TA   AGD_5841
418
.9
.T45
S65
1995

Thin-Film Deposition

Principles and Practice

Donald L. Smith

McGraw-Hill, Inc.
New York   San Francisco   Washington, D.C.   Auckland   Bogotá
Caracas   Lisbon   London   Madrid   Mexico City   Milan
Montreal   New Delhi   San Juan   Singapore
Sydney   Tokyo   Toronto

Library of Congress Cataloging-in-Publication Data

Smith, Donald L. (Donald Leonard), date.
   Thin-film deposition: principles and practice / Donald L. Smith.
      p.   cm.
   Includes index.
   ISBN 0-07-058502-4
   1. Thin films. 2. Vapor-plating. 3. Thin film devices.
   I. Includes index. II. Title
   TA418.9.T45S65   1995
   621.3815'2—dc20   94-47002
                                                              CIP

Copyright © 1995 by McGraw-Hill, Inc. All rights reserved. Printed in the United States of America. Except as permitted under the United States Copyright Act of 1976, no part of this publication may be reproduced or distributed in any form or by any means, or stored in a data base or retrieval system, without the prior written permission of the publisher.

1 2 3 4 5 6 7 8 9 0 DOC/DOC 9 0 0 9 8 7 6 5

ISBN 0-07-058502-4

The sponsoring editor for this book was Stephen S. Chapman and the production supervisor was Suzanne W. B. Rapcavage. It was set in New Century Schoolbook by J. K. Eckert & Company, Inc.

Printed and bound by R. R. Donnelley & Sons Company.

INTERNATIONAL EDITION

Copyright © 1995. Exclusive rights by McGraw-Hill, Inc. for manufacture and export. This book cannot be re-exported from the country to which it is consigned by McGraw-Hill. The International Edition is not available in North America.

When ordering this title, use ISBN 0-07-113913-3.

Cover photo: A thin film of the high-temperature superconductor $YBa_2Cu_3O_7$ is being deposited from a pulsed-laser vaporization source onto a ~750°C MgO substrate shown glowing orange at the top of the picture. Pulses from a UV (248 nm) KrF excimer laser enter the vacuum chamber from the right and impinge at 45° upon a sintered pellet of $YBa_2Cu_3O_7$ situated near the bottom where the white glow originates. Energy from the pulses electronically excites and partially ionizes both the vaporizing material and the 4 Pa of O_2 ambient gas, resulting in a spectacular plume of glowing plasma. The pulsed-laser deposition process is discussed in Sec. 8.4. This photo of Douglas Chrisey's apparatus was taken by M.A. Savell at the U.S. Naval Research Laboratories, Washington, D.C., and appeared on the cover of the *MRS Bulletin*, February 1992. (Used by permission of MRS and NRL.)

Information contained in this work has been obtained by McGraw-Hill, Inc., from sources believed to be reliable. However, neither McGraw-Hill nor its authors guarantee the accuracy or completeness of any information published herein and neither McGraw-Hill nor its authors shall be responsible for any errors, omissions, or damages arising out of use of this information. This work is published with the understanding that McGraw-Hill and its authors are supplying information but are not attempting to render engineering or other professional services. If such services are required, the assistance of an appropriate professional should be sought.

This work is dedicated to the memory of John A. Thornton,
1933–1987: magnetron pioneer, dedicated scientist,
inspiring teacher, and trusted counselor,
who left us suddenly in his prime.

ABOUT THE AUTHOR

Donald L. Smith has been on the research staff at Xerox's Palo Alto Research Center for 11 years, prior to which he was with Perkin-Elmer. He has worked with many thin-film processes in his 25-year career, including MBE, e-beam evaporation, sputtering, MOCVD, and PECVD. His current research includes stress control in metals and reactive evaporation of dielectrics. Over a period of several years preceding the writing of this book, he developed and taught a graduate course on thin-film deposition as an Adjunct Professor in Stanford University's Materials Science Department. He holds a B.S. from M.I.T. and a Ph.D. from the University of California at Berkeley, both in Chemical Engineering.

Contents

Preface   xi
Acknowledgments   xiii
Symbols and Abbreviations   xv

Chapter 1. Thin-Film Technology   1

1.1   Applications   1
1.2   Process Steps   3
1.3   Conclusion   7
1.4   Exercise   7

Chapter 2. Gas Kinetics   9

2.1   Vapors and Gases   9
2.2   Maxwell-Boltzmann Distribution   12
2.3   Molecular Impingement Flux   15
2.4   Ideal-Gas Law   17
2.5   Units of Measurement   20
2.6   Knudsen Equation   21
2.7   Mean Free Path   23
2.8   Transport Properties   25
     2.8.1   Diffusion   27
     2.8.2   Viscosity   28
     2.8.3   Heat conduction   29
2.9   Conclusion   32
2.10   Exercises   33
2.11   References   33
2.12   Recommended Readings   34

Chapter 3. Vacuum Technology   35

3.1   Pump Selection and Exhaust Handling   36
3.2   Problem Gases   40
3.3   Gas Throughput   41

3.4	Contamination Sources		45
	3.4.1	Oil backstreaming	46
	3.4.2	Gas evolution	50
	3.4.3	Dust	54
3.5	Pressure Measurement		55
3.6	Conclusion		60
3.7	Exercises		61
3.8	References		62
3.9	Recommended Readings		62

Chapter 4. Evaporation 63

4.1	Thermodynamics of Evaporation		63
4.2	Evaporation Rate		71
4.3	Alloys		74
4.4	Compounds		77
4.5	Sources		81
	4.5.1	Basic designs	81
	4.5.2	Contamination	84
	4.5.3	Temperature control	85
	4.5.4	Energy enhancement	90
4.6	Transport		94
4.7	Vapor Flux Monitoring		98
4.8	Deposition Monitoring		100
	4.8.1	Mass deposition	101
	4.8.2	Optical techniques	105
4.9	Conclusion		115
4.10	Exercises		115
4.11	References		117
4.12	Recommended Readings		118

Chapter 5. Deposition 119

5.1	Adsorption		120
5.2	Surface Diffusion		129
5.3	Nucleation		139
	5.3.1	Surface energy	139
	5.3.2	Three-dimensional nucleation	145
	5.3.3	Two-dimensional nucleation	154
	5.3.4	Texturing	156
5.4	Structure Development		159
	5.4.1	Quenched growth	161
	5.4.2	Thermally activated growth	170
	5.4.3	Amorphous films	177
	5.4.4	Composites	179
5.5	Interfaces		180
5.6	Stress		185
	5.6.1	Physics	186
	5.6.2	Problems	193
	5.6.3	Intrinsic stress	196
5.7	Adhesion		197

5.8	Temperature Control	200
	5.8.1 Radiation	202
	5.8.2 Gas conduction	208
	5.8.3 Measurement	211
5.9	Conclusion	215
5.10	Exercises	215
5.11	References	217
5.12	Recommended Readings	219

Chapter 6. Epitaxy — 221

6.1	Symmetry	221
6.2	Applications	226
	6.2.1 Semiconductor devices	230
6.3	Disruption	237
6.4	Growth Monitoring	241
	6.4.1 Electron spectroscopy	242
	6.4.2 Diffraction	246
6.5	Composition Control	259
	6.5.1 Precipitates	260
	6.5.2 Alloys	262
	6.5.3 Point defects and surface structure	263
	6.5.4 Gaseous sources	266
	6.5.5 Flux modulation	268
	6.5.6 Doping	273
6.6	Lattice Mismatch	279
6.7	Surface Morphology	293
	6.7.1 Two-dimensional nucleation	293
	6.7.2 Three-dimensional rearrangement	296
	6.7.3 Statistical roughening	299
6.8	Conclusion	300
6.9	Exercises	301
6.10	References	302
6.11	Recommended Readings	306

Chapter 7. Chemical Vapor Deposition — 307

7.1	Gas Supply	309
	7.1.1 Safety	311
	7.1.2 Flow control	312
	7.1.3 Contamination	315
7.2	Convection	317
	7.2.1 Laminar flow in ducts	318
	7.2.2 Axisymmetric flow	322
	7.2.3 Free convection	327
7.3	Reaction	330
	7.3.1 Chemical equilibrium	330
	7.3.2 Gas-phase rate	336
	7.3.3 Surface processes	344
7.4	Diffusion	353
	7.4.1 Diffusion-limited deposition	355

		7.4.2 Reactor models	358
		7.4.3 Temperature gradients	363
7.5	Conclusion		365
7.6	Exercises		365
7.7	References		367
7.8	Recommended Readings		369

Chapter 8. Energy Beams 371

8.1	Electron Generation		374
8.2	Electron Beams		382
8.3	Arc Plasmas		387
8.4	Pulsed Lasers		394
8.5	Ion Bombardment		400
	8.5.1	Surface effects	402
	8.5.2	Subsurface effects	411
	8.5.3	Bulk-film modification	423
	8.5.4	Sputtering	431
8.6	Conclusion		446
8.7	Exercises		447
8.8	References		449
8.9	Recommended readings		452

Chapter 9. Glow-Discharge Plasmas 453

9.1	Electron-Impact Reactions		454
9.2	Plasma Structure		462
9.3	DC Excitation		470
	9.3.1	Parallel-plate configuration	470
	9.3.2	Chemical-activation applications	473
	9.3.3	Sputtering	476
	9.3.4	Magnetrons	482
9.4	Frequency Effects		489
	9.4.1	Low-frequency regime	491
	9.4.2	Transition to high frequency	494
	9.4.3	RF bias	499
	9.4.4	Power coupling	503
9.5	Electrodeless Excitation		507
	9.5.1	Microwaves	508
	9.5.2	Helicons	518
	9.5.3	Inductive coils	521
9.6	Plasma Chemistry		524
	9.6.1	Kinetics	526
	9.6.2	Macroparticles	533
	9.6.3	Downstream deposition	536
	9.6.4	Case studies	541
9.7	Conclusion		549
9.8	Exercises		550
9.9	References		551
9.10	Suggested Readings		555

Chapter 10. Film Analysis — 557

- 10.1 Structure — 558
    - 10.1.1 Thickness — 558
    - 10.1.2 Surface topography — 560
    - 10.1.3 Bulk inhomogeneity — 563
    - 10.1.4 Crystallography — 564
    - 10.1.5 Bonding — 566
    - 10.1.6 Point defects — 568
- 10.2 Composition — 570
    - 10.2.1 Electrons and x-rays — 571
    - 10.2.2 Mass spectrometry — 572
    - 10.2.3 Rutherford backscattering — 572
    - 10.2.4 Hydrogen — 573
- 10.3 Properties — 574
    - 10.3.1 Optical behavior — 575
    - 10.3.2 Electrical behavior — 575
    - 10.3.3 Mechanical behavior — 581
- 10.4 Conclusion — 583
- 10.5 References — 583
- 10.6 Recommended Reading — 584

Appendix A Units 585
Appendix B Vapor Pressures of the Elements 587
Appendix C Sputtering Yields of the Elements 591
Appendix D Characteristics of Tungsten Filaments 593
Appendix E Sonic Orifice Flow 597
Index 599

Preface

This book evolved from a graduate course that I developed in the Materials Science and Engineering Department at Stanford University over a three-year period. One of the biggest challenges in this endeavor was finding suitable reading material for the students. Although many good book chapters and review articles were available on various thin-film topics, there remained untreated topics, inconsistencies, and even contradictions among these sources.

Here, I have thus endeavored to achieve the coherency, conciseness, and balanced perspective that single-authorship makes possible. Recognizing that the field is too vast to cover comprehensively in one volume, this book instead focuses on the basic features of the various deposition techniques, including their distinctions and especially their common underlying principles. In general, these principles have been under-represented in the literature, partly because much of thin-film technology has been developed empirically. Even so, as I have worked in the lab with many of the thin-film techniques, I have encountered common issues and explanations throughout the technology, and it is these elements which I have tried to develop and tie together.

Throughout the book, emphasis is on practical application of the basic principles which span the deposition techniques. These principles are developed from the level of freshman physics and chemistry, and almost all formulae presented are derived. Practical application is illustrated by frequent and diverse examples drawn from industrial thin-film technology.

The scope of the book embraces the vapor-phase deposition techniques, because they are closely related to each other and are also the most versatile. Omitted are the liquid-phase techniques (thermal and electrolytic), which are much more limited in adjustability of surface conditions, particularly temperature and ion bombardment. Emphasis is on the nature of the deposition process and how it determines film properties. Presented first are the principles that apply to all of the deposition techniques, such as gas kinetics, heat transfer, and vacuum technology. Then, specific techniques are grouped sequentially by the nature of the vapor phase over the substrate—vacuum, gas, or plasma. The final chapter is a brief survey of film characterization methods.

This treatment of thin-film deposition should be useful both to students and to professional scientists and engineers. The book may be used as the text for graduate courses in the curricula of materials science; applied physics; or electrical, mechanical, or chemical engineering, since all of these disciplines are involved in thin-film technology. To the professional, it provides an overall perspective useful in selecting pertinent research topics or in choosing a suitable deposition technique for a specific application. Finally, the book serves as a reference volume of useful formulae, data, explanations, and primary sources, accessed by an extensive index.

In a field that is rapidly evolving and in which application generally precedes understanding, a book such as this is likely to contain some misconceptions. Therefore, consider it as a starting point rather than as the last word. I welcome readers' comments and suggestions for use in the next edition.

Donald L. Smith

Acknowledgments

I am especially grateful to Stig Hagstrom for inviting me to participate in developing a course on thin-film deposition in the Materials Science and Engineering Department at Stanford, and for his guidance and encouragement throughout the project. As the subsequent process of transforming course notes into a textbook proceeded, many colleagues kindly read portions of the manuscript and provided valuable feedback, including David Graves, Mike Kelly, Sue Kelso, Bill Nix, Henry Weinberg, and finally, Dave Biegelsen, who actually read the entire manuscript. As questions arose during the writing, I gained understanding on numerous topics through discussions with the following colleagues, listed here in no intentional order and hoping that I have not overlooked anyone: John Arthur, Steve George, John Northrup, Conyers Herring, Dan Flamm, Rick Gottscho, Warren Jackson, Kit Bowen, Mark Perkins, Ross Bringans, Mike Kosterlitz, Robert Thornton, Dave Bour, Art Gossard, Dave Fork, Amy Wendt, Mike Walsh, Jerry Cuomo, Brian Chapman, Simone Anders, and Cal Quate. The staff of the Xerox PARC Technical Information Center was most responsive and persevering with my frequent requests, which were made most often of Sally Peters and Lisa Alfke. Dan Murphy of Xerox PARC rendered many challenging drawings, as did Alan Watson of Image Technology, and J.K. Eckert and Company gave the text a thorough copyediting. Finally, I am grateful for the considerable patience and support of my wife, Jane Mickelson (the Macintosh widow), my son Jared, and my editors Dan Gonneau and Steve Chapman.

Symbols and Abbreviations

Symbols

A	area
	optical absorbance
a	atomic spacing or diameter
a_B, a_C	activity coefficient of solute
a_o	lattice constant
(a)	adsorbed phase of a species
B	a constant
$\mathbf{B}$	magnetic flux density, $= \mu \mathbf{H}$
b	width
	magnitude of Burgers vector
$\mathbf{b}$	Burgers vector
C	capacitance
	flow conductance
C_A	flow conductance per unit area
C_f	fluid-flow conductance
C_m	molecular-flow conductance
c	particle speed
c_g	heat capacity per gram of liquid or solid
c_o	speed of light
c_p	heat capacity per mole of gas at constant pressure
c_v	heat capacity per mole of gas at constant volume
c^*	speed of sound
(c)	condensed phase of a species
D	diffusivity (mass diffusion coefficient)
D_T	coefficient of thermal diffusion (not thermal diffusivity, α)
d	distance
$\mathbf{E}$	electric field strength
E_a	activation energy
E_b	bond dissociation energy
	surface binding energy in sputtering
E_c	chemisorption energy
	conduction-band energy
E_d	adsorption energy
E_e	electron energy

E_F	Fermi level of energy
E_f	energy of Frenkel-pair formation
E_g	band-gap energy
E_i	threshold energy for ionization
E_k	activation energy of kth reaction
E_o	threshold energy of ion-activated process
E_p	potential energy
E_q	confinement energy of particle in quantum well
E_r	activation energy for chemisorption
	maximum reflected-ion energy
E_s	activation energy for surface diffusion
E_t	translational kinetic energy
E_v	valence-band energy
E_γ	kinetic energy of knock-on atom
E_+	Coulomb-potential energy
F	force
F_c	critical ratio of plasma power to reactant-gas flow
f	ion-gauge calibration factor
	fractional lattice mismatch
f_d	fractional depletion by diffusion
f_s	sputter-yield factor
f(x)	mathematical function of x
G	reciprocal-lattice vector
G	total Gibbs free energy of *system*
G_m	Gibbs free energy per mole
G^o	Gibbs free energy in standard state (10^5 Pa)
$\Delta_f G$	Gibbs-free-energy change of compound formation from the elements, per mole of compound
$\Delta_r G^o$	Gibbs free energy of reaction per mole of selected reactant, at p^o
G_s	shear modulus of elasticity
Gr	Grashof number
g	gravitational acceleration
(g)	gas phase of a species
H	magnetic field strength
H	total enthalpy of *system*
H_m	enthalpy per mole
$\Delta_a H$	heat of chemisorption per mole
$\Delta_c H$	heat of fusion per mole
$\Delta_f H$	heat of compound formation from the elements, per mole of compound
$\Delta_r H$	heat of reaction per mole of selected reactant
$\Delta_s H$	heat of sublimation per mole
$\Delta_v H$	heat of evaporation per mole
h	Planck's constant
h	film thickness
h_c	gas-film heat-transfer coefficient
h_r	radiative heat-transfer coefficient
I	current
	radiation intensity

Symbols and Abbreviations   xvii

i	$\sqrt{-1}$
	an integer
J	molecular flux
J_A	diffusion flux of component A
J_c	condensation flux
J_D	bulk diffusion flux
J_d	desorption flux
J_i	impinging flux
J_r	deposition flux
J_s	surface diffusion flux
J_v	evaporation flux
J_{vo}	maximum thermodynamic evaporation flux
j	current density
	an integer
K	chemical equilibrium constant
	curvature
Kn	Knudsen number
K_f	gas-flow calibration factor
K_T	thermal conductivity
$\mathbf{k}$	momentum vector
	wave vector
k	spring constant
k_B	Boltzmann's constant, $= R/N_A$
k_b	binary reaction rate coefficient
k_m	unimolecular reaction rate coefficient
k_t	ternary reaction rate coefficient
L	characteristic linear dimension
l	molecular mean free path
l_e	electron mean free path
l_+	ion mean free path
M	molar mass (molecular weight)
	bending moment
M_i	impinging-particle mass, in atomic mass units (u)
M_t	target-atom mass, in atomic mass units (u)
m	mass
m_e	electron mass
m_i	impinging-particle mass, kg
m_r	reduced mass
m_T	thermal load or mass
m_t	target-atom mass, kg
m_+	ion mass, kg
N	number of particles
N_A	Avogadro's number
N_m	number of moles
N_o	number of events
n	molecular concentration
$n_{A(B)}$	concentration of component A(B)
n_e	electron concentration
n_h	conducting hole concentration per unit volume
n_i	interstitial atom concentration per unit volume

Symbols and Abbreviations

n_m	molar concentration per unit volume
n_s	surface concentration per unit area
n_{so}	monomolecular-layer (monolayer) concentration per unit area
$\tilde{n}$	index of refraction, real part
$\hat{n}$	complex index of refraction
n_+	ion concentration per unit volume
n^*	concentration of critical nuclei per unit volume
P	power
Pe	Peclet number
p	impact parameter
	pressure
p_c	critical pressure
p_i	partial pressure of species i
$p°$	reference pressure (1 atm)
p_v	equilibrium vapor pressure
Q	molar (mass) flow rate
q	quantity of charge
	quantity of heat
R	gas constant
	reflectance
	resistance
R_d	desorption rate
R_r	chemisorption rate
R_H	hall coefficient
R_k	rate of kth chemical reaction per unit area or volume
Ra	Rayleigh number
Re	Reynolds number
$\Re$	real part of complex number
r	propagation distance
	radius
r_c	cyclotron radius
S	total entropy of *system*
S_c	molecular sticking coefficient
S_e	ion stopping power by electrons in matter
S_m	entropy per mole
S_n	ion stopping power by nuclei in matter
$\Delta_r S°$	entropy of reaction per mole of selected reactant, at $p°$
s	electrical conductivity
	steric factor in reaction rate
T	transmittance
	temperature
T_c	critical temperature
T_e	electron temperature in eV
T_m	energy transferred in a head-on binary collision
T_s	substrate temperature
T_v	vibrational period
$T(\theta)$	energy transferred in a binary collision at deflection angle θ
T_+	ion temperature
t	time
U	total internal energy of *system*

U_m	internal energy per mole
u	fluid-flow velocity
u_a	ambipolar diffusion velocity
u_D	diffusion velocity
u_d	drift velocity
V	voltage drop
	volume
V_b	plasma-sheath voltage drop
V_d	dc bias voltage
V_f	floating potential in a plasma
V_i	ionization potential (appearance potential)
V_m	molar volume
V_o	peak drive voltage
V_p	plasma potential measured from ground
ΔV_p	plasma potential measured from anode
V_s	surface potential
v	particle velocity
W	volume flow rate
w	quantity of work
	substrate thickness
X_C	capacitive reactance
X_l	load impedance
X_L	inductive reactance
X_s	source impedance
x	atomic fraction or mole fraction
	lateral distance
Y	Young's modulus of elasticity
Y_e	secondary-electron yield
Y_s	sputtering yield, atoms per ion
y	longitudinal distance
Z	atomic number
	partition function
z	acoustic impedance
	ion charge
	vertical distance

Greek Symbols

α	epilayer tilt angle
	mass function in sputtering-yield formula
	radiation absorptivity of a surface
	thermal diffusivity
α_λ	monochromatic absorptivity of a surface
α_r	light absorption coefficient of a solid
α_c	molecular condensation coefficient
α_v	molecular evaporation coefficient
α_T	linear thermal expansion coefficient
α_{TV}	volumetric thermal expansion coefficient
β	edge energy

Symbols and Abbreviations

γ	dislocation angle
	roughness exponent
	slip-plane angle
	surface tension
	surface energy per unit area
	thermal accommodation coefficient
	heat capacity ratio, c_p/c_v
Δ	ellipsometric phase difference
δ	incremental distance
	molecular trapping probability
	phase angle
δ_c	chemical boundary-layer thickness
δ_n	concentration boundary-layer thickness
δ_s	rf skin depth
δ_v	velocity boundary-layer thickness
ε	radiation emissivity of a surface
	reduced energy in ion-stopping theory
	strain
ε_d	dielectric constant
ε_k	oscillator strength of kth mode of a chemical bond
ε_o	electrical permittivity of vacuum
ε_m	total kinetic energy of a molecule
ε_p	potential energy of a molecule
ε_r	rotational kinetic energy of a molecule
ε_T	tetragonal strain
ε_t	translational kinetic energy of a molecule
ε_v	vibrational kinetic energy of a molecule
ε_λ	monochromatic emissivity of a surface
ζ	chemisorption reaction probability
	dimensionless height
η	vapor utilization fraction
	viscosity
Θ	fractional surface coverage
	dimensionless temperature
θ	angle
κ	solute segregation coefficient
Λ	diffusion length
λ	dislocation angle
	wavelength
λ_D	Debye length
λ_e	wavelength of fundamental optical absorption edge
λ_s	spatial wavelength
μ	chemical potential
	magnetic permittivity (= 1 in this book)
μ_e	electron mobility
μ_H	hall mobility
μ_h	hole mobility
μ°	chemical potential at 1 atm
ν	frequency
	kinematic viscosity

	Poisson's ratio
ν_A	stoichiometric coefficient of species A in a reaction
ν_e	electron collision frequency
ν_o	plasma drive frequency
	pre-exponential frequency factor
ν_r	resonant frequency of quartz crystal
ξ	fractional approach to completion of a process
ρ	resistivity
ρ_l	linear density of dislocations
ρ_m	mass density
ρ_q	space-charge density
ρ_s	sheet resistivity
σ	standard deviation
	Stefan-Boltzmann constant
	tensile stress
σ_i	ionization cross-section
σ_m	molecular cross-section
τ	relaxation time constant
	shear stress
Φ	heat flux
Φ_b	blackbody total radiative heat flux
Φ_u	universal electron screening function
$\Phi_{b\lambda}$	blackbody monochromatic radiative heat flux
ϕ	diameter
ϕ_B	Brewster's angle
ϕ_w	electron work function
ϕ_0	light incident angle
ϕ_1	light refracted angle
χ	Rutherford backscattering yield relative to randomly-orientedcrystal
ψ	ellipsometric angle
Ω	atomic volume
	number of quantum states
ω	azimuthal angle
	angular frequency
ω_c	electron-cyclotron angular frequency
ω_p	plasma angular frequency

Abbreviations

A	amperes
Å	angstroms
AES	auger electron spectroscopy
AFM	atomic-force microscope
AMLCD	active-matrix liquid-crystal display
APCVD	atmospheric-pressure chemical vapor deposition
at.	atomic
atm	atmospheres
bcc	body-centered cubic

C	centigrade
	coulombs
Chap.	chapter
CVD	chemical vapor deposition
dc	direct current
DLTS	deep-level transient spectroscopy
e	electron
ECR	electron cyclotron resonance
Eq.	equation
eV	electron-volts
F	farads
FTIR	fourier-transform infrared spectroscopy
fcc	face-centered cubic
g	grams
gb	grain boundary
GHz	gigahertz
h	hours
hcp	hexagonal close-packed
Hz	hertz (cycles per second)
HEMT	high-electron-mobility transistor
IR	infrared
J	joules
K	degrees kelvin
kg	kilograms
kHz	kilohertz
l	liters
L	langmuirs
LCD	liquid-crystal display
LEED	low-energy electron diffraction
LN_2	liquid nitrogen
LPCVD	low-pressure chemical vapor deposition
m	meters
MBE	molecular-beam epitaxy
mo	molecules
MHz	megahertz
min	minutes
ML	monolayer (monomolecular layer)
mol	moles
MSDS	material safety data sheet
MTP	multiple-twinned particle
N	newtons
nm	nanometers (10^{-9} m)
Pa	pascals
pF	picofarads (10^{-12} F)
ppb	parts per billion (1 in 10^9)
PECVD	plasma-enhanced chemical vapor deposition
PK	primary knock-on atom
ppm	parts per million (1 in 10^6)
psi	pounds per square inch
PVD	physical vapor deposition

R	hydrocarbon radical
RBS	Rutherford backscattering spectrometry
rf	radio frequency
RGA	residual gas analyzer (mass spectrometer)
RHEED	reflection high-energy electron diffraction
rms	root mean square $\left(\sqrt{\overline{x^2}}\right)$
RTP	rapid thermal processing
S	siemens (= $1/\Omega$)
s	seconds
sccm	standard (1 atm, 0° C) cm^3 per minute
Sec.	section
SEM	scanning electron microscope
SIMS	secondary-ion mass spectrometry
slm	standard (1 atm, 0° C) liters per minute
snafu	situation normal—all fouled up
STM	scanning tunneling microscope
stp	standard temperature (0° C) and pressure (1 atm)
T	tesla
TEM	transmission electron microscope
u	atomic mass units
UHV	ultra-high vacuum
UV	ultraviolet
VPE	vapor-phase epitaxy
W	watts
XPS	x-ray photoelectron spectroscopy
2D	two-dimensional
3D	three-dimensional

Chapter 1

Thin-Film Technology

1.1 Applications

Thin films are deposited onto bulk materials (substrates) to achieve properties unattainable or not easily attainable in the substrates alone. Thus, the first question to ask is, "What properties are required for the application at hand?" Table 1.1 divides these properties into five basic categories and gives examples of typical applications within each category. Examination of this table shows that the range of thin-film applications is very broad indeed. Often, multiple properties are obtainable simultaneously. For example, Cr coatings used on plastic parts for automobiles impart hardness, metallic luster, and protection against ultraviolet light. Titanium nitride (TiN) coatings on cutting tools offer hardness, low friction, and a chemical barrier to alloying of the tool with the workpiece (galling). They also offer a rich, gold color for decorative applications. The Cr coating on a plastic part achieves the functionality of the same part made from bulk metal, but at significant savings in cost and weight. The TiN coating achieves surface properties unattainable in a bulk material, since the bulk material must also offer high strength and toughness in the cutting tool application.

Additional functionality in thin films can be achieved by depositing multiple layers of different materials. Optical interference filters consist of tens or even hundreds of layers alternating between high and low indexes of refraction. When alternating layers are made using nanometer thicknesses of semiconducting materials such as GaAs and (AlGa)As, the result is a "superlattice" that has electrical properties governed by the constructed periodicity rather than by the atomic

2 Thin-Film Technology

TABLE 1.1 Thin-Film Applications

Thin-film property category	Typical applications
Optical	Reflective/antireflective coatings Interference filters Decoration (color, luster) Memory discs (CDs) Waveguides
Electrical	Insulation Conduction Semiconductor devices Piezoelectric drivers
Magnetic	Memory discs
Chemical	Barriers to diffusion or alloying Protection against oxidation or corrosion Gas/liquid sensors
Mechanical	Tribological (wear-resistant) coatings Hardness Adhesion Micromechanics
Thermal	Barrier layers Heat sinks

periodicity. Thus, multilayer thin films can behave as completely new, engineered materials unknown in bulk form. When multiple layering is combined with lithographic patterning in the plane of the films, microstructures of endless variety can be constructed. This is the basic technology of the integrated-circuit industry, and more recently it is being applied to optical waveguide circuitry and to micromechanical devices. The latter include such creations as rotary electrostatic motors tens of micrometers in diameter, which is a clear case of a device awaiting an application.

This book will examine the deposition of thin films from the vapor phase. Omitted will be the deposition of thick films as well as deposition from the liquid phase. The distinction between thin-film and thick-film technology is that the former involves deposition of individual molecules, while the latter involves deposition of particles. Examples of thick-film techniques are painting, silk screening, spin-on glass coating, and plasma spraying. The thick-film techniques are important and relatively inexpensive, but they do not offer the control or the material quality of the thin-film techniques. Note that films deposited by a thin-film technique can be thicker than those deposited by a thick-film technique, such as graphite plates several millimeters thick deposited by thermal decomposition (pyrolysis) of hydrocarbon vapors. Also omitted here will be films grown by reaction with the substrate rather than by deposition, such as silicon dioxide grown by the thermal reaction of water vapor with Si.

The vapor phase thin-film techniques have three significant advantages over the liquid-phase techniques: applicability to any material, wide adjustability in substrate temperature, and access to the surface during deposition. Temperature is a key variable in altering film properties. Access to the surface allows energy to be delivered by ion bombardment and allows surface analysis during deposition. We will see in Chap. 8 that ion bombardment can alter film properties dramatically. Surface analysis by electron diffraction during deposition has been central to research on epitaxial (single crystal) film growth behavior. Regarding the liquid-phase techniques, liquid-phase epitaxy is useful for low-cost production but does not have the control of the Chap. 6 processes. Electroplating from liquid solution is widely used, occasionally even for epitaxy. However, its technology is very different from the vapor-phase techniques to be examined here.

1.2 Process Steps

All thin-film processes contain the four (or five) sequential steps shown in Fig. 1.1. A source of film material is provided, the material is transported to the substrate, deposition takes place, sometimes the film is subsequently annealed, and finally it is analyzed to evaluate the process. The results of the analysis are then used to adjust the conditions of the other steps for film property modification. Additional process control and understanding are obtained by monitoring the first three steps during film deposition. In subsequent chapters, each of the deposition processes will be discussed in terms of the steps shown in Fig. 1.1, and various techniques for monitoring and controlling these steps will be examined. Annealing will not be discussed except to the extent that it occurs during deposition. Post-deposition annealing can be used to activate grain growth, alter stoichiometry, introduce dopants, or cause oxidation. However, this book confines itself to the deposition process itself. Below, we briefly elaborate on the steps shown in Fig. 1.1.

The **source** of the film-forming material may be a solid, liquid, vapor, or gas. Solid materials need to be vaporized to transport them to the substrate, and this can be done by heat or by an energetic beam of electrons, photons (laser ablation), or positive ions (sputtering). These methods are categorized as *physical* vapor deposition (PVD).

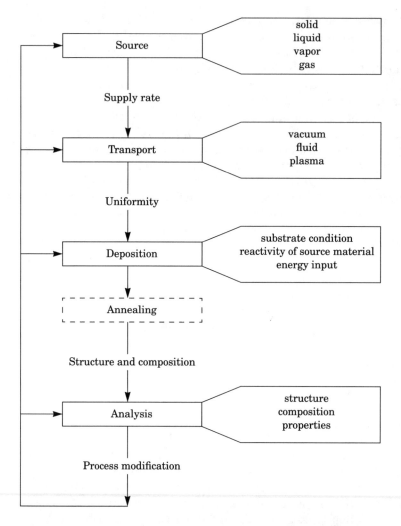

Figure 1.1 Thin-film process steps. In all steps, process monitoring is valuable, and contamination is a concern.

Occasionally, solid sources are instead *chemically* converted to vapors, such as Ga to GaCl. In other cases, the source material is supplied as a gas or as a liquid having sufficient vapor pressure to be transported at moderate temperatures. Thin-film processes that use gases, evaporating liquids, or chemically gasified solids as source materials are categorized as *chemical* vapor deposition (CVD). In both PVD and CVD, contamination and supply rate are the major source-material issues. Contamination is also an issue in the transport and deposition steps.

Supply rate is important because film properties vary with deposition rate and with the ratio of elements supplied to compound films.

In the **transport** step, the major issue is uniformity of arrival rate over the substrate area. The factors affecting this uniformity are very different, depending on whether the transport medium is a high vacuum or a fluid. (Here we are talking about gaseous fluids rather than liquid fluids.) In a high vacuum, molecules travel from the source to the substrate in straight lines, whereas in a fluid there are many collisions among molecules during the transport step. Consequently, in a high vacuum, uniformity of source-material arrival rate at the substrate is determined by geometry, whereas in a fluid it is determined by gas flow patterns and by diffusion of the source molecules through the other gases present. Often, the high-vacuum processes are equated with physical vapor deposition, and the fluid-flow processes are equated with chemical vapor deposition. However, this is not always a valid association. Although many physical deposition processes do operate in a high vacuum, others like laser ablation and sputtering often operate at higher pressures characteristic of a fluid. Similarly, although most chemical deposition processes operate at fluid pressures, chemical beam epitaxy operates in a high vacuum.

The high-vacuum transport medium has the important advantage of clear access to the deposition surface. This allows energy input from an ion beam and allows the use of analytical techniques involving electron beams, such as electron diffraction and Auger spectroscopy. On the other hand, the fluid medium has the advantage that it functions at atmospheric pressure or at easily-achieved moderate vacuum levels. In both media, contamination can be dealt with equally well and using similar techniques.

Many thin-film processes operate in a plasma, which is listed in Fig. 1.1 as a third transport medium. A plasma is a partially ionized gas and is often regarded as the fourth state of matter (solid, liquid, gas, and plasma). Plasmas contain a great deal of energy, which can activate film deposition processes at low temperature. The plasma operating pressure can be such that it behaves either as a fluid or as a high-vacuum medium.

The third step in the thin-film process is the actual **deposition** of the film onto the substrate surface. Deposition behavior is determined by source and transport factors and by conditions at the deposition surface. There are three principal surface factors which determine the deposition behavior, as seen in Fig. 1.1. These factors are substrate surface condition, reactivity of the arriving material, and energy input. Substrate surface condition includes roughness, level of contamination, degree of chemical bonding with the arriving material, and crystallographic parameters in the case of epitaxy. The reactivity factor refers to the probability of arriving molecules reacting with the surface and becoming incorporated into the film. This probability is known as the "sticking coefficient," S_c, and it can vary from unity to less than 10^{-3}. It is generally lower for CVD than for PVD processes, and low S_c aids in coating convoluted shapes and in deposition on selected areas.

The third deposition factor is energy input to the surface. It can come in many forms and has a profound effect on both the reactivity of arriving material and on the composition and structure of the film. Substrate temperature is the basic source of energy input, but there are many other sources. Photons are used in photoassisted and laser-assisted deposition. Positive ion bombardment carries very large amounts of energy. It is present in most plasma processes and in some high-vacuum processes. Chemical energy is carried by inherently reactive source molecules and by molecules that have been dissociated in the course of vaporization or plasma transport. In summary, the three deposition factors of substrate condition, reactivity, and energy input work together with the arriving fluxes to determine the structure and composition of the deposited film. The structure and composition in turn determine the various film properties listed in Table 1.1.

The final step in the deposition process is **analysis** of the film. One level of analysis consists of directly measuring those properties that are important for the application at hand, such as the hardness of a tool coating, the breakdown voltage of an insulator, or the index of refraction of an optical film. Many film deposition processes are optimized using the empirical approach of measuring key film properties as a function of the process variables that can be varied in the first three steps of the deposition. This approach is suggested by the "feedback" arrows in Fig. 1.1. A deeper level of analysis involves probing the film's structure and composition, since those are the factors that determine the observed properties. This type of analysis is generally more difficult, but it provides a bridge between the deposition step and the final film properties. This bridge leads to a better understanding of the overall process.

Analysis of the film after deposition can be thought of as the final stage of process monitoring. Monitoring is important at all steps in the thin-film process. The more monitoring that can be done during deposition, the better will be both the control and the understanding of the process. For example, the supply rate of each source material can be continuously metered and feedback-controlled. The composition of the transport medium can be analyzed for both reactant concentration and contaminants. Both the deposition rate and the film crystallography can be monitored in real time. Throughout this book, techniques for process monitoring will be introduced as appropriate for each deposition method.

1.3 Conclusion

This introductory chapter has presented many concepts in a general way to give the reader an overview and to provide a framework within which to discuss specific deposition techniques. These concepts may seem somewhat abstract at present, but they will take on more substance in the context of later discussion. It will be useful to refer back to this chapter from time to time, especially to Fig. 1.1, to see more clearly the parallels between the various deposition techniques and to develop a sense of the underlying principles which unify thin-film technology.

1.4 Exercise

1.1 List as many thick-film or thin-film applications as you can from among familiar industrial or consumer products. For each, list the properties required for the film to function effectively.

Chapter 2

Gas Kinetics

The behavior of vapors and gases on both the macroscopic and the molecular scale is fundamental to an understanding of thin-film deposition from the vapor phase. This chapter will examine the molecular interaction of gases and will relate that behavior to the macroscopic, measurable properties such as temperature and pressure.

2.1 Vapors and Gases

It is important at the outset to distinguish between the behaviors of vapors and gases. At a fixed temperature, T, a vapor can be condensed to a liquid or solid by compressing it sufficiently, whereas a gas cannot. This situation is illustrated by the pressure-volume-temperature (p-V-T) diagram of Fig. 2.1. The possible equilibrium states of any pure material can be represented by a surface in p-V-T space such as the one shown. Here, the V is that of a fixed amount of material which we will set at one mole (6.02×10^{23} molecules), so that V is the molar volume, V_m. Now visualize a cut through the p-V-T surface along a p-V plane positioned at temperature T_2. The cut intercepts the surface along the line a-b-c-d-e-f, so this line represents the relationship between p and V_m for that material at that T. At the lowest p, we are at point a, and the material exists in the vapor phase, as labeled. Reducing the volume of the container holding our mole of vapor raises its pressure, and we move along the line a-b. At point b, condensation to the liquid phase begins, and at point c, condensation is complete. During condensation, volume is decreasing at a fixed p and T; that is, the line b-c is perpendicular to the p-T plane. The p in this region, p_v,

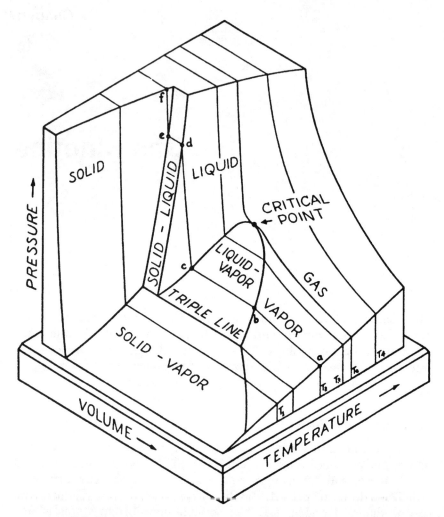

Figure 2.1 p-V-T diagram for a fixed amount of pure material. [Reprinted from Sears (1950) with permission of the publisher. © 1950, 1953 by Addison-Wesley Publishing Co., Inc.]

is called the saturation vapor pressure, or often just the vapor pressure. Of course, an abrupt decrease in V could be imposed during condensation, which would push the system to a p above the line b-c, but this would not be an *equilibrium* situation. In such an instance, the vapor is said to be "supersaturated." Supersaturation is an important driving force in the nucleation and growth of thin films, and it will be discussed in more detail in Chap. 5. The distance b-c along the V axis is the volume difference, ΔV, between the vapor and liquid phases at T_2. Note that at higher T values, ΔV becomes smaller and finally vanishes at the "critical" point. Every material has a single such point in p-V-T space, the coordinates of which are the critical p, V, and T (p_c, V_c, and T_c). Above T_c, there is no volume discontinuity upon compression and thus no distinction between the two phases. Instead, molar density ($1/V_m$) increases monotonically with increasing p, as it does at T_4. Above T_c, a material cannot be condensed to a liquid, and it is called a gas; below T_c, it can be condensed and is called a vapor.

A second important aspect of Fig. 2.1 for thin-film work is the p-T behavior of the surfaces labeled "liquid-vapor" and "solid-vapor," which represent the regions in which condensation is taking place. Since these surfaces, as well as the solid-liquid surface, are perpendicular to the p-T plane, their projections on that plane are lines, as shown in Fig. 2.2. Note that below a certain T, the liquid-phase region vanishes so that condensation occurs directly to the solid phase. The vaporization of material in this region is called sublimation. The point where the liquid phase vanishes is called the triple point because it is the only point where all three phases can exist simultaneously in equilibrium. The lines abutting the vapor phase are the vapor pressure (p_v) curves. These data are of fundamental importance in controlling thin-film processes, and they will be discussed quantitatively

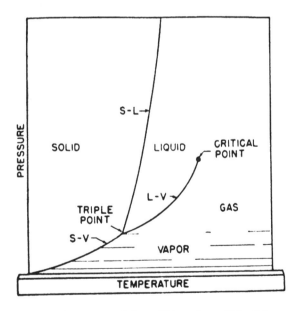

Figure 2.2 Projection of the p-V-T surface of Fig. 2.1 on the p-T plane. [Reprinted from Sears (1950) with permission of the publisher; "triple point" label added. © 1950, 1953 by Addison-Wesley Publishing Co., Inc.]

in Sec. 4.1. The p_v increases exponentially with T, and at its terminus at the critical point, p_v is well above one atmosphere. There, the vapor has to be compressed so much to make it condense that its concentration has approached the concentration of the liquid phase, and that is why the phase boundary vanishes. Thin-film processes are carried out at much lower p, sometimes above and sometimes below the p_v line. During film deposition, the amount by which the p of the source vapor exceeds p_v is the supersaturation, as mentioned above. In the first two steps of the thin-film process (source supply and transport to the substrate), good process control requires that one avoid condensation by keeping p below the p_v line. Condensation should also be avoided during the compression that occurs in vacuum pumps (more in Sec. 3.2).

By definition, all vapors are condensable, but sometimes a vapor will not be condensable under conditions encountered in thin-film work, these conditions usually being at or below one atmosphere of p and at or above room T (25° C). If the p_v of a particular vapor is > 1 atm. at 25° C, its condensation will not be encountered, and it may therefore be thought of as a gas for practical purposes. For example, nitrous oxide (N_2O) is, strictly speaking, a vapor at 25° C, because its T_c is 37° C; but since its p_v at 25° C is 55 atm, N_2O is considered to be a gas. In this chapter, we will be talking about gases and about vapors in the absence of condensation, so we will refer to them all as gases. Our discussion will also be restricted to *nonreactive* interaction of gas molecules with each other and with surfaces.

2.2 Maxwell-Boltzmann Distribution

Gas molecules are continually colliding randomly with each other and with containing surfaces. The energy exchange which takes place during these collisions leads to an equilibrium distribution of molecular speeds given by the Maxwell-Boltzmann formula:

$$\frac{dN/dc}{N} = 4\pi c^2 \left(\frac{m}{2\pi k_B T}\right)^{3/2} \exp\left(-\frac{\frac{1}{2}mc^2}{k_B T}\right) \qquad (2.1)$$

where N = total number of molecules in the distribution
  c = molecular speed, m/s
  dN/dc = incremental number of molecules dN within the speed increment dc
  m = molecular mass, kg
  k_B = Boltzmann's constant = 1.38×10^{-23} J/K
  T = absolute temperature, K

2.2 Maxwell-Boltzmann Distribution

Many important properties of gases follow from this formula, as we will see below. Its derivation is too involved to deal with here, but it is worthwhile reviewing its origin in statistical mechanics because of the utility of this thinking in many aspects of thin-film deposition. All particles have various modes, or degrees of freedom, of energy storage. For molecules, they are translational, rotational, vibrational, and electronic, listed here in order of increasing spacing between quantized energy levels. Electronic excitation requires the largest energy jump to reach the first level above ground state, so it is seen only at extremely high T or when there is a source of high-energy particles. In relation to thin-film processes, electronic and vibrational excitation are important in plasmas and in deposition activated by UV photons. Rotational excitation usually increases in degree over the T range of thermal film deposition, and this affects gas heat capacity, as we will see in Sec. 2.4. Translational energy levels have the closest spacing, and their distribution can be considered as a continuum at ordinary T values.

There are three translational-energy degrees of freedom or components, corresponding to the three orthogonal directions of velocity: v_x, v_y, and v_z. The total velocity, v, of a molecule is the vector sum of these three components, and it may be represented as a vector in velocity space as shown in Fig. 2.3. The absolute magnitude of v is called the molecular speed, c, which traces out a spherical surface of radius c as v changes direction. Thus, c is a scalar quantity that represents the magnitude of v independent of direction. The molecular translational energy of one degree of freedom (say, x) is given by

$$\varepsilon_{tx} = \frac{1}{2}mv_x^2 \qquad (2.2)$$

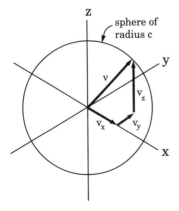

Figure 2.3 Representation of molecular velocity, v, in velocity space. Molecular speed is represented by c.

Now, at thermal equilibrium, it is known from Boltzmann statistics that, for each degree of freedom, the probability of finding a molecule at energy level ε is exponentially distributed in energy according to the Boltzmann factor, $\exp(-\varepsilon/k_B T)$, where k_B is Boltzmann's constant. This is the equilibrium distribution in ε because it provides the *maximum number of ways of arranging a group of molecules among the available quantum states to yield the same total energy for the group*. Each of the three translational-energy degrees of freedom is distributed in this way. The product of these terms gives the exponential term in Eq. (2.1), since we know from Pythagoras that $v^2 = v_x^2 + v_y^2 + v_z^2$, and we know furthermore that $v^2 \equiv c^2$. In addition, the probability of finding a molecule at speed c and energy $(1/2) mc^2$ will be proportional to the number of ways of distributing the energy among the x, y, and z components to arrive at this speed; that is, the probability is proportional to what is called the "degeneracy" of this translational energy level. The degeneracy is proportional to the surface area of the sphere of radius c in velocity space, $4\pi c^2$, which is the first term in Eq. (2.1). Finally, the middle term in Eq. (2.1) is a normalizing factor that makes the integral of the distribution over all speeds come out to unity.

The c^2 and exponential terms together lead to a peak in the speed distribution, as shown in Fig. 2.4 for Ar at 25° C. In this chapter, we

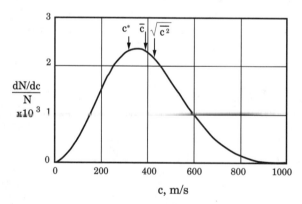

N = number of atoms in the distribution

$\bar{c}$ = mean speed

$\sqrt{\overline{c^2}}$ = root-mean-square (rms) speed

c^* = speed of sound

Figure 2.4 Maxwell-Boltzmann distribution of molecular speeds for Ar at room T.

will be expressing various properties of the gas in terms of the mean speed in the distribution,

$$\bar{c} = \sqrt{\frac{8k_BT}{\pi m}} = \sqrt{\frac{8RT}{\pi M}} \qquad (2.3)$$

or the root-mean-square (rms) speed,

$$\sqrt{\overline{c^2}} = \sqrt{\frac{3k_BT}{\pi m}} = \sqrt{\frac{3RT}{\bar{M}}} \qquad (2.4)$$

Here, the speeds have also been expressed in terms of the molar quantities R and M, where

R = $k_B N_A$ = gas constant = 8.31 J/mol·K or Pa·m^3/mol·K
   (J = joules = N·m)
N_A = Avogadro's number = 6.02 × 10^{23} mc/mol
   (mc = molecule)
M = mN_A = molar mass ("molecular weight") in kg

The above equations, obtained by integrating over the Maxwell-Boltzmann distribution, relate the molecular speeds to the measurable quantity T. The rms speed differs slightly from the mean speed because of the skew of the distribution. Both quantities are shown in Fig. 2.4, along with the speed of sound, c*, which is close in value because it is, after all, molecular collisions that carry sound pressure waves through a gas. All three speeds are seen to be near the peak of the distribution (see Exercise 2.2). Whether the mean or the rms speed is called for will depend on whether we are dealing with a quantity that is proportional to c or to c^2.

2.3 Molecular Impingement Flux

The molecular impingement flux at a surface is a fundamental determinant of film deposition rate. We can calculate this flux using the geometry of Fig. 2.5. There, a molecule at point A is shown moving at a velocity v and at an angle θ to the surface normal. We are considering a random distribution of molecular directions; that is, there is no bulk flow. This is always the case immediately above a surface, even if there is bulk flow farther away. With no bulk flow, half of the molecules have a positive value of the velocity component $v_x(\theta)$ in the direction of the surface. For a given value of v_x, the corresponding molecular impingement flux (mc/m^2·s) would be given by

$$J_i = \frac{1}{2}nv_x \quad (2.5)$$

where n is the molecular concentration (mc/m³). Now we need to determine the average value of v_x as θ ranges from 0° to 90°. For this purpose, imagine that the molecules originating at point A and having positive $v_x(\theta)$ pass through the surface of a hemisphere of unit radius (r = 1) having its origin at A. For random motion, the same molecular flux will pass through every differential area on the surface of this hemisphere. At an angle θ, the x component of velocity is given by $v_x(\theta) = v\cos\theta$. Molecules traveling at that angle pass through the dashed circle shown in Fig. 2.5, which has circumference (2π sin θ) and differential area (2π sin θ dθ). Integrating over the hemisphere and normalizing by dividing by the area of the hemisphere, A = 2π, we have for the average value of v_x over the hemisphere of impingement directions:

$$\bar{v}_x = \frac{\int v_x(\theta)\,dA}{\int dA} = \frac{1}{2\pi}\int_0^{\frac{\pi}{2}} (v\cos\theta)(2\pi\sin\theta)\,d\theta = v \cdot \frac{1}{2}\sin^2\theta \Big|_0^{\frac{\pi}{2}} = \frac{1}{2}v$$

(2.6)

Substituting into Eq. (2.5) gives

$$J_i = \frac{1}{4}nv$$

We must now average over the Maxwell-Boltzmann distribution of speeds. First, we may substitute c for v, since the distribution of magnitudes of these two quantities within the hemisphere of positive x is

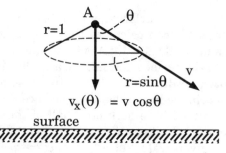

Figure 2.5 Geometry of molecular impingement.

the same. Next, since J_i is linearly proportional to c, we may average over c simply by substituting the mean speed, $\bar{c}$. Thus,

$$J_i = \frac{1}{4} n \bar{c} \tag{2.7}$$

where J_i (mc/m²·s) is the molecular impingement flux, and $\bar{c}$ is the mean molecular speed from Eq. (2.3). We next need an expression for the molecular concentration, n, in terms of p and T, to be able to express J_i entirely in terms of macroscopic quantities; so we will proceed to derive one.

2.4 Ideal-Gas Law

The ideal-gas law is an "equation of state" which expresses the relationship between pressure, volume, and temperature (p, V, and T) in the high-volume vapor and gas regions of Fig. 2.1. We can approach its derivation by considering the meaning of pressure. Pressure is the force per unit area exerted on a surface by the momentum reversal that occurs when molecules impinge on it and bounce back off. For specular reflection, the momentum change per molecule would be $2mv_x$, using the geometry of Fig. 2.5. Even though molecules bounce off of surfaces in random directions rather than specularly, the momentum reversal has the same magnitude when averaged over the hemisphere, because the molecules are *approaching* in random directions as well. Newton tells us that the force exerted in a collision process is equal to the *rate* of momentum change, which is proportional to the impingement flux given by Eq. (2.5); that is,

$$F/A = J_i(2mv_x) = nmv_x^2 \tag{2.8}$$

where F/A = force/area. Now, since $v^2 = v_x^2 + v_y^2 + v_z^2$, and because the distribution of directions is random, the three orthogonal components on the right are all the same. Thus,

$$v_x^2 = \frac{1}{3} v^2 \tag{2.9}$$

and

$$F/A = \frac{1}{3} nmv^2$$

To finally obtain the pressure, we need to average this expression for F/A over the Maxwell-Boltzmann distribution. Since F/A is proportional to v^2 and thus to c^2, we average by substituting the mean square speed, which is the square of Eq. (2.4). Thus,

$$p = \frac{1}{3}nm\overline{c^2} = nk_BT = n_m RT = \frac{N_m RT}{V} = \frac{RT}{V_m} \qquad (2.10)$$

where p = pressure, N/m² [N/m² = Pa (pascals); N = newtons = kg·m/s²]
  n = molecular concentration, mc/m³ = N/V
  n_m = molar concentration, mol/m³
   = n/N_A (= 2.4×10^{14}p mc/cm³ at 25° C)
  N_m = N/N_A = number of moles of gas
  V = volume, m³
  V_m = molar volume = 22,400 cm³/mol at 0° C and 1 atm ("standard" conditions)

The last equality is the familiar form of the ideal gas law.

The ideal gas law is obeyed when the volume of the *molecules* in the gas is much smaller than the volume of the gas and when the cohesive forces between the molecules can be neglected. Both of these assumptions are valid when the concentration n is low enough that the spacing between molecules is much larger than the molecular diameter. This is always the case in thin-film deposition from the vapor phase, where T is at or above room T and p is at or below 1 atm. These conditions place molar concentration, n_m, well away from the critical point on Fig. 2.2, since most materials have $p_c \gg 1$ atm, and those that do not, have $T_c \ll 25°$ C.

We now consider the ideal gas law in connection with the energy content of the gas. Note that Eq. (2.10) contains the molecular translational energy, $\varepsilon_t = (1/2)m\overline{c^2}$. In fact, inspection shows that ε_t is proportional to the pressure-volume product pV, which has the units of energy (N·m). By rearranging Eq. (2.10), we obtain

$$\varepsilon_t = \frac{3}{2}k_B T \qquad (2.11)$$

This important result is an expression of the equipartition theorem of classical statistical mechanics, and it will be very useful when we are discussing chemical reactions. The classical (as opposed to the quantum mechanical) statistical treatment assumes that the quantized energy levels of molecules are close enough together so that they may be approximated as a continuum. This is a good assumption for *translational* energy when T is well above absolute zero K, although it is not a good one for rotational, vibrational and electronic energy in gases. According to the equipartition theorem, energy is distributed equally among all degrees of freedom, and each one contains $(1/2)k_BT$ of energy. There are three degrees of freedom of translational energy, corresponding to the x, y, and z directions; thus we obtain Eq. (2.11). For *atomic* gases, ε_t also represents the total kinetic energy content. For *molecular* gases, there is additional kinetic energy in various rotational degrees of freedom at ordinary T and also in vibrational degrees of freedom at very high T. This energy must be included when estimating gas heat capacity, c_v, which is the rate of increase in total molecular kinetic energy with T. Thus,

$$\frac{c_v}{N_A} = \frac{d\varepsilon_m}{dT} = \frac{d(\varepsilon_t + \varepsilon_r + \varepsilon_v)}{dT} \qquad (2.12)$$

where c_v = molar heat capacity at constant volume, J/mol·K
ε_m = total kinetic energy content per molecule
$\varepsilon_r, \varepsilon_v$ = rotational, vibrational energy content

Thus, it is found experimentally that c_v for *atomic* gases is equal to $(3/2)R$ and is independent of T. Conversely, c_v for *molecular* gases is larger, the amount depending on the complexity of the molecule. For small diatomic molecules at room T, the two rotational degrees of freedom are excited, but the vibrational ones are not, so that $c_v = (5/2)R$. For all molecules, c_v increases with T as more of the rotational and vibrational quantum levels become accessible.

The heat capacity of any gas is larger when measured at constant pressure (c_p) rather than at constant volume (c_v), because then the heat input is doing pdV work on the surroundings in addition to adding kinetic energy to the molecules. From the ideal gas law, it is seen that this work is RdT per mole, so that the added heat capacity is R; thus,

$$c_p = c_v + R \qquad (2.13)$$

where c_p is the molar heat capacity at constant pressure (J/mol·K). The distinction between these two heat capacities may also be expressed in thermodynamic terms; thus,

$$c_v = \left(\frac{\partial U_m}{\partial T}\right)_V \tag{2.14}$$

where U_m = internal energy per mole, = $\varepsilon_m N_A$ [see Eq. (2.12)] except at extreme T when electronic excitation is also involved. The notation in Eq. (2.14) indicates a partial derivative, ∂, at constant V. On the other hand, at constant p we have

$$c_p = \left(\frac{\partial H_m}{\partial T}\right)_p \tag{2.15}$$

where H_m = *enthalpy* per mole. The enthalpy function is defined as

$$H_m = U_m + pV_m \tag{2.16}$$

Therefore, by differentiation,

$$\left(\frac{\partial H_m}{\partial T}\right)_p = \left(\frac{\partial U_m}{\partial T}\right)_p + p\left(\frac{\partial V_m}{\partial T}\right)_p \tag{2.17}$$

For the ideal-gas case, the second term on the right is just R, and the first term on the right is c_v, since for that case U_m is a function only of T. Thus, Eq. (2.17) is equivalent to Eq. (2.13).

In dealing with heat *transfer* in gases, c_v is the quantity to use, even though the system under study may be operating at constant p. This is because heat is transferred only by way of the internal energy; that is, pV does not contribute to the thermal conductivity. However, the heat *content* of a gas being raised in T at constant p is measured by c_p. Note also that here we have been working with heat capacity per *mole*, whereas heat capacity per *gram* is the quantity usually reported in data tables. Finally, the term "specific heat" refers to c_p per gram relative to water. It is important to keep in mind the distinctions between these various expressions of heat capacity.

2.5 Units of Measurement

In all of the above, we have used SI (Système International) units, which are based on m, kg, and s. This is the easiest way to avoid errors in any calculation, since these units and those derived from them (such as J, N, Pa) are self-consistent; that is, all conversion factors are unity. To demonstrate this point, insert SI units for the terms on the right-hand side of Eq. (2.10) and show that they cancel to yield pressure in Pa.

Unfortunately, other units are frequently encountered in thin-film technology either for historical reasons or for convenience of size. Pressure is frequently measured either in torr, where 1 torr = 133 Pa = 1 mm of rise on a mercury manometer (mm Hg); in millitorr or microns (μm Hg); or in bar, where 1 bar = 750 Torr = 1.0×10^5 Pa = 0.99 standard atmosphere (atm). (The difference between 1 bar and 1 atm is negligible for practical calculations in thin-film work.) Also, cm and cm^3 are often used because they are more convenient in size than m and m^3.

The "standard" conditions of T and p ("stp") are 0° C and 1 atm (760 torr). From the ideal gas law, Eq. (2.10), it may be found that at stp the molecular density, n, is 2.69×10^{19} mc/cm^3, and the molar volume, V_m, is 22,400 cm^3. Note that these conditions are different from the standard conditions to which thermodynamic data are referenced: those are 25° C and 1 bar.

The term usually encountered in gas supply monitoring is the "mass" flow rate measured in standard cm^3 per minute or per second or liters per minute ("sccm," "sccs," or "slm"), "standard" meaning at 1 atm and 0° C. Mass flow rate thus has the units of (pressure × volume/time), which can be seen from Eq. (2.10) to be actually proportional to *molecular* or *molar* flow rate, dN/dt, rather than to "mass" flow rate, the conversion being 1 sccm = 4.48×10^{17} mc/s. In vacuum pumping, mass flow rate (pump "throughput") is often measured in torr·l/s or Pa·l/s (l = liters).

2.6 Knudsen Equation

So far we have developed expressions for molecular energy, speed, concentration and flow rate in terms of macroscopic quantities. We can do the same for the molecular impingement flux given by Eq. (2.7), now that we have expressions for both the mean speed, $\bar{c}$, in Eq. (2.3) and the molecular concentration, n, in Eq. (2.10). Thus, in SI units, we have:

$$J_i\left(\frac{mc}{m^2 \cdot s}\right) = \frac{N_A p}{\sqrt{2\pi MRT}} \tag{2.18}$$

However, it is more common to use cm^2 rather than m^2 for flux, and g rather than kg for molecular weight, in which case we obtain

$$\boxed{J_i\left(\frac{mc}{cm^2 \cdot s}\right) = 2.63 \times 10^{20} \frac{p}{\sqrt{MT}}} \tag{2.19}$$

where the units of p, M, and T are Pa, g, and K. For p in torr, the proportionality factor changes to 3.51×10^{22}.

Equation (2.19), the Knudsen equation, is one of the most important relationships in thin-film technology. It is valid in both the vacuum and fluid regimes of operating p, and it can also be used to calculate effusion flux through an orifice if p is low enough, as will be discussed in Sec. 3.3. To get a feeling for impingement flux magnitudes, consider that for an M of 40, 25° C and 10^{-3} Pa, $J_i = 2.4 \times 10^{15}$ mc/cm^2·s. If all of this impinging gas is adsorbing and sticking on the surface in a thin-film deposition process, how rapidly does it accumulate? For a typical molecular diameter of 0.3 nm, there are about 10^{15} molecules per monomolecular layer ("monolayer" or ML), so the accumulation rate is 2.4 ML/s or 2.6 μm/h, which is a typical thin-film deposition rate (although rates can be much faster). This means that only 10^{-8} atm of reactant gas p is needed to achieve this deposition rate if all the gas impinging on the surface deposits. It also means that if this gas is a background impurity in the deposition chamber rather than a constituent of the desired film, the film is likely to incorporate a high level of contamination. To deposit a 99.9 percent pure film at 2.6 μm/h, the p of reactive impurity gases must be kept below 10^{-6} Pa, which is considered to be in the range of "ultra-high" vacuum (UHV). On the other hand, the p of *inert* background gases or of carrier gases used in CVD, such as H_2, can be much higher without compromising film purity. Indeed, CVD processes are often run at 1 atm. The impurity impingement flux is determined by the *partial* p of the impurity gas(es), not the total p. The key quantity to keep in mind is the *ratio* of film molecular deposition flux to reactive impurity impingement flux, or what one might think of as the "goodies-to-crud" ratio or the purity ratio. To improve film purity, this ratio needs to be increased, either by increasing deposition rate or by decreasing impurity partial p values.

In any thin-film process, the linear deposition rate, dh/dt, is easily determined from film thickness, h, and deposition time. The molecular deposition flux may be found from this rate if the film density is known. In SI units,

$$J_r = \frac{dh}{dt}\left(\frac{\rho_m N_A}{M}\right) \qquad (2.20)$$

where J_r = molecular deposition flux, mc/m^2·s

ρ_m = film density, kg/m^3

or in more conventional units,

$$J_r\left(\frac{mc}{cm^2 \cdot s}\right) = 1.67 \times 10^{16} \frac{\frac{dh}{dt}(\mu m/h) \rho_m(g/cm^3)}{M(g/mol)} \quad (2.21)$$

It is worth remembering that, for most solids, the molecular concentration, $(\rho_m N_A/M)$, is of the order of 5×10^{22} mc/cm^3.

Another quantity encountered in the study of gas adsorption is the langmuir, L, which refers to the amount of exposure to a gas that a surface has seen, in mc/cm^2. Exposure is proportional to pt, and 1 L = 10^{-6} torr·s. If all the impinging gas adsorbs, 1 L represents a surface accumulation of about 1/3 monolayer. Note that the term "*ad*sorption," which refers to gas accumulation on a surface, differs from the term "*ab*sorption," which refers to the takeup of a gas into the bulk of a liquid or solid.

2.7 Mean Free Path

The mean free path, l, is the mean distance that a particle travels in a gas before encountering a collision with a gas molecule. We need to estimate l for gas molecules and also for electrons and ions, since all of these particles are involved in various deposition processes. We begin with the simplest case—the electron. Consider an electron passing through an array of gas molecules of effective collision diameter a as shown in Fig. 2.6a, directly into the plane of the drawing. Since electrons are much smaller than molecules, the collision cross section, σ_m, is just the projected area of the gas molecule, $(\pi/4)a^2$. For a gas concentration of n mc/cm^3, the total projected area of gas molecules per cm^2 of gas cross section and per cm length of electron travel is nσ_m. This is the fractional area that is "filled" by collisions per cm of travel, and its inverse is the distance that the electron needs to travel before the projected area of collision cross section just fills one cm^2. This distance is the electron's mean free path, l_e. Thus,

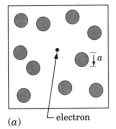
(a) electron

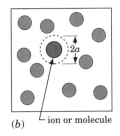
(b) ion or molecule

Figure 2.6 Geometry for mean-free-path calculation showing a particle traveling into the plane of the drawing and encountering gas molecules of diameter a; the particle is an electron is part (*a*) and an ion in (*b*).

$$l_e = \frac{1}{\sigma_m n} = \frac{1}{(\pi/4)a^2 n} \qquad (2.22)$$

(This equation is an approximation, since σ_m is actually a function of electron kinetic energy, as we will see in Sec. 9.1.) In considering instead the passage of a molecular ion, which has a radius about the same as that of a molecule, the collision diameter becomes doubled to the dashed circle shown in Fig. 2.6b, because the ion and molecule need only touch at their peripheries to collide. Thus,

$$l_i = \frac{1}{\pi a^2 n} \qquad (2.23)$$

In the above two cases, we have been able to assume that the molecules are standing still while the electrons or ions are moving through them, because the charged particles encountered in thin-film work travel much faster than molecules. However, in the case of molecule-molecule collisions, the molecules are sometimes heading straight at each other and sometimes sideswiping each other, so that on average they approach each other at 90°. Thus, the mean speed of mutual approach is $\sqrt{2}\,\bar{c}$ rather than $\bar{c}$, which shortens l by $1/\sqrt{2}$; thus,

$$l = \frac{1}{\sqrt{2}\pi a^2 n} \qquad (2.24)$$

Let us estimate the mean free path of a typical 0.3 nm diameter molecule at 1 Pa and 25° C. We first need to find n from the ideal gas law, Eq. (2.10). Working in SI units, we have

$$n = \frac{pN_A}{RT} = \frac{1 \times 6.02 \times 10^{23}}{8.31 \times 298} = 2.43 \times 10^{20} \text{ mc/m}^3$$

Inserting this into Eq. (2.24) gives

$$l = \frac{1}{\sqrt{2}\pi \left(0.3 \times 10^{-9}\right)^2 \left(2.43 \times 10^{20}\right)} = 1.03 \times 10^{-2} \text{ m} = 1.03 \text{ cm}$$

Unless T is extremely high, p is the main determinant of l, and the above example shows that $l \propto 1/p$. It is worth remembering that the mean free path at 1 Pa and room T is about 1 cm for small molecules. The order of magnitude of l is very important in film deposition, because it determines whether the process is operating in the high-vacuum regime or the fluid-flow regime. The regime is determined by the value of the Knudsen number,

$$\boxed{Kn = l/L} \qquad (2.25)$$

where L is a characteristic dimension in the process, such as the distance between the source and the substrate. For Kn > 1, the process is in high vacuum. This is also known as the molecular flow regime, since the molecules flow independently of each other and collide only with the walls. Conversely, for Kn < 0.01, the process is in fluid flow. Intermediate values of Kn constitute a transition regime where the equations applicable to either of the limiting regimes are not strictly valid. Plasma processes often operate in the transition regime. High-vacuum processes require $p < 10^{-2}$ Pa for typical chamber sizes to ensure Kn > 1.

Kn is the first and most elementary of several dimensionless numbers that we will encounter in this book. Dimensionless numbers are very useful for characterizing a process, because a particular magnitude of such a number has the same meaning irrespective of the scale or operating conditions of the process. Thus, a Kn of 1 represents the high-vacuum edge of the transition regime whether one has a 1 meter chamber at 10^{-2} Pa or a 1 cm chamber at 1 Pa.

2.8 Transport Properties

The remaining macroscopic properties that we wish to relate to molecular interaction are the transport properties, which quantify the rate of transport of mass (diffusion), momentum (viscous shear), and energy (heat conduction) through a fluid, where here again we mean a gaseous fluid. The transport to be discussed here occurs by random molecular motion through a gas which has no bulk flow in the direction of the transport. Mass and heat can be transported by bulk flow as well, but these topics will be deferred to Chap. 7. Table 2.1 summarizes the quantities and equations describing the transport properties, and they will be dealt with in turn below. Also shown in the table are the quantities relating to the flow of electricity, because that is sometimes a helpful analogy.

Transport is always described by an equation of the form

$$\boxed{(\text{flux of A}) = -(\text{proportionality factor}) \times (\text{gradient in A})} \qquad (2.26)$$

TABLE 2.1 Gas Transport Properties*

Transported quantity	Describing equation	Proportionality factor: Derivation from elementary kinetic theory	Proportionality factor: Typical value at 300 K, 1 atm
Mass	Diffusing flux = $J_A\left(\frac{mc}{cm^2 \cdot s}\right) = -D_{AB}\left(\frac{dn_A}{dx}\right)$ (Fick's law)	Diffusivity = $D_{AB}\left(\frac{cm^2}{s}\right) = \frac{1}{4}\bar{c}l \sim \frac{T^{7/4}\left(\frac{1}{M_A}+\frac{1}{M_B}\right)^{1/2}}{p\,(a_A + a_B)^2}$	Ar-Ar: 0.19 cm^2/s Ar-He: 0.72
Momentum	Shear stress = $\tau\left(N/m^2\right) = \eta\frac{du}{dx}$	Viscosity = $\eta\,(\text{Poise})^{\dagger} = \frac{1}{4}nm\bar{c}l \sim \frac{\sqrt{MT}}{a^2}$	Ar: 2.26×10^{-4} Poise[†] He: 2.02×10^{-4}
Energy (heat)	Conductive heat flux = $\Phi\left(\frac{W}{cm^2}\right) = -K_T\frac{dT}{dx}$ (Fourier's law)	Thermal conductivity = $K_T\left(\frac{W}{cm \cdot K}\right) = \frac{1}{2}n\left(\frac{c_v}{N_A}\right)\bar{c}l \sim \sqrt{\frac{T}{M}}\,\frac{c_v}{a^2}$	Ar: 0.176 mW/cm·K He: 1.52
Charge	Current density = $j\left(\frac{A}{cm^2}\right) = \frac{-1}{\rho}\,\frac{dV}{dx} = -s\,\frac{dV}{dx}$ (Ohm's law)		

*cm-g-s (cgs) units appear here instead of SI units when they are commonly used for these quantities.
†1 Poise = 1 g/cm·s = 0.1 kg/m· s or N·s/m^2 or Pa·s.

The general form of such equations is three-dimensional, but in this book we will always be using the simpler one-dimensional form, where the flux is in the x direction and the gradient is dA/dx. To make an example of the electrical analog, current density, $j(A/cm^2)$, is driven by a gradient in electrical potential or voltage, dV/dx (V/cm), and the proportionality factor is the electrical conductivity, s, or inverse resistivity, $1/\rho$ ($1/\Omega\cdot$cm or A/V·cm), as shown in Table 2.1. This is just the familiar Ohm's law.

2.8.1 Diffusion

The behavior of molecular diffusion will be described using Fig. 2.7a. Here, the concentration of the black A molecules in a mixture with the B molecules is decreasing from n_A to $(n_A - \Delta n_A)$ in the x direction over a distance of one mean free path, l. Diffusion of A occurs in the direction of decreasing n_A. A rough estimate of the diffusion flux can be made by calculating the net flux through an imaginary slab of thickness l, using Eq. (2.7) for the fluxes in opposite directions. Thus,

$$J_A \text{ (net)} = J \text{ (down at } x) - J \text{ (up at } x+l) = \tfrac{1}{4}\Delta n_A \bar{c}$$

Since $\Delta n_A = l(-dn_A/dx)$, we have

$$J_A = -\left(\tfrac{1}{4}\bar{c}l\right)\frac{dn_A}{dx} = -D_{AB}\frac{dn_A}{dx} \tag{2.27}$$

which shows that the diffusivity or diffusion coefficient of A through B, D_{AB}, is proportional to $\bar{c}$ and l. We will see that this is also the case for momentum and energy transport.

Inserting expressions for $\bar{c}$ and l from Eqs. (2.3) and (2.24), we find that D_{AB} should be proportional to $T^{3/2}/\sqrt{M}a^2p$. Empirically, the T dependence is found to be $T^{7/4}$, which is given in Table 2.1 along with

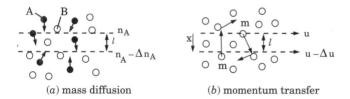

(a) mass diffusion (b) momentum transfer

Figure 2.7 Illustration of transport properties on a molecular scale. Flux is downward in the x direction.

the M and a averaging factors that account for the A-B mixture. The numerical constant is not given, because the quantitative values of D_{AB} need to be determined empirically. An empirical correlation for many binary mixtures is available [1]. For rough calculations satisfactory for thin-film deposition work, the 1 atm values given for Ar-Ar and Ar-He in Table 2.1 can be extrapolated to other T, p, and M with factor-of-two or so accuracy. T and p are the main determinants of D_{AB}. p is particularly important, since it can be lower than 1 atm by many orders of magnitude in deposition processes, with a correspondingly large increase in D_{AB}. When p is reduced so far that Kn > 1, diffusion no longer occurs. Instead, molecules travel from wall to wall in molecular flow, without encountering each other in the gas phase.

Note that D_{AB} is not dependent upon the ratio of n_A to n_B, which may seem strange at first. However, consider a very dilute mixture of A in B. Total p is constant along the n_A gradient, meaning that (n_A + n_B) does not vary with x. Therefore, dn_B/dx must be equal in magnitude and opposite in sign to dn_A/dx. In addition, for (n_A + n_B) not to increase with time at one end or the other, the diffusion flux of B through A must be equal and opposite to that of A through B; that is, $J_A = -J_B$. Thus, in accordance with the transport equation [Eq. (2.27)], $D_{BA} = D_{AB}$ even though $n_B \gg n_A$. On the other hand, the coefficient of *self*-diffusion of B through B (D_{BB}) is different from D_{BA} and D_{AB}. This is because D_{BB} involves the scattering of molecules of B by other molecules of B with no regard for what the dilute A constituent is doing.

2.8.2 Viscosity

Gas viscosity is the result of molecular momentum transport along a gradient in bulk flow velocity u. In Fig. 2.7b, gas is flowing to the right with velocity u at level x and $(u - \Delta u)$ at level $(x + l)$. The u gradient is along the x axis and perpendicular to the bulk flow. This situation is encountered in fluid flow whenever the flow stream approaches a boundary, since u must go to zero at the boundary. Superimposed upon the bulk flow is the random molecular motion at velocity $\bar{c}$, which causes molecules to continually cross up and down between flow streams separated by l, as shown in Fig. 2.7b. Those moving upward will gain momentum $m\Delta u$ upon colliding with molecules in the faster flow stream, thus exerting a drag force on that flow stream. Similarly, those moving downward will exert an accelerating force on the slower flow stream. These forces are equal and opposite in a steady-state situation. They appear as a viscous shear stress, $\tau(N/m^2)$, between the two flow streams, and τ is equal to the rate of momentum transfer per unit area perpendicular to x. This rate is equal to the molecular flux times the momentum gain/loss per molecule; thus,

$$\tau = J(m\Delta u) = \frac{1}{4}n\bar{c}ml\frac{du}{dx} = \eta\frac{du}{dx} \qquad (2.28)$$

where η = viscosity (kg/m·s = N·s/m^2 = Pa·s in SI units; g/cm·s = Poise in cgs units).

Substituting for n from the ideal gas law and for $\bar{c}$ and l as we did for diffusion, we find that $\eta \propto \sqrt{MT}/a^2$. There are two surprises here. First, η *increases* with T, which is the opposite of liquid behavior. Second, η is *independent* of p. The latter is because, although the molecules are traveling longer distances along the u gradient at lower p, there are fewer of them doing so, so these trends cancel out. As in the case of diffusion, viscosity has no meaning for Kn > 1, where molecules no longer collide with each other. In that regime, flow is still limited by the proximity of boundaries, but not due to viscous drag. Flow is limited instead by the fact that molecules bounce off of surfaces randomly rather than in specular reflection, so that they are just as likely to be scattered back in the direction from which they came as to be scattered forward. This behavior accounts for pumping in the "molecular drag" vacuum pump, which operates across the transition flow regime (see Sec. 3.1).

2.8.3 Heat conduction

Gaseous heat conduction occurs by transfer of energy in molecular collisions downward along a gradient in molecular kinetic energy ε_m. Its equation is analogous to that for momentum transfer, with $m\Delta u$ replaced by $2\Delta\varepsilon_m$. The reason for the 2 is that the average energy of molecules crossing a plane is twice the average over all directions [2], because the faster ones cross the plane more often. Thus,

$$\Phi = \frac{1}{4}n\bar{c}\left(l\frac{d\varepsilon_m}{dx}\right) \qquad (2.29)$$

where Φ = heat flux, J/s·cm^2 (= W/cm$^{2)}$. We can substitute T for ε_m using Eq. (2.12) for the heat capacity; thus,

$$\frac{d\varepsilon_m}{dx} = \frac{d\varepsilon_m}{dT}\frac{dT}{dx} = \frac{c_v}{N_A}\frac{dT}{dx} \qquad (2.30)$$

Inserting this into Eq. (2.29) gives the expression in Table 2.1 for the thermal conductivity, K_T. Values of K_T are tabulated in various handbooks [3]. As in the case of D and η, the derivation of K_T here has been simplified to focus on the physical process rather than on obtaining the exact quantitative value. Consequently, the fractions appearing in the equations for these quantities, such as the 1/4 in Eq. (2.29), are not to be considered exact, although the dependencies on n and the other molecular quantities are quite close to what is observed experimentally.

Note that in terms of the macroscopic quantities, we have $K_T \propto c_v/\sqrt{M}a^2$. Thus, small, light molecules generally have higher K_T, although this trend is sometimes reversed by the higher c_V of more complex molecules, which have more rotational and vibrational modes of energy storage. Note also that K_T, like η, is independent of p, and for the same reasons: as p decreases, the molecular flux is lower, but the molecules travel farther between collisions. However, as p decreases into the molecular flow regime, the situation changes, as we shall see below.

Consider heat transfer by gas conduction between two parallel plates as shown in Fig. 2.8. This is a common situation in film deposition, where one plate is a heated platform at temperature T_h, and the other is a substrate for film deposition that is being raised to T_s by heat transfer from the platform. For a gap of b between the plates, the appropriate Knudsen number is l/b. At higher p where Kn << 1 (Fig. 2.8a), the heat flux is given by

$$\Phi = -K_T \frac{dT}{dx} = \frac{K_T}{b}(T_h - T_s) \tag{2.31}$$

Now consider instead the case where Kn > 1 (Fig. 2.8b). Here, gas molecules are bouncing back and forth from plate to plate without encountering any collisions along the way, so the use of K_T, which is a bulk fluid property, is no longer appropriate. Instead, the heat flux between plates is proportional to the flux of molecules across the gap times the amount of heat carried per molecule, or

$$\Phi = \left[J_i \gamma' \frac{c_v}{N_A}\right](T_h - T_s)$$

$$= \left[4.37 \times 10^{-4} \frac{p}{\sqrt{MT}} \gamma' c_v\right](T_h - T_s) = h_c(T_h - T_s) \tag{2.32}$$

where J_i = molecular impingement flux, mc/cm²·s, from Eq. (2.19)
γ' = thermal accommodation factor (≈ unity except for He)
h_c = heat transfer coefficient, W/cm²·K

Here we have introduced two new quantities, γ' and h_c. γ' will be discussed below. The heat transfer coefficient, h_c, is given by the

2.8.3 Heat conduction

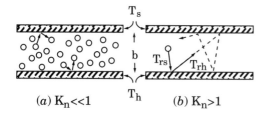

(a) $K_n \ll 1$ (b) $K_n > 1$

Figure 2.8 Gas-conductive heat transfer between parallel plates at (a) low and (b) high Knudsen numbers, Kn.

quantity in square brackets in Eq. (2.32). Note that h_c contains *no dependence on the distance* over which heat is being transferred. Note also that h_c appears, rather than K_T, whenever heat transfer is taking place across an interface rather than through a bulk fluid or other material. This applies to the situation of Fig. 2.8b, because there are no gas-gas collisions.

We can see that there are two fundamental differences between the Eq. (2.31) and Eq. (2.32) expressions for Φ. In the first (Kn << 1), Φ is independent of p, because K_T is independent of p; also, Φ is inversely proportional to b. In the second (Kn > 1), Φ is proportional to p and independent of b. These relationships and the transition region between them are illustrated in Fig. 2.9 for the case of Ar at 25° C. Here, the effective h_c in the Kn << 1 regime is K_T/b. One important conclusion which can be drawn from Fig. 2.9 is that heat transfer to a substrate from a platform can be increased by increasing the gas p between them, but only if the gap is kept small enough that Kn > 1.

Helium is often chosen to improve heat transfer to substrates because of its high K_T, but in fact it is not the best choice when Kn > 1, because of the thermal accommodation factor in Eq. (2.32). Referring again to Fig. 2.8, consider the molecule approaching the heated platform. It has the temperature T_{rs}, which it acquired by being reflected from the substrate. Upon being reflected from the platform, it will have temperature T_{rh}, and the thermal accommodation *coefficient* is defined as

$$\gamma = \frac{T_{rs} - T_{rh}}{T_{rs} - T_h} \qquad (2.33)$$

Gamma (γ) represents the degree to which the molecule accommodates itself to the temperature T_h of the surface from which it is reflected. For most molecule-surface combinations, γ is close to unity, but for He it is 0.1 to 0.4, depending on the surface [4]. If γ is less than unity and

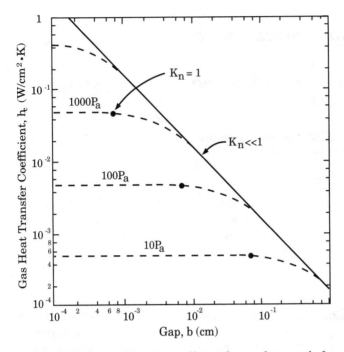

Figure 2.9 Gas-film heat-transfer coefficient, h_c, vs. plate gap, b, for various pressures of Ar. For $Kn \ll 1$, $h_c = K_T/b$, where K_T = bulk thermal conductivity of the gas.

is the same at both surfaces, it can be shown (Exercise 2.10) that the overall reduction in heat flux represented by γ' in Eq. (2.32) is given by

$$\gamma' = \frac{\gamma}{2-\gamma} \quad (2.34)$$

Thus, if $\gamma = 0.2$, then $\gamma' = 0.11$. From Eq. (2.32), it can be seen that the best choice for a heat-transfer gas is one having low molecular mass to give high J_i, while also having many rotational modes to give high c_v per Eq. (2.12). Choices will usually be limited by process chemistry, of course.

2.9 Conclusion

We now have all the basic gas-kinetics concepts that we need to examine the interaction of gases and vapors with surfaces in the deposition process. It is useful to keep in mind this motion of individual molecules even though one is controlling the film-deposition process using macroscopic quantities such as T and p.

2.10 Exercises

2.1 List six molecules that are gases (excluding the noble gases) and six that are vapors at room temperature. Distinguish the vapors listed as to whether their vapor pressures are above or below 1 atm at room T.

2.2 (a) Calculate the ratio between the mean molecular speed and the rms speed in the Maxwell-Boltzmann distribution. (b) Calculate the ratio between the mean speed and the most probable speed (the peak of the distribution).

2.3 Show that 1 sccm = 4.48×10^{17} mc/s.

2.4 Show that the proportionality factor is indeed 2.63×10^{20} in Eq. (2.19).

2.5 An Al film is being deposited at 5 μm/h in a background of 1×10^{-7} torr of O_2. What will be the maximum atomic percent oxygen in the deposit?

2.6 For a gas of 0.3 nm diameter molecules at 1 bar pressure and 25° C: (a) What fraction of the total volume is occupied by the molecules? (b) What is the ratio of the mean free path to the molecular diameter?

2.7 Show how R enters into $c_p = c_v + R$ by differentiating the ideal gas law.

2.8 Show that the molecular concentration is 2.69×10^{19} mc/cm^3 at stp.

2.9 Single-crystal Si is being deposited at 1.0 μm/h onto one face of a 4 inch diameter heated substrate in a CVD process by pyrolysis (thermal decomposition) of the reactant dichlorosilane (SiH_2Cl_2) flowing at 10 sccm. What fraction of the reactant is being utilized in the deposition?

2.10 Show that $\gamma' = \gamma/(2 - \gamma)$ for the geometry of Fig. 2.8, assuming that γ is the same at both surfaces. To simplify the algebra, normalize the temperatures to $T_h = 1$ and $T_s = 0$.

2.11 References

1. Perry, R.H., et al. (eds.). 1984. *Chemical Engineers' Handbook*, 6th ed. New York: McGraw-Hill, 3–285.
2. Fowler, R., and E.A. Guggenheim. 1952. *Statistical Thermodynamics*. London: Cambridge University Press, 124.
3. For K_T, c_v, and η, see, for example, *CRC Handbook of Chemistry and Physics*. Boca Raton, Fla.: CRC Press. For more comprehensive listings, see Touloukian, Y.S., et al. (eds.) 1970. *Thermophysical Properties of Matter*. New York: Plenum Press.
4. Ho, C.Y., S.C. Saxena, and R.K. Joshi (eds.). 1989. *Thermal Accommodation and Adsorption Coefficients of Gases*, part A. New York: Hemisphere Publishing.

2.12 Recommended Readings

Glang, R. 1970. Chap. 1, "Vacuum Evaporation." In *Handbook of Thin Film Technology*, L. I. Maissel and R. Glang (eds.). New York: McGraw-Hill.

Sears, F.W. 1950. *An Introduction to Thermodynamics, the Kinetic Theory of Gases, and Statistical Mechanics*, Chap. 12–13. Cambridge, Mass.: Addison-Wesley.

Chapter 3

Vacuum Technology

Most of the film deposition processes to be discussed in this book operate under some degree of vacuum. Only atmospheric-pressure CVD does not, but the same vacuum techniques of contamination reduction and process control still apply to it. Vacuum technology is a large topic which is well treated in textbooks such as those in the recommended readings list at the end of this chapter. Our purposes here are more specific: first, to become oriented to the general topic and, second, to examine certain aspects of vacuum technology that are particularly relevant to film deposition and deserve special emphasis. As we know, "Nature abhors a vacuum," so good equipment and techniques are needed to create one.

Figure 3.1 is a schematic diagram of a typical vacuum system for thin-film deposition. The purpose and functioning of the components shown will be elaborated upon in the subheadings below. Sometimes not all of these components will be required for a particular process. As shown, the substrate is introduced through a "load-lock" chamber to allow the main process chamber to remain under vacuum, because this reduces contamination and shortens substrate turnaround time. The roughing pump evacuates the load-lock chamber from atmospheric pressure after the substrate has been loaded into it and before the valve is opened into the process chamber. Once the substrate is in the process chamber, it is heated and controlled at the film deposition temperature. Process gases and vapors are metered into the chamber through mass flow-controlled supply lines, which are discussed more in Sec. 7.1.2. Process pressure is measured by a vacuum gauge that can be coupled to a motor-driven throttle valve in the pump throat for pressure control. Sometimes, pressure is controlled instead by cou-

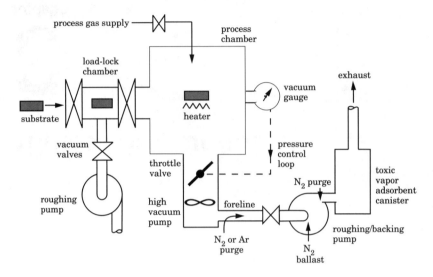

Figure 3.1 Typical vacuum-system components for thin-film deposition.

pling the vacuum gauge to the gas-supply metering valve, but that technique does not allow independent control of gas flow rate and pressure. Finally, process and impurity gases are evacuated through a high-vacuum pump followed by a "backing" pump which often serves to rough out the process chamber from atmosphere as well. For process vacuums above 10 Pa, only one stage of pumping is needed. The foreline and "ballast" nitrogen/argon purges which are shown are often required for reducing process and pump contamination, respectively. The exhaust nitrogen purge and the vapor-adsorbent canister provide safe disposal of flammable and toxic process vapors, respectively.

3.1 Pump Selection and Exhaust Handling

The principal types of vacuum pumps are listed in Table 3.1, along with their key characteristics. The choice of pumps will depend on the process vacuum level and on the properties of the vapors to be handled. Pumps fall into two categories by pumping principle: those that *displace* gas from the vacuum chamber and exhaust it to atmosphere, and those that *trap* it within the pump itself. Displacement pumps are often oil lubricated, which means that great care must be taken to avoid contaminating the process chamber with oil. On the other hand, they can pump large gas flows continuously without becoming saturated like trapping-type pumps do. Trapping pumps of the cryogenic variety are not recommended when pumping flammable vapors,

TABLE 3.1 Vacuum Pump Characteristics

Pressure ranges	Name	Category	Approx. $/(l/s)	Backing pump req'd?	Oil present? Inlet	Oil present? Outlet	Problematic gases and vapors	Other comments
	Dry rotary	Displacement	1000	No	No	Yes	Condensables require gas ballasting; see text	
	Oil-sealed rotary	Displacement	300	No	Yes	Yes		Common for roughing/backing
	Roots blower	Displacement	70	Yes	No	Yes		Oil contam. unless foreline purged
	Molecular drag	Displacement	35	Yes	No	Yes*	Low compression ratio for H_2 and He	
	Turbo-molecular	Displacement	40	Yes	No	Yes*		
	Oil diffusion	Trapping	5	Yes	Yes	Yes		Greatest risk of oil contam.
	Cryosorption	Trapping	450	No	No	(No outlet)	Explosion danger with flammables	For dry roughing
	He-cycle cryopump	Trapping	7	No	No	No†		Low capacity for He, H_2
	Sputter-ion	Trapping	25	No	No	(No outlet)	Poor for inerts	

Pressure ranges axis: $-10, -5, 0, +5$ Pa / $-10, -5, 0$ Torr, 1 atm

Legend: ultimate limits (dashed), process operation (solid)

molecular ↔ transition ↔ fluid

Flow regime for 5 cm diameter tube

log of pump inlet pressure

*except magnetically levitated bearing types, which use no lubrication
†Purge roughing pump line to avoid oil contamination during warmup regeneration cycle.

37

because usually air is pumped too. When the saturated pump is warmed up to regenerate it, the released gas can form an explosive mixture. Sputter-ion pumps are trapping pumps that do not release chemically active trapped gases, but they can periodically release bursts of inert gases during operation. Unlike the cryogenic trapping pumps, they cannot be regenerated when saturated but must have their internal parts replaced.

The next consideration is that the pump operate well at the desired process pressure. Notice in the table that the recommended pressure limits for process operation are much narrower than the ultimate pressure limits. The ultimate upper limit is the maximum pressure at which the pump can be started, whereas the process upper limit is the maximum pressure at which it will operate well on a continuous basis. Excessive operating pressure causes overheating or stalling, as well as rapid saturation in the case of trapping pumps. If the maximum starting pressure is less than 1 atm or if a turbomolecular or molecular drag pump is being used, the system must first be roughed out with another pump. The lower pressure limit of process operation is the pressure below which pumping speed drops off, except in the case of the oil-sealed rotary pump, which is widely used for both roughing and backing. There, the problem at < 10 Pa is contamination from pump oil back-diffusing (backstreaming) into the process chamber, as will be discussed in Sec. 3.4.1. The other two choices for roughing and backing avoid the oil problem but have other drawbacks. The dry rotary pumps are larger, noisier, and more expensive. The cryosorption pumps require frequent refills with liquid nitrogen and frequent regeneration due to saturation with pumped gas.

Another consideration is the cost factor. Approximate purchase costs per unit of pumping speed for medium-sized pumps are listed in Table 3.1. There is a large difference among pumps, but in many cases the higher cost is justified by improved performance. Moreover, there are other cost factors involved, such as maintenance, service life, and whether a backing or roughing pump is required. The pumps used for backing or roughing are the dry and oil-sealed rotary pumps and the cryosorption pump. These all have relatively high cost per unit of pumping *speed* in l/s. However, they require a much smaller speed than the high-vacuum pump on a given chamber, because of their higher operating pressure [see Eq. (3.3) below].

The molecular-drag pump is the most recent addition to the selection of available pumps. Sometimes, it is integrated with a turbomolecular pump in a single unit, thus increasing the upper operating pressure of the latter. By itself, it still has a wide operating pressure range that spans the transition-flow regime. This range encompasses all of the glow-discharge plasma deposition processes as well as some

CVD processes, so this pump is particularly useful for thin-film work. Moreover, its principle of operation is a classic illustration of gas kinetic behavior. As illustrated in Fig. 3.2, the pump incorporates a channel between the stationary surface of a "stator" and the surface of a "rotor," which is moving to the right at a supersonic velocity u_r. In construction, one of the two surfaces is a spiral channel, and the other is a cylinder surrounding it and almost touching the channel rim. In the molecular-flow regime, molecules bounce from surface to surface without encountering collisions along the way. They approach the rotor surface with mean thermal velocity $\bar{c}$ and bounce off of it having the sum of two velocities, namely: $\bar{c}$, which is randomly directed; and u_r, which is directed to the right. Thus, molecules are pumped to the right. This behavior is a consequence of the strong interaction between gas molecules and surfaces, which causes them to become trapped temporarily and thus lose memory of the direction from which they approached. (Helium may be an exception, at least on very smooth surfaces.)

The molecular-drag pump can also operate at the lower end of the fluid-flow regime. There, a velocity profile from u_r to zero develops across the channel as shown in Fig. 3.2b. In the absence of pressure gradients, this profile would be linear in accordance with a force balance using Eq. (2.28) and as discussed in fluid-mechanics texts. However, the pumping action causes pressure to decrease from right to left, which causes some back flow in that direction and somewhat "collapses" the velocity profile, as shown. Now, the mass pumping rate is proportional to molecular concentration, n, and thus to pressure, p,

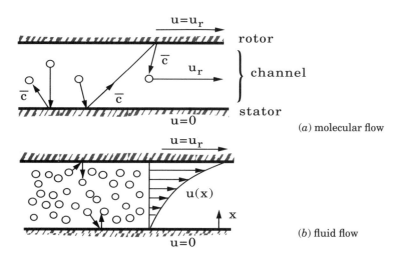

Figure 3.2 Molecular-drag pump operation in the two flow regimes.

whereas it will be seen in Sec. 3.3 that the mass flow rate induced by a p drop in the fluid regime is proportional to p^2. This means that as p is raised, back flow eventually becomes much larger than pumping rate, and this sets an upper limit to pump operating p.

3.2 Problem Gases

Some gases present special pumping problems. H_2 and He have high molecular speeds because of their low mass [Eq. (2.3)]. This limits their compression in molecular-drag, turbomolecular, or oil-diffusion pumps, all of which use supersonic velocity of the pumping medium to push molecules along. Reactive, condensable or toxic gases require special handling whatever pump is used, and we will discuss these three categories in turn below.

Reactive gases cause several problems. Acidic ones such as the chlorinated gases can decompose hydrocarbon pump oil and corrode metal pump parts. Oxidizing ones such as O_2 can explode the pump-oil vapor in the exhaust. Flammable ones such as methane can explode when mixed with air in the exhaust. Decomposition and explosion of the oil are avoided by using perfluorinated pump oils. These are organic molecules that have all of their H replaced by F, which makes them chemically inert due to the high strength of the C-F bond. Even with these, acidic gases dissolved in the oil can corrode the pump, and flammable gases can explode in the exhaust system. These two problems are avoided by N_2 purging at the pump ballast port and at the pump exhaust casing, respectively, as shown in Fig. 3.1. Sufficient and reliable N_2 purging can also prevent decomposition and explosion of hydrocarbon pump oil as an alternative to using the very expensive perfluorinated oil.

N_2 purging at the ballast port not only sweeps dissolved corrosives out of the pump oil, but also prevents **vapor condensation**. Condensates emulsify the oil and thus destroy its sealing and lubricating properties. The ballast port addresses this problem by injecting N_2 into the vapor as it is being compressed, which lowers the partial pressure of the vapor. If the dilution is sufficient, the vapor's partial pressure drops below its saturation vapor pressure, at which point no condensation can occur. In addition, the equilibrium concentration of corrosive gas dissolved in the oil drops roughly in proportion to the drop in its partial pressure over the oil. One example of a condensable and corrosive vapor situation is the $TiCl_4 + H_2O$ mixture sometimes used in the CVD of TiO_2. The saturation vapor pressures of these two reactants at room T are about 1400 and 3000 Pa, respectively. To prevent their condensation upon compression of the process stream to 1 atm (10^5 Pa), a minimum dilution of $10^5/1400 = 70/1$ is necessary. If the dilution provided in the process-gas supply and in the pump foreline is less than 70/1, the difference must be made up at the ballast port.

When **toxic gases** are involved, special procedures are necessary to protect personnel and the environment. For example, arsine (AsH_3), which is used in CVD of the important semiconductor GaAs, is one of the most toxic gases known. Oil removed from pumps for disposal or recycling should be assumed to contain some amount of any toxic gases which have been pumped, and it must be handled and labeled accordingly. The dissolved concentration can be reduced, though not necessarily to negligible levels, by a long purge with N_2 before the oil is removed. Regardless of whether any pumps are employed, the process exhaust stream must be tightly sealed and must be treated to reduce the concentration of toxics to negligible levels before release to the environment. Treatment methods include adsorption (usually on activated charcoal) as shown in Fig. 3.1, washing in a scrubber, and decomposition by burning or by catalysis. Different methods are appropriate for each vapor. The chemical manufacturer and local government authorities need to be consulted prior to selection of process equipment. In all cases, continuous environmental-monitoring sensors are advised at the gas supply, at the process chamber, and downstream of the exhaust treatment equipment.

3.3 Gas Throughput

The required sizes of the pump and its connecting line to the process chamber are determined by the gas load imposed by the film deposition process. The essential features of this situation are illustrated in Fig. 3.3. The mass flow of gas must satisfy the continuity equation, which is an expression of the law of mass (and also energy) conservation. This equation is fundamental to all flow and film-deposition pro-

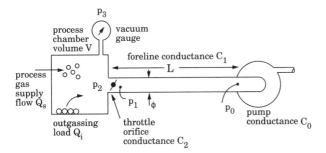

Figure 3.3 Geometry for gas throughput calculations.

cesses, and it will come up again and again throughout the book. It has the general form

$$\boxed{\text{input} + \text{generation} = \text{output} + \text{accumulation}} \quad (3.1)$$

That is, whatever mass or energy is introduced into a given space must either come out again or build up there. Here, the space is the process chamber's gas phase. The generation term typically arises with chemical reactions (mass) or heat production (energy), and it is zero here. We will first consider operation at a constant process-chamber pressure, p_2, so the accumulation term is also zero. There are two input terms: the mass flow rate of process gas supplied, Q_s, and "outgassing" from the chamber walls, Q_i (neglecting unwanted leaks, which should have been plugged). Outgassing is evolution of gaseous contaminants from chamber materials, and its minimization will be discussed in Sec. 3.4.2. The supply flow, Q_s, is often expressed in sccm, which is proportional to mc/s by the ideal-gas law, Eq. (2.10).

The output term in Eq. (3.1) is the flow of gas toward and through the pump(s). In Fig. 3.3, there are three elements in this flow path, each of which has a certain "conductance," C (liters/s, = l/s), for the gas. There is a throttle restriction (C_2), a pumping line restriction (C_1), and the pump itself (C_0). The throttle is needed only when p_2 is larger than the maximum operating p of the pump (see Table 3.1), as in the case of low-p CVD or glow-discharge processes run using a turbomolecular pump. The mass flow or "throughput," Q, past these elements is usually expressed in (Pa·l/s) or (torr·l/s), which is also proportional to mc/s by the ideal-gas law. For the two flow restrictions, C_1 and C_2, C is defined by

$$Q = C\Delta p \quad (3.2)$$

where Δp is the pressure difference across the element and is the driving force for the gas flow. For the vacuum pump, throughput is given by

$$Q_0 = C_0 p_0 \quad (3.3)$$

where the conductance C_0 is known as the pumping "speed" (l/s). Usually, C_0 varies little with p_0 over the pump's operating range, meaning that pump throughput is proportional to its inlet pressure, p_0. Setting all of these Q values equal in accordance with Eq. (3.1), and taking the case where $Q_i \ll Q_s$ (it had better be!), we have

$$Q_s + Q_i \approx Q_s = C_2(p_2 - p_1) = C_1(p_1 - p_0) = C_0 p_0 \quad (3.4)$$

3.3 Gas Throughput

For a given Q_s, the three p values will adjust themselves to satisfy these equalities. Then, control of the process pressure, p_2, is accomplished by throttling down C_2 as shown in Fig. 3.1.

Conductances depend on geometry and flow regime. The simplest case is an **orifice in molecular flow,** meaning that the molecular mean free path is larger than the orifice diameter (Kn > 1). In this case, the flux through the orifice in each direction (downstream and upstream) is equal to the impingement flux at the plane of the orifice as given by Eq. (2.18). It is an important property of molecular flow that these two fluxes are independent of one another. This is because of the fact that the molecules cross paths without colliding. (We will see in Sec. 3.4.1 how this causes oil backstreaming.) The orifice throughput is the difference between the downstream and upstream fluxes through the orifice times its area, A; thus

$$Q_2 = (J_{i2} - J_{i1})A = \left[\frac{N_A}{\sqrt{2\pi MRT}}\right] A(p_2 - p_1) = C_2(p_2 - p_1) \quad (3.5)$$

The term in square brackets is the orifice conductance per unit area, C_A. For air at room temperature, conversion from SI units to liters and cm gives $C_A = 11.6$ $l/\text{s·cm}^2$, a very useful quantity to remember, since any restriction in the vacuum plumbing can be modeled approximately as an orifice. The conductance of an **orifice in fluid flow** is much more complicated to analyze, but it is also much higher. Therefore, it usually is not the throughput-limiting element in Fig. 3.3 unless it is made so deliberately, as in the case of the throttle valve. Appendix E gives the conductance of a fluid-flow orifice in the limit of sonic velocity in the orifice ("choked" flow).

For **long tubes in molecular flow,** the conductance in l/s for air at room temperature is

$$C_m = 12.3\phi^3/L \quad (3.6)$$

where ϕ = tube diameter, cm
L = tube length, cm; L $\gg$ ϕ

Note that C_m is proportional to ϕ^3, versus ϕ^2 for the orifice in Eq. (3.5). Two powers of ϕ come from the area factor as in the case of the orifice, and the third power comes from the fact that the axial distance traversed by a molecule between bounces off the wall is proportional to ϕ. For gases other than air, C_m scales according to Eq. (2.18). For **long tubes in fluid flow,**

$$C_f = 1.41\phi^4 \bar{p}/L \tag{3.7}$$

where $\bar{p}$ is the average pressure from end to end, in Pa. Note that C is independent of p in molecular flow but proportional to p in fluid flow. Thus, in fluid flow, Q is proportional to p^2. One power of p comes from the concentration of the fluid, and the second comes from the Δp driving force for flow. The third and fourth powers of ϕ here come from the viscous drag force, which is proportional to the radial gradient in axial flow velocity [Eq. (2.28)]. This velocity is inversely proportional to tube area for a given volumetric flow rate. For gases other than air, C_f scales inversely with viscosity. In the transition region (0.01 < Kn < 1), both of the above equations will estimate C somewhat low.

The minimum size of foreline tube and pump for a given process can be calculated from the anticipated gas load, Q_s. The tube should be sized so that its C is at least as large as that of the pump, since tubes are less expensive than pumps. However, depositing films often become contaminated by the outgassing load, Q_i. In such cases, a larger pump than the minimum is desirable. For processes operating without a throttle, faster pumping (larger C_0) reduces the partial p of these contaminant gases, p_i, in accordance with Eq. (3.3): that is,

$$p_i = Q_i/C_0 \tag{3.8}$$

For processes operating at some fixed pressure p_2 with a throttle, faster pumping instead allows the supply gas flow, Q_s, of (presumably pure) process gas to be higher, which dilutes the flow of outgassing contaminant, Q_i, and thus reduces its p_i by

$$\frac{p_i}{p_2} = \frac{Q_i}{Q_i + Q_s} \approx \frac{Q_i}{Q_s} \tag{3.9}$$

This equation assumes that the gases mix well in the chamber and that the ideal-gas law holds. Note here that p_i also scales with p_2, whereas in the unthrottled case of Eq. (3.8), p_2 does not appear. It does appear in the throttled case because it is presumed that Q_s is already set at the maximum which the pump can handle. Thus, p_2 cannot be increased by increasing Q_s, which would have diluted the contaminant. Instead, p_2 can only be increased by closing the throttle to reduce C_2, which also raises p_i. For this reason, film purity is improved by reducing process pressure, p_2, in cases where the pump must be throttled. Alternatively, a pump operable at higher p_0 can be selected. The molecular-drag pump is attractive here, because it also minimizes oil contamination. Equation (3.9) also applies to atmospheric-pressure CVD, where Q_s *can* be increased to reduce p_i, because there is no pump throughput limitation.

The rate of pumpdown from 1 atm can also be calculated from the continuity equation, Eq. (3.1). Here we are considering only evacuation of the air in the process-chamber volume, V; we are not supplying any additional gas. Therefore, the input term is zero. The accumulation term is the evacuation rate. We also assume that the output term is limited by the pump throughput, since (1) the throttle valve will be open, and (2) the tube throughput is proportional to p^2 and will be relatively high at the high-p end. Thus, using the notation of Fig. 3.3, we have $p_2 = p_0$, and Eq. (3.1) becomes

$$0 = C_0 p_2 + V\frac{dp_2}{dt} \qquad (3.10)$$

where both terms have the units of throughput (Pa·l/s). Rearranging, integrating, and applying the initial condition that $p_2 = p_{2_0}$ at time t = 0, we have

$$p_2 = p_{2_0} \exp\left(\frac{-t}{V/C_0}\right) \qquad (3.11)$$

This is a classic exponential-decay situation and is analogous to discharging a capacitor through a resistor. V/C_0 is the time constant for the process, and for a typical case of a 100 l chamber and a 10 l/s roughing pump, the time constant is 10 s. This means that evacuation to 10^{-5} Pa would take place in a mere 4 min. In practice, however, Eq. (3.11) is obeyed only down to 10 Pa or so. At lower p, evacuation slows down progressively as the added load from outgassing of the chamber walls now dominates the situation. Outgassing is one of the major contamination sources that we will discuss in the next section.

3.4 Contamination Sources

The sensitivity of a deposition process to contamination varies greatly with the materials involved and depends on whether the depositing film incorporates or rejects the contaminants arriving at its surface. This, in turn, depends on the chemical reactivity of the surface with the contaminants. For example, gold is not very reactive and can be deposited with high purity in a poor vacuum, whereas aluminum reacts with and incorporates almost all arriving contaminants except the inert gases. The degree to which contamination needs to be controlled must be assessed separately for each material system by analysis of the resulting film. Whenever film properties are poorer than desired, the question should be asked: to what degree are contaminants responsible for this result?

As discussed in Chap. 1, contaminants can be introduced in all four steps in the deposition process: source, transport, deposition, and analysis. The contaminants associated with the source and analysis steps will be addressed in Chaps. 4 and 10, respectively. Here, we are concerned with contaminants entering the vapor environment between the source and the substrate. These adsorb on the substrate before film deposition commences, and they mix with the transporting source material during deposition. Their reduction requires the application of good vacuum practice, whether the process is operating in ultra-high vacuum (UHV) or at atmospheric pressure. Substrates also can be contaminated before introduction into the process chamber, and this problem will be addressed in Chaps. 5 and 6. The principal sources of contamination entering the process vapor environment will be dealt with in turn below; they are: oil backstreaming from the pumps, gas evolution from chamber materials, and dust stirred up from surfaces.

3.4.1 Oil backstreaming

Oil backstreaming into the process chamber can occur whenever oil is used as the pump operating fluid or lubricant. The rate at which this occurs is much higher than one would predict from the relatively low room-temperature vapor pressures (p_v) of oils used in vacuum technology. This is because pumps usually run hot, and the p_v of any material rises very steeply (exponentially) with T. For example, a typical rotary pump oil has a p_v of 10^{-4} Pa at 25° C but a p_v of 10^{-1} Pa at a pump operating temperature of 85° C.

Backstreaming can occur in both the molecular-flow and fluid-flow regimes, but the mechanisms and remedies differ with regime. Molecular flow is illustrated in Fig. 3.4a for the tube between the process chamber and an oil diffusion pump. Despite the flow of gas toward the pump, oil molecules bounce freely backward toward the process chamber without encountering any resistance from the countercurrent gas flow, because the gas and oil molecules do not collide with each other when Kn > 1. Some but not all of the oil will condense on the room-temperature walls of the tube before reaching the chamber. The remedy is to place in the tube a baffle which is optically opaque, meaning that there is no line-of-sight path for an oil molecule through it without encountering at least one surface (two is better). Figure 3.4a shows the traditional "chevron" baffle or trap, named after the look-alike sergeant's stripes. The trap is cooled with chilled water or, better yet, liquid nitrogen (LN$_2$), so that the oil condenses on it and stays

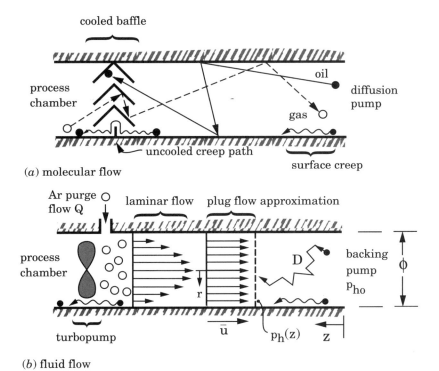

Figure 3.4 Oil backstreaming behavior in the two flow regimes, showing both gas-phase and surface paths

there. However, the trap will be warmed up on occasion, either for cleaning or due to coolant failure. A valve can be used to block off the process chamber on such occasions, but then oil still reaches the downstream face of the valve and is bound to work its way into the chamber eventually. Gas purging of the warm trap can stop the backstreaming, as we will see below, but this is getting complicated. It is clear why oil diffusion pumps are listed in Table 3.1 as having the highest risk of oil backstreaming.

A more insidious backstreaming mechanism which is independent of flow regime is surface diffusion (or "creep," or migration), as shown in Figs. 3.4a and b. It is possible to stop this with a trap having no surface pathway through it that does not encounter an LN_2-cooled surface, but most traps do not have this feature. Another possible but undocumented remedy is to heat the surface so that the oil evaporates from the surface and into the cold trap or, in fluid flow, into the gas stream. Even turbomolecular pumps, which are generally accepted as being clean due to the very high compression ratio for heavy molecules such as oil, would seem to be susceptible to surface creep.

Turbopumps and molecular drag pumps lose their high compression ratio when stalled, such as during a power failure or following bearing seizure. Then, rapid backstreaming can occur through the gas phase if the pressure in the pump remains low enough, so these pumps should be interlocked to vent automatically on such occasions.

Unless one is careful, molecular-flow backstreaming also will occur in the foreline between the high-vacuum pump and the backing pump and in the roughing lines to all chambers and pumps. It will happen whenever the pressure in these lines is allowed to drop into the transition flow regime (Kn > 0.01). Thus, a 5 cm diameter tube should never be allowed to drop below 20 Pa. Oil roughing pumps should be valved off below this pressure, and backing-pump forelines should be held in the fluid-flow regime with a purge of N_2 or Ar unless oil already exists by design at the outlet of the high-vacuum pump. Alternatively, there are foreline traps available which use a canister of zeolite pellets to trap the oil. Zeolite, or "molecular sieve," is a ceramic with extremely fine porosity which gives the pellets an enormous internal surface area of about 1000 m^2/g. The chemically active surface holds onto oil that adsorbs on it until the trap becomes saturated and is regenerated by baking. Like the chevron baffles, these traps are risky. They can become saturated without anyone realizing it, or they may get baked without the required continuous purge.

Foreline purging is the most reliable way to prevent backstreaming. Figure 3.4b shows an Ar purge flow, Q, entering just downstream of a turbomolecular pump. The same technique can be used for Roots blowers and molecular-drag pumps. N_2 can be used instead when the presence of its backpressure in the vacuum chamber is not a contamination issue for the deposition process at hand. Again, pressure in the line must be held high enough so that Kn < 0.01, either by raising Q or by throttling the pumping line at the backing-pump end. Sometimes, process-gas flow alone is enough to keep the pressure up. Even with the purge, backdiffusion of oil will occur to some extent against the downstream flow of gas. However, diffusion is driven by a concentration gradient in accordance with Eq. (2.27). Thus, the partial pressure of oil vapor, $p_h(z)$ (h denoting hydrocarbon oil) decreases with increasing distance z upstream from the backing pump and eventually becomes negligible. The profile of $p_h(z)$ in fact reaches a steady state at which the upstream diffusional mass flow due to the p_h gradient is exactly balanced by the downstream bulk flow of oil vapor diluted in purge gas. This balance applies to every cross-sectional plane along the z direction, one such plane being indicated by the vertical dashed line in Fig. 3.4b.

3.4.1 Oil backstreaming

The solution to the above problem sets the criterion for sufficient purging and also illustrates a classic differential mass-balance situation which occurs elsewhere in film-deposition work. To solve it, we will make a simplifying assumption regarding the radial velocity profile, u(r), of the purge gas. The actual profile for laminar flow is parabolic, as shown in Fig. 3.4b and as derived later as Eq. (7.6). However, instead of this profile, we will assume "plug flow," in which u is constant across the tube diameter and is equal to the area-weighted average value of u(r), $\bar{u}$, as also shown in Fig. 3.4b. This assumption is valid for tube length L >> ϕ (diameter), because each backdiffusing oil molecule moves sideways in r as well as axially in z, so it soon samples all values of u(r). Now, setting the upstream diffusional flux equal to the downstream bulk flux for oil vapor of concentration $n_h(z)$, we have

$$-D\frac{dn_h}{dz} = |\bar{u}|n_h \left(\frac{mc}{cm^2 \cdot s}\right) \tag{3.12}$$

This integrates to

$$\frac{n_h}{n_{ho}} = \frac{p_h}{p_{ho}} = \exp\left(-\frac{|\bar{u}|z}{D}\right) \tag{3.13}$$

where n_{ho} and p_{ho} are the oil vapor concentration and partial pressure at the backing pump inlet. We can find $\bar{u}$ from

$$\bar{u} = \frac{W}{A} = \frac{Q/p}{\frac{\pi}{4}\phi^2} \tag{3.14}$$

where W = volume flow rate, cm^3/s
        A = tube cross-sectional area, cm^2
        Q = gas mass flow rate, $Pa \cdot cm^3/s$ at 25° C, = sccm × 1837
        p = *total* pressure in the tube

As a conservatively high value of D, we take D (Ar-Ar) from Table 2.1. Scaling it for p, we have

$$D\ (cm^2/s) = 0.19 \times 10^5/p \tag{3.15}$$

Thus, the exponential factor in Eq. (3.13) becomes

$$\frac{\bar{u}z}{D} = 0.123\ Qy/\phi^2 \tag{3.16}$$

for units of sccm and cm. Note that there is no dependence on p, provided that the tube remains in fluid flow. For long, narrow tubes, very little flow is required to attenuate p_h. For example, with 5 sccm flowing through a 2 cm-diameter, 200 cm-long tube, $p_h/p_{ho} = 4\times10^{-14}$.

Thus, foreline and roughing-line purging can be very effective at preventing oil backstreaming through the gas phase. Surface creep still may occur, however, as shown in Fig. 3.4b. In critical applications, the only complete solution is to use an all-dry (that is, oil-free) pumping system.

3.4.2 Gas evolution

All materials constantly evolve gases and vapors, a behavior known as outgassing. Consequently, the materials used in thin-film deposition systems must be chosen and treated carefully to minimize contamination from this source. These contaminant gases come both from within the bulk of the material and from its surface, and they enter the process environment at a rate which is little affected by the pressure of the process gases. This means that the same problem exists and the same remedies apply whether one is dealing with UHV evaporation at 10^{-8} Pa or atmospheric-pressure CVD at 10^5 Pa. A few materials that evolve much less gas than others have become generally accepted for thin-film equipment, and these will be highlighted below, following some general comments about surface versus bulk sources of gas.

All surfaces that are exposed to ambient air or handled become covered with contaminants. These consist largely of oil and water, nature's two basic and ubiquitous solvents. The oil comes largely from machining and from fingerprints, and it should be removed with degreasing solvents such as acetone or trichloroethane. The water comes from adsorption of atmospheric moisture. The high polarity of the water molecule causes the first monolayer to bond quite strongly to chemically active surfaces such as metal, glass, and ceramic. After a few monolayers of adsorption, the chemical influence of the substrate dies out, and the water vapor behaves as it does over the liquid. That is, it does not condense if its partial pressure over the surface is lower than its saturation vapor pressure at that T (about 3×10^3 Pa at 25° C). In other words, it does not condense if the relative humidity is < 100 percent. But the first few monolayers do adsorb on the surface, and they stay there until the moisture-containing air is pumped away or purged out with a flow of dry process gas. Then, this water slowly "desorbs" into the process environment. ("Desorb" is the converse of adsorb.)

Although a monolayer does not seem like much, in a film-deposition environment it can be quite significant. A monolayer contains about

3.4.2 Gas evolution 51

10^{15} mc/cm^2 (mc = molecule), so a chamber having 1 m^2 of internal surface area will need to dispose of at least 10^{19} mc of adsorbed water, which is 40 Pa·l. Suppose that the room-temperature desorption rate is such that a 100 l/s pump can maintain 10^{-4} Pa during the desorption, a typical situation. Since V∝1/p, the 40 Pa·l expands to 4×10^5 l at this pressure and will take 4000 s to pump away at 100 l/s. In practice, the desorption rate and water pressure decrease gradually over a longer period of time. Alternatively, consider the same 40 Pa·l desorbing into an atmospheric-pressure process-gas flow of 10 slm (= standard liters per minute or atm · l/m). If the desorption rate is such that it adds 1 ppm of water to this flow stream, the desorption rate is 10^{-2} cm^3/m, and it will take 40 m to purge out the monolayer of water. Again, in practice the desorption rate and the ppm level decrease gradually with time. Thus, background water levels on the order of 10^{-4} Pa in a high-vacuum process and 1 ppm in an atmospheric-pressure process are to be expected upon first startup, and in many cases these levels must be reduced considerably before good thin films can be deposited. In the above calculation, we have assumed a surface area equal to the macroscopic geometrical surface area. The microscopic surface area will be much larger if the surface is rough or porous, or if the native oxide on a metal surface is porous or dusty rather than impervious and adherent (see Exercise 3.8). Poorly finished or heavily oxidized metals will adsorb substantially more water.

The desorption rates of water and other surface contaminants can be speeded up by baking, which uses thermal energy to overcome the surface bonding. Desorption rates increase exponentially with T, so 150° C for a few hours is usually sufficient and still stays below the T tolerance limit of Viton, the usual elastomeric sealing material. The same degree of adsorbate removal can be achieved by desorption at room T over a longer period of time, typically a few days. Alternatively, exposure to UV light or to plasma can speed up desorption, but these methods only affect that part of the surface area which is directly exposed, whereas baking reaches into all the nooks and crannies, too. The amount of water introduced upon substrate loading from atmosphere can be reduced considerably by use of the load-lock of Fig. 3.1. An enormous pumping speed for water can be obtained at low cost by the use of a LN$_2$-cooled shroud within the process chamber (a Meissner trap). The vapor pressure of water at the 77 K temperature of boiling LN$_2$ is about 10^{-17} Pa (see discussion of Fig. 4.4), so all water that hits the shroud freezes out and stays there. This results in a water pumping speed of 11.6 l/s·cm^2, the same value as for orifice conductance [Eq. (3.5)]. Of course, if process gases are being used which also condense at this temperature, such as silane (SiH$_4$), this approach cannot be employed.

Outgassing contaminants come not only from the surface but also from the bulk of process-containment materials, especially elastomers and other polymers. Elastomers are resilient synthetic rubbers used for gasket seals (O-rings). Viton is a perfluorinated elastomer which has a relatively low solubility for vapors as well as high chemical inertness, and it is the material generally used for vacuum seals. When particularly aggressive reactants are present, the much more expensive Kalrez may be needed. More rigid polymers such as Teflon and Delrin are used for insulators in fixturing. All polymers are particularly susceptible to *ab*sorption of water and solvent vapors into the bulk. This can involve much larger quantities than just *ad*sorption onto surfaces. Prolonged contact with common degreasing solvents should be avoided because of solvent absorption; instead, a quick swipe with a methanol-dampened lint-free cloth is recommended for the degreasing of polymer surfaces.

Because the outside of an O-ring seal is always exposed to atmosphere, it continually absorbs moisture from the air. This water diffuses through the O-ring and outgasses into the process chamber. Since the water is being replenished continually, it never goes away. Consequently, water is always the dominant background gas in a clean, elastomer-sealed vacuum chamber, and it limits the vacuum level to the 10^{-6} Pa range. Elastomers used only on internal seals not exposed to atmosphere eventually do lose their water, so if these are the only elastomer seals present, vacuum levels of $<10^{-8}$ Pa can be attained.

Metals outgas both high-vapor-pressure (p_v) alloying elements and dissolved gases. Common alloying elements to be avoided because of high p_v include Zn (in brass), Cd (in electroplated hardware), Pb and Sb (in soft solder), P (in phosphor-bronze bearings), S and Se (in type 303 stainless steel), and Te. Appendix B gives p_v data for the elements. Stainless steel types 304(L) and 316(L), which are Fe-Ni-Cr alloys, are the most common vacuum metals. They are free of S and Se, nonmagnetic, easily welded, and corrosion-resistant. They form a nonporous Cr oxide surface, and the L-designated versions have low carbon content, which makes them less susceptible to destruction of this protective layer by Cr carbide formation during welding. For parts which need to be heated above 450° C or so, stainless steels are not recommended because of metallurgical degradation and because of volatilization of their small Mn content. Mn evolution from hot stainless steel was a major problem in the early days of GaAs molecular-beam epitaxy, because it forms an electrical-charge-carrier trap in this semiconductor, so ppm concentrations of Mn can destroy the electrical conductivity of GaAs. The refractory (high-melting) metals W, Ta, and Mo are recommended for high-T applications. Among the three, Ta has the advantage of not becoming brittle after incandescent (glowing) heating. W has the highest melting point and lowest p_v of any metal, which is why it is used in light-bulb filaments. For solder-joining or sealing of vacuum equipment, the one low-T alloy which is suitable is the Sn-4 at.% Ag eutectic, which melts at 221° C. Sn has an extremely low p_V despite its low melting point.

Gas evolution from stainless steel consists mainly of CO, followed by H_2, CO_2, and CH_4. These gases come from reactions of dissolved O and H with the C in the steel. Outgassing rate can be reduced to the order of 10^{-10} Pa·l/cm·s by baking. Aluminum alloys can have even lower outgassing rates [1], but special surface treatment is required to prevent porous-oxide formation, welding is more difficult, seals are more easily damaged, and Al is subject to attack by elements often encountered in thin-film work, such as Ga and Cl.

Good choices for electrically insulating components in film-deposition systems include quartz, the alkali-free borosilicate glasses such as Pyrex, nonporous grades of silica-alumina ceramic, and pyrolytic boron nitride (PBN). Boron nitride has the valuable advantage of being easily machinable, and PBN is nonporous, unlike the less expensive hot-pressed variety of BN. These inorganic insulators are less gassy but more expensive to fabricate than the polymeric ones mentioned above.

Another source of gas evolution is the "virtual" leak. The one illustrated in Fig. 3.5 was produced by capturing a volume of gas between a substrate and its heater enclosure. The gas in this volume has only crevices through which to pump out, so evacuation will take a long time. Such situations often are misinterpreted as real leaks from atmosphere. The remedy is to provide a pumpout hole into the gas cavity. Another common place for virtual leaks is at the bottom of blind-tapped holes used for mounting internal hardware; here, one should use vented screws, which have holes drilled through them. Even vented cavities can be problematic when running processes in the fluid flow regime, because they act as "dead spots" in the gas flow

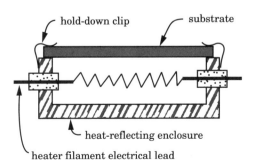

Figure 3.5 Virtual-leak problem caused by poor substrate heater design. Cross section through an enclosed cylindrical cavity is shown.

pattern. For example, if the gas composition in a CVD process is changed at one point to initiate the deposition of a second layer, the first gas will take a while to diffuse out of the cavity and get carried away. While doing so, it will contaminate the deposition of the second layer. If the dead spot cannot be avoided, pressure cycling of the second gas can be used to speed up the rate of gas exchange into the cavity by causing it to "breathe."

3.4.3 Dust

In many thin-film applications, contamination of the substrate surface with even submicron ($<10^{-4}$ cm) particles can destroy the process results. For example, it causes polycrystalline defects in epitaxial layers, electrical shorts between levels of integrated circuitry, and excessive light scattering in optical coatings. Ironically, thin-film deposition systems are copious generators of such particles. This is because films always deposit not only on the substrate of interest but also on other surfaces in the process chamber. The films on these other surfaces eventually begin to flake off (spall) due to a combination of poor adherence and stress buildup. They are also likely to deposit as loose particles because of gas-phase nucleation (more in Sec. 7.3.2). The resulting dust is easily transferred to the substrate before and during deposition. In high-vacuum processes, replaceable shielding around the source can be used to reduce deposition on chamber components. Minimizing T cycling of substrate heaters and of LN_2 shrouding reduces flaking caused by thermal stress. (Film stress is discussed in Sec. 5.6.)

Since pumpdown from atmosphere is very effective at stirring up dust, the load-lock of Fig. 3.1 is a helpful remedy; however, it is not a complete one. Even the load-lock will accumulate some particles by transfer both from atmosphere and from the process chamber. Moreover, while under vacuum, particles of previously deposited film can be *thrown* from surfaces by stress release, and these can reach the substrate even if it is facing downward. They will be held to the substrate by electrostatic forces, which are far stronger than the gravitational pull for such small particles, since they scale with area while the gravity force scales with volume. Also, the velocity of inrushing process gases can release particles and stir them up, even at process pressures of <100 Pa. Consequently, it is often necessary to periodically clean the process chamber of deposits by plasma etching, chemical etching, or mechanical scrubbing. The first technique has the advantage of not requiring disassembly, but it does require that the etching chemistry, reactant gases, and plasma equipment be available. Since larger particles are affected by gravity, it is preferable to deposit onto substrates held vertically whenever possible, so that these particles will fall neither onto the substrate nor into the source of depositing material.

Regardless of whether a load-lock is used, particle stirring can be minimized by "soft" (slow) pumpdown and venting. The conventional criterion for these operations is to avoid turbulence in the gas flow. However, the ease with which particles are lifted off of surfaces scales more closely with the gas *velocity* than with the particular gas flow pattern [2]. This velocity is kept to a minimum by using large-diameter inlet and outlet tubes and by keeping mass flow rates low. Low mass flow rate is especially important on the low-pressure ends of the cycles, where flow velocity is much higher for a given mass flow rate. It is not possible to quantify the maximum tolerable velocity, because this will vary with particle adherence and size as well as with chamber geometry. The best way to assess the situation is to examine the substrate by high-intensity light scattering after a pumpdown/venting cycle to determine the extent of particle accumulation.

When the dust from deposits is toxic, opened chambers should be dealt with cautiously to avoid breathing the dust and contaminating the surroundings. Some materials, such as As, produce a toxic reaction when they are absorbed into the bloodstream through the lung tissue or digestive tract. Others, such as Be and SiO_2, remain in the lungs and over a period of years become encased in scar tissue. Since the scar tissue does not exchange O_2, suffocation eventually results. Wearing of respirators and use of "HEPA"-filtered vacuum cleaners and wet wiping are among the precautions that should be taken in working around toxic dust.

3.5 Pressure Measurement

The above techniques have permitted us to achieve an appropriate level of vacuum to operate a deposition process. It is now necessary to measure the process pressure. In flow processes, it is usually necessary to *control* the pressure as well, by the throttle feedback technique illustrated in Fig. 3.1. Many different vacuum gauges have been developed, but we will discuss only the few most widely used in thin-film work.

The **thermocouple gauge** and the **Pirani gauge** are simple, inexpensive instruments whose operation is based on heat removal by the gas. The thermocouple gauge is illustrated in Fig. 3.6a. A constant current, I, is passed through a resistive wire, which heats up. The wire temperature, T, is monitored as a millivolt (mV) reading on a thermocouple attached to it. When no gas is present, the wire reaches a

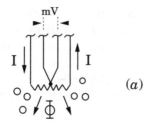

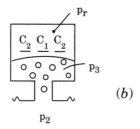

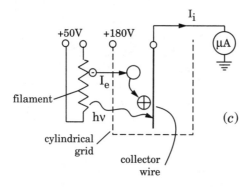

Figure 3.6 Vacuum-gauge operating principles: (a) thermocouple gauge, (b) capacitance diaphragm gauge, and (c) ion gauge.

steady-state T which is determined by radiative heat loss and by heat conduction along the four wires. In the molecular-flow regime, background gas removes additional heat in proportion to its heat-transfer coefficient, h_c, as defined in Eq. (2.32). Since $h_c \propto p$, the wire T drops with increasing p. However, when p rises into the fluid-flow regime, gas heat removal is governed instead by the bulk thermal conductivity, K_T, which is independent of p. From Fig. 2.9, it can be seen that gauge sensitivity is going to die out above 100 Pa or so for reasonable heat-transfer gap lengths. For p < 0.1 Pa, sensitivity is again lost because heat transfer through the gas becomes negligible compared to radiative and conductive losses from the wire. The thermocouple gauge also has the disadvantages of nonlinearity, calibration dependence on gas composition, and drift in the zero reading, which is the wire T at zero p.

The best gauge for the above p range is the **capacitance diaphragm gauge** or capacitance "manometer," which is illustrated in Fig. 3.6b. Here, a thin metal diaphragm separates a region at pressure p_3 from a sealed-off reference vacuum at $p_r \ll p_3$. Here, p_3 is the gauge p, which is also referred to in Fig. 3.3. The mechanical force exerted on the diaphragm by the p difference across it deflects it as shown in Fig. 3.6b and thus increases the capacitance, C_1, between the diaphragm and a fixed metal disc. The capacitance, C_2, to a fixed metal annulus increases less, and this C difference is converted to p. The use of such a differential measurement is a very effective way of decreasing drift and noise in many instruments. Here, it cancels much of the drift resulting from thermal expansion of the components. This drift can be reduced further by T control of the gauge body. Operating the gauge at an elevated T has the additional advantage of avoiding condensation of process gases which are operating near their saturation vapor pressures.

Residual T drift limits the sensitivity of the capacitance diaphragm gauge to about 10^{-3} Pa, which is actually quite remarkable when one considers that this p corresponds to a diaphragm deflection of less than one atomic diameter! The upper p limit is determined by the strength of the diaphragm or by collapse of the capacitance gap. For a given diaphragm, the dynamic range can be as high as 10^5. The gauge is usually linear to better than 1 percent, and its calibration is independent of gas composition. It is the best choice for reactive gases, because only corrosion-resistant alloys are exposed to the process and because there are no hot wires to decompose these gases. The manufacturer's calibration is usually reliable, but it can be checked against atmospheric (barometric) pressure for high-range gauges. Lower-range gauges then can be calibrated against the high-range one. The one problem with capacitance gauges is zero offset. This offset must be balanced electronically under conditions where p_3 is known to be less than the uncertainty desired in the process p reading. To do this, a pump must be available which can achieve this low a p_3, and when there is any uncertainty about the p achieved, p_3 needs to be verified using a gauge that does not have a significant zero offset, namely the ionization gauge discussed below. When oil is present at the inlet of the pump being used, time spent at low p_3 must be minimized to avoid oil backstreaming.

Still lower pressures are measured by ionizing the gas and measuring the collected ion current I_i, as shown in Fig. 3.6c. This is called the Bayard-Alpert or **ion gauge**. Ionization is accomplished by acceleration of electrons emitted from a heated filament. Electron-impact ionization is used in many areas of thin-film technology and is discussed further in Sec. 8.1. The filament is biased at about +50 V so that the electrons will be repelled from the chamber wall, which is at ground potential. The electrons are accelerated toward the cylindrical grid, which is biased at about +180 V. I_i is proportional to the electron emission current, I_e, the ionization cross section, σ_i, and the gas *concentration*, n. I_e is regulated at a specified value, typically 4 mA. The value of σ_i varies with gas composition and electron energy. Small molecules with high ionization potentials have lower σ_i. Helium has the lowest σ_i, at one-sixth that of N_2. The lower p limit of the ionization gauge is determined by the flux of high-energy photons (shown as hv in Fig. 3.6c) irradiating the collector wire from the filament; this is called the "x-ray limit." The resulting photoelectrons emitted from the collector produce a residual collector current which is equivalent to about 10^{-9} Pa. The upper p limit of this gauge is reached at about 1 Pa, where the electron mean free path, l_e, becomes too small to allow the electrons to reach the space between the grid and the collector. Between these limits, the gauge is very linear. The hot filament causes considerable outgassing from itself and surrounding materials, and this can make the local pressure, p_3, higher than the chamber pressure, p_2, by as much as 10^{-4} Pa. Most gauge controllers have a provision for degassing the gauge at a higher T than the measuring conditions. If the chamber is to be baked, degassing is most effective if carried out during bakeout.

Since the x-ray limit of an ion gauge is much lower than the sensitivity of a capacitance gauge, the ion gauge can be used to verify the low p_3 that is needed to correct the zero offset error, p_e, of the capacitance gauge. Having made this correction, the capacitance gauge's indicated pressure, p_c, will be the true p. Then, the ion gauge's indicated p, p_i, can be calibrated against p_c for each gas of interest. These procedures are illustrated in Fig. 3.7. A sufficiently sensitive capacitance gauge will be accurate at a low enough p so that the ion gauge

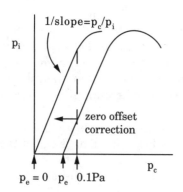

Figure 3.7 Calibration of an ion gauge (p_i) against a capacitance diaphragm gauge (p_c); p_e = zero offset error.

will be linear over some finite overlapping range of the two gauges. Once this range of overlapping linearity has been verified by the Fig. 3.7 plot, the resulting calibration factor, $f = p_c/p_i$, can be used to extrapolate p readings accurately down to the x-ray limit of the ion gauge. Sometimes, ion-gauge calibration factors are provided by the manufacturer, but it is best to verify them by the above procedure when accuracy of better than 50 percent is desired.

A valuable extension of the ion gauge involves separating the ions by mass-to-charge ratio, m/q, in a mass spectrometer. The type of mass spectrometer which uses a quadrupole electrostatic rf field rather than a magnetic field to accomplish the separation is the most convenient and is known as a **residual gas analyzer** (RGA). Examination of the gas spectrum distinguishes readily between water outgassing, pump oil, air leaks, and leaks through faulty process gas valves. Leaks from atmosphere can be located by setting the RGA to He and then blowing He gas around the outside; this is known as a He leak detector. If an *extremely low* flow of He through a fine tube is used, the leak point can be located within 1 mm. The RGA can also be used to monitor process-gas purity. However, for those impurities which also outgas from the RGA itself, such as water and CO, sensitivity is limited. The maximum operating p of the RGA is 10^{-2} Pa, because the mean free path, l_i, of the ions must be long enough for them to pass through the 20 cm-long quadrupole mass filter. Gases at higher p levels can be analyzed by separately (differentially) pumping the RGA and metering the gas into it through an orifice or adjustable leak valve. However, much better sensitivity to impurities in process gases is obtained by first ionizing the gas at the higher p and then extracting the ions into the differentially-pumped mass filter through an orifice, thus avoiding ionization of the impurities which are outgassing from the mass filter and its vacuum chamber. This technique can be extended to **atmospheric-pressure ionization** by using a corona plasma discharge from a high-voltage tip for ionization. Here, *preferential* ionization of the impurities is also obtained when the primary gas has a higher "ionization potential" [defined after Eq. (8.13)] than do the impurities, and then sub-ppb (parts per billion, or one in 10^9) sensitivity is often achieved [3]. This is a powerful technique for monitoring process-gas purity.

With any of the above vacuum gauges, T differences between the gauge (at p_3, T_3) and the process chamber (at p_2, T_2) produce an error in p measurement due to "thermal transpiration" when the gauge and chamber are separated by a restriction or tube as shown in Fig. 3.3 and when Kn > 1. In steady state, the molecular fluxes must be the same going both ways through the tube, and they will be proportional to the fluxes impinging upon each end of the tube, which are given by Eq. (2.19). This means that

$$\frac{p_3}{\sqrt{MT_3}} = \frac{p_2}{\sqrt{MT_2}} \qquad (3.17)$$

or

$$\frac{p_2}{p_3} = \sqrt{\frac{T_2}{T_3}} \qquad (3.18)$$

Thus, a gauge operating at 100° C will read 11 percent high in measuring the p of a 25° C high-vacuum chamber. Conversely, a 25° C gauge connected to a 600° C hot-wall CVD reactor will read only 58 percent of the reactor p if the connecting tube is operating at Kn > 1. The thermal-transpiration error dies out for Kn << 1, because then the p levels equilibrate by the fluid flow which develops in any p gradient. A solution to thermal transpiration has also been derived for the more complicated transition-flow regime [4]. T-stabilized capacitance gauges sometimes come with calibrations already corrected for thermal transpiration (these assume a room-T chamber).

3.6 Conclusion

The choice of vacuum equipment for film deposition depends on the level of vacuum and process cleanliness required and on the type and quantity of gas being pumped. Different deposition processes have widely different operating-pressure ranges, as we will see in subsequent chapters. Whatever the pressure, a good deal of care may be required to achieve a sufficiently clean process environment, and much of vacuum technology addresses this issue. The *tolerable* level of impurities depends on the composition of both the impurities and the material being deposited. Susceptibility of a film to contamination depends largely on its chemical reactivity, as pointed out in Sec. 3.4 for Au versus Al. It also depends on the composition of the contaminant compared to that of the film. For example, water will not contaminate the CVD of SiO_2 from SiH_4 (silane) and O_2, because water is a product of the film-forming reaction anyway. It is best to assess at the outset which contaminants are likely to be especially troublesome and which are of little concern, and then to design the system accordingly. Then, the system must be *operated* carefully to avoid introducing contamination from the atmosphere, backstreaming, outgassing, and dust. One can further reduce film contamination by predepositing chemically-active film constituents on vacuum-chamber surfaces or by purging a CVD system with reactive gases. These procedures serve to "getter" impurities out of the system just prior to deposition. Impurities in the process gases and vacuum chamber can be monitored by mass spectrometry, and impurities in the film can be analyzed by the techniques described later in Sec. 10.2.

3.7 Exercises

3.1 Show that the conductance per unit area of a molecular-flow orifice is 11.6 $l/s \cdot cm^2$ for air at 25° C.

3.2 It is desired to operate a plasma CVD process at 100 Pa with a supply flow of 200 sccm SiH_4. The chamber is being pumped through a 2 m-long, 3 cm-diameter tube. (a) What is the maximum allowable pressure at the pump inlet? (b) What is the minimum speed of the pump?

3.3 The molecular-flow and fluid-flow expressions for tube conductance should give the same value of C somewhere around the middle of the flow transition region. For what value of Kn do Eqs. (3.6) and (3.7) give the same C, using ϕ as the characteristic dimension in Kn?

3.4 A process chamber of 1 m^2 surface area is outgassing at a rate of 10^{13} mc/$cm^2 \cdot$s. For a process-gas flow of 200 sccm and a total pressure of 100 Pa, what is the partial pressure of the outgassing impurities?

3.5 List the advantages of N_2 purging of pumping systems.

3.6 What are the pump choices for a process operating at 10^{-2} Pa, and what are the relative advantages and disadvantages of each?

3.7 A pump foreline 5 cm in diameter and 300 cm long is being purged with Ar. (a) How many sccm of Ar flow are required to attenuate the pump-oil partial pressure by 10^{10} over the length of the tube? (b) What will be the total pressure, p, at the upstream end if the pumping speed is such that p at the downstream end is 50 Pa?

3.8 Suppose that the inside surface of an Al vacuum chamber is coated with 100 nm of poorly formed anodic oxide having 20 percent porosity consisting of 2-nm-diameter cylindrical pores. (a) What is the ratio of the total internal surface area of the pores to the macroscopic area of the Al surface? (b) If two monolayers of water are adsorbed on all of the internal surface area, how many scc are adsorbed per m^2 of macroscopic area?

3.9 How many seconds will it take for a 1 cm^3 virtual leak to evacuate from 10^2 to 10^{-6} Pa through a 10^{-2} cm-diameter, 1 cm-long capillary?

3.10 Water diffusing through an elastomer O-ring seal from atmosphere at a rate of 0.01 sccm is the principal background gas in a particular vacuum chamber being pumped at 1000 l/s. What is the partial pressure, in Pa, of water in the chamber?

3.11 An ion gauge which indicates pressures p_i is being calibrated against a capacitance diaphragm gauge which indicates pressures p_c and which has a zero offset of p_e. The following three data points (p_i, p_c) are taken, in units of Pa: (0.12, 0.04), (0.10, 0.03), and (0.06, 0.01). (a) What is p_e? (b) What is the ion gauge calibration factor $f = p_c/p_i$?

3.12 List the factors affecting the level of water background in a vacuum chamber.

3.8 References

1. Ishimaru, H. 1990. "Developments and Applications for All-Aluminum Alloy Vacuum Systems." *Mater. Research Soc. Bull.* July:23.
2. Strasser, G., H.P. Bader, and M.E. Bader. 1990. "Reduction of Particle Contamination by Controlled Venting and Pumping of Vacuum Loadlocks." *J. Vac. Sci. Technol.* A8:4092.
3. Siefering, K., and H. Berger. 1992. "Improved APIMS Methods." *J. Electrochem. Soc.* 139:1442.
4. Poulter, K.F., M.-J. Rodgers, P.J. Nash, T.J. Thompson, and M.P. Perkin. 1983. "Thermal Transpiration Correction in Capacitance Manometers." *Vacuum* 33:311.

3.9 Recommended Readings

Fraser, D.B. (ed.). 1980. *Pumping Hazardous Gases*. New York: American Vacuum Society.

Hablanian, M.H. 1990. *High Vacuum Technology, a Practical Guide*. New York: Marcel Dekker.

O'Hanlon, J.F. 1989. *User's Guide to Vacuum Technology*, 2nd ed. New York: John Wiley & Sons.

Santeler, D.J. 1990. Topics in Vacuum System Gas Flow Applications. *J. Vac. Sci. Technol.* A8:2782.

Tompkins, H.G. 1991. *Pumps Used in Vacuum Technology*. New York: American Vacuum Society.

Chapter 4

Evaporation

In the previous two chapters, we examined the foundations of vapor deposition in gas kinetics and vacuum technology. Now we can begin to examine specific deposition processes. In this chapter, we will address the thermal evaporation of source material and its transport to the substrate within a high-vacuum environment. This is the basic physical vapor deposition (PVD) process. In the next two chapters, we will look at the deposition behavior of the vapor once it arrives at the substrate. In subsequent chapters, we will discuss other ways of supplying vapor, including the use of chemical vapors (CVD) and nonthermal methods of vaporization: that is, energy beams and sputtering.

4.1 Thermodynamics of Evaporation

We begin the discussion of evaporation by reviewing some basic principles of thermodynamics and chemical kinetics which will be useful in many aspects of thin-film work. Evaporation is a classic illustration of these principles, so it provides a convenient context within which to develop them. Our present objective is to predict the evaporation rate of a material from available data such as the boiling point and heat of evaporation. Readers familiar with the concept of entropy as disorder, with the significance of the Gibbs free energy and the chemical potential in determining equilibrium, and with the origins of the Clausius-Clapeyron equation may wish to skip this section.

The vapor-liquid or vapor-solid equilibrium situation is shown in Fig. 4.1a. Here, we are considering either a single element or a pure compound which does not dissociate upon evaporation. Alloys, mix-

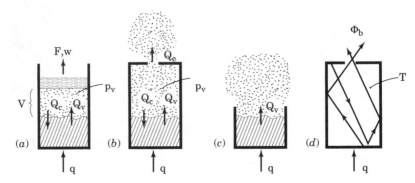

Figure 4.1 Evaporation situations: (a) vapor-liquid (or solid) equilibrium in a closed system, (b) Knudsen-cell effusion, (c) vacuum evaporation, and (d) radiation-blackbody analogy.

tures, and dissociatively evaporating compounds will be dealt with in Secs. 4.3 and 4.4. The condensed phase of our material, liquid or solid, is in equilibrium with its pure vapor in a closed isothermal container which will be the system under study. At equilibrium, the pressure in the vapor phase is the (saturation) vapor pressure, p_v, as discussed in Sec. 2.1. It is important to note that this equilibrium is a *dynamic* situation in which the molar condensation rate, Q_c, and the molar evaporation rate, Q_v, are balanced.

The system of Fig. 4.1a is closed; that is, no mass crosses its boundaries. However, energy in the form of heat, q, may be added, causing evaporation, and then energy in the form of mechanical work, $w = p_v \Delta V$, is removed from the resulting vapor as it pushes back the surrounding atmosphere or pushes on the piston shown, exerting force F. This is the principle of the steam engine, from which Josiah Willard Gibbs gained much of the inspiration for his development of the theory of thermodynamics. Not all of the heat input can be transformed into work, however. Most of it is consumed in increasing the energy of the molecules *within* the system, which is called the "internal" energy, U. In our earlier discussion of heat capacity leading to Eq. (2.14), we were concerned only with the molecular *kinetic* energy components of U. Now, we must include the much larger molecular *potential* energy component, ε_p, which accompanies the removal of a molecule from the condensed phase into the vapor phase. This removal process is illustrated in the potential-energy diagram of Fig. 4.2. The molecule in the condensed phase sits near the bottom of a potential well created by its bonds to neighboring molecules situated at the material's surface (z = 0). Evaporation involves removing the molecule to $z = \infty$ by breaking these bonds and raising ε_p. In the above interchange of heat and work accompanying evaporation and piston-pushing, total energy must be

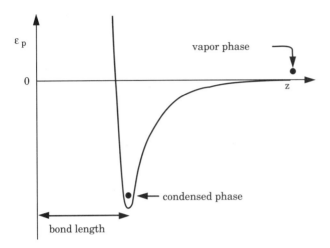

Figure 4.2 Molecular potential-energy diagram for evaporation and condensation.

conserved in accordance with the first law of thermodynamics, which is expressed in differential form as

$$dU + dw = dq \qquad (4.1)$$

That is, an increment of heat input must show up either as internal energy stored or as work output.

To further discuss equilibrium, we need to invoke the more subtle concept of entropy. Consider a process involving changes in temperature and pressure (T and p) to our system. Let these changes be carried out slowly—close to equilibrium—and let the system be brought back to its original state at the conclusion. This is called a reversible process. The classic example used in thermodynamics texts is the Carnot engine cycle. Analysis of such processes shows that they obey the second law of thermodynamics, namely,

$$\oint \frac{dq}{T} = \oint dS = 0 \qquad (4.2)$$

where the circle indicates integration around a closed loop, and T is in K. This equation defines the entropy, S; that is, $dS = dq/T$, where dq is the heat flow in a process being carried out reversibly. For reversible processes, total S of the system plus the surroundings is constant. For irreversible processes, total S increases. Entropy increase is a measure of the degree of randomization of energy that was initially in a form out of which work could be extracted, such as a difference in p, T or concentration. For example, if the force exerted by the piston in

Fig. 4.1a were not used to store mechanical energy, but rather were allowed to dissipate its energy as heat (such as by friction), total S would increase. In processes carried out near equilibrium, S does not increase because it is already as high as it can be given the amount of energy available.

The following example of entropy increase involving a concentration difference shows the connection between randomization and equilibrium. Consider the eight molecules in Fig. 4.3a, four black ones in one box (such as a crystal lattice) and four white ones in another, with a barrier between the boxes (such as an activation energy for diffusion). There are only four positions available for the four black molecules in the top box, and since the black molecules are indistinguishable from one another, there is only one way to arrange them (one "quantum state"). Now, we know that removal of the barrier causes the molecules to mix, just as salt dissolves when dropped into water. After mixing, there are most likely to be two black and two white molecules in the top box, and now there are *six* ways to arrange them, as shown in Fig. 4.3b. It may be shown by statistical arguments that

$$S = k_B \ln \Omega \qquad (4.3)$$

where k_B = Boltzmann's constant
Ω = number of quantum states

Thus, irreversible mixing has increased the entropy of the molecules in the top box from zero to 1.79 k_B. By comparison, recall from the discussion of Sec. 2.2 that the Maxwell-Boltzmann distribution is the equilibrium distribution of molecular speeds because it is the distribution that provides the maximum number of ways of arranging a system of molecules both along the ladder of quantized translational energy levels (the Boltzmann factor) and among the quantum states available at each energy level (the degeneracy), given a fixed total energy for the system. The reason for the appearance of the natural logarithm in Eq. (4.3) is that S appears in an energy term, as we will see below, and energy terms are additive, whereas probability terms such as Ω are multiplicative. The above example has been an admit-

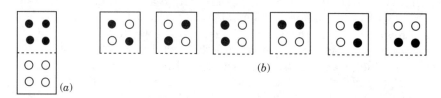

Figure 4.3 Ways of arranging four molecules in a box: (*a*) before mixing, (*b*) after mixing.

tedly cursory but necessarily brief introduction to the subject of statistical mechanics. It is included because it is important to think in terms of the degree of molecular randomization in examining chemical processes. For a more rigorous and comprehensive discussion, see Sears (1950) or other texts. There will be further discussion of these concepts in Sec. 5.2 in the context of surface diffusion.

We have shown above that at fixed energy, entropy tends toward a maximum as a system approaches equilibrium. Conversely, at fixed entropy, energy tends toward a minimum as a system approaches equilibrium. For example, entropy considerations aside, the molecule in Fig. 4.2 would fall to the bottom of the potential well in moving toward equilibrium, just as all the stuff in your house tends to end up on the floor if you don't keep picking it up. In evaporation, and in many other processes, both energy and entropy are varying. Removal of the molecule from the potential well into the vapor phase increases its potential and kinetic energy. Removal also increases its entropy, since clearly a molecule in the vapor phase has more quantum states available to it than in the condensed phase, both in position ("configurational" entropy) because it is in free space, and in energy level ("thermal" entropy) because the translational-energy quantum states are so closely spaced.

The equilibrium relationship between the competing factors of energy and entropy is seen by making the following substitutions in Eq. (4.1): dw = pdV and dq = TdS [from Eq. (4.2)]. Rearranging the result, we have

$$dU + pdV - TdS = dG = 0 \quad (4.4)$$

Here, we have also introduced G, the Gibbs free energy, often simply called the free energy. G is defined as

$$G = (U + pV) - TS = H - TS \quad (4.5)$$

Here, H is the enthalpy, which appeared earlier in connection with heat capacity at constant p [Eq. (2.15)] and which will be useful below. H is the energy term to use for constant-p processes, where pdV work is being done on the surroundings, while U would be the term to use for constant-V processes. Differentiating Eq. (4.5), we have

$$dG = dU + pdV + Vdp - TdS - SdT \quad (4.6)$$

which reduces to Eq. (4.4) for processes carried out at constant T and p, such as evaporation and many other processes in thin-film work. Thus, we see that G provides a concise definition of equilibrium; namely, a system held at constant T and p is at equilibrium when dG =

0 for any disturbance, such as the evaporation of dN_m moles of condensate. This implies that G for the system is at a minimum. Mathematically, G also could be at a maximum (unstable equilibrium), but we know that moving toward equilibrium involves decreasing H and increasing S, which means that G is *decreasing* toward a minimum by Eq. (4.5). Note that T appears in the entropy term in the above equations and gives it the units of energy by the definition of entropy in Eq. (4.2). This means that the entropy term becomes more important in proportion to T. That is, increasing the thermal motion promotes randomization, as one would expect, just as you know that it is more difficult to put your house in order when there is a party going on. During evaporation, H is raised, but this is compensated by an increase in S so that G remains at the minimum. At higher T, more H can be compensated because of the larger TS increase, so p_v increases with T, as we well know. Also note that evaporation can proceed by *absorption* of heat from the surroundings (evaporative cooling), which creates a T difference. This process is driven by the increase in S accompanying the evaporation, and it is an example of an "endothermic" (heat absorbing) reaction.

At this point, it is useful to introduce another function which expresses the incremental change in G for addition of material to a phase at constant T and p. This function also may be applied to multicomponent mixtures by defining it as

$$\mu_i = \left(\frac{\partial G}{\partial N_{mi}}\right)_{T, p, N_{mj}} \qquad (4.7)$$

where μ_i is called the chemical potential of component i, and N_{mi} is the number of moles of that component. Different components consist of different kinds of molecules. The partial differential symbol, ∂, is used above because G also is a function of T, p, and N_{mj}, where j denotes the other components; but here, the amounts of the other components are being held constant. For a single-component system such as our evaporating pure condensate, μ is just the free energy per mole of condensate, G_{mc}. We have shown that, at vapor-liquid equilibrium, G for a closed system (vapor plus liquid) does not change as evaporation proceeds at constant T and p. In practice, the system must be *slightly* removed from equilibrium so that the heat of evaporation can flow into it along a T gradient. In any case, moving dN_m moles of material from the condensate into the vapor involves no change in the G of the total system, and thus,

$$\mu_c = \mu_v \qquad (4.8)$$

where c and v denote the condensed and vapor phases, respectively. This is another way of stating the equilibrium condition, and it applies whenever the number of moles of each component is constant (that is, no reactions). The difference in μ from one phase to another when the two are *not* at equilibrium may be thought of as the driving force for motion toward equilibrium; hence the term "chemical potential" for μ. Material will move from phases of high μ to those of low μ until all the μ values are equal.

We now have all the tools we need to quantify the dependence of p_v on T. We know that at any point along the vapor-liquid (or solid) equilibrium curve of Fig. 2.2, Eq. (4.8) holds. As we move up the curve, the μ values of both phases increase, and they must both increase at the same rate in order for Eq. (4.8) to continue to hold; that is, $d\mu_c = d\mu_v$, or for our pure material, $dG_{mc} = dG_{mv}$. Since we are at equilibrium, Eq. (4.4) holds; that is, $dU = TdS - PdV$. Substituting this expression for dU into Eq. (4.6), we have

$$dG = Vdp - SdT \quad \text{or} \quad dG_{mi} = V_{mi}dp - S_{mi}dT \qquad (4.9)$$

where i denotes condensate (c) or vapor (v), and m denotes "per mole." Thus, at equilibrium between the two phases,

$$V_{mc}dp - S_{mc}dT = V_{mv}dp - S_{mv}dT$$

or

$$\frac{dp_v}{dT} = \frac{S_{mv} - S_{mc}}{V_{mv} - V_{mc}} = \frac{\Delta S_m}{\Delta V_m} \qquad (4.10)$$

Since $\Delta G_m = 0$ for evaporation, we have from Eq. (4.5) that $\Delta_v H = T\Delta S_m$, where $\Delta_v H$ is the "latent heat" (enthalpy change) of vaporization per mole. Note that the heat of vaporization is an H term rather than a U term because it is measured at constant p, not at constant V, and thus it includes the pΔV work of expansion. Substituting $\Delta_v H/T$ for ΔS_m in Eq. (4.10) leads to the Clausius-Clapeyron equation,

$$\frac{dp_v}{dT} = \frac{\Delta_v H}{T\Delta V_m} \qquad (4.11)$$

The volume term may be eliminated as follows. Consider V_{mc} to be negligible, since it is typically 1/1000 of V_{mv} at 1 atm. For ideal gases, $V_{mv} = RT/p$ from Eq. (2.10). Thus,

$$\frac{dp_v}{p_v} = \left(\frac{\Delta_v H}{RT^2}\right) dT \qquad (4.12)$$

and this integrates to

$$\ln p_v = 2.3 \log_{10} p_v = \left(\frac{-\Delta_v H}{R}\right)\left(\frac{1}{T}\right) + B \qquad (4.13)$$

where B is the constant of integration. Here, we have assumed that $\Delta_v H$ is independent of T, which is not exactly so but nevertheless is reasonable in regions far from the critical point (Fig. 2.2). Alternatively, we may express Eq. (4.13) in exponential form:

$$p_v = B' \exp(-\Delta_v H/RT) \qquad (4.14)$$

The exponential factor here is the familiar one which appears in all thermally activated chemical processes, evaporation being an example of such a process. This factor will be derived in Sec. 5.2 Most of the energy in $\Delta_V H$ goes into raising the condensate molecule out of the potential well of Fig. 4.2, and the remainder goes into the resulting vapor molecule as random translational kinetic energy (ε_t) and as the $p\Delta V$ work which it does on the surroundings (see Exercise 4.1).

Note that a plot of $\ln p_v$ or $\log_{10} p_v$ versus $1/T$ (T in K) is a straight line according to Eq. (4.13). This is a useful way to estimate the p_v of materials for which p_v versus T data are not directly available (see Exercise 4.2). Almost always, one has available [1, 2] the boiling point of a liquid, or equivalently, the sublimation T of a solid, and the $\Delta_v H$ at that T. For compounds, often just the molar enthalpies of formation from the elements, $\Delta_f H$, are listed for the two phases, from which one can calculate $\Delta_v H = \Delta_f H_v - \Delta_f H_c$. Figure 4.4 shows the Eq. (4.13) plot for water as a dashed line extrapolated from the boiling point downward using the slope calculated from $\Delta_v H$ at 100° C (40.6 kJ/mole). The solid curve shows the actual p_v data. At a p_v level of 1 Pa, which is typically encountered in source-material evaporation for thin-film processes, the linear extrapolation has overestimated p_v by ×4. This is not so bad for a ×10^5 extrapolation in p_v, but clearly it is not good enough for setting source-material flux in a thin-film process. For this and other reasons which we will examine in the next two sections, source flux must always be measured directly. Nevertheless, even an order-of-magnitude estimate of p_v is useful for several reasons. It can predict the source T that will be required fairly accurately; in the case of water, the predicted T for 1 Pa is low by only 15 K. It can predict whether a material is going to have to be cooled or capped off to pre-

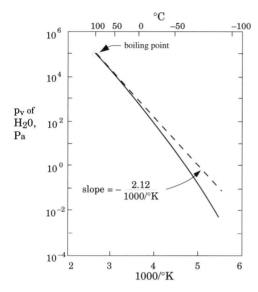

Figure 4.4 Vapor-pressure behavior of water. (Solid line = measured data.)

vent its evaporation at room T or at vacuum-chamber bakeout T; for example, the Group II (Zn, etc.) and VI (S, etc.) elements are problematic in this respect. It can predict how cold a cryogenic trap has to be to pump a material. For example, over the liquid nitrogen-cooled trap discussed in Sec. 3.4.2, the extrapolated p_v of water is about 10^{-17} Pa, which is totally negligible. Finally, it can predict how hot the walls of a CVD chamber have to be to prevent vapor condensation.

The p_v is the first piece of information needed to determine evaporation rate, Q_v, which we will discuss next. The p_v data for elements are given in Appendix B. For compounds, p_v sometimes can be found in the literature [3]; otherwise, it can be estimated as discussed above.

4.2 Evaporation Rate

When evaporation is taking place close to vapor-liquid (or solid) equilibrium, the effusion rate, Q_e, from a vapor source can be found from the vapor pressure, p_v, alone. This is the case for the vapor source shown in Fig. 4.1b. Here, the closed isothermal system of Fig. 4.1a has been modified by opening an orifice in the top which is small enough so that the resulting effusion does not significantly reduce the pressure, p, in the vapor phase; that is, $p \approx p_v$. Suppose that the orifice is small enough so that Kn > 1, and the orifice is therefore operating in the molecular-flow regime. Here, Kn is calculated from Eq. (2.25) using the orifice diameter as the characteristic dimension, L. If in addition, the orifice length is appreciably less than the orifice diameter, then the orifice is termed "ideal," and the effusion rate, Q_e, is given by Eq. (3.5). In that equation, the downstream pressure, p_1, may be taken as zero, since we are talking here about effusion into high vacuum. This situation is known as a "Knudsen cell."

To avoid reducing p significantly, the Knudsen-cell orifice has to be small enough that $Q_e \ll Q_v$, where Q_v is the evaporation rate from the surface of the condensed phase. At this point, we know Q_e, but we do not know Q_v. The determination of Q_v from first principles has not yet been accomplished, but we can reliably determine an *upper limit* to Q_v as follows. A mass balance around the vapor phase in the cell, using the steady-state form of Eq. (3.1), gives

$$Q_v = Q_c + Q_e \qquad (4.15)$$

where Q_c is the condensation rate of the vapor. Neglecting Q_e, which we have specified as relatively small, we have remaining the expression for vapor-liquid equilibrium in terms of a balance of condensation and evaporation rates. We can write the same balance in terms of fluxes per unit area of condensate surface (J = Q/A). We know the *impinging* flux, J_i, from kinetic theory [Eq. (2.18)]. Upon impingement, there is a range of degrees of interaction with the condensate surface which will be analyzed more thoroughly in Sec. 5.1. At the lower limit of this range, the vapor molecule bounces off after a single collision. At the upper limit, it becomes bonded strongly enough to the surface to become incorporated into the condensed phase. The portion of J_i that condenses is denoted as J_c. *If* all of the impinging flux does condense ($J_i = J_c$), then the vapor-liquid equilibrium ($J_c = J_v$) may be written as

$$J_i = J_{vo} \qquad (4.16)$$

where subscript o denotes the upper limit to J_v that corresponds to all of the impinging flux condensing. Moreover, we would expect J_{vo} to be unchanged if we were to remove the vapor phase and were to consider evaporation from an open crucible or surface as shown in Fig. 4.1c. We would expect this because the evaporation of an individual molecule of condensate from the surface is not retarded by the impingement of vapor molecules elsewhere on the surface. Molecular potential interactions can extend only over a few atomic distances, and the probability of a vapor molecule landing that close to a surface molecule in the course of its evaporation is very small even at p_v = 1 atm (see Exercise 4.3).

It has been verified experimentally that evaporation occurs at this upper-limit value ($J_v = J_{vo}$) for those metals that have atomic vapors, as most metals do. (The vapor compositions of the elements are given in Appendix B.) For other materials, J_v generally is lower than J_{vo}, and this is accounted for empirically by introducing an evaporation coefficient α_v, defined by

$$J_v = \alpha_v J_{vo} \qquad (4.17)$$

There is a corresponding condensation coefficient α_c defined by

$$J_c = \alpha_c J_i \qquad (4.18)$$

At equilibrium, where $J_c = J_v$, it follows that $\alpha_c = \alpha_v$. Away from equilibrium, the coefficients can differ from each other and can have differing p and T dependencies. The coefficient α_v has been determined only for a few materials; for example, it has a very low value of about 10^{-4} for As [4]. However, α_v generally is not known, and the factors that influence it are not well understood. This means that unless evaporation is being carried out from a Knudsen cell ($Q_e \ll Q_v$), effusion rate cannot be predicted accurately and must instead be measured directly in thin-film deposition processes. For the Knudsen cell, it must be verified that the orifice is small enough to avoid depleting the vapor phase (see Exercise 4.4). Even for metals having atomic vapors, where $\alpha_v \approx 1$, a melt surface that is partially obstructed from evaporating freely, such as by an oxide skin, will have an evaporation rate lower than that calculated using the macroscopic melt area.

It is useful to observe that α_c and α_v are analogous to the absorptivity, α, and emissivity, ε, of a surface for electromagnetic radiation (heat and light). Moreover, the behavior of heat radiation will be important later on for the analysis of T control in vacuum, so we digress to discuss it now. For radiation impinging on a surface, the fraction absorbed is given by α; the remainder is reflected (or transmitted). Absorptivity is a function of wavelength, λ, and of incident angle, just as α_c is expected to be a function of molecular kinetic energy and incident angle. The value of α at a given λ is called the monochromatic absorptivity, α_λ. There is a maximum radiation flux that can be emitted from a surface, just as there is a maximum evaporation flux, and both are found from the second law of thermodynamics. The maximum radiation flux is that from a so-called *blackbody* and is given by the Stefan-Boltzmann blackbody radiation law:

$$\boxed{\Phi_b = \tilde{n}^2 \sigma T^4} \qquad (4.19)$$

where Φ_b = blackbody radiation flux, W/cm^2
σ = Stefan-Boltzmann constant = 5.67×10^{-12} W/cm$^2 \cdot$K^4
$\tilde{n}$ = index of refraction of the medium over the surface ($\approx$ 1 for vapors at <1 atm)

This radiation is spectrally distributed according to Planck's law; likewise, evaporating molecules are distributed in energy. The radiation flux per unit wavelength interval is $\Phi_{b\lambda}$. The fraction of this spectral blackbody flux that is emitted by surfaces not completely black in a given λ interval is accounted for by the emissivity at that λ, ε_λ; thus

$$\Phi_\lambda = \varepsilon_\lambda \Phi_{b\lambda} \tag{4.20}$$

Moreover, it can be proven thermodynamically that $\alpha_\lambda = \varepsilon_\lambda$ (Kirchhoff's law), just as $\alpha_c = \alpha_v$ for molecules. The conventional way of constructing an ideal blackbody emitter from materials for which ε_λ is not unity is to use a cavity such as that shown in Fig. 4.1d. Although ε_λ for the internal surfaces may be <1, the radiation flux from the orifice still is given by Eq. (4.19). This is because, looking into the orifice, one sees the sum of the reflected and the emitted radiation. In effect, the radiation field within the cavity builds up to the blackbody value, just as the pressure within a Knudsen cell builds up to p_v and results in a maximum value for the effusion flux.

4.3 Alloys

So far, we have considered the evaporation only of single-component materials; that is, the elements and those compounds that do not dissociate upon evaporation. With multicomponent materials, we must deal with an additional complication, namely that the composition of the vapor phase generally differs from that of the condensed phase. We will address alloys in this section and dissociatively evaporating compounds in the next section. The distinction between the two is important. The term alloy is used to designate either a solid solution or a mixture of solid phases, and its composition is variable over a wide range. A familiar example of the solid-solution type is the solder alloy Pb-Sn or Pb$_x$Sn$_{1-x}$, where x is the mole fraction of Pb. Conversely, a compound has a specific ratio of elements, such as the semiconductor GaAs or the dielectric SiO$_2$. That is, compounds have a specific "stoichiometry." It is also possible to have an alloy of compounds, such as the laser-diode alloy (AlAs)$_x$(GaAs)$_{1-x}$, more often written as Al$_x$Ga$_{1-x}$As or (AlGa)As. Finally, there are other three-element solids that are not alloys of binary compounds but are ternary compounds; that is, they have a specific ratio of all three elements, such as the solar cell material $CuInSe_2$. Each of these categories of material behaves differently in evaporation, and it is important to know which behavior is operative to control composition during thin-film deposition.

Consider a generalized binary metal alloy B_xC_{1-x} whose component elements B and C are completely miscible at the evaporation T; that is, the atomic fraction x can vary from 0 to 1 without precipitating a second solid phase. We will consider evaporation from a well mixed liquid phase to avoid the complication of composition gradients developing within a solid condensate as evaporation proceeds. The total equilibrium vapor pressure over the melt will be the sum of the component p_v values: p_B and p_C. These p_v values will be lower than those of the pure elements because of the dilution of one element by the other; thus,

$$p_B = a_B x p_{vB} \quad \text{and} \quad p_C = a_C(1-x)p_{vC} \qquad (4.21)$$

where p_{vB} and p_{vB} are the p_v values of the pure elements and the a values are the "activity coefficients." For simplicity, we will assume "Raoult's law" behavior, where the a values are unity, even though they generally deviate somewhat from unity due to differences between B-C versus B-B and C-C bond strengths. If the evaporation coefficients are unity, as is common for metals, the evaporation flux of each element is given by the Knudsen equation, Eq. (2.18), as discussed in the previous section. Then the ratio of these two fluxes is

$$\frac{J_{vB}}{J_{vC}} = \frac{x}{1-x} \frac{p_{vB}}{p_{vC}} \sqrt{\frac{M_C}{M_B}} \qquad (4.22)$$

Thus, the vapor flux will be richer than the melt in the more volatile element for any composition x, so the melt will continue to deplete in that element as evaporation proceeds and will never reach a steady-state flux ratio. This effect is used to advantage in purging a melt of volatile contaminants during a "soak" period prior to deposition, but it is not suitable for the deposition of an alloy film.

There are two ways to achieve a steady flux ratio in such a situation. One is to use separate sources of the two pure elements, operating at different T levels. Here, it is important either to monitor each flux separately and continuously as discussed in Sec. 4.7, or to use isothermal Knudsen-cell or crucible sources whose effusion can be reliably set by T control. In either case, flux ratio will vary over the deposition area due to the separate locations of the two sources, which is undesirable unless one is specifically studying film properties versus composition. This variation can be quantified by the information given in Sec. 4.6.

The second way to achieve a steady flux ratio is to feed an alloy wire or rod steadily into the melt during evaporation, as shown in Fig. 4.5 for a feed alloy $B_y C_{1-y}$. If the volume feed rate, W, is adjusted to hold the melt volume V constant, mass conservation dictates that a steady state must eventually be reached in which the evaporating atomic flux ratio of the two elements in Eq. (4.22) is equal to their atomic feed ratio, $y/(1-y)$. During the approach to steady state, the melt composition becomes depleted in the more volatile element until the two ratios become equal at a steady-state melt composition x_s, where

$$x_s = \frac{1}{1 + \left(\frac{1-y}{y}\right)\left(\frac{p_{vB}}{p_{vC}}\right)\sqrt{\frac{M_C}{M_B}}} \quad (4.23)$$

An estimate of the time required to reach steady state can be made by applying a transient mass balance [Eq. (3.1)] to one of the elements. Doing so for B in units of atoms/s, we have:

$$\underbrace{Wyn}_{\text{(input)}} = \underbrace{Wyn\frac{x}{x_s}}_{\text{(output)}} + \underbrace{Vn\frac{dx}{dt}}_{\text{(accumulation)}} \quad (4.24)$$

where n is the atomic density (atoms/cm^3). Below, we make the reasonable assumption that n does not vary much with composition, so it may be cancelled out. Note that the output (evaporation rate) has been expressed as a fraction of the steady-state input value, Wyn, rather than directly as $J_{vC}A$ (A = melt surface area), to simplify the expression. Note also the further simplification of treating V as a constant, whereas in fact V will decrease during the transient period if W is held constant. Integration of Eq. (4.24) from an initial condition of x = y at t = 0 gives

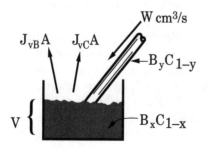

Figure 4.5 Alloy evaporation with continuous feed.

$$\xi = \frac{x_s - x}{x_s - y} = \exp\left(\frac{-t}{Vx_s/Wy}\right) = e^{-t/\tau} \tag{4.25}$$

where ξ is the fraction remaining from the initial state to the steady state at $x = x_s$. The denominator in the exponential term has the units of t and is in fact the time constant, τ, for the exponential approach to steady state. This solution is not exact because of the constant-V assumption, but the important point is that relaxation time scales as V/W. For exponential relaxation problems of this sort in which V is constant, $\tau = V/W$ exactly. This is the case for another common situation in which there is a step change in the concentration of a reactant flowing into a well mixed CVD reactor. The problem is also similar to the chamber-evacuation situation of Eq. (3.11).

For typical values of V and W in alloy evaporation, the approach to steady state can take many hours and consume much source material (see Exercise 4.7). This transient can be avoided by starting with a melt having the steady-state composition. This composition can be estimated from Eq. (4.23), but it may need to be adjusted based on actual film composition measurements if the value of $a_{B,C}$ [activity coefficients of Eq. (4.21)], α_v [evaporation coefficient of Eq. (4.17)], or S_c (sticking coefficient of Fig. 5.1) is not the same for each element. In special cases where the elemental vapor pressures are similar, x_s will be so close to y that the drift of x from y to x_s may not matter. This is the case for Al-2% Cu, which is used to prevent electromigration in integrated-circuit metal lines.

4.4 Compounds

Compounds behave very differently from alloys during evaporation. Some compounds evaporate as molecules that have the composition of the condensed phase. These compounds may be regarded as single-component materials, since they have only one p_v to consider. Ionically bonded compounds, including alkali halides such as the infrared-optics material KI and alkaline-earth halides such as the antireflection coating MgF_2, fall into this category. Oxides vary from one another in behavior, from no dissociation to partial to complete dissociation. There are limited data and some discrepancies about the degree of oxide dissociation, so it is wise to anticipate an oxygen-deficient deposited oxide film. Sometimes, reduction from a higher to a lower oxide occurs during evaporation. For example, SiO_2 evaporates as the monoxide SiO in the presence of a reducing agent M, where M is Si, C, or H_2. Since SiO has a much higher p_v than does SiO_2, these reactions have been used to thermally evaporate native oxide from Si substrates prior to film deposition. Also falling into the category of partial dissociation are the Group IV-VI compounds (the "four-sixes"), which are used for infrared detectors and diode lasers. Thus, PbTe evaporates mostly as PbTe, but partly as Pb + (1/2)Te$_2$. Some compounds dissociate completely upon evaporation, and we will examine this behavior below. Among these materials are the Group III-V semiconductors (the "three-fives") such as GaAs, which are used for light-emitting diodes (LEDs), lasers, and microwave devices. Also completely dissociating are the II-VI semiconductors such as ZnSe, which have similar but less well developed applications. Both the III-Vs and the II-VIs sublimate, and their vapors usually consist of atoms of the metallic element, which we will call M, and dimers of the nonmetallic element, which we will call Y$_2$. The III-Vs and II-VIs differ from each other in that the III metals are much less volatile than the II metals, which has consequences to be seen below.

The behavior of compounds in equilibrium with their constituent elements is described by the "phase diagram" of the system [5]. Fig. 4.6a shows a relatively simple phase diagram for a binary mixture of M and Y, which happens to form only one compound, MY. The coordinate x is the atomic fraction of Y in the mixture. The existence of MY is conventionally indicated by a vertical line at x = 0.5, but in fact there is always a narrow *range* of x over which the compound can exist

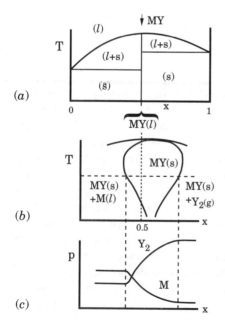

Figure 4.6 General form of phase equilibria for nonstoichiometric binary semiconductors M$_{1-x}$Y$_x$: (a) T-x projection, (b) T-x projection near x = 0.5, and (c) p-x projection. (Source: Reprinted from Ref. 6 with permission. Copyright © 1979, Pergamon Press.)

without precipitating a second phase of M or Y. This range is known as the single-phase field. A typical situation is shown in Fig. 4.6b, which is greatly expanded along the x axis. The distance in x between the M-rich and Y-rich boundaries shown here for the compound is usually less than 10^{-4}. Because it is so difficult to chemically determine such small changes in stoichiometry (see Sec. 10.2), the positions of these boundaries are not known for most compounds; but that is not important for the present discussion. The compound accommodates nonstoichiometry within the single-phase field by generating native point defects within its crystal lattice. The three types of such defects and their effect on stoichiometry are listed in Table 4.1. Usually, not all types are present in a given compound. The term "native" distinguishes these defects from point defects caused by impurity atoms. The nonstoichiometric single-phase compound may be viewed as a solid solution of native point defects in the lattice. Precipitation of M or Y occurs when the solubility range of the defects is exceeded at the M-rich or Y-rich boundary. Note that the two boundaries converge at the melting point in Fig. 4.6b; this is a manifestation of Gibbs' phase rule. Note also that the range of x does not always include the stoichiometric composition.

The important consequence of this range of x for compound evaporation and deposition is that the p_V values of M and Y_2 vary by orders of magnitude from one boundary to the other. This behavior is shown in Fig. 4.6c for a fixed T, which is indicated by the horizontal dashed line on Fig. 4.6b. The p_V of each element increases with the atomic fraction of that element, as one would expect. In each case, upon reaching the phase boundary, the p_V of the enriched element levels out at the value for the pure element, because that is when the pure element begins to precipitate out. Meanwhile, the p_V of the other element over the compound has become much less than that over the pure element. These

TABLE 4.1 Native Point Defects in Binary Compounds MY

Defect type	Symbol*	Stoichiometry
Vacancy (V)	V_M	Y-rich
	V_Y	M-rich
Interstitial	M_i	M-rich
	Y_i	Y-rich
Antisite	M_Y	M-rich
	Y_M	Y-rich

*Subscript denotes lattice location; i = interstitial site.

curves will both shift up with T, and for each element, the p_v behavior may be visualized as a sheet in p_v-T-x space. At the T shown in Fig. 4.6b, the p_v levels happen to cross each other within the single-phase region. If the crossover is large enough so that at some point before reaching the M-rich boundary, the evaporation flux of M becomes twice as large as that of Y_2, the surface composition of the compound will adjust itself by evaporation until that point is reached. From then on, evaporation will proceed "congruently;" that is, in the stoichiometric proportion. This is always the case for the II-VI compounds, and therefore they may be evaporated to completion congruently without forming precipitates. This is a good way to maintain a constant and known II/VI flux ratio at the substrate. If this ratio does not happen to be the one which yields the desired film properties, it can be adjusted by adding a second source of the pure II or VI element.

The III-V compounds present a different situation, because the p_v of M is often lower than that of Y_2 even at the M-rich boundary. In that case, M enrichment continues at the boundary by accumulation of a second condensed phase of pure M precipitate (which may also have a small amount of Y dissolved in it). Figure 4.7 shows the p_v versus T data for Ga and As_2 along each phase boundary. The upper curve for each element is the one corresponding to the phase boundary rich in that element. The point where the two curves converge at the left-hand edge is the melting T of GaAs, 1,238° C. Note that for T levels above about 680° C, $p_v(As_2) > p_v(Ga)$, even at the Ga-rich boundary, so that at high T, congruent evaporation is not possible. The maximum T for congruent evaporation depends on the evaporation coefficients, α_v, as well as on the p_v crossover point, and for GaAs that T is just over 600° C. However, at 600° C the evaporation fluxes, J_v, of Ga and As_2 are insufficient for film deposition at reasonable rates, so GaAs is deposited instead from separate Ga and As elemental sources. This is the case for the other III-V compounds as well. Elemental As evaporates as As_4, which has a lower S_c than As_2 in the deposition of GaAs, so the As_4 vapor often is superheated in a small furnace over the source crucible to crack it into As_2.

Clearly, considerable care must be taken to obtain constant and appropriate evaporation rates of the various species when dealing with dissociatively evaporating compounds. When these data are not available, one needs to do flux monitoring of each species (Sec. 4.7) and compositional analysis of both the depleted source material and the deposited film (Sec. 10.2) to define the situation. The dependence of solid-phase existence and composition on elemental p_v levels is also an important factor in determining the composition of the deposited film, and this relationship will be discussed more fully in Sec. 6.5.1. Basically, there are two major effects. One is that the native-point-

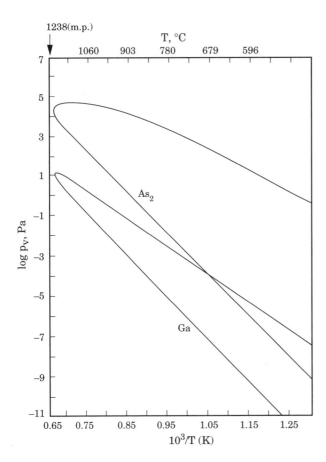

Figure 4.7 Equilibrium vapor pressures of Ga and As_2 along the GaAs liquidus, as functions of T. (Source: Reprinted from Ref. 7 with permission. Copyright © 1979, Pergamon Press. As_1 and As_4 curves deleted.)

defect distribution of the compound film shifts with the impingement-flux ratio of the constituent elements. Another is that if the impingement flux of Y_2 is too low and M is not volatile enough at the deposition T, the film will contain M precipitate as well as MY.

4.5 Sources

4.5.1 Basic designs

The common types of thermal evaporation sources are shown in Fig. 4.8. The simplest are the twisted-wire coil and the dimpled sheet-

82 Evaporation

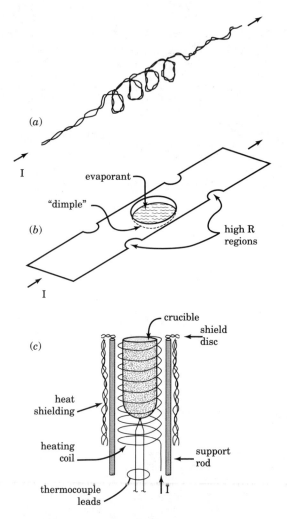

Figure 4.8 Evaporation sources: (a) twisted-wire coil, (b) dimpled boat, and (c) heat-shielded crucible.

metal "boat," both made from one of the refractory metals, W, Ta, and Mo, which were discussed in Sec. 3.4.2. These are heated by passing current of the order of I = 100 A through them (joule heating). The amount of heat generated is I^2R, where R is the parallel resistance of the source/evaporant combination at the evaporation temperature, T. T is nonuniform because of heat conduction down the current contacts. Therefore, if evaporation rate control is important, continuous flux monitoring is necessary, preferably with feedback control of the source I. The wire can be used only for metals that wet the wire. In the spe-

4.5.1 Basic designs

cial case of Cr, which sublimes, sources are available in the form of a thick electroplated coating on W rod. Although boats can also be used for evaporants which do not wet them, thermal coupling to the evaporant is poor because it is by radiation only, so the boat will have to be much hotter than the evaporant. The thermal coupling problem in vacuum is discussed more thoroughly in Sec. 5.8.1. To prevent wetted evaporant from spreading away from the hot zone of the boat, a narrowed region is often provided on each end of this zone as shown in Fig. 4.8b. The resulting higher R increases heating enough in those regions so that evaporation rate exceeds spreading rate.

Alloying of the evaporant with the wire or boat metal may result in emrittlement or melting. If there are no metals suitable for contact with the desired evaporant, ceramic-coated boats or ceramic crucibles can be used. Information on boat materials and configurations suitable for various evaporants is available from their manufacturers. With evaporants which are very reactive at the evaporation T, such as Si, even ceramic sources will dissolve, thus contaminating the melt and eventually disintegrating. In such cases, evaporation must be carried out by focussing energy into the center of the melt, as discussed in Chap. 8.

In applications where simultaneous flux control of more than one evaporant is required, one must provide either separate flux monitoring of each evaporant or closely T-controlled sources which can maintain and reproduce the desired fluxes. Since the former method is relatively awkward and expensive, the latter is employed more often. Because of the exponential dependence of p_v on T [Eq. (4.14)], T control needs to be very good. Typically, p_v increases about 20 percent per 1 percent rise in evaporant T, or 2 percent per 0.1 percent rise; that is, $dp/p \sim 20 \, dT/T$ in Eq. (4.12). Crucible sources provide the best T uniformity and control, especially if configured as shown in Fig. 4.8c, a design that has evolved from the stringent requirements of molecular-beam epitaxy (MBE; see Chap. 6). Key design features are as follows.

The Ta heater coil is minimally supported within notched ceramic support rods to reduce heat sinking. Alternatively, the coil can be hermetically sealed in a metal can to reduce O_2 outgassing from metal reaction with the ceramic at the high T of the coil. The coil extends well below the crucible bottom so that the thermocouple touching the crucible bottom is immersed within the same heating environment as the crucible and thus stabilizes at the same T. The thermocouple leads are made of fine wire to reduce heat sinking and are made of refractory alloys such as W-Re. Proper thermocouple readout wiring is discussed in conjunction with T measurement in general in Sec. 5.8.3. The heater coil is of finer pitch (denser) near the crucible mouth, so that the T in that location is higher than the T of the evaporant melt.

This prevents condensation of evaporant droplets at the mouth. Such droplets are undesirable because, when they fall back down into the melt, they cause the ejection ("spitting") of macroparticles that end up as defects in the film. The higher mouth T also prevents evaporants which wet the crucible from migrating out of it, as did the notches on the boat source. The radiation heat shielding consists of several wraps of corrugated Ta foil. Ta is used because it is the most ductile of the refractory metals. The shielding reflects radiation from the coil and crucible, thereby improving source-T uniformity and reducing heating and outgassing of nearby hardware. The corrugations minimize thermal contact between successive layers. (Principles of radiative heat transfer which are relevant to the behavior of heat shielding are discussed in Sec. 5.8.1.) The shield disc at the top provides additional heat shielding and also blocks the emission of outgassing contaminants in the direction of the substrate.

The crucible source can be converted to a Knudsen cell by adding to the crucible a closely fitting cover having a thin-lipped (ideal) orifice operating at Kn > 1. The crucible sources used for MBE are often referred to as Knudsen cells, but usually they do not satisfy the Kn > 1 criterion. If, in addition, one wants the Knudsen cell to operate at the saturation p_v of the evaporant, it must satisfy the criterion that $Q_e \ll Q_v$ in Eq. (4.15).

4.5.2 Contamination

Source structural materials and evaporants evolve contaminant vapors both from their surfaces and from the bulk. The vapor evolution problem was discussed in connection with vacuum materials in Sec. 3.4.2. The same remedies apply here, namely: degrease all materials, minimize surface area, avoid volatile constituents, and employ baking.

Considering source structural materials, surface area is minimized by using nonporous grades of ceramic for crucibles. In the case of graphite or BN, this means the "pyrolytic" grade. Pyrolytic crucibles are formed on the surface of a core or "mandrel" by the thermal decomposition (pyrolysis) of the gases CH_4 or $B_2H_6 + NH_3$. (These are examples of thin-film CVD processes in which the deposit—the crucible—is actually very thick.) Much of the volatile contaminant can be removed from source crucibles and associated components by "firing" the source at a higher T than the desired evaporation T, prior to filling the crucible with evaporant. Crucible materials should be chosen whose residual contaminants are the least harmful to the film being deposited. For instance, BN is a good choice for III-V semiconductor film deposition even though it is not one of the most refractory ceramics. It is a good choice because BN is itself a III-V compound, and therefore its elements do not form charge-carrier traps within the band gap of a III-V semiconductor. The most refractory ceramics are alumina and zirconia, but both evolve some oxygen.

Considering evaporant materials, chemical analysis should be obtained to determine if there are contaminants present that will be harmful to the film being deposited. Be aware that generally, only the metallic impurities are reported by the manufacturer. That is, a 99.99 percent pure ("four-nines" pure) material will have <0.01 percent (100 ppm) total *metallic* impurity content, but it also may contain a great deal more of O, C, Cl, and so forth. The evaporant should be obtained in large pieces rather than in powdered form to minimize surface area. A powder of 10 μm particle size has a very large surface area of about 0.3 m^2/cm^3. The amount of outgassing of adsorbed water from such large areas is considerable and was estimated in Sec. 3.4.2 and Exercise 3.8.

Much of the volatile impurity content in the evaporant can be removed before film deposition. This includes adsorbed gases and also dissolved elements of higher p_v than the evaporant. Removal is accomplished by "soaking" the source at a T somewhat below the T for film deposition, both before substrate loading and again afterwards while blocking the source from the substrate with a shutter; the second soak removes water adsorbed during the loading process. The dissolved impurities having high p_v will progressively deplete relative to the evaporant as described by Eq. (4.25). Conversely, impurities with p_v significantly *less* than that of the evaporant will selectively remain behind in the crucible, as described by the same equation. This phase separation by p_v is essentially the process of purification by distillation, as is done commercially for many materials. However, as the evaporant charge becomes depleted, the fraction of low-p_v impurities in it increases, and therefore so does the fraction of impurity in the evaporant flux. Therefore, it is unwise to evaporate the source material to completion when film purity is important. The only advantage of evaporating to completion is that a film of known thickness can be obtained simply by preweighing the evaporant charge, without having to measure the evaporation flux or time. The third group of impurities, those with p_v *similar* to that of the evaporant, cannot be selectively removed and will remain part of the evaporation flux.

4.5.3 Temperature control

The close control of T in evaporation sources, substrates, and CVD reactors is crucial to the reliable operation of film deposition processes. Here, we discuss the principles of T control within the context of evaporant heating.

To maintain constant crucible temperature, T, in the Fig. 4.8c source, the heater coil's power supply is driven by a feedback control loop which compares the thermocouple voltage with a set-point voltage. The simplest kind of T control is "on-off," such as the thermostat in your house. There, the heater power is fully on when T is below set point and fully off when it is above set point, so that T continues to oscillate about the set point. For most thin-film applications, this type of control is inadequate, and more sophisticated controllers are employed instead. The common "PID" type have three modes of control: *p*roportional, *i*ntegral or "reset," and *d*erivative or "rate." If the thermocouple T is lower than the set point by ΔT, the proportional control signal driving the power supply is a fraction of "full on." The fraction is given by $\Delta T/T_b$, where T_b is called the proportional bandwidth. T_b is inversely proportional to the "gain" of the feedback loop. On-off control is essentially proportional control with infinite gain. With proportional control alone, the power will remain "on" enough to maintain steady T only if there remains a finite offset ΔT from set point. The integral control mode gradually removes this offset by adding a control signal proportional to $\int_t \Delta T dt$. Finally, the derivative mode reduces T overshoot past the set point on initial heat-up by adding a third control signal proportional to minus the rate of approach of T to set point, $d(\Delta T)/dt$. The need for derivative control depends on the degree of overshoot and whether this would result in excessive evaporant loss or other problems.

The maximum speed of response to a step change in set point depends on the thermal time constant of the evaporation source or other thermal system. Response speed is important in applications involving multilayer films, where it is desired to quickly change the T of a source or of the substrate between layers or to deposit a layer of graded composition. A shutter may be closed between source and substrate to interrupt the deposition during this transient period. However, this allows background contaminants to build up on the substrate surface, as pointed out in Sec. 2.6. To avoid this problem in multilayer film deposition, separate sources of the same evaporant operating at the fluxes desired for each layer are often employed in conjunction with synchronized shutters. This technique is used, for example, in the deposition of quantum-well structures by MBE, where fluxes must be changed within one monomolecular layer of deposition.

The thermal time constant, τ, for T-control response may be determined by considering that heat must obey the continuity equation [Eq. (3.1)]: input − output = accumulation. To make the solution tractable, we will make various simplifying assumptions which are quite acceptable given that we are only seeking a rough estimate of τ. Thus, in the evaporation-source geometry of Fig. 4.9, we are looking down into a

4.5.3 Temperature control

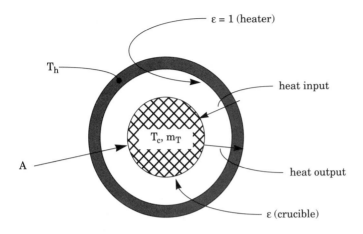

Figure 4.9 Geometry for crucible heating calculation; view looking into crucible mouth.

cylindrical crucible of surface area A and at T_c. It is completely surrounded by a cylindrical heater at T_h. Under vacuum, heat transfer is by radiation and follows Eqs. (4.19) and (4.20). For the crucible, we will make the so-called "gray-body" assumption that ε_λ and α_λ are independent of λ over the λ range of concern here; thus, $\varepsilon = \alpha$ by Kirchhoff's law. For the heater, we will assume that $\varepsilon = \alpha = 1$. For the system consisting of the crucible and its contents, the input term in the continuity equation is the heat radiated from the heater and absorbed by the crucible, and the output term is the heat radiated by the crucible. The accumulation term is the rate of heat buildup in the crucible and its contents. We will assume that the crucible and the evaporant are well coupled thermally. (This assumption is true for a melt; however, the T of a sublimating evaporant would lag that of the crucible considerably.) We may now write

$$\underbrace{A\alpha\sigma T_h^4}_{\text{(input)}} - \underbrace{A\varepsilon\sigma T_c^4}_{\text{(output)}} = \underbrace{(c_{gc}m_c + c_{ge}m_e)\frac{dT_c}{dt}}_{\text{(accumulation)}} = m_T\frac{dT_c}{dt} \qquad (4.26)$$

where the c_g and m values are the heat capacities (J/g·°C or J/g·K) and masses of the crucible and evaporant, m_T is the resulting overall "thermal load" or "thermal mass," and the other symbols are as defined previously. We may now substitute ε for α, combine the two terms on the left, and factor the polynomial to give

$$A\varepsilon\sigma\left(T_h^4 - T_c^4\right) = A\varepsilon\sigma\left(T_h^3 + T_h^2 T_c + T_h T_c^2 + T_c^3\right)(T_h - T_c)$$

$$\approx A\left(4\varepsilon\sigma T^3\right)(T_h - T_c) = Ah_r(T_h - T_c) \quad (4.27)$$

where $T = (T_h + T_c)/2$

$h_r = 4\varepsilon\sigma T^3$ = radiative heat transfer coefficient

The T^3 approximation has changed an equation in T_c^4 into an equation linear in T_c and having an h_r of the same units as the h_c for gas conduction in Eq. (2.32). For typical situations in thin-film deposition, the error introduced by this approximation is negligible, as seen in Exercise 4.9. We now have in Eqs. (4.26) and (4.27) a differential equation which we can solve readily to give

$$\frac{T_h - T_c(t)}{T_h - T_c(t=0)} = \exp\left(\frac{-t}{m_T/Ah_r}\right) = \exp\left(\frac{-t}{\tau}\right) \quad (4.28)$$

where τ is the time constant for heat-up or cool-down. The expression for τ in the denominator of the exponential is completely analogous to that in Eq. (3.11) for vacuum pumpdown and to that in Eq. (4.25) for alloy composition relaxation. In the electrical analogy, all three of these τ values consist of a capacitance (C) term divided by a conductance (1/R) term, which gives an "RC" time constant. For crucible heat-up, low thermal mass and high thermal conductance or coupling result in rapid response (small τ). Note that according to Eq. (4.28), $T_c \to T_h$ as $t \to \infty$, whereas in practice $T_h > T_c$ at steady state. This discrepancy is due to the simplifying assumption that the heater completely surrounds the crucible. To estimate T and h_r, consider that during heat-up, we have $T_h > T_c$, and during cool-down, we have $T_h < T_c$. Thus, we will take $T \approx T_c$ to give an overall average h_r for the system.

Let us now calculate τ for a typical situation consisting of a 500° C crucible with $A = 4$ cm^2, containing 5 g of Sb ($c_g = 0.23$ J/g·K). If the mass of the crucible can be neglected, then $m_T = 1.15$ J/K. If $\varepsilon = 1$, then $h_r = 0.011$ J/s·cm^2·K. Thus, $\tau = 26$ s, which would be a fairly long time to have to interrupt deposition waiting for T_c to reach steady state. The T_c^3 factor in h_r means that a more volatile evaporant will have much slower control response, since it will be operating at a lower T_c. It also means that controllers need to be retuned if T_c is changed significantly, because the control-feedback gain must be matched to τ. The same considerations apply to substrate T control. It does not help

4.5.3 Temperature control

to increase the heater power to give a higher T^3 for heat-up, because the T^3 for cool-down will still be determined by T_c. Instead, it is best to set "full-on" heater power so that the heat-up rate is the same as the cool-down rate in the neighborhood of the desired T_c. Heater power can be set higher for faster heat-up response, but derivative control may be needed to prevent excess overshoot.

Once heater power has been set, each of the three control modes needs to be tuned. If there is too much gain in any mode, T_c will oscillate at an angular frequency of roughly $1/\tau$. If there is too little proportional gain (too much proportional bandwidth), T_c will drift, and response will be sluggish. These behaviors are shown in Fig. 4.10. The proper tuning procedure is to turn off reset and rate action and then tune the proportional band to just short of the setting which causes oscillation when a step change in set point is applied. Then, reset is added to just short of oscillation. Finally, enough rate action is added to minimize overshoot without making control response too sluggish. The thermal load and the heat supply are analogous to the familiar mass-and-spring system, whose resonant frequency increases with spring stiffness and decreases with increasing mass. Critical damping in either system means having just enough resistance to response so that oscillation is avoided, as with the shock absorbers on your car. With proper tuning, 0.1° C stability of thermocouple T usually can be attained in thin-film applications. Keep in mind, however, that this stability only applies to the steady state. Any disturbance in heat flow conditions, such as the opening of a shutter over the crucible and the accompanying increase in radiative heat loss, will cause a control response transient before re-equilibration.

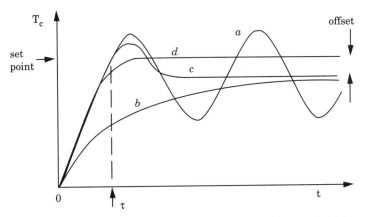

Figure 4.10 Control response to a step increase in set point at $t = 0$: (*a*) too much gain, (*b*) not enough gain, (*c*) critical damping, and (*d*) critical damping with reset and rate action added.

4.5.4 Energy enhancement

In many thin-film deposition processes, energy input to the film surface is required in order to increase the reaction rate of depositing species with each other or with the substrate, or to assist the depositing species in moving about on the surface. The effects of this energy will be discussed frequently throughout the book, since its control is a major factor in obtaining good films. In thermal evaporation, the energy content of the evaporant is much smaller than typical chemical bond strengths, as can be seen from Table 4.2, so it does not have much effect on film growth. In this table, the physics energy units of electron volts (eV) are given as well as the chemistry units of kilojoules because they relate more easily to the use of electrical means of energy enhancement: that is, electron bombardment and plasmas. One eV is the energy gain of a particle having one electronic charge upon passing through a potential drop of one volt.

There are many ways to enhance evaporant energy so as to modify film growth. Some involve nonthermal methods of source-material volatilization, such as laser ablation and sputtering (Chap. 8). Others use a glow-discharge plasma in the transport step, such as activated reactive evaporation (Sec. 9.3.2). The term "energetic condensation" is applied to any film-deposition process in which the kinetic energy of the depositing species is well above thermal but still below the onset of crystallographic damage and implantation (~30 eV; see Sec. 8.5.2.1). The present discussion will be limited to nonplasma ways of enhancing the energy of the basic thermal-evaporation source of Fig. 4.1b.

We first examine the supersonic nozzle. Consider a source having an orifice which is effusing vapor into vacuum, so that the pressure, p, is negligible downstream. We have seen that for such an orifice in molec-

TABLE 4.2 Characteristic Energies in Thin-Film Deposition

	eV/mc	kJ/mol
ε_t^* at : 298 K	0.038	3.71
: 2200 K	0.28	27.4
Xe in H_2, 2200 K[†]	14.3	1380
100 mc, 1 keV cluster ion	10	965
Laser-ablated atoms	40	
Sputtered neutrals	5	
Bond energy: Si-Si	3.29	318
: N≡N	9.83	949

* Mean thermal translational energy of gas molecules, per Eq. (2.11)
[†]Supersonic nozzle beam [8]

4.5.4 Energy enhancement

ular flow (Kn > 1), the throughput Q_e (mc/s) is proportional to upstream p [Eq. (3.5)]. Conversely, in the fluid-flow regime (Kn << 1), $Q_e \propto p^2$. One factor of p comes from the vapor concentration increase with p via the ideal-gas law, as in molecular flow. The second p comes from the fact that *fluid* flow velocity through an orifice increases with the Δp driving force. From the molecular viewpoint, collisions between molecules in the approach to the orifice serve to convert random thermal translational energy into directed kinetic energy moving downstream. This conversion occurs because molecules in the orifice are being hit more often from the high-p upstream side than from the low-p downstream side, and thus they accelerate. The limit to this acceleration occurs when the flow velocity in the orifice throat reaches the speed of sound, c^* (Fig. 2.4). This situation is known alternately as sonic or choked or critical flow. However, downstream of the orifice, further acceleration into the supersonic regime can occur, and this effect can be enhanced by incorporating a conical nozzle as shown in Fig. 4.11. Such nozzles are commonly employed in rocket engines to maximize thrust. The cone angle is made as large as possible to minimize friction, but not so large as to cause separation of the laminar-flow boundary layer from the cone wall. Separation would lead to randomization of the kinetic energy in eddy currents generated adjacent to the wall. Typical conditions for present applications would be a p of 3 atm, an orifice diameter of 0.3 mm, and a cone included angle of 15°. Note that in the subsonic regime upstream of the orifice, flow velocity, u, is increasing with decreasing cross-sectional area for flow, A; in the throat, $u = c^*$ and is thus independent of A; and in the expansion nozzle, u is increasing with *increasing* A.

Two characteristics of supersonic nozzles are important for film-deposition applications. One is that considerable vapor cooling takes place due to the transfer of kinetic energy from thermal to directed. Cooling

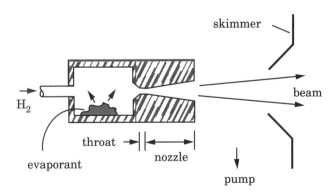

Figure 4.11 Supersonic-nozzle evaporation source.

assists in cluster formation, as we will see later. Cooling is efficient for two reasons. First, because u is so high, little heat is received from the orifice wall; that is, the expansion is "adiabatic." In addition, for an optimized cone angle, there are negligible frictional losses, which would transfer kinetic energy back to thermal energy; that is, the process is "isentropic" (constant entropy). The second important characteristic of these nozzles arises from the fact that all components of a gas mixture get accelerated to the same u in any fluid flow stream. Since $\varepsilon_t = (1/2)mv^2$, the evaporation of a small fraction of high-mass vapor into a low-mass carrier gas such as H_2 results in an ε_t for this "seeded" vapor which is increased by the ratio of masses over what it would be for expansion of the pure vapor. For example, measurements of Xe in H_2 gave ε_t well above chemical bond energies, as seen in Table 4.2. The mass ratio for this combination is 66. This technique requires a large flow of carrier gas which must be pumped away. Most of the pumping is done between the nozzle and "skimmers" (Fig. 4.11) which serve both to isolate the deposition chamber from the higher-p region and to deflect carrier gas out of the supersonic stream to minimize collisional losses. A supersonic-nozzle beam of Ge_2H_6 gas in He has been used to deposit single-crystal Ge film onto GaAs at 50 nm/s, much higher than the rate using thermal effusion [9]. The rate enhancement was attributed to activation of Ge_2H_6 dissociation at the growth surface by the high molecular kinetic energy.

Ionization and acceleration constitute a widely-used method of enhancing the energy of vapor molecules. Ionization of molecules evaporating from a crucible can be done by electron impact from a filament source, as was used in the ion gauge (Sec. 3.5). The ions are then accelerated using an electrical bias between source and substrate. For a typical design, see Ref. [10]. Low ionization rate limits the ion flux of such sources, so they are mostly useful for enhancing the incorporation of minor constituents (dopants) into a depositing film. Much larger ion fluxes are obtainable by using electron-beam or cathodic-arc evaporation (Chap. 8) or by incorporating a glow-discharge plasma into the transport space (Chap. 9).

About two decades ago, Takagi of Kyoto University made the intriguing proposal that film deposition be carried out using accelerated *cluster* ions. With sufficient collisions and cooling, vapor from an orifice is known to condense into clusters of tens or hundreds of atoms or molecules. It was postulated that, upon colliding with the film surface, such a cluster would disintegrate and transfer its perpendicular kinetic energy into energy directed laterally along the surface and distributed among the scattered molecules. Thus, a 100 mc cluster accelerated to 1 keV would result in 10 eV molecules skittering along the film surface (see Table 4.2). This is just the direction and level of energy that is desired for the modification of film deposition behavior. However, it has not yet been demonstrated that this sort of energy transfer from clusters actually takes place. Both the production and verification of clusters are nontrivial, and most of the work issuing from the cluster-ion proposal has been done with uncharacterized cluster masses and with collision rates within the nozzle which were much too low to lead to substantial cluster formation. Much of this work does show evidence of film improvement when the vapor flux is ionized and accelerated versus when it is not, and it is likely that these results are due to bombardment of the surface by single-molecule vapor ions along with a larger flux of neutral thermal molecules. Ion bombardment of a depositing material is widely used to improve film properties, as will be discussed in Sec. 8.5.3.

In the meantime, there has been considerable progress in the understanding of and equipment for cluster formation. Expanded interest in clusters has arisen from the discovery of "magic" numbers of atoms (corresponding to completed shells of conduction electrons) and from the potential of nanoparticles as new materials, such as the 60-atom polyhedral carbon clusters known fondly as Buckminster-Fuller-ene or "buckyballs." A good sampling of recent cluster work can be found in the particle-symposium proceedings listed at the end of the chapter. The basic requirements for cluster formation are that there be enough vapor-vapor molecular collisions to grow the clusters and that there be a sink for the considerable heat of condensation, which would otherwise prevent the cluster from nucleating in the first place. The sink is provided by contact with a cold "third body" during nucleation, this contact being either with the orifice wall (heterogeneous nucleation) or with other molecules of the vapor or the carrier gas (homogeneous nucleation). Under optimized conditions in a supersonic nozzle, there is sufficient cooling and collision rate for substantial cluster generation; for example, Zn beams with 2200-atom average cluster size have been obtained [11]. Alternatively, in the gas-condensation technique, the vapor flows through a chilled tube in an inert carrier gas which can be held at much lower p than in the supersonic nozzle. This technique reduces gas load on the pumps but would be expected to result in substantial cluster loss to the walls. A third mechanism has been proposed to explain observations of clusters in some of the cluster-ion film-deposition work done under conditions unlikely to generate them homogeneously. This mechanism [12] invokes heterogeneous nucleation on a nonwetting orifice wall, followed by entrainment of the clusters in the vapor stream. While a likely explanation, it is unlikely that this mechanism would be controllable or would result in clusterization of a substantial fraction of the vapor.

It may be concluded with regard to cluster-ion deposition that the original proposal remains intriguing but untested. With the present improved state of cluster-generation technology, it is now possible to study the use of cluster ions in film deposition more rigorously.

4.6 Transport

At this point in our development of the principles of thin-film deposition processes, we have completed the first of the four process steps shown in Fig. 1.1, namely the supply of source material. The main issues addressed have been supply rate and contamination. Thus far, these issues have been addressed only for thermal evaporation as a supply method. They will be revisited in Chaps. 7 and 8 for other supply methods. The second of the four process steps is transport of the source material to the substrate. Here, the main issues are arrival-rate uniformity at the substrate and, again, contamination. Contamination entering the process during the transport step has been addressed in Sec. 3.4. Here, we will examine uniformity. The factors determining arrival-rate uniformity at the substrate are very different, depending on whether the medium is a high vacuum or a gaseous fluid, as pointed out in Sec. 1.2. Transport in the fluid medium depends on flow patterns and diffusion, and these factors will be examined in Chap. 7 in the context of CVD. Here, we will examine transport in the high-vacuum medium. For present purposes, high vacuum means that the background pressure is low enough that the Knudsen number, $Kn = l/L$ [Eq. (2.25)], is >1 for L taken as the distance from source to substrate. Under these conditions, transport is along the line of sight from source to substrate, because the probability is small for an evaporant molecule to collide with a background-gas molecule along the way. Consequently, arrival-rate uniformity across the substrate is determined solely by geometrical factors, and we will proceed to examine these below.

Consider the evaporation geometry of Fig. 4.12, where material is being evaporated from a source centered at point B onto a flat substrate situated at a perpendicular distance r_0 away. Several source shapes are shown. The circular disc, which is emitting material from the top surface only, well represents the boat, the Knudsen-cell orifice, and the filled crucible. The sphere roughly approximates the wire-coil source. The collimated source represents a partly filled crucible or a nonideal orifice—an orifice whose length is significant compared to its diameter. We first want to know the evaporant flux J_θ at radius r_0 from the source and at some angle θ from the perpendicular (say at point R), compared to the flux J_0 at the perpendicular (point P). If you

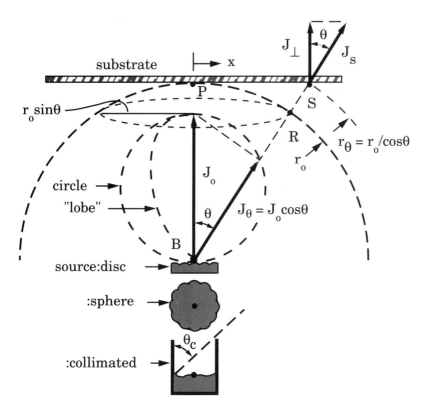

Figure 4.12 Geometry of vacuum evaporation. The flux ration J_θ/J_o shown is for the "cosine distribution" characteristic of the disc-shaped source centered at point B; alternate source shapes shown below would have different flux distributions when positioned at B.

think of yourself as an observer looking into the source from radius r_o, it is clear that J_θ will be proportional to the projected area of the source in that direction. For the sphere, this area is independent of θ, so for a total evaporation rate of Q mc/s, we have simply

$$J_\theta = J_o = Q/4\pi r_o^2 \qquad (4.29)$$

where Q (mc/s) = $J_v A$ = total evaporation rate from a source of area A emitting flux J_v from its surface.

More commonly, the source is a disc. Looking into this source from r_0, the projected area changes from a circle to an ellipse to a line as one moves away from the perpendicular; in fact,

$$J_\theta = J_o \cos\theta \qquad (4.30)$$

This is known as a cosine flux distribution. In Fig. 4.12, the flux vectors J_o and J_θ have been displaced from their proper locations at points P and R, and to point B, to illustrate that the vector J_θ traces a circle over θ. This is because J_o, J_θ, and θ always form a right triangle, as shown. To calculate the J values, consider that J_θ will pass through the r_o sphere in an annulus having radius $r_o \sin\theta$, circumference $2\pi r_o \sin\theta$, and differential width $r_o d\theta$. Knowing this, we can integrate J_θ over the hemisphere of evaporation, and by mass conservation, this integral must equal the total evaporation rate; that is,

$$Q = \int_0^\pi J_o \cos\theta \cdot 2\pi r_o \sin\theta \cdot r_o d\theta = \pi r_o^2 J_o \qquad (4.31)$$

or

$$\boxed{J_o = \frac{Q}{\pi r_o^2}} \qquad (4.32)$$

Note that Eq. (4.31) is the same integral as Eq. (2.6) for the average perpendicular component, $\bar{v}_x$, of molecular velocity, v, over the hemisphere. There, we found that

$$\bar{v}_x = \frac{1}{2} v$$

Here, the average flux over the hemisphere would be $Q/2\pi r_o^2$, which is $(1/2) J_o$ by Eq. (4.32).

Looking now into the collimated source from r_o, we see that the projected area drops off much more rapidly with increasing θ than for the disc-shaped source, due to shadowing from the lip, and this leads to the "lobe"-shaped flux distribution shown in Fig. 4.12. Although there is some angle, θ_c, above which the evaporant surface is not visible at all, there will still be some flux at higher angles because of evaporant scattering from the collimating sidewalls. Indeed, if all of this evaporant were to reflect at the specular angle, θ would never change upon such reflection, and therefore the cosine distribution would be preserved. However, molecules do not do this, but instead they scatter randomly, as discussed in Sec. 2.8.2 on viscosity. Some of these scattered molecules return to the crucible rather than escape from it, and this results in the lobed distribution. Calculations of such distributions have been reported [13]. The lobe can be broadened by using conical crucibles. Lobed distributions also are seen from *disc*-shaped sources when they are heated in one spot by a focused beam of electrons or laser light. These energy-beam techniques are discussed in Chap. 8. Supersonic nozzles produce sharply lobed distributions. Deposition uniformity over large substrates is a major concern with lobed flux distributions.

Let us consider uniformity across a flat substrate for the simpler case of a cosine distribution. There are two factors that attenuate the flux with increasing θ besides the projected-source-area factor. One is that the substrate point S in Fig. 4.12 is at radius r_θ rather than at r_0 (point R) from the source, and $r_\theta = r_0/\cos\theta$. Since J is inversely proportional to r^2 by Eqs. (4.29) and (4.31), we have for the flux through the r_θ sphere at substrate point S,

$$J_s = J_\theta \cos^2\theta \tag{4.33}$$

Now, the flux that determines deposition rate is the flux perpendicular to the substrate at point S, $J_\perp$, and that flux is reduced from J_s by another $\cos\theta$ factor, as shown in Fig. 4.12. Combining all of these factors, we have, finally,

$$J_\perp = J_0 \cos^4\theta = \frac{Q\cos^4\theta}{\pi r_0^2} \tag{4.34}$$

where Q (mc/s) is the total evaporation rate into the hemisphere from a disc-shaped source. The $\cos^4$ factor causes, for example, a 10 percent reduction in $J_\perp$ from center to edge for a substrate whose radius is one-fourth of the source-to-substrate distance r_0. Consequently, there is always a trade-off between nonuniformity at short distances and evaporant waste at large distances. By the way, the flux $J_\perp$ is sometimes expressed in terms of a "beam-equivalent pressure" by using the Knudsen equation, Eq. (2.18).

Various refinements of the Fig. 4.12 geometry can be incorporated to improve flux uniformity. Rotation of a substrate about its own axis, coupled with off-axis orientation of a single source, improves uniformity considerably [14, 15] compared to Eq. (4.34). Numerous small substrates such as lenses can be coated uniformly from a single source with low material waste by placing the substrates on a "planetary" fixture. There, the substrates rotate simultaneously about two different axes so that over the course of the deposition, their complex orbits cover a large fraction of the hemisphere. Note that rotating substrates always incur an oscillatory deposition rate that can affect film properties, especially when the process involves co-evaporating different materials from different sources, since the oscillations of the different material fluxes will not be in phase. For large-area substrates, multiple, distributed sources of the same material and rate can be used, but rate balancing can be difficult. Alternatively, several sources are distributed in a line, and the substrate is passed by perpendicular to the line as shown in Fig. 4.13. This geometry is also used in the sputter-deposition process to be discussed in Sec. 9.3.4, in which case one uses a single source of long rectangular shape, and uniformity is better than 2 percent.

The example of Fig. 4.13 also illustrates the "web" coating process, in which a flexible substrate is supplied from one roll, taken up by a second, and suspended as a web between the two rolls during coating. This process is widely used to apply magnetic metal to videotape, Al or SiO_2 moisture-barrier layers to plastic film for food packaging, and metal electrodes to Mylar film for capacitors. The rolls may be located either in the vacuum, or outside with the web being passed through multiple, differentially pumped sliding seals.

Uniform coating of large *rigid* substrates is also achieved by transporting the substrate past a distributed vapor source as in the web process. For high throughput, entrance and exit load-locks (Fig. 3.1) are used to supply and remove the substrates. Multilayer coatings can be applied by placing sources of different materials in succession along the direction of substrate motion. With these techniques, coatings can be applied uniformly to substrates of many m^2 area, as with heat-control coatings on window glass.

4.7 Vapor Flux Monitoring

Often, it is not possible to predict accurately the flux arriving at the substrate, due to uncertainty in the evaporation rates of the various species or due to the presence of lobed flux distributions, as discussed

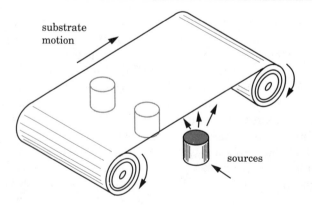

Figure 4.13 Web coating process for flexible substrates.

above. When the arriving flux at least is reproducible for successive depositions, a few measurements of film thickness are sufficient to set the source operating conditions. This can be done for sources having good T control, unity evaporation coefficients, and cosine flux distributions. In other cases, it is difficult to reproduce arriving flux, and therefore flux monitoring is needed. Flux can be monitored either as the concentration of source vapor near the substrate or as actual deposition rate. Monitoring both is especially useful, because then one can calculate the fraction of arriving vapor which becomes incorporated into the deposit—the sticking coefficient, S_c. Deposition monitoring will be discussed in the next section. Here, we will address vapor monitoring.

The basic technique for vapor flux monitoring is to excite the vapor molecules and detect their response. Excitation with a beam of electrons or light raises electrons in the molecules into excited states. This is followed by ionization if the excited electron leaves the molecule, or by photon emission (or Auger-electron emission) if it relaxes back toward its ground state. The simplest type of flux monitor is the ion gauge, which was discussed in Sec. 3.5. Recall that the ionization rate and resulting collected ion current are proportional to the concentration of vapor molecules within the gauge, $n\,(\text{mc/cm}^3)$. The flux of any stream of particles is related to its concentration by

$$\boxed{J = nv} \qquad (4.35)$$

where v is the particle velocity in the direction of the flux. For flux from thermal evaporation, one can take v as roughly the mean speed, $\bar{c}$, of a Maxwellian distribution at the T of the source, as given by Eq. (2.3). For particles vaporized by laser ablation or ion sputtering, v is much higher.

The ion-gauge flux monitor must be shielded as shown in Fig. 4.14 so that evaporant does not react with the filament. Electrical connectors also must be shielded from accumulating conducting deposits which would short them out, and the grid must be apertured to avoid accumulation of insulating deposits which would charge up. Simultaneous monitoring of multiple evaporants can be achieved by shielding each monitor from the other sources as shown in Fig. 4.14. However, it is more convenient to use a single species-specific monitor. One such device is the mass spectrometer mentioned in Sec. 3.5, which is basically an ion gauge followed by a mass-selective filter. Another is the electron-impact emission spectrometer [16], in which the light emission from the relaxation of electron-impact-excited molecules is spectrally analyzed to separate the emission lines characteristic of each

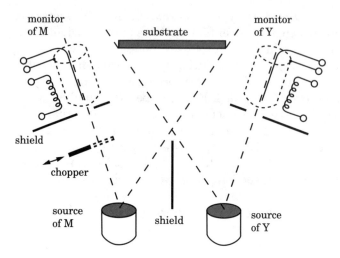

Figure 4.14 Separate monitoring of vapor fluxes from two sources using ion gauges.

element. With any of the above devices, the mechanical beam chopper shown in Fig. 4.14 is useful for distinguishing beam from background species when the background n is a significant fraction of the beam n, as in the case of volatile evaporants such as As.

The above electron-beam techniques are usable only when the electron mean free path is larger than the monitor diameter, which means below 1 Pa or so [Eq. (2.22)]. They also are not usable in a plasma environment because of interference from plasma electrons and ions, although light emission from molecules excited by *plasma* electrons is an important analytical tool in plasmas. For the shorter mean free paths of the fluid-flow processes, or in plasmas, optical excitation and detection techniques must be used to monitor vapor concentration.

Closed-loop control can be set up for flux just as it was for T in Sec. 4.5.3. The flux-monitoring signal is compared to a set point, and the difference is used to drive the source power through a proportional controller. As with T control, the thermal time constant of the source determines maximum response speed. The same control loop can be used with the quartz-crystal deposition monitor which is to be discussed next.

4.8 Deposition Monitoring

We now discuss direct monitoring of deposition flux, which is the vapor flux arriving at the substrate times the sticking coefficient, S_c. Though deposition monitoring is really part of the deposition step of the thin-film process, its close relationship to vapor flux monitoring makes it appropriate to discuss here in the context of the transport step. There are several approaches to deposition flux monitoring, principally:

1. mass deposition on a vibrating quartz wafer adjacent to the substrate
2. change in optical reflection intensity or phase from the film itself with increasing thickness
3. periodicity in the reflection high-energy electron diffraction (RHEED) signal from epitaxial films

Since RHEED is specific to epitaxy, it will be discussed in Chap. 6. The mass and optical techniques are discussed below. All three techniques are extremely sensitive, with submonolayer resolution being achievable routinely. It is important to keep in mind that none of them measures film thickness directly. Thickness must by calculated by knowing or assuming a film property, namely, for the cases listed above, (1) density, (2) index of refraction (except in ellipsometry), and (3) monomolecular layer (monolayer) thickness.

4.8.1 Mass deposition

The vibrating quartz crystal mass-deposition monitor, or quartz crystal microbalance, is one of the most powerful and widely used diagnostic instruments in thin-film technology. It uses the resonant crystalline quartz wafers that were developed for frequency control in radios and are used also for timing in computers and watches. Crystalline quartz is piezoelectric, so a quartz wafer generates an oscillating voltage across itself when vibrating at its resonant frequency, and this voltage can be amplified and fed back to drive the crystal at this frequency. Electrical coupling is done with thin-film metal electrodes deposited on opposite faces of a thin quartz wafer having the proper crystallographic orientation, as shown in Fig. 4.15. For deposition monitoring, one electrode is exposed to the vapor flux and proceeds to accumulate a mass of deposit. This mass loading reduces the crystal's resonant frequency, v_r. Comparison of the loaded v_r with the v_r of a reference crystal located in the instrument's control unit is used to calculate the mass of deposit. Measurement of any quantity relative to a reference value of similar magnitude is always much more accurate than making an absolute measurement, and here it accounts for the submonolayer resolution of the instrument. The v_r of about 5 MHz, when loaded, generates a beat frequency with the reference crystal which is equal to their v_r difference. The resonant frequencies are very stable, so one can easily measure a beat frequency of a few hertz,

102 Evaporation

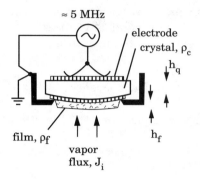

Figure 4.15 Quartz-crystal deposition monitor in cross section.

which corresponds to a sensitivity of <1 ppm in mass, or one monolayer on a typical 350 µm-thick crystal.

The resonant crystal may be crudely modeled as a familiar mass-spring oscillator, for which

$$v_r = \frac{1}{T_v} = \frac{1}{2\pi}\sqrt{\frac{k}{m}} \qquad (4.36)$$

where v_r = resonant frequency, Hz
T_v = vibrational period, s
k = spring constant (stiffness), N/m (= kg/s^2)
m = mass, kg

For a mass-loaded crystal, we may then write

$$T_v = T_{vo} + \Delta T_v \propto \sqrt{m_q + m_f} = \sqrt{m_q\left(1 + \frac{m_f}{m_q}\right)} \approx \sqrt{m_q}\left(1 + \frac{1}{2}\frac{m_f}{m_q}\right) \qquad (4.37)$$

which means that

$$\Delta T_v \propto m_f \propto \rho_f h_f \qquad (4.38)$$

where ΔT_v = change in T_v due to the loading
m_q = mass of the unloaded crystal
m_f = mass of the deposit
ρ_f = mass density of the deposit, g/cm^3
h_f = thickness of the deposit

The last equality on the right of Eq. (4.37) is valid for $m_f \ll m_q$ and is sufficiently accurate in practice for $m_f/m_q < 0.1$. Thus, for h_q = 350 µm, a deposit thickness of about $35(\rho_q/\rho_f)$ µm can be tolerated. To further increase the tolerable loading and thereby increase the useful life of a crystal, a correction must be applied to account for the fact that k as well as m is changing in Eq. (4.36). These factors are accounted for together by the acoustic impedance, z, of the crystal-deposit combination, since the resonance is essentially an acoustic standing wave which pervades both the crystal and the deposit. Most instruments allow the programming in of a z-correction factor, the value of which will depend on the film being deposited. This correction extends the tolerable loading to $m_f \approx 0.7\ m_q$. An extensive analysis of the z correction and tabulation of its value for various materials has been reported [17]. Note, however, that the z of a material in thin-film form may be different from the z in bulk because of differences in structure, bonding, ρ_f, and composition. For accurate work, therefore, it is necessary to determine z for the film being deposited by comparing actual film thickness with crystal readings. Deviation of ρ_f from the bulk value also changes the film thickness, which is inferred from the measured mass using Eq. (4.38).

The quartz crystal monitor gives an absolute reading of m_f, and therefore an absolute measurement of arriving vapor flux, if S_c is unity. From this measurement, one can calculate the vapor pressure, p_v, of the source material using the cosine law [Eq. (4.34)] and the Knudsen equation [Eq. (2.19)], provided that the cosine law is valid at the location of the crystal and that a Knudsen-cell source (Fig. 4.1b) is employed. This is a convenient way to determine the p_v values of evaporants.

Since the crystal and substrate positions relative to the source are different, the ratio of the vapor fluxes at the two positions must either be determined empirically or estimated using the information in Sec. 4.6. If the cosine-effusion law is being obeyed, the calculation of this geometrical correction factor is straightforward; however, in the case of a sharply lobed or fluctuating flux distribution, large errors can result. Locating the crystal as close as possible to the substrate minimizes geometrical errors. But in that case, too much radiation from a hot substrate can cause drift in the crystal reading, for the reason explained below.

Crystal T control is important. Figure 4.16 shows how v_r varies with T for the standard "AT-35°" cut of crystallographic orientation. For this cut, v_r is independent of T near room T, but not at other temperatures. Thus, if the crystal heats up during a deposition measurement due to radiation from the substrate or the source, v_r will increase and will appear as a *decrease* in mass. If the radiation is coming from the source, transient errors due to quartz T restabilization will occur upon source heat-up or shutter opening at the start of deposition, and again

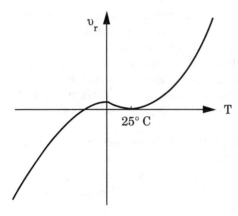

Figure 4.16 Thermal drift in quartz resonant frequency, v_r, for AT-35° cut of crystal.

upon cool-down at the end, as shown in Fig. 4.17. These transients can be recognized most easily by recording h_f versus time. The actual deposition rate is the slope in the linear region shown in Fig. 4.17. Also illustrated is an abrupt jump in reading due to the crystal spontaneously hopping from one vibrational mode to another. The convex shape of the crystal shown in Fig. 4.15 and the placement of electrode spring contacts off center help stabilize the fundamental vibrational mode, but uneven deposits or high mass loading can cause "mode hopping." Frequency jumps will also occur if the deposit begins to flake off. An additional increase in mass may be seen upon admitting air to the deposition chamber, or admitting any gas which reacts with the film. This increase is an indication of film porosity: the internal surface area along the pores is adsorbing water or becoming oxidized. In such

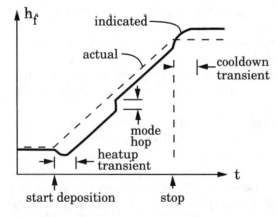

Figure 4.17 Quartz monitor film thickness, h_f, versus time, t (indicated versus actual).

cases, the quartz crystal is a very sensitive microbalance (±0.01 μg) for studying film porosity and reactivity.

Surface T also affects the sticking coefficient, S_c, of many materials. If one wants to monitor the flux actually depositing at the substrate T rather than that depositing at room T, it is possible to operate the crystal at the elevated T despite the steeper dependence of v_r on T which is shown in Fig. 4.16. In my lab, I have found that satisfactory stability of v_r can be achieved by applying the feedback T-control techniques discussed in Sec. 4.5.3 to the crystal holder. Above 350° C or so, the piezoelectric effect weakens, and it is difficult to maintain oscillation. If instead of duplicating substrate T, one wants to increase the S_c of a vapor flux toward unity, the crystal can be cooled, or a reactive material can be codeposited. For example, in the monitoring of (AlGa)As deposition, the S_c of As_4 vapor can be increased by codepositing a stoichiometrically excess flux of Al. In the case of the Group II-VI semiconductors, an excess flux of either one of the elemental vapors drives the S_c of the other one toward unity [18].

Loss of oscillation will occur if a liquid is deposited, such as Ga or Hg, or if the film reacts with or dissolves the electrode material, thus destroying the electrical contact. Pt is the most robust of the commonly available electrodes. Quartz monitors can even be used in plasma environments, provided that the holder is well grounded and that the 5 MHz wire is well shielded all the way to the vacuum feedthrough. Materials exposed to plasma or to ion beams can *lose* mass due to ion-sputter removal of surface atoms, as discussed in Sec. 8.5.4. In such cases, the quartz crystal, predeposited with the material of interest, can be used as a sensitive measure of sputtering rate.

4.8.2 Optical techniques

The optical techniques of deposition flux monitoring have several advantages: they are applied directly to the depositing film, yet they involve too low an energy density to perturb the deposition process, and they can be used in high-pressure or plasma environments. On the other hand, their correct interpretation sometimes requires considerable knowledge of the optical properties of film and substrate, and the film must have some transparency. Here, we will briefly review the physics of light interaction with matter and apply it to the two basic rate-monitoring techniques: interference oscillations and ellipsometry. Light *scattering* can be used to monitor film surface roughness during deposition, and this will be discussed in Sec. 5.4.1.

Light (and electromagnetic radiation of other frequencies) consists of sinusoidally oscillating electric and magnetic fields which are oriented transverse to the direction of ray travel and at right angles (orthogonal) to one another. Interaction with matter is almost entirely due to the electric field, which causes charges in the matter to oscillate. In Chap. 9, we will see how radio-frequency and microwave-frequency electromagnetic fields can energize plasmas by causing their free electrons to oscillate. Here, we begin by considering in-phase light rays a and b of the plane wavefront in Fig. 4.18 as they approach the surface of a transparent solid (or liquid) medium from the gas phase. The incident, reflected, and refracted rays all lie in the "plane of incidence" (the plane of the figure), which is always perpendicular to the solid surface. The electric-field vector, $\mathbf{E}$, may be oriented parallel to the plane of incidence, as shown by $\mathbf{E}_p$, or perpendicular to it, which would be denoted by $\mathbf{E}_s$, since perpendicular is senkrecht in German. Other orientations of $\mathbf{E}$ may always be described in terms of its $\mathbf{E}_p$ and $\mathbf{E}_s$ components. The $\mathbf{E}_p$ orientation is known as TM (transverse magnetic) polarization, and the $\mathbf{E}_s$ as TE (transverse electric).

The speed of light in vacuum or gas, c_0, slows down to speed c_1 upon refraction into medium 1 due to the electric polarization which $\mathbf{E}$ induces in the medium, and the speed ratio, c_1/c_0, is called the index of refraction, $\tilde{n}$. Since the frequency of the light, ν, must remain constant, and since

$$\nu = \frac{c}{\lambda} \tag{4.39}$$

the wavelength is proportionately shortened upon refraction; that is,

$$\lambda_1 = \lambda_0 \tilde{n}_0 / \tilde{n}_1 \tag{4.40}$$

where λ_0 and λ_1 are the wavelengths in gas and in medium 1, respectively. It is therefore convenient to think of an "optical path length," L,

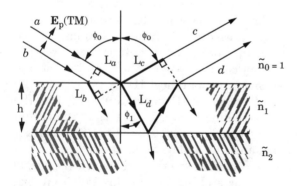

Figure 4.18 Geometry of refraction and interference.

measured in wavelengths, for light traveling in a refractive medium; thus, $L \propto 1/\tilde{n}$. For the rays a and b to continue along in phase as a plane wavefront after being refracted into medium 1, the optical path lengths L_a and L_b, shown by the heavy lines in Fig. 4.18, must be equal. This leads directly to Snell's law:

$$\tilde{n}_0 \sin \phi_0 = \tilde{n}_1 \sin \phi_1 \quad (4.41)$$

where the angles of incidence, ϕ_0, and of refraction, ϕ_1, are measured from the perpendicular to the surface. Another part of ray a is shown being reflected from the surface as ray c. Energy imparted to ray c comes from charge oscillations in medium 1 that are induced by the refracted part of ray a, so these oscillations will be perpendicular to the refracted ray. At a particular value of ϕ_0, ray c will *also* be perpendicular to the refracted ray, and under this condition there will therefore be no component of charge oscillation perpendicular to ray c for TM-polarized light. This means that the energy delivered to ray c will be zero, so no light can be reflected. This particular angle of incidence is known as Brewster's angle, ϕ_B, and it can be shown by trigonometry and Snell's law that

$$\phi_B = \tan^{-1}(\tilde{n}_1/\tilde{n}_0) \quad (4.42)$$

Note that reflection does *not* go to zero for TE polarization, and thus ϕ_B is known also as the "polarizing angle." We will see that the value of ϕ_0 relative to ϕ_B is very important in optical measurements on thin films.

If medium 1 is a thin film of thickness h on a substrate of index $\tilde{n}_2$, light refracted into it will continue to bounce back and forth across h with some reflection and refraction at each interface as shown for ray a in Fig. 4.18. The refracted light leaving medium 1 as ray d after one bounce traverses an optical path length of $2L_d$ within this medium during the time that the reflected ray c traverses a shorter path L_c in the gas before meeting up with ray d in the outgoing beam. The particular film thickness, h_o, for which the path length difference is exactly one wavelength is called the "thickness period," and it can be shown by trigonometry and Eqs. (4.40) and (4.41) that

$$h_o = \frac{\frac{1}{2}\lambda_0}{\sqrt{\tilde{n}_1^2 - \sin^2 \phi_0}} \quad (4.43)$$

The same value of h_o is obtained for rays leaving the medium after any two successive bounces. If all of these rays are in phase, the intensity of the reflected beam is a maximum; whereas if successive rays are 180° (π radians) out of phase, the rays interfere destructively, and the reflected intensity is a minimum. A phase shift of 2π is induced by the path through a film of thickness h_0, and there can be additional phase shifts upon reflection at each interface depending on conditions, as specified in Table 4.3. There is never a phase shift upon refraction. Note that the reflection situation is less complicated when $\phi_0 < \phi_B$. In that instance, the phase shift is the same at both of the interfaces of Fig. 4.18 when $\tilde{n}_2 > \tilde{n}_1 > \tilde{n}_0$. Then, maxima are obtained when $h = jh_0$, where j is any integer, and minima are obtained when $h = (j + 1/2)h_0$. The same formulae apply to both polarizations. For perpendicular ("normal") incidence ($\phi_0 = 0°$), the first minimum (j = 0) occurs at $h = (1/4)\lambda_0/\tilde{n}_1$, and this is the situation for the familiar quarter-wave antireflection (AR) coating used on lenses. If in addition, $\tilde{n}_1 = \sqrt{\tilde{n}_2}$, the reflection from the AR coating is exactly zero. In thin-film monitoring, one can also have $\tilde{n}_2 < \tilde{n}_1$. In that case, the phase shifts at the 0-1 and 1-2 interfaces differ from each other by π for both polarizations, as seen in Table 4.3. Then, the conditions for maxima and minima as described above are reversed.

Deposition rate monitoring by recording the periodic intensity oscillations of monochromatic light is particularly appropriate for optical thin films such as AR coatings, since the optical thickness is the main property of interest. The He-Ne laser ($\lambda_0 = 632.8$ nm) is an ideal light source for films transparent at that wavelength. The technique can be applied to any film which has some transparency. If the film has some absorptivity, the oscillation amplitude will attenuate with increasing thickness as the contribution from the refracted ray becomes weaker. In addition, the phase relationships become more complicated with absorption. In such cases or when the film thickness to be measured is

TABLE 4.3 Phase Shifts upon Reflection of Light from an Interface

Polarization	ϕ_0	$\tilde{n}_b > \tilde{n}_a$*	$\tilde{n}_b < \tilde{n}_a$*
$\mathbf{E_s}$ (TE)	$< \phi_B$	π	0
	$> \phi_B$	π	$0 \to \pi$ as $\phi_0 \to \pi/2$
$\mathbf{E_p}$ (TM)	$< \phi_B$	0	π
	$> \phi_B$	π	$0 \to \pi$ as $\phi_0 \to \pi/2$

*Light is in medium of refractive index $\tilde{n}_a$; interface is with medium $\tilde{n}_b$; both media are nonabsorbing.

much less than a quarter wave, the more sophisticated optical technique of ellipsometry is preferred.

In ellipsometry, one extracts much more information from the reflected light beam by measuring the amplitude ratio and phase relationship of the $\mathbf{E}_p$ and $\mathbf{E}_s$ components; that is, by measuring the polarization ellipse of the reflected light. This concept is illustrated in Fig. 4.19, which shows the E-vector oscillations of light waves of both polarizations as they travel in the +r direction. The oscillation of each is described by its variation with time, phase shift, and position:

$$\mathbf{E}_k = \mathbf{E}_{ok} \cos(\omega t + \delta_k - 2\pi r \tilde{n}/\lambda_0) \tag{4.44}$$

where subscript k = p or s polarization
$\quad\quad\quad\mathbf{E}_{ok}$ = amplitude (maximum value) of $\mathbf{E}_k$
$\quad\quad\quad\omega$ = angular frequency = $2\pi\nu$
$\quad\quad\quad\delta$ = phase angle

The last term may also be written as $\mathbf{kr}$, where the "wave vector" $\mathbf{k} = 2\pi/\lambda = 2\pi\tilde{n}/\lambda_0$. For the two waves shown in Fig. 4.19, inspection shows that $\delta_s = 0$ and $\delta_p = \pi/2$. At any particular t and r, the resultant E vector obtained by the superposition of these two waves will have an amplitude and direction equal to their vector sum, $\mathbf{E} = \mathbf{E}_p + \mathbf{E}_s$. Now visualize the evolution of this sum in amplitude and direction as the two waves pass to the right through the r = 0 plane, and visualize the projection of $\mathbf{E}$ on the p-s plane at r = 0. This projection is shown to the left of Fig. 4.19 for t = 0 and looking into the beam. As t increases and

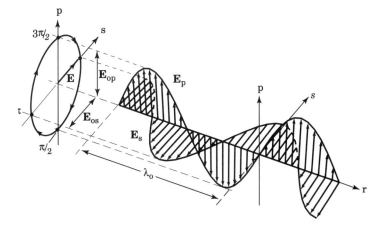

Figure 4.19 Superposition of two light waves to form the polarization ellipse. Here, the p-polarized wave leads the s-polarized wave by $\pi/2$, resulting in right-hand elliptical polarization.

the waves pass to the right, E_s decreases from its maximum value toward zero, while E_p, starting from zero, becomes negative. For equal amplitudes, E_{ok}, this progression results in E tracing a circular path in a clockwise direction as shown, and this is called right-hand circular polarization. Had δ_s been $-\pi/2$, the path would have been counterclockwise and the polarization left-hand. If the waves are instead *in phase* ($\delta_p = \delta_s$), the path becomes a line bisecting the top-right and bottom-left quadrants; if they are out of phase ($\delta_p = \delta_s \pm \pi$), the path is a line bisecting the other two quadrants. Also, if one of the amplitudes is zero, a line is obtained. These three latter cases are examples of *linear* polarization. Finally, for other phase and amplitude relationships, the path is an *ellipse* whose major axis is inclined at some azimuthal angle θ as shown in Fig. 4.20. The maximum excursions of the ellipse in the p and s directions are the amplitudes of the electric fields in those directions. The significance of the angle ψ will be clear later. Note that monochromatic light is required to obtain a time-invariant phase relationship between the two waves. The smaller the bandwidth, Δλ, of the light, the longer is the distance of travel in r before the phase relationship is lost. This distance is the coherence length, Δr, and

$$\Delta r = \lambda \left(\frac{\lambda}{\Delta \lambda} \right) \qquad (4.45)$$

A good monochromator or a laser is required for Δr to be long enough for ellipsometry.

In ellipsometry, we are concerned with the phase of the reflected light but not with the time dependence, which is simply sinusoidal. The time dependence can be factored out of Eq. (4.44) by switching to exponentials using Euler's formula: $\cos x + i \sin x = e^{ix}$ (which is obtained by comparing the series expansions of these functions). If $\Re$

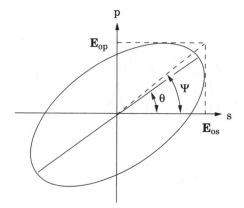

Figure 4.20 Geometry of the polarization ellipse of the reflected light. The ellipsometric angle, ψ, is oriented as shown only when the *incident* amplitudes are equal.

4.8.2 Optical techniques 111

denotes the real part of these complex numbers, then $\cos x = \Re(e^{ix})$, and

$$E_k = \Re\left(e^{i\omega t} \cdot e^{i\delta_k} \cdot e^{-i \cdot 2\pi r \hat{n}/\lambda_0}\right) \quad (4.46)$$

It turns out that operating on the complex exponential terms is equivalent to operating on the real part only, at least for present purposes, so the $\Re$ is dropped henceforth. The substitution of $\tilde{n}$ by $\hat{n}$ indicates that the refractive index is also in general a complex number,

$$\hat{n} = \tilde{n} + i\kappa \quad (4.47)$$

Here, κ is the extinction coefficient, which accounts for light absorption and scattering in the film. The reason that the extinction term is imaginary becomes clear upon factoring the $\hat{n}$ exponential term in Eq. (4.46):

$$e^{-i \cdot 2\pi r \hat{n}/\lambda_0} = e^{-i \cdot 2\pi r \tilde{n}/\lambda_0} \cdot e^{-(2\pi\kappa/\lambda_0) r} \quad (4.48)$$

The second term on the right is a *real* number, and it describes the exponential attenuation of light passing through an absorbing medium. Photodetectors actually measure light *intensity*, $I = |E_k|^2$, and applying this relationship to the attenuation term leads to the familiar Beer's law:

$$\boxed{\frac{I(r)}{I(r=0)} = e^{-(4\pi\kappa/\lambda_0) r} = e^{-\alpha_r r}} \quad (4.49)$$

where α_r is the light absorption coefficient. Note that α_r is different from the radiation *absorptivity* discussed in Sec. 4.2. The absorptivity equals the fraction of incident light that is absorbed in the *total* thickness of the medium rather than being reflected from its surface or transmitted through to the other side, whereas α_r expresses attenuation per unit of thickness. In ellipsometry, the effect of extinction in a thin film shows up as a reduced amplitude for light reflected from the substrate interface. Additional loss of reflected amplitude occurs due to refraction into the substrate. There are also phase changes due to reflection from interfaces and due to path length differences, as discussed above.

The amplitude and phase changes upon reflection of each polarization component (k = p or s) are accounted for together by the complex reflection coefficients

$$r_k = |r_k|e^{i\delta_k} \qquad (4.50)$$

where $|r_k|$ is the reflected/incident amplitude ratio, and where it has been assumed that the phases of both *incident* components are zero. Here, we have dropped both the t and the r factors of Eq. (4.46) and are considering only the δ factor at a fixed t and r. This expression is called the "phasor" representation of a periodic function, which should not be confused with the energy-beam sidearm carried by Enterprise crew members. It is common in ellipsometry to direct at the surface linearly polarized light which is inclined at an azimuthal angle of 45° to the plane of incidence, so that the amplitudes and phases of each component are equal. The light is polarized by passing it through a "dichroic" material such as a Polaroid sheet. Dichroic materials cause polarization by having a much higher extinction coefficient for one polarization than for the orthogonal one. Using such light, the ratio of the complex reflection coefficients is obtained directly by measuring the polarization ellipse of the reflected light. These are the basic data which are obtained from ellipsometry, and the ratio is expressed as:

$$\rho = \frac{r_p}{r_s} = \frac{|r_p|}{|r_s|}e^{i(\delta_p - \delta_s)} = \tan\Psi e^{i\Delta} \qquad (4.51)$$

In the case of equal amplitudes for the incident light, ψ is also given by the ratio of the reflected *amplitudes* as shown in Fig. 4.20, which also shows the azimuthal orientation angle of the ellipse, θ. Since ψ can vary over 90° and Δ over 360°, the data space forms a hemisphere in polar coordinates.

In practice, the reflection ellipse is mapped by passing the reflected light through a spinning polarizer into a photodetector. From the ellipse shape and azimuth, the ellipsometric angles, ψ and Δ, can be calculated. These angles uniquely determine the $\hat{n}$ of a bare substrate. Upon film deposition, the new ψ and Δ plus the previously measured $\hat{n}$ of the substrate uniquely determine both the $\hat{n}$ and the thickness, h, of the film. The complex dielectric constant of the film, $\varepsilon_d = \varepsilon_1 + i\varepsilon_2$, can also be calculated, since it is related to $\hat{n}$ by $\varepsilon_d = \hat{n}^2$. The imaginary part of ε_d is associated with optical attenuation in the film, as was the case with $\hat{n}$. The mathematics of transforming the raw ellipsometric data into ψ and Δ and thence into film properties is very complicated; but fortunately, good software is available. It is possible, for example, to plot ψ and Δ in real time during film deposition. In this mode, if the incident angle, ϕ_0, is set near to the ϕ_B [Eq. (4.42)] *of the film* which is being deposited, Δ can detect less than one monolayer of

4.8.2 Optical techniques

deposition! This seems surprising, since the light wavelength amounts to thousands of monolayers, but it comes about as follows. Recall that there is no reflection of the p component at ϕ_B. Therefore, only the s component is reflected from the surface of the film, while the p component is completely refracted through the film and thus becomes retarded in phase by the added optical path length. We know that one thickness period of film, h_0 in Eq. (4.43), corresponds to a 360° phase retardation, so for a typical h_0 of 100 nm, the h sensitivity of this technique is 0.3 nm per degree of Δ. The choice of ϕ_0 is important, because if it is too close to ϕ_B, too much sensitivity results, meaning that small errors in measuring ϕ_0 will result in large errors in calculated $\tilde{n}$ and h. Another characteristic of ellipsometry to be wary of is that the sensitivity of the ellipsometric angles to changes in the optical properties of the film varies widely with conditions. For example, sensitivity to the $\tilde{n}$ and h of a film are low when h is near the thickness period of Eq. (4.43) (Tompkins, 1993).

Figure 4.21 shows results obtained [19] for the ψ, Δ trajectory in depositing $Al_{0.25}Ga_{0.75}As$ onto a GaAs substrate by CVD at 600° C. Point 1 corresponds to the ψ and Δ for the GaAs substrate before deposition. With increasing deposition, the trajectory follows a logarithmic

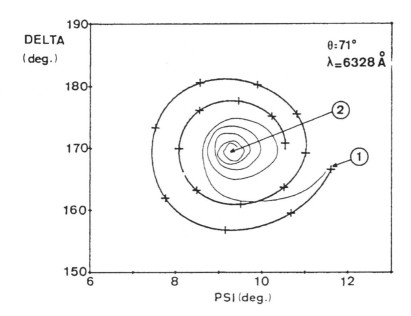

Figure 4.21 Experimental Ψ–Δ trajectory during CVD of $Al_{0.25}Ga_{0.75}As$ onto a GaAs substrate (heavy line with + marks) and calculated trajectory using abrupt-interface model (light line). (Source: Reprinted from Ref. 19 by permission. © 1981 by Annual Reviews, Inc.)

spiral as indicated by the heavy line, which has been marked with crosses at 10 nm increments of h. The spiral would eventually converge to point 2, which corresponds to the ψ and Δ for bulk $Al_{0.25}Ga_{0.75}As$. If the film were completely transparent to the 632.8 nm He-Ne light used, the spiral would not converge, but would repeatedly circulate back on itself, passing through point 1 after the deposition of each thickness period, h_0. The observed convergence results from the attenuation, or extinction, of the refracted ray as h increases. Comparison of these data with a model calculation provides additional information about the depositing film. The narrow-line trajectory plotted in Fig. 4.21 is the theoretical trajectory for the deposition of a smooth film having a uniform ñ equal to that of $Al_{0.25}Ga_{0.75}As$. The considerable deviation from the data indicates that the ñ of the film is *not* initially that of $Al_{0.25}Ga_{0.75}As$, but instead it gradually changes from that of GaAs to that of $Al_{0.25}Ga_{0.75}As$ with increasing h. Intermixing of interfaces is a common problem in thin-film deposition, especially in CVD, and its measurement in real time is a powerful tool in pursuing its solution.

The above results at the same time point up the principal shortcoming of ellipsometry, which is that the calculated film properties must be based on an assumed model for the optical structure. Thus, the more that is known about the film(s), the more likely it is that a correct and unambiguous interpretation of the data can be made. For example, the development of roughness in the depositing film can be misinterpreted as a reduced ñ. This is because roughness on a scale much smaller than λ_0 will not cause detectable scattering, but rather will appear optically as an "effective medium" composed of film material plus gas inclusions. This too can be modeled, but it needs to be known whether roughness is indeed the cause of the reduced ñ. In addition, effective-medium theory is complicated, and most of the effective-medium models, including the popular Bruggeman approximation, assume isotropic inclusions. Unfortunately, rough or porous thin films often have a columnar structure rather than an isotropic one. These ambiguities, like all ambiguities, are best resolved by collecting more data, and the easiest way to do this in ellipsometry is by varying λ_0; that is, by employing *spectroscopic* ellipsometry. Clearly, fitting a model to the ellipsometric data over a wide λ range is a much more rigorous test of it than fitting it at only one λ. Moreover, it usually happens that a certain λ is particularly responsive to the film layers under study, in which case that λ can be selected for real-time monitoring and feedback control of deposition rate and composition [20]. Spectroscopic ellipsometry is also useful for determining the h and ñ values of multilayer film structures *after* deposition, as will be discussed in Sec. 10.1.1.

4.9 Conclusion

In this chapter, we have examined the supply of source material by thermal evaporation and the transport of the resulting vapor to the substrate within a high-vacuum environment. Although a material's p_v versus T behavior can be estimated by thermodynamics knowing only the boiling point and heat of evaporation, its net evaporation *rate* can be calculated from p_v only in the cases of true Knudsen-cell sources or atomic vapors of metals. In other cases, p_v only gives an *upper limit* to net evaporation rate. This points up a general fact of chemistry, namely that the kinetics of a process are always more difficult to estimate than the equilibrium situation. The evaporation of alloys and compounds introduces the additional complication of vapor component ratios drifting with time due to changing source composition. Materials knowledge, source design, and T stability are all important in controlling supply rate. Once the vapor has been generated, it must be transported uniformly to the substrate. In that step, key factors are the geometry between source and substrate and the monitoring of vapor flux or of deposition rate. We have seen that there are some very sensitive monitoring techniques available, which can also be used for feedback control of supply rate. Having now dealt with evaporation and transport, we are ready to examine the deposition itself.

4.10 Exercises

4.1 For Al evaporating at 1100° C, $\Delta_v H$ = 318 kJ/mole. Assuming atomic vapor, what fraction of this energy goes into pΔV, what fraction into kinetic energy, and what fraction into potential energy?

4.2 The sublimation T of TiF$_4$ is 284° C. At that T, $\Delta_f H_c$ = −1639 kJ/mole and $\Delta_f H_v$ = −1551 kJ/mole. (a) Write an equation for p_v (Pa) with T (in K) as the only unknown. (b) What is p_v at room T (25° C)? (c) What is the TiF$_4$ effusion rate at room T from an ideal Knudsen cell having a 1 mm-diameter orifice?

4.3 For water in equilibrium with its vapor at 100° C: (a) What is average time between arrivals of vapor molecules within a radius of three atomic distances (about 1 nm) from a particular surface site? (b) How much time does it take the average arriving molecule to traverse the last three atomic distances above the surface, neglecting acceleration due to potential interaction? (c) What is the probability that there will be an impinging molecule within three atomic distances of a particular surface site?

4.4 The p_v of arsenic vapor over the elemental solid is 1 Pa at the operating T of a particular Knudsen cell. If the evaporation coefficient of arsenic is 10^{-4}, the Knudsen-cell orifice has an area of 0.01 cm^2, and the surface area of the evaporating arsenic is 100 cm^2, what is the steady-state pressure of arsenic vapor in the cell?

4.5 Show that Eq. (4.23) for the steady-state alloy melt composition follows from the Knudsen equation, stating assumptions needed.

4.6 Derive Eq. (4.25) from Eq. (4.24).

4.7 An alloy is being deposited by wire feed into a crucible melt having V = 10 cm^3. The deposition rate on a substrate 30 cm directly above the crucible is 10 µm/h. (a) Assuming cosine-law effusion, what is the volume feed rate of wire, W, required to hold V constant after the melt reaches steady-state composition? (b) How many hours will it take for the melt to drift 90 percent of the way from initial to steady-state composition, assuming that τ = V/W?

4.8 The compound MY sublimes into M plus Y$_2$, with unity α_V values. At the particular T in the crucible used for subliming it, the p_V values of the two vapor species over MY at the phase boundaries of solid MY are: $p_V(M) = 10^{-1}$ Pa and $p_V(Y_2) = 1$ Pa at 50.1 percent M; $p_V(M) = 10^{-3}$ and $p_V(Y_2) = 10^3$ at 49.98 percent M. Describe the evolution with time of the composition in the crucible as sublimation proceeds, starting from exactly stoichiometric MY.

4.9 (a) Show that the following expression for the blackbody radiative heat-transfer coefficient

$$h_r = \sigma\left(T_h^3 + T_h^2 T_s + T_h T_s^2 + T_s^3\right)$$

follows from Eq. (4.19). (b) For $T_h = 800°$ C and $T_s = 500°$ C, what percent error is introduced by the approximation $h_r = 4\sigma T^3$?

4.10 Ag having 1 at.% Cu impurity is being evaporated at 1000° C. Assuming unity evaporation coefficients, what is the atomic fraction of Cu in the evaporating flux, (a) initially and (b) after 90 percent of the Ag has been evaporated? (An approximate answer without using calculus is being sought here; state your assumptions.)

4.11 Show by integration that, for a cosine distribution of evaporant, the evaporant flux perpendicular to the source plane is twice the average flux over the hemisphere.

4.12 A flat, 10 cm-diameter substrate is situated with its center along the axis of an ideal-orifice evaporant source and 20 cm away from it. The substrate is tilted at 30° from being perpendicular to the axis. Assuming that all material arriving at the substrate condenses on it, by what percents do the deposition rates at the substrate points nearest to and farthest from the source differ from the rate at the center of the substrate?

4.13 How many cm^3 of MgF_2 must be loaded into a crucible to deposit 10 μm of film on a substrate placed perpendicular to and centered on the axis of the crucible at a distance of 30 cm, assuming cosine effusion and unity S_c, and consuming only 80 percent of the MgF_2?

4.14 A molecularly evaporating compound of molecular weight 70 is effusing from the 1.0 mm-diameter ideal orifice of a 700° C Knudsen cell and is depositing with unity S_c and at 1.3 μm/h, assuming ρ_f = 2.8, onto a quartz-crystal monitor located on the axis of the cell, perpendicular to it, and 10 cm away. The compound has been progressively ground up until the deposition rate has stopped increasing with further grinding. (a) What does the last statement imply about thermodynamic conditions within the cell? (b) What is the p_v of this compound at 700° C?

4.15 Derive Eq. (4.42) for Brewster's angle.

4.16 Derive Eq. (4.43) for the thickness periodicity, h_o.

4.11 References

1. Chase, M.W., et al. (eds.). 1985. *JANAF Thermochemical Tables*, 3rd ed. Washington, DC: American Chemical Society.
2. Wagman, D.D., et al. (eds.). 1982. "NBS Tables of Chemical Thermodynamic Properties." *J. Phys. Chem. Ref. Data* 11, suppl. no. 2. Washington, DC: American Chemical Society.
3. *CRC Handbook of Chemistry and Physics*. Boca Raton, Fla.: CRC Press.
4. Rosenblatt, G.M. 1976. "Effect of Incident Flux on Surface Concentrations and Condensation Coefficients when Growth and Vaporization Involve Mobile Surface Species." *J. Chem. Phys.* 64:3942.
5. Baker, H. 1992. "Introduction to Alloy Phase Diagrams." Sec. 1 in vol. 3, *ASM Handbook, Alloy Phase Diagrams*. Materials Park, Ohio: ASM International.
6. Smith, D.L. 1979. "Nonstoichiometry and Carrier Concentration Control in MBE of Compound Semiconductors." *Prog. Crystal Growth and Characterization* 2:33.
7. Arthur, J.R. 1967. "Vapor Pressures and Phase Equilibria in the Ga-As System." *J. Phys. Chem. Solids* 28:2257.
8. Winters, H.F., H. Coufal, C.T. Rettner, and D.S. Bethune. 1990. "Energy Transfer from Rare Gases to Surfaces." *Phys. Rev. B* 41:6240.
9. Eres, D., D.H. Lowndes, J.Z. Tichler, J.W. Sharp, T.E. Haynes, and M.F. Chisholm. 1990. "The Effect of Deposition Rate on the Growth of Epitaxial Ge on GaAs(100)," *J. Appl. Phys.* 67:1361.
10. Hasan, M.-A., J. Knall, and S.A. Barnett. 1987. "A Low-Energy Metal Ion Source for Primary Ion Deposition and Accelerated Ion Doping during Molecular Beam Epitaxy." *J. Vac. Sci. Technol. B* 5:1332.
11. Urban III, F.K., S.W. Feng, and J.J. Nainaparampil. 1993. "Study of Zinc Films Formed using Large Clusters in the Ionized Cluster Beam Deposition Technique." *J. Vac. Sci. Technol. B* 11:1916.
12. Knauer, W. 1987. "Formation of Large Metal Clusters by Surface Nucleation." *J. Appl. Phys.* 62:841.

13. Adamson, S., C. O'Carroll, and J.F. McGlip. 1989. "Monte Carlo Calculations of the Beam Flux Distribution from Molecular Beam Epitaxy Sources." *J. Vac. Sci. Technol. B* 7:487.
14. Shiralagi, K.T., A.M. Kriman, and G.N. Maracas. 1991. "Effusion Cell Orientation Dependence of Molecular Beam Epitaxy Flux Uniformity." *J. Vac. Sci. Technol. A*, 9:65.
15. Wasilewski, Z.R., G.C. Aers, A.J. SpringThorpe, and C.J. Miner. 1991. "Studies and Modeling of Growth Uniformity in Molecular Beam Epitaxy." *J. Vac. Sci. Technol. B* 9:120.
16. Manufactured by Leybold Inficon, Inc.
17. Benes, E. 1984. "Improved Quartz Crystal Microbalance Technique." *J. Appl. Phys.* 56:608.
18. Smith, D.L., and V.Y. Pickhardt. 1975. "Molecular Beam Epitaxy of II-VI Compounds." *J. Appl. Phys.* 46:2366.
19. Theeten, J.B., and D.E. Aspnes. 1981. "Ellipsometry in Thin Film Analysis." *Ann. Rev. Materials Science* 11:97.
20. Quinn, W. E., D.E. Aspnes, M.J.S.P. Brasil, M.A.A. Pudensi, S.A. Schwarz, M.C. Tamargo, S. Gregory, and R.E. Nahory. 1992. "Automated Control of III-V Semiconductor Composition and Structure by Spectroellipsometry." *J. Vac. Sci. Technol. B* 10:759.

4.12 Recommended Readings

Aspnes, D.E., and R.P.H. Chang. 1989. "Spectroscopic Ellipsometry in Plasma Processing." In *Plasma Diagnostics, vol. 2*, ed. O. Auciello and D.L. Flamm. Boston, Mass.: Academic Press.

Herman, M.A., and H. Sitter. 1989. *Molecular Beam Epitaxy*, Chap. 2. Berlin: Springer-Verlag.

1991. Proceedings of the Fifth International Symposium on Small Particles and Inorganic Clusters (Konstanz, Germany). *Zeitschrift fur Physik D*, vol. 19 and 20 (in English).

Sears, F.W. 1950. *An Introduction to Thermodynamics, the Kinetic Theory of Gases, and Statistical Mechanics*. Cambridge, Massachusetts: Addison-Wesley.

Stoecker, W.F. 1989. *Design of Thermal Systems*, 3rd ed, Chap. 2. New York: McGraw-Hill.

Tompkins, H.G. 1993. *A User's Guide to Ellipsometry*. Boston, Mass.: Academic Press.

Chapter

5

Deposition

We now come to the heart of the thin-film process sequence. Deposition may be considered as six sequential substeps, and we will examine these one by one in this chapter. The arriving atoms and molecules must first (1) adsorb on the surface, after which they often (2) diffuse some distance before becoming incorporated into the film. Incorporation involves (3) reaction of the adsorbed species with each other and the surface to form the bonds of the film material. The (4) initial aggregation of the film material is called nucleation. As the film grows thicker, it (5) develops a structure, or morphology, which includes both topography (roughness) and crystallography. A film's crystallography may range from amorphous to polycrystalline to single-crystal. The last is obtained by epitaxy—that is, by replicating the crystalline order of a single-crystal substrate. Epitaxy has special techniques and features which are dealt with separately in Chap. 6. Finally, (6) diffusional interactions occur within the bulk of the film and with the substrate. These interactions are similar to those of post-deposition annealing, since they occur beneath the surface on which deposition is continuing to occur. Sometimes, after deposition, further heat treatment of a film is carried out to modify its properties. For example, composition can be modified by annealing in a vapor, and crystal growth can be achieved by long annealing or by briefly melting. These post-deposition techniques will be mentioned only in passing in this book.

In the latter part of the chapter, we will examine three key practical aspects of deposition: the development of mechanical stress, adhesion of the film to the substrate, and substrate T control. The issue of composition control during deposition will be reserved for Chap. 6, since it is more easily studied under epitaxial growth conditions.

For the present, we will consider that only thermal energy is being supplied to the surface except where energy enhancement is specifically noted. Adding energy to the deposition process by nonthermal means is an important process technique which can activate deposition reactions and alter film structure. It was introduced in Sec. 4.5.4 and will be treated extensively in Chaps. 8 and 9.

5.1 Adsorption

Consider a molecule approaching a surface from the vapor phase, as shown in Fig. 5.1. Upon arriving within a few atomic distances of the surface, it will begin to feel an attraction due to interaction with the surface molecules. This happens even with symmetrical molecules and with inert gases, neither of which have dipole moments. It happens because even these molecules and atoms act as *oscillating* dipoles, and this behavior creates the induced-dipole interaction known as the van der Waals force or London dispersion force. Polar molecules, having permanent dipoles, are attracted more strongly. The approaching molecule is being attracted into a potential well like the one that was illustrated in Fig. 4.2 for condensation. Condensation is just a special case of adsorption in which the substrate composition is the same as that of the adsorbant. This is sometimes the case in thin-film deposition and sometimes not. In either case, the molecule accelerates down the curve of the potential well until it passes the bottom and is repelled by the steeply rising portion, which is caused by mutual repulsion of the nuclei (more on this in Sec. 8.5.2.2). If enough of the

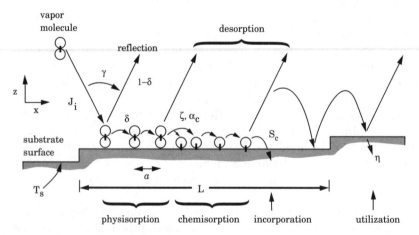

Figure 5.1 Adsorption processes and quantities. α_c is used only for condensation (adsorption of a material onto itself). A vertical connecting bar denotes a chemical bond.

molecule's perpendicular component of momentum is dissipated into the surface during this interaction, the molecule will not be able to escape the potential well after being repelled, though it will still be able to migrate along the surface. This molecule is trapped in a weakly-adsorbed state known as physical adsorption or physisorption. The fraction of approaching molecules so adsorbed is called the trapping probability, δ, and the fraction escaping (reflecting) is $(1 - \delta)$ as shown in Fig. 5.1. The quantity δ is different from the thermal accommodation coefficient, γ, which was defined by Eq. (2.33). In general, a molecule is at least partially accommodated thermally to the surface temperature, T_s, even when it is reflected without having been trapped.

The physisorbed molecule is mobile on the surface except at cryogenic T, so it is shown hopping (diffusing) between surface atomic sites on Fig. 5.1. It may desorb after a while by gaining enough energy in the tail of the thermal energy distribution, or it may undergo a further interaction consisting of the formation of chemical bonds with the surface atoms; that is, chemisorption. If both adsorption states exist, the physisorbed state is called the precursor state. Chemisorption involves the sharing of electrons in new molecular orbitals and is much stronger than physisorption, which involves only dipole interactions. These two types of adsorption can be distinguished in almost all vapor-surface combinations, so they constitute a valuable model with which to analyze any surface process. This model has long been applied to heterogeneous catalysis, thin-film deposition, and condensation of molecular vapors. Recent theory indicates that even the condensation of a monatomic vapor such as Al can involve both adsorption states, the precursor state in that case being an Al-Al dimer whose bonding to the bulk Al is inhibited by the existence of the dimer bond [1]. In such a case, and in the case of condensing *molecular* vapors such as As_4, the vapor would not be considered actually *condensed* until it had become fully incorporated into the solid phase by chemisorption. Thus, the condensation coefficient, α_c, defined by Eq. (4.18) is that fraction of the arriving vapor that becomes not only trapped but also chemisorbed, as indicated in Fig. 5.1. However, the term α_c is not used in the case of chemisorption on a foreign substrate. Then, we speak of the chemisorption reaction probability, ζ, which will be derived later. The precursor model may also be applied to cases where both of the adsorption states involve chemical bonding, but where the bonding in one state is weaker than in the other.

Since some of the physisorbed species eventually escape back into the vapor phase, a third term, called the sticking coefficient, S_c, is used to denote that fraction of the arriving vapor that remains adsorbed *for the duration of the experiment*. Since this duration is arbitrary, S_c has less of a fundamental meaning than δ and α_c, which are determined solely by chemistry and energy. Nevertheless, S_c is very useful in thin-film deposition, since it is equal to the fraction of arriving vapor which becomes incorporated into the film. That is, this fraction becomes adsorbed and then buried before it can desorb. Note that in the limiting case of vapor-solid equilibrium, $S_c \to 0$ whereas $\alpha_c > 0$. One more fraction, which is useful in CVD practice, is the utilization fraction, η, of a chemical vapor. Chemical-vapor molecules diffuse around in the deposition chamber and can hit the film surface many times before finally being swept downstream in the flow, as we will see in Chap. 7 and as shown in Fig. 5.1. Thus, η can approach unity even when S_c is very low.

There are many examples of precursor adsorption in thin-film technology. The precursor phase exists any time there is both a weak and a strong bonding state of the adsorbing vapor with the substrate or film surface. Here are some examples:

1. In most CVD reactions, the feed vapors adsorb as molecules and then undergo the reactions which break their molecular bonds and form new bonds to the film surface; thus,

$$SiH_4(g) \to \cdots \to SiH_4(a) \to Si(c) + 2H_2(g)$$

where (g), (a), and (c) denote the gas, adsorbed, and condensed phases, and where $(\cdots)$ denotes a series of intermediate reaction steps.

2. In deposition of compounds from separate vapor sources of each element, adsorbing vapor bonds much more strongly to those surface sites occupied by the other element; thus,

$$Zn(g) + Se(a) \to ZnSe(c) \quad \text{and} \quad Se(g) + Zn(a) \to ZnSe(c)$$

3. Silicon that is chemically passivated by an atomic layer of H reacts with adsorbates mainly at those few sites that are missing an H atom. On the H-passivated sites, adsorbates remain only physisorbed.
4. Atomically flat surfaces often bond more strongly with adsorbates at atomic steps such as those shown at a spacing of L in Fig. 5.1.
5. Upon adsorption, atoms of low-reactivity metals often bond much less easily to nonmetallic substrates than to those sites containing another metal atom.

We will revisit these examples after examining the energetics of the precursor adsorption model in more detail.

Consider a hypothetical diatomic gas-phase molecule $Y_2(g)$ adsorbing and then dissociatively chemisorbing as two Y atoms. Figure 5.2

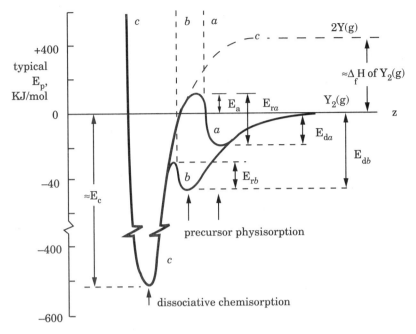

Figure 5.2 Energetics of the precursor adsorption model. Energy scale is typical only.

shows a diagram of the potential energy versus molecular distance, z, from the surface. This is similar to Fig. 4.2 for condensation except that we have changed from the molecular (ε_p) to the molar (E_p) quantities of potential energy which are more conventional in chemistry. The energy scales shown represent typical bond strengths. Three curves are shown: two alternate ones for the precursor state (a and b), and one for the chemisorbed state (c). By convention, the zero of E_p is set at the E_p of the element Y in its thermodynamic standard state, which we specify for this element to be the diatomic molecule in the gas phase. In fact, all gaseous elements except the inert gases have diatomic standard states. Note that lifting *atomic* Y out of its potential well along curve c results in a much higher E_p in the gas phase, which corresponds roughly to the heat of formation, $\Delta_f H$, of 2Y(g) from Y_2(g). [$\Delta_f H$ usually can be found in thermodynamic tables (Refs. 1–3, Chap. 4).] The result of this high E_p for Y(g) is that curves a and c intersect at positive E_p, meaning that there is an activation energy, E_a, to be overcome for Y_2(g) to become dissociatively chemisorbed. For the deeper precursor well, b, chemisorption is not "activated," though there still is a barrier, as shown. The level of E_{ra} or E_{rb}, and hence of E_a, is determined by the degree to which the bonds within both the precursor and the surface must be strained from their relaxed condition before new bonds can be formed between the precursor and the surface.

There are two ways in which vapor can arrive at the surface having an $E_p > 0$. Gaseous molecules have their E_p raised by becoming dissociated. Solids and liquids have it raised by evaporating, as discussed in Sec. 4.1. If the E_p of the arriving vapor is high enough, curve c is followed, and direct chemisorption can occur without involving the precursor state. In the language of surface chemistry, direct reaction between an incoming species and a surface site or adsorbate is called the Eley-Rideal mechanism, whereas reaction among surface species is called the Langmuir-Hinshelwood mechanism.

A principal advantage of the energy-enhanced deposition processes is that they can provide enough energy so that the arriving molecules can surmount the E_a barrier and adsorb directly into the chemisorbed state. In other words, the arriving molecules immediately react with the surface to deposit the film. In sputter deposition, species arrive having *kinetic* energies of around 1000 kJ/mol as well as having $E_p > 0$ by having been vaporized. In plasma-enhanced deposition, vapor molecules become dissociated in the plasma and thus arrive along curve c, above the E_a barrier. Thus, an energy-enhanced process can supply E_a to the arriving species either as kinetic energy of accelerated molecules or as potential energy of dissociated ones.

Conversely, in thermally controlled deposition processes such as evaporation and CVD, the vapor often adsorbs first into the precursor state; that is, it falls to the bottom of the well on curve a or b. Thence, it may either chemisorb by overcoming the barrier $E_{r(a,b)}$ shown in Fig. 5.2, or it may desorb by overcoming the heat of physisorption, which is roughly $E_{d(a,b)}$. The competition between these two reactions results in a net rate of chemisorption whose behavior we would like to describe, since it is the basic film-forming reaction. We start with the conventional expression for the rate of a first-order chemical reaction, first-order meaning that rate is proportional to the concentration of one reactant; thus,

$$R_k = k_k n_s = k_k n_{so} \Theta \tag{5.1}$$

where R_k = rate of the kth surface reaction per unit surface area, mc/cm²·s

k_k = rate constant, s^{-1}

n_s = surface concentration of reactant, mc/cm²

n_{so} = monolayer surface concentration, mc/cm²

Θ = fractional surface coverage by reactant

Rate and concentration here are in surface units; for volume reactions such as in CVD vapor phases, they would instead be in mc/cm^3·s and mc/cm^3, respectively (Sec. 7.3.2). The rate constant follows the Arrhenius equation, which we will derive in the next section:

$$k_k = \nu_{ok} e^{-E_k/RT} \quad (5.2)$$

where ν_{ok} = frequency factor or pre-exponential factor
E_k = reaction activation energy, kJ/mol

We now make the simplifying assumptions that n_s is constant over time (steady state) and that the chemisorption reaction only occurs in the forward direction. In film deposition practice, chemisorption reversal occurs only when surface T is so high that the film is beginning to decompose. We may now write a mass balance [Eq. (3.1)] for the physisorbed precursor:

$$J_i \delta (1 - \Theta) = R_r + R_d = (k_r + k_d) n_{so} \Theta \quad (5.3)$$

where J_i = molecular impingement flux, mc/cm^2·s [Eq. (2.18)]
R_r = reaction (chemisorption) rate
R_d = desorption rate
n_s = surface concentration of the precursor

Here, we have also made the reasonable assumption that adsorption does not occur on the area already occupied by adsorbate (Θ). Rearranging this expression, we have

$$\Theta = \frac{J_i \delta / n_{so}}{J_i \delta / n_{so} + k_r + k_d} \quad (5.4)$$

and substituting into the chemisorption rate expression [Eq. (5.1)],

$$R_r = k_r n_{so} \Theta = \frac{J_i \delta k_r}{J_i \delta / n_{so} + k_r + k_d} \quad (5.5)$$

With this, we may now define the sticking coefficient in film deposition more precisely as

$$S_c = R_r / J_i \quad (5.6)$$

Note that S_c depends on both J_i and the rate constants; this dependency has consequences for CVD film conformality over topography, as will be discussed further in Sec. 7.3.3.

In the case of small J_i and thus small Θ, Eq. (5.5) simplifies (Weinberg, 1991) to

$$R_r = \frac{J_i \delta}{1 + k_d/k_r} = J_i \left[\frac{\delta}{1 + \dfrac{\nu_{od}}{\nu_{or}} e^{-(E_r - E_d)/RT_s}} \right] = J_i \zeta \qquad (5.7)$$

This equation defines ζ, the chemisorption reaction probability—the fraction of that vapor impinging on *bare* $(1 - \Theta)$ sites that becomes chemisorbed rather than being reflected or desorbed. For $\Theta \ll 1$, $S_c \approx \zeta$; but for larger Θ, $S_c < \zeta$. It is important not to confuse these two terms. For the special case of film deposition from a single vapor having the same composition as the film, $\zeta \equiv \alpha_c$, the condensation coefficient. For the more complicated case of compound-film deposition from multicomponent vapors, the assumption of first-order kinetics on which Eqs. (5.5) and (5.7) depend is not always valid, because more than one reactant is involved. This case will be discussed more in Sec. 7.3.3.

The quantity R_r governs the rate of film deposition when k_r is the same from site to site along the surface and when T_s is not so high that decomposition or re-evaporation of the film is occurring. By Eq. (5.7), R_r is going to increase as the exponential energy term, $(E_r - E_d)$, decreases. If this term is positive, there is an activation energy, E_a, for chemisorption, as shown for curve *a* in Fig. 5.2, where $E_a = E_{ra} - E_{da} > 0$. If E_a is high enough, the film will fail to deposit unless T_s is raised to make the exponential term smaller. On the other hand, when chemisorption is not activated ($E_r < E_d$) as on curve *b*, R_r *decreases* with increasing T_s. Thus, R_r can go either way with T_s depending on the energetics at the surface. The activated case is very common in CVD. For example, Si deposits from silane gas (SiH_4) at elevated T but not at room T. Of course, if T_s becomes *too* high, the evaporation flux of the Si itself [J_v from Eq. (4.17)] will exceed R_r, and Si again stops depositing. The net deposition flux of Si is thus given by

$$J_r = R_r - J_v \qquad (5.8)$$

For this case, there is a T_s *window* between reaction activation and re-evaporation, within which deposition can be achieved. The opposite case of R_r increasing with *decreasing* T_s is more difficult to identify, because other factors can cause J_r to increase with decreasing T_s even after J_v vanishes, such as increasing *nucleation*.

Nucleation is a complication that must often be added to the above model, which assumed identical kinetics for all surface sites. We will examine the nucleation process in more detail in Sec. 5.3, but it needs some introduction now. When nucleation is important, $J_r > 0$ in Eq. (5.8) only on certain active substrate-surface sites called nucleation sites or on nuclei of film material which have spontaneously accumulated. There are many ways in which nucleation sites may arise. In deposition examples (3) and (5) mentioned earlier in this section, R_r was higher at unpassivated Si surface atoms and at metal sites, respectively, because of a lower E_r there. In example (2) on the deposition of compounds from separate vapors, chemisorption only occurred at sites containing the other element, and in the atomically-flat-surface example (4), it only occurred at atomic steps. In examples such as (2) through (4), exploiting the nucleation phenomenon can result in films which have less roughness and defects, as we will see on many occasions later on. Note that the precursor bonds more readily to the surface than to itself in all three of these latter examples.

In other cases, where the precursor bonds more readily to itself than to the surface, nucleation limitations result in undesirably nonuniform deposition. One such case is example (5) for deposition of metals on a nonmetallic substrate, particularly weakly reacting metals such as Zn and Cd on glass or on an ionically bonded substrate such as NaCl or CaF_2. The activation energies (E_a) for chemical bonding of Zn and Cd to these substrates are very high because of the high bond strength between the elements making up the substrate. Consequently, the metals remain in the physisorbed state, from which they easily desorb, unless they encounter other adsorbed metal atoms with which to bond and form a nucleus. When the nucleus becomes large enough to behave like bulk metal, it will not evaporate as long as T_s is low enough so that J_v for the bulk metal is negligible in Eq. (5.8). Thus, at T_s well below where bulk Zn and Cd evaporation becomes significant, such as room T, one still obtains a most undesirable "splotchy" deposit on such substrates, consisting of islands of metal separated by areas devoid of deposition where nuclei have not yet had a chance to form. These bare areas represent a situation where $R_r \ll J_i\delta$ in Eq. (5.7).

Note that we have now seen two opposite examples for Zn vapor: Zn adsorbing onto a Se-covered surface, in which the bonding to the surface is much stronger than that of Zn to itself; and Zn on glass, in which the bonding is much weaker than that of Zn to itself. It is important to keep in mind that the re-evaporation rate of any species adsorbing onto a foreign surface will usually be vastly different from the evaporation rate of that species in its pure bulk form, because the rate is dominated by the degree of interaction with the foreign surface.

Faster adsorption onto a foreign surface is exploited in atomic-layer epitaxy (Sec. 6.5.5), and slower adsorption is exploited in selective deposition (Sec. 7.3.3).

Another dramatic example of nucleation-induced nonuniformity is diamond-film CVD, which is carried out using carbon-containing precursors such as methane (CH_4). Figure 5.3 shows diamond nuclei growing on the edges of etch pits in a Si substrate. The deposition rate on the diamond nuclei is high, and elsewhere it is zero. It is not clear why diamond nucleation occurs only at these particular sites, and this is a subject of intense current research. In many other cases, energy-enhanced techniques can activate bonding between adsorbing species and substrates; this enhances both nucleation and film adhesion to the substrate. Thus, sputter-deposited films often have better adhesion than those deposited from thermal evaporation sources.

The existence of certain sites which are more active for adsorption than the rest of the surface is common in thin-film processes. In such cases, access of the precursor to these favored sites can dominate the deposition kinetics. Access sometimes occurs mainly from the vapor

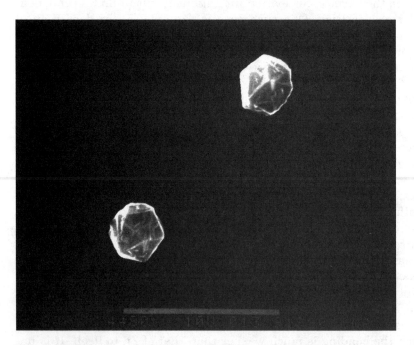

Figure 5.3 SEM photograph of two diamond nuclei growing on a patterned single-crystal Si substrate. The CVD of diamond from 1% CH_4 in H_2 at 4000 Pa was activated by a ≈2000° C Ta filament positioned 8 mm above the 900–1000° C substrate. (Previously unpublished photo courtesy of Paul A. Dennig from the laboratory of David A. Stevenson, Stanford University.)

phase (Eley-Rideal mechanism), but often surface diffusion (Langmuir-Hinshelwood mechanism) is the dominant route. Surface diffusion and nucleation will be addressed in the next two sections. These two complicating factors along with the possibility of non-first-order CVD reactions make the kinetic analysis of thin-film deposition very difficult, and consequently, such analysis has not often been performed. Nevertheless, simple models such as the one presented above for adsorption can provide a useful framework within which to think about deposition processes, provided that one remains aware of the limitations of the models being employed.

5.2 Surface Diffusion

Surface diffusion is one of the most important determinants of film structure because it allows the adsorbing species to find each other, find the most active sites, or find epitaxial sites. Various methods have been applied to measuring surface diffusion rates of adsorbed molecules, but most of this work has been done on chemical systems relevant to heterogeneous catalysis rather than to thin-film deposition. The role of surface diffusion in thin films has mainly been inferred from observations of film structure. However, the recent advent of the scanning tunneling microscope (STM) gives us the extraordinary power to directly observe individual atoms on surfaces in relation to the entire array of available atomic surface sites (Lagally, 1993). STM observation of the diffusion of these atoms should ultimately provide a wealth of data relevant to thin-film deposition.

We will develop here an expression for the rate of surface diffusion using absolute-reaction-rate theory. Although this approach cannot provide a quantitative estimate of the diffusion rate, it will provide valuable insight into what factors determine this rate. Figure 5.2 showed that adsorbed atoms or molecules reside in potential wells on the surface, but it did not consider the variation in well depth with position, x, along the surface. Figure 5.4a shows that this depth is periodic, or corrugated, with a potential-energy barrier of height E_s between surface sites. The top of the barrier is considered to be the "transition state" between surface sites, in the language of reaction rate theory. Figure 5.4b illustrates a typical adsorbate situation leading to this corrugation. It is a hexagonally close-packed surface lattice on which the adsorption sites are the centers of the triangles of surface atoms, and the transition state is the "saddle point" between them. Other bonding situations can lead to the adsorption sites being other points, such as the centers of the surface atoms. The surface diffusion process requires partly breaking the bond(s) between the adsorbate and the surface site so that the adsorbate may move to the

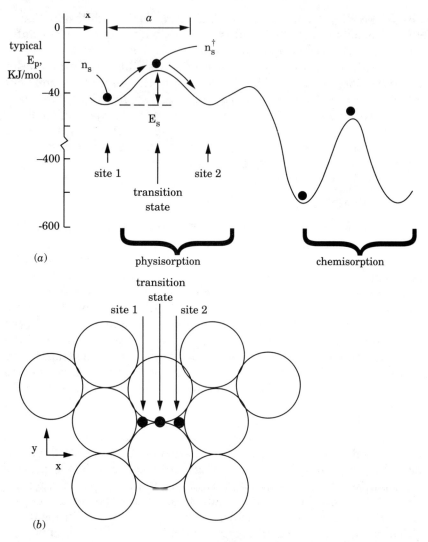

Figure 5.4 Surface diffusion: (a) potential energy vs. position x along the surface, and (b) typical adsorption sites on a surface lattice.

neighboring surface site and form new bonds there. This process may be viewed as an elementary form of chemical reaction, because any reaction involves the partial breaking of reactant bonds and partial formation of product bonds during motion of the atoms through a transition state. Thus, the principles to be discussed below apply to any chemical reactions, including those occurring in CVD.

There will be some flux, J_s (mc/cm·s), of adsorbate across the E_s barrier between sites 1 and 2 in the x direction of Fig. 5.4b. The flux here is in surface units, which are per linear cm of crosswise distance, y, instead of the previously encountered volume flux units, which are per cm^2 of cross-sectional area. If the distance between sites is a, then the *rate* of barrier crossing by transition-state molecules, per unit area of surface, is

$$R_s = J_s/a \text{ (mc/cm}^2\text{·s)} \tag{5.9}$$

Considering the adsorbate to be a two-dimensional gas at thermal equilibrium, the Maxwell-Boltzmann distribution applies to these translating molecules. Thus, we may use Eq. (2.7) for the flux of molecules impinging on the barrier and Eq. (2.3) for the mean speed. Here, for simplicity, we ignore the small changes in the numerical proportionality factors that arise in going from a three-dimensional to a two-dimensional situation. (It turns out that these factors cancel each other, anyway.) Inserting these equations into Eq. (5.9), we have

$$R_s = \frac{1}{4}n_s^\dagger \bar{c}/a = \frac{n_s^\dagger}{a}\sqrt{\frac{RT}{2\pi M}} = \frac{n_s^\dagger}{a}\sqrt{\frac{k_B T}{2\pi M}} \tag{5.10}$$

where $n_s^\dagger$ (mc/cm^2) denotes the surface concentration of adsorbate residing in the transition state. Now we must find the relation between $n_s^\dagger$ and n_s, the latter being the concentration of molecules in adsorption sites. At thermal equilibrium, statistical mechanics says that the concentration of molecules in a given state is proportional to the total number of ways of distributing the available thermal energy around a large system of molecules in that state. For each type of kinetic energy contributing to the thermal energy, the number of ways, Z, is equal to the sum over all of the quantized energy levels, ε_j, of the following products: the Boltzmann factor for each energy level times the number of ways of distributing energy at that level (the degeneracy of the level, g_j). Thus,

$$Z = \sum_j g_j e^{-\varepsilon_j/k_B T} \tag{5.11}$$

This type of summing was done for translational kinetic energy in the discussion of the Maxwell-Boltzmann distribution function (Sec. 2.2). The quantity Z is called the "partition function," and the product of the Z quantities for all the types of energy involved in a given state will be proportional to the concentration of molecules in that state; thus, for *any* two states,

$$\frac{n_s^\dagger}{n_s} = \frac{Z_r^\dagger Z_v^\dagger Z_t^\dagger}{Z_r Z_v Z_t} e^{-E_s/RT} \quad (5.12)$$

where r, v, and t denote the rotational, vibrational, and translational kinetic energies, each of which has various directional components. The final Boltzmann factor in Eq. (5.12) accounts for the *potential* energy difference between the adsorption-site state (n_s) and the transition state ($n_s^\dagger$), as shown in Fig. 5.4a. We neglect electronic excitation (Z_e), which occurs only at extremely high T.

We now need to evaluate the Z ratio in order to know $n_s^\dagger$. Because the molecule's rotational modes are eliminated or at least "frustrated" in their free motion by the adsorption, we may write $Z_r^\dagger = Z_r = 1$. For vibrational energy, the partition function derived from quantum mechanics for a harmonic oscillator is

$$Z_{vk} = \frac{1}{1 - e^{-h\nu_k/k_B T}} \quad (5.13)$$

where **h** = Planck's constant = 6.63×10^{-34} J·s

ν_k = frequency of the kth vibrational mode

[Often, Z_{vk} is written with the oscillator's "residual" or "zero-point" energy factor, $\exp(-\mathbf{h}\nu_k/2k_B T)$, included in the numerator, but here this is accounted for in the potential energy factor of Eq. (5.12).] Optical absorption-band wavelengths for the vibrations of adsorbate bonds are in the infrared. For a typical value of $\lambda = 30$ μm or $1/\lambda$ (wavenumbers) = 333 cm^{-1}, we have $\nu_k = c_0/\lambda = 10^{13}$ s^{-1}, which means that at typical film deposition T, $\mathbf{h}\nu_k/k_B T > 1$, and Z_v is near unity. In other words, the vibrational modes are mostly in their ground states, because the excited states are just beginning to be accessible at ordinary T.

In all reactions, the one component of vibration that is aligned with the reaction coordinate (x, here) is transformed into a translational component crossing the barrier. The loss of this transformed vibrational component from $Z_v^\dagger$ in the numerator of Eq. (5.12) makes no difference in our calculation, since it is near unity anyway. Conversely, the newly created $Z_t^\dagger$ will increase the value of the numerator, since translational-energy quantum levels are much more closely spaced and therefore much more accessible at ordinary T. The translational-energy partition function is

$$Z_{tx}^\dagger = a \frac{\sqrt{2\pi m k_B T}}{\mathbf{h}} \quad (5.14)$$

Note here that the value of $Z^\dagger_{tx}$ is proportional to the linear dimension a. If the x component is the only translational component that increases in moving to the transition state, then Eq. (5.12) becomes

$$n^\dagger_s = n_s Z^\dagger_{tx} e^{-E_s/RT} \tag{5.15}$$

Upon inserting these two equations into Eq. (5.10), we have, finally

$$R_s = n_s \left(\frac{k_B T}{h}\right) e^{-E_s/RT} = n_s v_{os} e^{-E_s/RT} = n_s k_s \tag{5.16}$$

which is the main result of absolute-reaction-rate theory. Note that in the second equality we have arrived at the Eq. (5.2) Arrhenius expression for the rate constant, k_s (s^{-1}), of a chemical reaction. For this particular reaction, the pre-exponential factor is $k_B T/h$, which is 2×10^{13} s^{-1} at 960 K, for example. The rate constant here represents the frequency with which an individual adsorbate molecule "hops" to an adjacent site.

The above important result gives us some insight into the meaning of v_{os}. In particular, note that v_{os} is *not* the frequency of any vibrational component, v_k, of the adsorbate, although such an implication is often made in the literature. That would be the case only in the event that $k_B T \gg hv_k$ in Eq. (5.13), in which case Z_v would become equal to $k_B T/hv_k$. Then, since there is one more vibrational component in the adsorbed state than in the transition state, this $k_B T/hv_k$ would end up in the denominator of Eq. (5.12), cancelling $k_B T/h$ and leaving the pre-exponential factor equal to v_k. However, at any reasonable T we do *not* have $k_B T \gg hv_k$, as pointed out after Eq. (5.13). Note also that Eq. (5.16) assumed that the only translational component acquired upon entering the transition state was the one in the direction of the reaction coordinate. This is a good assumption if sites 1 and 2 are identical, as in Fig. 5.4a. However, if the adsorbate is moving from a chemisorption site to a physisorption site in the course of diffusing, there may be an increase of as much as one additional translational component, which would raise v_{os} by as much as 10^3 by evaluation of Eq. (5.14). Thus, v_{os} may vary from 10^{13} to 10^{16} and may *not* simply be assumed to be $\sim 10^{13}$ s^{-1} as is so often done.

The rate of surface diffusion also increases exponentially with T and with decreasing E_s, as seen in Eq. (5.16). E_s, the activation energy for surface diffusion from Fig. 5.4a, is always considerably lower than the desorption activation energy, which is E_c or E_d, depending on whether the species in question is chemisorbed or physisorbed, respectively (see Fig. 5.2). E_s is lower because the bonds are being only partially broken in diffusion, whereas they are completely broken in re-evaporation. Therefore, at film deposition T approaching the onset of re-evaporation, where $\exp(-E_c/RT)$ is becoming significant, one expects a high rate of surface diffusion. This is one of the principal ways in which substrate T affects film structure, as we will see in Sec. 5.4. The ratio E_s/E_c is sometimes known as the "corrugation ratio," and it is lower for metals than for semiconductors because of the absence of bond directionality in metals.

We must now relate the molecular hopping rate, k_s, from Eq. (5.16) to the distance which an adsorbate molecule travels during film deposition. This is an adaptation of the classic random-walk problem. Since each hop is equally likely to be forward or backward in any given direction on the surface, there is no net motion in any one direction. However, as time passes, the molecule is more likely to be found further from its starting point. This is equivalent to saying that if one carries out a large number of trials of starting a molecule diffusing from a single point at $t = 0$, then with increasing t the final locations of these molecules become more widely dispersed from the starting point. In fact, for t corresponding to a large number of hops N_o ($N_o = k_s t \gg 1$), the molecules will be dispersed in a Gaussian (normal) distribution whose median is at the starting point. The width of a Gaussian distribution is characterized by its standard deviation, σ, which is the root-mean-square (rms) deviation from the median and is also the half-width of the bell-shaped curve at its inflection point. For the diffusing molecule, if r is the per-hop rms change in distance from the starting point, then it turns out that $\sigma = r\sqrt{N_o}$. For diffusion on a given two-dimensional surface *lattice*, r is related to the hop distance, a, by a geometric factor, β, which depends upon the angles between the possible hopping directions on the lattice, but for present purposes we will assume $\beta \approx 1$. If we now consider σ as a measure of the diffusion length, Λ, of the molecule in time t, we can write

$$\Lambda = r\sqrt{N_o} = \beta a \sqrt{N_o} \approx a\sqrt{N_o} = a\sqrt{k_s t} \qquad (5.17)$$

To get a feeling for the magnitudes involved here, assume that Eq. (5.16) holds and that the substrate T is 960 K, so that $v_{os} = k_B T/h = 2 \times 10^{13}$ s^{-1}. If the molecule is physisorbed and thus has a relatively low E_s of, say, 20 kJ/mol, then $k_s = 1.6 \times 10^{12}$ s^{-1}. For a typical a value of 0.3 nm and a t of 1 s, $\Lambda = 380$ μm, which is very large compared to the typical dimensions of thin-film topography. Conversely, for a *chemi*sorbed molecule having an E_s of, say, 200 kJ/mol, $k_s = 2.6 \times 10^2$ s^{-1}, and $\Lambda = 4.8$ nm, which is only tens of atomic diameters. Clearly, Λ varies enormously with bonding conditions at the surface.

We have arbitrarily chosen t = 1 s in the above example, but t will vary considerably with deposition conditions. Two regimes need to be considered separately: one in which t is the time between adsorption and burial by the next depositing monolayer, and another in which the adsorbate is more likely to desorb than to be buried. For the burial case,

$$t = \frac{n_o}{J_r} \quad (5.18)$$

where n_o = adsorption sites/cm^2
J_r = deposition flux, mc/cm$^2 \cdot$s

Inserting this and the expression for k_s [Eq. (5.16)] into Eq. (5.17), we have

$$\Lambda = a \sqrt{\frac{\nu_{os} n_o}{J_r}} e^{-E_s/2RT} \quad (5.19)$$

That is, Λ increases exponentially with T. This behavior appears as a straight line with a negative slope of $-E_s/2R$ on the Arrhenius plot of ln Λ versus 1/T in Fig. 5.5. Conversely, when T is high enough that film re-evaporation is becoming significant, diffusing species are likely to desorb before they become buried. Then, t is instead the adsorption

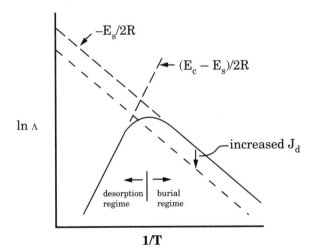

Figure 5.5 Behavior of surface diffusion length, Λ, with substrate T.

lifetime. To simplify its estimation, we neglect the precursor state, whose concentration is likely to be negligible at this high a T anyway, and we consider only desorption from the chemisorbed state. In this case,

$$t = \frac{1}{k_c} = \frac{1}{\nu_{oc}} e^{E_c/RT} \qquad (5.20)$$

where subscript c denotes the chemisorbed state as in Fig. 5.2. Inserting this and the expression for k_s [Eq. 5.16)] into Eq. (5.17), we have

$$\Lambda = a \sqrt{\frac{\nu_{os}}{\nu_{oc}}} e^{(E_c - E_s)/2RT} \qquad (5.21)$$

Because $(E_c - E_s)$ is always positive, as pointed out earlier, the slope of the Arrhenius plot is *positive* in this regime, as shown in Fig. 5.5. Note that this leads to a *maximum* in Λ at a T just below the onset of significant film re-evaporation. We will see in Sec. 5.4 that long Λ leads to films which are smoother, more homogeneous, and lower in crystallographic defects. Therefore, it is found empirically that the highest film quality is often obtained just below the re-evaporation point.

The above derivation of surface diffusion quantities has been based on examining the motion of individual adsorbed molecules. It is useful now to see how these quantities relate to familiar *macroscopic* quantities encountered in transport theory and thermodynamics. First of all, surface diffusion may be expressed in the form of the transport equation [Eq. (2.26)]; that is,

$$J_s = -D \frac{dn_s}{dx} \qquad (5.22)$$

With J_s and n_s in surface units of mc/cm·s and mc/cm^2, respectively, D here has the same units of cm^2/s as for the volume diffusion case. For the volume case, we found in Sec. 2.8.1 that

$$D = \tfrac{1}{4} \bar{c} \, l \qquad (5.23)$$

In adapting to the surface case, we again ignore the accompanying small changes in numerical factors as we did for Eq. (5.10). The mean free path, l, here becomes the hop distance, a; and the mean speed, c, will be $k_s a$. Thus,

$$D = \tfrac{1}{4} k_s a^2 \qquad (5.24)$$

and

$$\Lambda = a\sqrt{k_s t} = 2\sqrt{Dt} \tag{5.25}$$

Now, using Eq. (5.16) for k_s, we can express D in Arrhenius form:

$$D = \tfrac{1}{4}v_{os}a^2 e^{-E_s/RT} = D_o e^{-E_s/RT} \tag{5.26}$$

D is usually the quantity which appears in adsorbate diffusion experiments, while Λ is the quantity of interest for thin-film deposition.

The partition-function ratio of Eq. (5.12), which determines the preexponential factor for surface diffusion, can be related to the macroscopic thermodynamic concept of entropy which was discussed in Sec. 4.1. Indeed, this important relationship exists for any chemical reaction, and we will now develop it for the case of surface diffusion. Assuming thermal equilibrium as we did above, Eq. (4.9) for dG holds, and from it we may write for our pure adsorbate that

$$\left(\frac{\partial G_m}{\partial p}\right)_T = \left(\frac{\partial \mu}{\partial p}\right)_T = V_m = \frac{RT}{p} \tag{5.27}$$

where the last equality also assumes the ideal-gas law, Eq. (2.10). A *mobile* adsorbate may in fact be thought of as a two-dimensional (2D) gas. We have essentially postulated a 2D gas already in writing Eqs. (5.10) and (5.24) above, and we will assume here that this gas obeys a 2D ideal-gas law. The dependence of the chemical potential, μ, on p at constant T may be found by integration of Eq. (5.27) starting from some standard reference state ($\mu°$, $p°$), where $p° = 10^5$ Pa $\approx$ 1 atm by convention; thus,

$$\mu = \mu° + RT \ln \frac{p}{p°} \tag{5.28}$$

Now, for a system in equilibrium, the μ values of all components are equal as long as a shift in equilibrium involves no change in the number of moles of material present. (Such molarity changes can occur by reaction, as we will see in Sec. 7.3.1.). So, for our simple diffusion-activation reaction situation, $\mu = \mu^\dagger$, and using Eq. (5.28) we have

$$\ln\left(\frac{p^\dagger}{p}\right) = \ln\left(\frac{n_s^\dagger}{n_s}\right) = \frac{-\left(\mu^{°\dagger} - \mu°\right)}{RT} = -\frac{\Delta_r G°}{RT} \tag{5.29}$$

In the first equality, we have used the ideal-gas law to switch from p to n_s units. The last equality defines the Gibbs free energy of reaction per mole at standard pressure, $\Delta_r G°$. This important equation relates the ratio of the concentrations of species in *any* two bonding configurations to their reference-pressure chemical potentials (free energies per mole) in those bonding configurations. The second and third equalities apply also to adsorbates of limited mobility, which are not behaving as 2D gases, provided that $n_s \ll 1$ ML as assumed previously, so that $n_s \propto p$ (more on this in Sec. 7.3.3).

Also, at equilibrium, forward rate equals reverse rate, so for this first-order reaction [Eq. (5.1)] we have

$$R_s = R_{-s} \quad \text{or} \quad n_s k_s = n_s^\dagger k_{-s} \tag{5.30}$$

where $-s$ denotes the reverse reaction from the transition state back to adsorption site 1 in Fig. 5.4a. Rearranging this equation gives the definition of the equilibrium constant:

$$K = \frac{k_s}{k_{-s}} = \frac{n_s^\dagger}{n_s} = e^{-\Delta_r G°/RT} \tag{5.31}$$

where we have used Eq. (5.29) to obtain the last equality. Again, this expression is valid for any first-order reaction, not just for surface diffusion. Except for the $n_s^\dagger/n_s$ equality, the expression is valid for reactions of any order.

Now, using the definition of the Gibbs free energy in Eq. (4.5), we can write that

$$\Delta_r G° = U° + p° \Delta_i V° - T \Delta_r S° \tag{5.32}$$

The first term on the right is essentially the height of the potential-energy barrier, E_S. The second term is negligible for condensed-phase reactions. Upon inserting these results into Eq. (5.31), we have

$$\frac{n_s^\dagger}{n_s} = e^{-\Delta_r S°/R} \cdot e^{-E_s/RT} \tag{5.33}$$

Comparing this with Eq. (5.12) shows that

$$\Delta_r S° = R \ln \left(\frac{Z_r^\dagger Z_v^\dagger Z_t^\dagger}{Z_r Z_v Z_t} \right) \tag{5.34}$$

This result is equivalent to the expression for S presented in Eq. (4.3) and gives further insight into the meaning of entropy. We see from this that the pre-exponential term in the Arrhenius equation [Eq. (5.2)] is an entropy term; that is, it increases with the degree of delocalization or the degree of access to different quantum states in the numerator. It is useful to think of the Arrhenius pre-exponential term from this perspective. Note, however, that there may be other factors in the pre-exponential term as well. For example, if E_s has some T dependence, say $E_s = E_s^\circ - BT$, then a multiplier of $e^{B/R}$ will enter the pre-exponential [2].

Having now developed some ideas about the mechanisms and behavior of surface diffusion, we can proceed to discuss this key phenomenon in the context of thin-film nucleation and structure development.

5.3 Nucleation

5.3.1 Surface energy

To understand nucleation, we need to introduce the concept of surface energy. The familiar experiment of drawing a liquid membrane out of soapy water on a wire ring is illustrated in Fig. 5.6. The force required to support the membrane per unit width of membrane surface is known as the surface tension, γ, expressed as N/m in SI units, or more commonly as dynes/cm in cgs units. For a wire of circumference b, the width of surface is 2b, since the membrane has both an inner and an outer surface. Thus, the total force required to support the membrane is $F = 2b\gamma$. As the membrane is extended upward in the x direction, work $F\Delta x$ (N·m or J) is done to create the new surface, and the surface area created is $A = 2b\Delta x$, assuming for simplicity a constant membrane circumference. The work is stored as surface *energy* (as in stretching a spring), so the surface energy per unit area of surface is

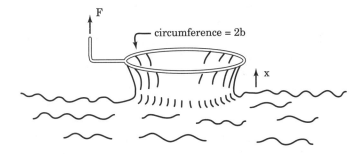

Figure 5.6 Surface tension of a liquid membrane.

$$F\Delta x/A \; (\text{N·m/m}^2) = (2b\gamma)\Delta x/2b\Delta x = \gamma \; (\text{N/m or J/m}^2). \tag{5.35}$$

Thus, surface tension (N/m) and surface energy per unit area (J/m^2) are identical, at least for liquids. For solids at T > 0 K, surface Gibbs free energy is reduced by an entropy factor [Eq. (4.5)] which depends on the degree of surface disorder (Williams, 1994). For solids, there is also a quantity called surface stress, which differs from surface energy by a surface elastic-strain term (more in Sec. 5.6.1). Liquids cannot support such strain, because the atoms just rearrange to relax it.

Surface energy exists because the molecules of a condensed phase are attracted to each other, which is what causes the condensation. The creation of surface involves the removal of molecular contact from above that surface (bond-breaking), and thus involves energy input. Consequently, to the extent that motion within the condensed phase can occur, such motion will proceed so as to minimize the total surface energy, γA. In the liquid-membrane case, where γ is fixed, this means minimizing A. Thus, when the wire is lifted far enough, the membrane snaps taut into the plane of the ring; and when a bubble is blown, it becomes spherical. In the case of solids, surface energy proceeds to minimize itself by surface diffusion, and this process is fundamental to the development of structure in thin films. In thin-film growth, both A and γ are varying. Area, A, depends on surface topography, and γ depends on many properties of the exposed surface, including chemical composition, crystallographic orientation, atomic reconstruction, and atomic-scale roughness. In materials that have no orientation dependence of a particular property, that property is said to be isotropic. However, in most crystalline solids, γ is not isotropic, but is *an*isotropic.

Examining the anisotropy of γ calls for a brief digression to review crystallographic nomenclature. Some readers may be able to skip this paragraph, and others may need refer to a more thorough description in a solid-state physics or crystallography text. Figure 5.7 shows a crystal unit cell. There are fourteen variations of this cell (the fourteen "Bravais lattices") having varying ratios of the three sides, *a*, *b*, and *c*, varying angles between the sides, and varying positions of atoms supplementing the eight corner atoms. In the cubic lattice, the sides are of equal length and the angles are all 90°. The face-centered cubic (fcc) lattice shown has an additional atom in the center of each of the six faces; this structure is also known as cubic close-packed, and it is very common. The orientation of an atomic plane in the unit cell is described by its Miller indices, (jkl). These are obtained by taking the reciprocals of the plane's intersection points with the x, y, and z axes, measured in integral numbers of unit cells. Thus, the plane containing the atoms labeled 1, 2, 3, and 4, which intersects the axes at 1, ∞, and

5.3.1 Surface energy 141

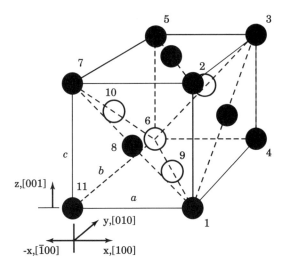

Figure 5.7 Geometry of a face-centered cubic crystal.

∞, has indices (100). Other planes with the same (100) symmetry in this crystal are (010) and (001), and the set of them is denoted using different brackets: that is, {001}. Similarly, the plane containing atoms 1, 2, 5, and 6, which intersects the axes at 1, 1, and ∞, has indices (110), and the set of planes with this symmetry is denoted as {011}. Finally, the set of planes like the one containing atoms 1, 6, and 7 is the {111} set. These are the three primary sets of planes in a cubic lattice.

We will see below that a crystal tends to expose surfaces having low Miller indices, and these exposed surfaces are known as "facets." The use of reciprocals in the indices comes from x-ray diffraction work, where the spacing of diffraction spots is reciprocally related to the spacing of atomic planes (more in Sec. 6.4.2). In the hexagonally close-packed or "hcp" lattice, a and b of Fig. 5.7 are at 120° in the so-called "basal" a-b plane; also, $c \perp a$-b, and $c \neq a = b$. There is a third direction in the basal plane, oriented at 120° to a and b and symmetrically equivalent to a and b, so *four* indices are commonly used to describe a plane in an hcp crystal, even though the fourth index is redundant. Thus, the basal plane is (0001). The basal plane in an hcp crystal has the same symmetry as the {111} planes in a cubic crystal, and this can be used to advantage in the epitaxial growth of one type of crystal upon the other. *Directions* with respect to the axes of a crystal are denoted by the x, y, and z components of their vectors, again measured in unit cells. Thus, the x direction in Fig. 5.7 is the [100] direction, and the set of directions having this symmetry is ⟨001⟩, where again the type of bracket identifies the meaning. When the axis angles are all 90°, a direction is always perpendicular to a plane having the same indices. Negative-pointing directions or negative-facing exposed planes of a crystal are denoted by a bar over that index, as shown for [$\bar{1}00$] in Fig. 5.7 and for various faces later in Fig. 5.11.

Returning now to the anisotropy of γ, one source is the anisotropy in bonding directions within the lattice. For example, note in Fig. 5.7 that the atoms in the (111) plane of the fcc lattice are hexagonally close-packed. That is, they would all have six nearest neighbors in the plane if the lattice were extended to the neighboring unit cells. This arrangement maximizes bonding possibilities within the plane and consequently minimizes them perpendicular to the plane. Hence, an exposed {111} face of an fcc crystal has fewer unsatisfied bonds ("dangling" bonds) sticking out of the surface than the other faces have, and it therefore has a lower γ than the other faces. In crystals having ionic bonds, such as CaF_2, or polar bonds, such as GaAs, γ also tends to be lower for faces containing equal numbers of cations (Ca, Ga) and anions (F, As), since this results in charge neutrality at the surface. Consequently, these are the *nonpolar* faces. In materials having a "layered" structure, such as graphite and MoS_2, there are *no* chemical bonds between the atomic layers of the basal plane. The sliding of these planes past each other accounts for the performance of layered materials as dry lubricants. The low-energy facets for various crystal structures are listed in Table 5.1. Aside from hcp, all of those listed are cubic and are distinguished from each other only by the positions of the atoms within the unit cell.

For most faces of crystals in general, γ is actually lower than one would predict from the dangling-bond density of the separated bulk lattice. This is because the dangling bonds and their atoms at the surface become distorted from their bulk lattice angles and positions, respectively, to cross-bond with each other and thereby reduce surface energy. The result is a "reconstructed" surface having patterns of atom positions and surface bonds that are different from those in the bulk and having γ reduced by as much as half. Further γ reduction may re-

TABLE 5.1 Facets of Lowest Surface Energy for Various Crystal Structures

Structure	Examples	Low-γ facets
Body-centered cubic (bcc)	Cr, Fe	{110}
Face-centered cubic (fcc)	Au, Al	{111}
Hexagonal close-packed (hcp)	Zn, Mg	{0001}
Diamond	Si, Ge	{111}
Zinc blende	GaAs, ZnSe	{110}
Fluorite	MgF_2, CaF_2	{111}
Rock salt	NaCl, PbTe	{100}

sult from the adsorption of a "passivating" monolayer of an element with which the dangling bonds react to become terminated bonds. This is more effective than reconstruction, because less bond strain is involved, and it can thus take the place of and prevent the reconstruction. For example, Si can be passivated by H [3] or by As [4], and GaAs can be passivated by S [5]. By the way, reconstruction and passivation can each generate energy barriers to adsorbing vapor becoming incorporated into the bulk, and this makes the precursor-adsorption model of Sec. 5.1 applicable even to the condensation of pure materials onto themselves. The experimental and theoretical study of surface reconstruction is a major branch of surface science, and it will be discussed further in Sec. 6.5.3 because it has a profound effect on epitaxy.

For deposition onto a foreign substrate, nucleation behavior is strongly influenced by the γ of the substrate. Here, we need to consider the γ both of the substrate free surface, γ_s, and of the substrate-film interface, γ_i, relative to that of the film free surface, γ_f. All three γ values will in general depend on crystallographic orientation, passivation, and sometimes other factors. Assume that there is enough surface diffusion so that the depositing material can rearrange itself to minimize γ; that is, assume that the nucleation is not kinetically limited and can approach equilibrium. For this, we must have $\Lambda \gg a$ [see Eq. (5.17)]. In the opposite case, $\Lambda < a$, every atom sticks where it lands, and the growth behavior is "quenched." Quenched growth develops its own characteristic film structure which we will examine in Sec. 5.4. With our assumption of $\Lambda \gg a$, there are two nucleation situations on the bare substrate, as shown in Figs. 5.8a and b. In a, the film spreads across or "wets" the substrate because

$$\gamma_f + \gamma_i < \gamma_s \tag{5.36}$$

That is, total surface energy is lower for the wetted substrate than for the bare one. This leads to smooth growth, atomic layer by layer, which is sometimes referred to as the Frank-van der Merwe growth mode. For this mode to occur, there must be strong enough bonding between film and substrate to reduce γ_i to where Eq. (5.36) holds. If there were no such bonding at all, we would have $\gamma_i = \gamma_f + \gamma_s$, so spreading the film across the substrate would always increase total surface energy by $2\gamma_f$, as in the case of the free-standing liquid membrane of Fig. 5.6. Thus, with insufficient substrate bonding, Eq. (5.36) ceases to hold, and the film does not wet the substrate but instead forms three-dimensional (3D) islands, shown in Fig. 5.8b and sometimes referred to as the Volmer-Weber growth mode. There is a third growth mode, Stranski-Krastanov, shown in Fig. 5.8c, in which the growth changes from layer to island after a monolayer or two due to a change in the

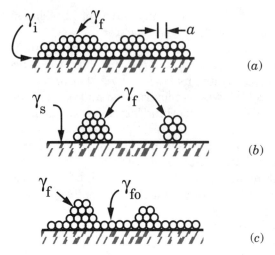

Figure 5.8 Film growth modes: (a) Frank-Van der Merwe (layer), (b) Volmer-Weber (island), and (c) Stranski-Krastanov.

energy situation with successive monolayers. This mode is associated with epitaxy and will be discussed in Sec. 6.7. For liquids contacting solids, the degree of wetting is most easily observed by the rise or depression of a liquid column in a narrow tube (a capillary). Thus, film nucleation analysis in terms of degree of wetting is known as the "capillarity" model. Nucleation and the three growth modes of Fig. 5.8 are further discussed by Venables (1984).

Three-dimensional nucleation is usually undesirable, since it leads to rough, nonuniform films, an extreme example being the diamond nuclei of Fig. 5.3. Often, it can be changed to 2D growth by manipulating one of the γ terms in Eq. (5.36) to satisfy the inequality. For example, γ_i decreases with increased film-substrate bonding. This bonding depends on the chemical reactivity of the film and substrate and on the similarity of the bonding within the two materials. There are three types of chemical bonding: (1) covalent, in which valence electrons are shared between atoms, as in Si and GaAs; (2) ionic, in which valence electrons are completely transferred from the cation to the anion, as in CaF_2 and NaCl; and (3) metallic, in which the valence electrons wander through the crystal, and the metal ions exist as an array of positive charges in a sea of negative charge (the "jellium" model). In general, interfacial bonding is stronger between materials having the same type of bonding. Bonding will still occur across types if there is enough chemical reactivity. For example, the chemically-active metal, Cr, will bond to glass by breaking an O-Si bond and forming a Cr-O or Cr-Si bond. Conversely, the noble metal, Au, cannot do this and does not bond well to glass. Yet Au does form a strong metallic bond to clean (unoxidized) Cr. Thus, by the use of an intermediate "glue" layer which bonds well to both the film and the substrate, such as Cr between Au and glass, γ_i can be reduced and wetting accomplished. Another good bonding material is Ti. Only enough bonding material is needed to ensure a continuous layer—typically 10 nm. Alternatively, energy-enhanced techniques, particularly ion bombardment and sputtering (Chaps. 8 and 9), can provide the activation energy for bonding between film and substrate and thereby reduce γ_i. Ion bombardment can also break up the nuclei of 3D islands, thus *counteracting* the equilibrium tendency (more in Sec. 6.7.2). Energy-enhanced techniques are very powerful in modifying film structure and are being employed more and more routinely in a variety of applications.

Equation (5.36) can also be satisfied in principle by reducing γ_f or increasing γ_s. The former is the familiar "surfactant" effect known in liquids such as soap. However, I do not know of a case where this effect has been used to assist wetting in thin-film deposition. There has been a so-called surfactant effect reported in the suppression of 3D nuclei in epitaxy (Sec. 6.7.2), but it is not the result of γ_f reduction. In another approach to wetting, γ_s can be increased by activating the substrate surface through energetic irradiation. For example, polymers are treated with plasma exposure to increase bonding to metal films. Presumably, the irradiation creates dangling bonds at the surface.

5.3.2 Three-dimensional (3D) nucleation

Despite the above techniques to encourage smooth film growth, 3D nucleation is often a problem, so it is important to understand its behavior to deal with it effectively. There are two ways in which 3D nuclei can form. The more common one is when bonding initiates at active surface sites such as atomic steps, crystal defects, or impurities. At these nucleation sites, the activation energy for bonding is less than elsewhere, as discussed in Sec. 5.1; or equivalently, γ_i is lower in Eq. (5.36) because of the interfacial bonding which develops. The vapor may arrive at these sites either by surface diffusion or by direct impingement. The latter is the dominant mode when the vapor has a very low trapping probability, δ, elsewhere on the surface or when the desorption rate of the adsorbed precursor, R_d [Eq. (5.5)], is so high that most of it never reaches a nucleation site. However, it is more common for nuclei to accumulate by surface diffusion.

5.3.2.1 Classical nucleation. Even if there are no active nucleation sites on the surface, 3D nuclei can still form at random surface locations by the spontaneous accumulation of mobile adsorbed atoms plus arriving vapor into "critical" nuclei which are big enough to be stable. This is the classic nucleation problem (Lewis, 1978). We will present below an elementary treatment of this difficult problem, using the "capillarity" approach, which is based on surface-energy minimization. This will serve to give a qualitative idea of the important factors which determine nucleation behavior. Quantitative evaluation of nucleation behavior in a particular deposition situation is more reliably done by direct microscopic observation of the nuclei using transmission electron microscopy (TEM) or scanning tunneling microscopy (STM).

Figure 5.9 illustrates the geometry of nucleus formation. In this example, wetting to the substrate is poor, so that the contact area, A_i, and its effect on total surface energy can be neglected. If anisotropy of the surface energy, γ_f, can also be neglected, the nucleus will be spherical as shown. Vapor arrives at the surface of the nucleus directly from the gas phase as well as by surface diffusion of adsorbed "monomers" (single atoms or molecules). The nucleus can also lose material by evaporation back into the gas phase or into the 2D gas of adsorbed monomers. Recall from the Sec. 4.1 discussion of evaporation that at vapor-liquid or vapor-solid equilibrium, the chemical potentials of the two phases are equal [Eq. (4.8): $\mu_v = \mu_c$] and the vapor is at its saturation vapor pressure, p_v. If the p of the vapor is increased above p_v, μ_v increases in accordance with Eq. (5.28) (for ideal gas), which we applied there to the 2D gas situation. Thus,

$$\mu_v - \mu_c = RT \ln \frac{p}{p_v} = RT \ln \frac{J_c}{J_v} \tag{5.37}$$

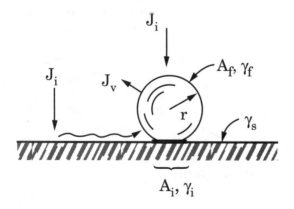

Figure 5.9 Geometry of condensate nucleus formation on nonwetting substrate.

5.3.2 Three-dimensional (3D) nucleation

Here, equality between the ratio of p values and the ratio of condensing and evaporating molecular fluxes, J_c and J_v, has been obtained by using Eqs. (2.18), (4.17), and (4.18) and by assuming that we are close enough to equilibrium so that $\alpha_c = \alpha_v$. This ratio is known as the "supersaturation" ratio. If the ratio were <1, the system would be called "undersaturated." For single-component systems, as we have here, $\mu = G_m$, the Gibbs free energy per mole. The difference in μ or G_m between the two phases is the driving force for condensation. Processes always proceed in the direction of decreasing G_m, which is toward condensation in the supersaturated case.

The quantity γ was not involved in the Sec. 4.1 treatment of vapor-condensate equilibrium, because there the condensate area, A, and therefore the total surface energy, γA, were constant. In that case, an infinitesimal amount of supersaturation was enough to drive the condensation. Conversely, in the nucleation situation of Fig. 5.9, nucleus surface area, A_f, is *increasing* with condensation, so there is a $\gamma_f A_f$ contribution to the total G of the nucleus. Consequently, the G change *per nucleus* (not per mole) for forming a nucleus of radius r and volume V from the vapor becomes

$$\Delta G = -(\mu_v - \mu_c)\frac{V}{V_{mc}} + \gamma_f A_f = -\left(RT \ln \frac{p}{p_v}\right) \cdot \frac{(4/3)\pi r^3}{V_{mc}} + \gamma_f \cdot 4\pi r^2 \tag{5.38}$$

where V_{mc} is the molar volume of the condensate. The γ_f term is positive and causes ΔG to *rise* with increasing r at small r. The behavior of this function is shown in Fig. 5.10 for water at two p/p_v ratios. For water, the surface tension γ_f = 73 dynes/cm = 0.073 J/m^2 at 20° C. nuclei having r less than that at which ΔG goes through a maximum will spontaneously decompose, since that is the direction of decreasing ΔG. Larger nuclei are stable and will spontaneously grow. The critical nucleus radius, r^*, at which the slope of $d(\Delta G)/dr$ changes sign is found by setting $d(\Delta G)/dr = 0$. Operating on Eq. (5.38) in this way, we obtain

$$r^* = \frac{2\gamma_f}{\left(\frac{RT}{V_{mc}}\right)\ln\left(\frac{p}{p_v}\right)} \tag{5.39}$$

and

$$\Delta G^* = \frac{(16/3)\pi\gamma_f^3}{\left[\left(\frac{RT}{V_{mc}}\right)\ln\left(\frac{p}{p_v}\right)\right]^2} \tag{5.40}$$

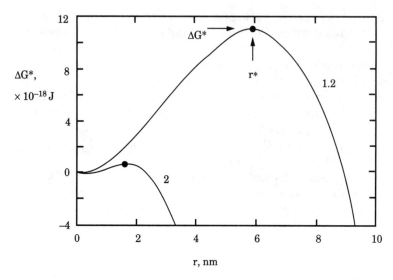

Figure 5.10 Classical nucleation behavior of water for two values of the super-saturation ratio (• denotes critical-cluster condition).

Here, ΔG^* is the energy barrier to nucleation and is analogous to the energy barrier for a chemical reaction.

The accumulation of individual adatom monomers into a nucleus may indeed be considered as a chemical reaction. Since nuclei with $r > r^*$ are stable and proceed to grow spontaneously, the reaction of adatoms with such nuclei is an irreversible one and leads rapidly to coalescence of the nuclei with each other to form a continuous film. Conversely, for $r < r^*$, the reaction is reversible because such nuclei spontaneously disintegrate. We may therefore view the population of nuclei with $r < r^*$ as being in a state of equilibrium with p, albeit a metastable one because of the ever-expanding "sink" provided by nuclei having $r > r^*$. Equation (5.31) relating reactant concentrations to ΔG now applies, with one modification. There, the reaction to the excited state was first-order, whereas here the reaction to a critical nucleus containing j atoms is j^{th} order. Thus, the equilibrium condition becomes

$$\mu_j = j\mu_1 \qquad (5.41)$$

where μ_1 denotes the chemical potential of the adsorbed monomers. This leads to

$$\frac{n_j}{n_1^j} = \frac{n^*}{n_1^j} = e^{-\Delta G_m^*/RT} = e^{-\Delta G^*/k_B T} \qquad (5.42)$$

5.3.2 Three-dimensional (3D) nucleation

where n_j is the surface concentration (cm^{-2}) of nuclei containing j monomers, and n_1 is the surface concentration of adsorbed monomers. Note that here G_m denotes a mole of *nuclei*, not a mole of atoms or molecules, and thus we obtain the last equality, where ΔG^* is per nucleus and is given by Eq. (5.40). Here, ΔG^* is equivalent to the standard-state $\Delta G°$ from Eq. (5.31), because the concentration chosen for the standard state is arbitrary and will not affect the difference of $\mu°$ terms or the ratio of n values significantly.

The concentration of critical nuclei, n*, will be determined by both n_1 and ΔG^* in Eq. (5.42). The behavior of n_1 depends on whether the adatom desorption rate is appreciable. If it is, then n_1 will reach a nearly steady-state fractional surface coverage given by Eq. (5.4). It will not be precisely steady-state, because there will always be some finite rate of critical-nucleus generation, and these nuclei will ultimately take over the surface. Nevertheless, before this happens, n_1 is largely determined by the vapor arrival rate, J_i, which is proportional to p by the Knudsen equation [Eq. (2.18)]. Since increasing p also raises n* by reducing ΔG^* [Eq. (5.40)], n* becomes a very strong function of p, such that nucleation appears to initiate suddenly as p is raised. The dependence of n* on T is more complicated. With increasing T at fixed p, p_v increases exponentially by Eq. (4.14), but this effect is more than cancelled by the RT factor in Eq. (5.40), so that ΔG^* decreases. But meanwhile, n_1 is decreasing exponentially by Eq. (5.4), and thus it is experimentally observed that n* decreases with increasing T. For example, it is the difficulty of nucleating Zn and Cd on glass before monomer desorption which accounts for the "splotchy" appearance of thin deposits of these metals despite high supersaturation during deposition, as discussed in Sec. 5.1. The situation can be remedied by cooling the substrate to inhibit desorption.

If precursor desorption is negligible, then n_1 does not approach steady state, but increases steadily with time at a constant arrival rate (constant supersaturation) of vapor, J_i. In this case, n* rises much more rapidly than when precursor desorption is appreciable. In either case, the rate at which n* rises determines the "coarseness" of the nucleation. When the process objective is to produce a smooth, uniform film, a large n* and small r* are preferred; that is, less coarse nucleation. One way to achieve this is by using a very high vapor arrival rate (high supersaturation), at least until the nucleation phase is over and the film is continuous. At sufficient supersaturation, the number of atoms in the critical nucleus approaches unity, and ΔG^* becomes negligible. For example, this is believed to be the situation for the classic nucleation study case of Au on NaCl, which is characterized by poor adherence. Coarsening will still occur even with one-atom critical nuclei, to the extent that the atoms and nuclei are mobile on the surface. Mobility allows the nuclei to migrate and coalesce with each other, becoming larger and fewer with time. It also allows transfer of atoms from smaller nuclei to larger ones, a process driven by the higher γ_f of the former which results from the curvature effect to be discussed in Sec. 5.4.2. These processes are analogous to the "Ostwald ripening" process of bulk metallurgy, in which similar crystallite (grain) growth occurs with annealing. Coarsening will continue to occur even if deposition is stopped, which is why the arrival rate needs to remain high until the film is continuous, if smoothness is the objective. Sometimes, it is instead desired to obtain polycrystalline films with large grain size and a correspondingly low concentration of grain boundaries. Grain boundaries can degrade film performance by acting as electrical-carrier traps in semiconductors or as channels for diffusion through chemical-barrier layers. In such cases, coarser nucleation may be preferable.

5.3.2.2 Kinetics vs. thermodynamics. Another approach to achieving smooth growth is to lower the substrate T to inhibit surface diffusion and thus "freeze out" the nucleation and coalescence processes. If the arriving species do not have enough thermal energy to either desorb or diffuse, they remain where they land, which leads to the quenched growth mentioned earlier. In this case, the nucleation process is *kinetically inhibited* by the surface-diffusion activation-energy barrier, E_s, in Eq. (5.16). This is also the case for ion-bombardment dissipation of 3D nuclei, as mentioned at the end of Sec. 5.3.1: the nuclei do not have time to reassemble themselves by surface diffusion before they are buried by depositing material.

The question of whether a process is approaching equilibrium or is instead limited by kinetics is an important one, and it arises often in thin-film deposition. Process behavior and film properties are profoundly affected by the degree to which one or the other situation dominates. The answer is not always apparent in a given process, and this often leads to confusion and to misinterpretation of observed phenomena. Therefore, to elaborate briefly, the generalized mathematical representation of this dichotomy is embodied in Eq. (5.30) describing the rate balance of a reversible reaction and Eq. (5.31) defining its equilibrium constant. Approach to equilibrium requires the forward and reverse rates to be fast enough so that they become balanced *within the applicable time scale*, which may be the time for deposition of one monolayer, for example. Then, the concentrations of reactant and product species are related by the difference in their free energies, $\Delta_r G°$. If, on the other hand, the forward rate is so slow that the product concentration does not have time to build up to its equilibrium level within this time scale, then the product concentration is determined not by $\Delta_r G°$ but, instead, by the forward rate. This rate is governed by Eqs. (5.1) and (5.2), in which E_k/T plays the dominant role. So it is, that reactions can be frozen out and equilibration avoided if so desired, by lowering the T.

The difficulty of answering the question of kinetics versus thermodynamics arises from the fact that the applicable rate constants, k_k, are often unknown or not known accurately enough. The measurement of k_k is much more difficult than just measuring equilibrium concentration, both because it is a dynamic measurement and because it must be made in the absence of the reverse reaction. The problem is perhaps most troublesome in CVD, where many reactions are involved, as we will see in Sec. 7.3.

5.3.2.3 Other complications. The assumption of sufficient surface diffusion for approach to equilibrium is inherent in the above treatment of nucleation. Other simplifying assumptions have also been made, and it is important to be aware of these in order to recognize the limitations of the model. As mentioned earlier, active sites for nucleation have deliberately not been included in order to examine spontaneous nucleation on a homogeneous surface. Nevertheless, in almost all practical situations, nucleation will occur *predominantly* at active sites. The extent to which this mechanism increases nucleation density will depend on the activity (γ_s) and concentration of such sites on the particular substrate involved, so there is no way to construct a general model. Remaining simplifying assumptions relate to surface energy, γ. We have neglected any reduction of total surface energy arising from reduced γ_i over the contact area, A_i, in Fig. 5.9. Such reduction will reduce ΔG^* rapidly through the γ^3 factor in Eq. (5.40). The concept of γ also implies that the surface is a continuum, whereas in fact when the nuclei become smaller than perhaps 10^3 atoms (radius r = 6 atoms), the surface is more accurately treated as an array of discrete atoms. Also, when r is small, γ_f becomes larger than it is on a plane surface, because the atoms on a convex surface are more exposed and less well connected to the bulk. This convexity effect is very important to structural development, as we will see in Sec. 5.4.2. Finally, we have ignored the crystallographic anisotropy of γ_f, which we will now discuss.

The shape of a nucleus will try to adjust itself to minimize total surface energy; that is,

$$\boxed{\sum_k \gamma_k A_k = \text{minimum}} \quad (5.43)$$

Subscript k denotes terms corresponding to the nucleus free surface, the interface to the substrate, and the substrate free surface. In the case of liquid or amorphous nuclei, which have no γ anisotropy, there is only one term, $\gamma_f A_f$, for the nucleus free surface. In the more common case of crystalline nuclei, these surface terms include all of the various exposed atomic planes or facets, and the nucleus shape-adjustment process is called facetting. Consider first the case of poor wetting ($A_i \to 0$ in Fig. 5.9), so that the nucleus free-surface terms are the only significant terms in Eq. (5.43). The "Wulff theorem" states that when total surface energy of a crystal of constant volume is minimized per Eq. (5.43), then it will also hold that

$$\gamma_k / r_k = \text{constant} \tag{5.44}$$

where r_k is the perpendicular distance from the center of the crystal to the kth facet. This situation is easier to visualize in two dimensions, so it is illustrated in Fig. 5.11 for the case of a needle crystallite oriented along the z axis, perpendicular to the paper. For this hypothetical crystallite, we have specified that γ on the {110} facets (γ_1) is 20 percent higher than γ on the {100} facets (γ_0), so $r_1 = 1.2 r_0$ as shown. The facets of lowest γ for various crystal structures are listed in Table 5.1. These are the preferred faces of exposure and therefore the largest ones on an equilibrated crystallite. Facetted crystals are abundant in Nature. Even grains of common table salt show up as perfect little cubes under the microscope. Covalent and ionic crystals have much larger γ anisotropy than metallic crystals, because they have strongly

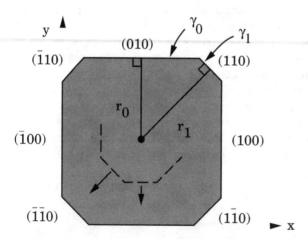

Figure 5.11 Wulff construction for a needle crystallite oriented along the z axis (perpendicular to the paper).

5.3.2 Three-dimensional (3D) nucleation

preferred bond directionality and bond polarity, respectively, while metals have neither. Polar covalent crystals and ionic crystals have, in addition, a preference for exposing charge-neutral (nonpolar) facets. In metals, γ anisotropy arises from anisotropy in the surface packing density of atoms and is typically only a few percent, but often this is still enough to result in preferred planes of exposure.

There are two ways in which a growing crystallite evolves in shape: deposition anisotropy and surface diffusion. Deposition rate from the vapor is often higher on facets of high γ because bonding is stronger and therefore δ or S_c is higher (see Fig. 5.1). It can be seen from the arrows in Fig. 5.11 leading from the nonequilibrated dashed-line facets that such deposition-rate anisotropy will cause the facets of high γ to become proportionately smaller as growth proceeds. Meanwhile, surface diffusion will redistribute mobile atoms so as to minimize total surface energy whether growth is occurring or not. Without surface diffusion, crystallites would evolve toward a "growth shape" which is in general different from the equilibrium "Wulff shape." Not all nuclei are single crystals, however, and this leads to shapes other than the Wulff shape even with surface diffusion occurring. In particular, there are multiple-twinned particles (MTPs). The twin of a crystal is its mirror image. For example, in the crystal of Fig. 5.7, if the pyramidal portion enclosed by the origin and atoms 1, 6, and 7 in the (111) plane were to form a twin on the (111) plane, the result would be two pyramids with a common base on (111) and six exposed {100} facets. MTPs consist of many single-crystal facetted particles nested together along twin planes to form one particle. In summary, we have seen that γ anisotropy in thin-film nuclei will cause them to develop facets and shapes that are governed by the anisotropy, by the dynamics of the deposition process, and by the extent of twinning.

In the case of liquid or amorphous nuclei, Eq. (5.43) still applies and in general has three terms—$γ_f$, $γ_i$, and $γ_s$—where $γ_f$ is now isotropic and therefore constitutes only one term. At substrate T above 2/3 of the absolute melting T of the film material in bulk form, small nuclei are likely to be molten [6], so the liquid case is encountered more often than one might expect. In previously discussing 3D nucleation, we neglected for simplicity the effect of $γ_i$ and $γ_s$ on total surface energy. Accounting for them now in Eq. (5.43) leads to the spherical-cap nucleus shown in Fig. 5.12, and leads to the "wetting angle," θ, with the substrate (Exercise 5.5). Angle θ may also be obtained by considering the γ values as surface-tension vectors as shown and by writing the force balance known as Young's equation:

$$γ_i + γ_f \cos θ = γ_s \qquad (5.45)$$

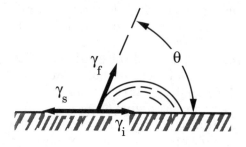

Figure 5.12 Wetting angle of a liquid nucleus.

The poorer the wetting, the larger θ becomes. When there is no wetting, θ = 180°, and γ_i is just $\gamma_s + \gamma_f$. When wetting is complete, θ = 0°, and Eq. (5.45) turns into the inequality of Eq. (5.36).

5.3.3 Two-dimensional nucleation

When wetting is complete and Eq. (5.36) holds, the adsorbing atoms do not accumulate into 3D islands but, instead, spread out on the surface in a partial monolayer as shown in Fig. 5.8a. Because total surface energy is *reduced* rather than increased by this process, there is no nucleation barrier in going from the vapor state to the adsorbed state; that is, the γ term in Eq. (5.38) is negative when the interfacial area is included. This means that deposition can proceed even in *under*saturated conditions. A familiar example of this situation is the oxidation of a metal, which involves O_2 adsorbing into a chemisorbed state (that is, reacting with the metal) at well below its saturation vapor pressure, p_v. Also in Example 2 of Sec. 5.1 on ZnSe deposition from Zn and Se vapor, each element will chemisorb onto the other at well below the element p_v values.

Assuming, as we did for 3D nucleation, that there is sufficient surface diffusion for equilibration, the partial monolayer of adsorbed atoms will behave as a 2D gas. By analogy to a 3D gas condensing into 3D nuclei, the 2D gas then condenses into 2D nuclei as illustrated in Fig. 5.13. Here, only the top monolayer of atoms is drawn. The "atomic terrace" to the left represents a monolayer which is one atomic step (a) higher than the surface to the right. But unlike the 3D nucleation case, 2D nucleation from a 2D gas involves no change in any of the γ values, so one might expect there to be no nucleation barrier. However, the chemical potential, μ, of a 2D nucleus is higher than that of a continuous monolayer because of the exposed *edge*. This situation may be viewed in terms of an excess edge energy, β (J/m), which is analogous to the surface energy, γ, of the 3D case. The surface concentration of the 2D gas for which its μ is the same as that on the straight terrace

5.3.3 Two-dimensional nucleation

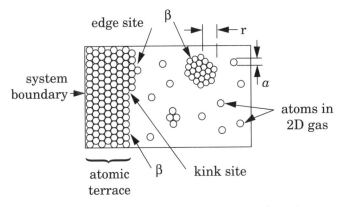

Figure 5.13 Geometry of 3D nucleation, looking down at the surface.

edge of a continuous monolayer (μ_c) may be thought of as the 2D saturation vapor concentration, n_v (mc/m^2). If n_s is the actual concentration of the 2D gas, then (n_s/n_v) becomes the 2D supersaturation ratio. By the same procedures as in the 3D case, we may then derive expressions for the critical nucleus which are analogous to Eqs. (5.39) and (5.40):

$$r^* = \frac{\beta}{a\left(\dfrac{RT}{V_{mc}}\right)\ln\left(\dfrac{n_s}{n_v}\right)} \quad (5.46)$$

and
$$\Delta G^* = \frac{\pi\beta^2}{a\left(\dfrac{RT}{V_{mc}}\right)\ln\left(\dfrac{n_s}{n_v}\right)} \quad (5.47)$$

Here, a is the monolayer thickness.

Once supercritical nuclei form, the 2D gas continues to attach to their edges until coalescence occurs and the monolayer is complete. Meanwhile, the next monolayer is beginning to form, and the film continues to build up in this way, atomic layer by layer. In the special case of single-crystal film deposition (epitaxy), the surface may contain many atomic terraces with straight edges as shown in Fig. 5.13. Their importance for epitaxy will be discussed in Sec. 6.7. The "kink" sites shown in Fig. 5.13 are also important surface features. Attachment of a 2D gas atom to a random site on the *straight* edge involves an increase in total edge energy, because it increases the length of the edge. Conversely, attachment to the kink site makes no change in the length of the edge; this is therefore an energetically preferred site, and edge growth can most easily occur by attachment-driven motion of these kink sites along the edge.

We now see that surface energy is determined not only by facet orientation as discussed in Sec. 5.3.1, but also by the densities of steps and kinks (Williams, 1994). The equilibrium densities of these two features increase with T because of their associated entropy (disorder), S. That is, when the TS term in Eq. (4.5) for the Gibbs free energy, G, becomes larger, the internal energy term, U, also becomes larger to minimize G, and U here mostly consists of the potential energy of step and kink formation. This is the same T-driven tendency toward disorder that causes vapor pressure to rise with T (see Sec. 4.1).

During film deposition, if the surface diffusion rate is high enough and n_s is low enough so that the 2D gas atoms are more likely to attach to an edge than to form a critical nucleus within an atomic terrace, then edge attachment becomes the dominant growth mode; that is, we have $\Lambda > L$, where is the surface diffusion length from Eq. (5.25) and L is the distance between terraces. This is called the "continuous" growth mode, as opposed to the nucleated mode. The continuous mode of 2D growth is analogous to the type of 3D nucleation in which nucleation is more likely to occur at active surface sites than by spontaneous nucleation elsewhere on the surface. Active sites and step edges, especially kinked edges, break the nucleation barrier by providing wetting at those sites.

Two-dimensional nucleation is usually preferred to 3D because it leads to smooth growth. In nonepitaxial growth, large grain size (coarse nucleation) may be desired in addition to smoothness. Unlike in the 3D nucleation case, here large grain size and smoothness are not incompatible. That is, if adatom mobility on the substrate is sufficient, large 2D nuclei will form before the first monolayer coalesces, and then subsequent monolayers will grow epitaxially on those nuclei. But there is another problem. High adatom mobility requires a low surface-diffusion activation energy, E_s, in accordance with Eq. (5.16), but E_s tends to increase with the strength of the adsorption, E_d or E_c, as suggested in Fig. 5.2b. At the same time, good wetting requires low γ_i and therefore requires strong adsorption. As a result, it will not always be possible to achieve strong enough adsorption for wetting without immobilizing the adsorbate and preventing grain growth. Even so, small-diameter grains can become wider as the film grows thicker, as we will see in Sec. 5.4.2.

5.3.4 Texturing

Here, we are referring to *crystallographic* texturing rather than to surface topography, although they are often correlated. The degree of texturing is the degree to which the crystallites in a polycrystalline film are similarly oriented. In one limit, there is random orientation (no texturing), and in the other limit, there is the single crystal. A material in which the crystallites are *nearly* aligned in all three dimensions is called a "mosaic," and the limit of a perfect mosaic is a single crystal. The degree of texturing is best measured by x-ray techniques, as discussed in Sec. 6.4.2. Texturing can occur in one, two, or three dimensions. Epitaxy is the best way to achieve perfect three-dimensional texturing. Epitaxy occurs when the bonds of the film crystal align with the bonds of the substrate surface, making the interfacial energy, γ_i, very low—zero in the case of homoepitaxy, which is when the film material is the same as the substrate material. In other cases, when no such alignment is operative, the most common form of thin-film texturing is a two-dimensional one in which the crystallite planes are aligned with respect to rotation about the two axes which lie in the plane of the substrate. This means that the film has a preferred growth plane parallel to the substrate but has random orientation with respect to rotation about the axis perpendicular to the plane of the substrate—the "azimuthal" axis.

One frequently wants to deposit a film onto a substrate to which no crystallographic alignment is possible, such as an amorphous substrate (glass, for example) or one which has a crystal symmetry or lattice dimension very different from that of the film. The achievement of 2D texturing in such cases can be very desirable when the desired film property is also crystallographically anisotropic. For example, ZnO exhibits its largest piezoelectric effect along the [0001] axis, which also happens to be its preferred growth plane, so texturing is very beneficial in that case. The magnetic saturation properties of ferromagnetic polycrystalline films used in memory discs are also highly anisotropic. The *thermodynamically* driven orientation of nuclei into a texture in nonepitaxial situations requires two conditions: (1) surface energy anisotropy, so that there will be a preferred facet; and (2) adatom mobility, so that the adatoms can arrange themselves to minimize surface energy. Since these requirements are often met, some degree of texturing is commonly observed, but the degree varies greatly with materials and deposition conditions. It is also possible that texturing can be *kinetically* driven, if the deposition rate on certain facets is higher than that on others due to a difference in S_c [Eq. (5.6)]. The kinetically favored texture is likely to be different from the thermodynamically favored one, as is the case for shape evolution in bulk crystals (Sec. 5.3.2.3).

Even with good 2D texturing, grain boundaries can be a problem. In the electronics industry, thin-film polycrystalline semiconductor materials are often used to obtain electrical isolation of transistors built on an insulating sublayer or in order to avoid the cost of single-crystal bulk material. The latter is particularly important in large-area applications such as solar cells and flat-panel display screens. The performance of these polycrystalline semiconductors is degraded from that of single-crystal ones, because the dangling or strained bonds present at grain boundaries act as traps and recombination centers for the charge carriers. Consequently, the achievement of single-crystal semiconductor growth on inexpensive substrates has been a long-standing goal. More recently, work on high-T superconductors has shown that grain boundaries lower the current-carrying capacity by orders of magnitude [7]. Avoidance of grain boundaries requires not only a preferred growth plane (two-axis rotational alignment) but also rotational alignment with respect to the azimuthal axis. Attempts have been made for about two decades to encourage rotational alignment in this dimension by generating on an amorphous substrate a pattern matching the film's crystallographic symmetry—a template [8]. Presumably, nuclei should preferentially align to the edges of the template to minimize their edge energy, β, thus achieving what is called "graphoepitaxy." These attempts have had varying degrees of success, but a reliable, large-scale process has yet to emerge. Some reasons are discussed below.

For a nucleus to diffuse to and align itself with an edge of the template pattern, it must be mobile on the surface. The mobility of a nucleus decreases rapidly as its radius increases, for several reasons: (1) increased contact area and bonding to the substrate; (2) increased mass and therefore lower velocity for the same thermal energy; and (3) reduction of the translational-energy fluctuations that result from impact by adatoms of the 2D gas (2D Brownian motion). Not much is known about nucleus mobility, and it is often ignored in models of nucleus growth. In graphoepitaxy attempts, it is likely that a principal cause of failure relates to the scale of the template relative to the scale of the largest mobile nuclei. This situation is illustrated in Fig. 5.14, in which we are looking down onto a patterned substrate containing various square nuclei. There are two matters of scale involved here. First, if nucleation is occurring on terraces between adjacent pattern edges, as in the case of the large nucleus in (a), we must have $\Lambda > L$ for the largest nuclei, where Λ is the nucleus diffusion length and L is the pattern spacing. If instead, the large nucleus in (a) becomes immobilized where it is shown, graphoepitaxy will fail. The second matter of scale involves the radius, r, of nuclei forming on or diffusing to the pattern edge relative to the spatial wavelength, λ, of the edge roughness. For $r < \lambda$, nuclei will misalign at the edge as shown in (a); for $r > \lambda$, they will align well as shown in (b). In a relatively successful recent attempt [9], 85 percent alignment of ZnS was achieved in a pattern of

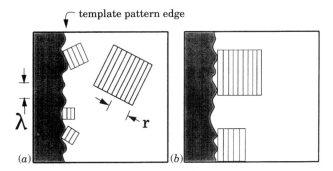

Figure 5.14 Graphoepitaxy on an irregular template, when the larger nuclei are (a) immobile and (b) mobile on the surface.

pyramidal pits transferred to polyimide from Si(100) which had been etched into {111} pyramids. Success was attributed to several factors: (1) strongly preferred {111} planes of exposure for ZnS; (2) smoothness of the template pattern; (3) nonwetting of the nuclei to the substrate, which increased their freedom to rotate; and (4) selective nucleation in the bottoms of the pyramidal pits.

5.4 Structure Development

Upon coalescence of the surface nuclei to form a continuous film, the nucleation step of film deposition is complete, and the fourth step begins: development of the bulk film structure. The form of this structure changes dramatically with the amount of thermal motion taking place during film growth, which scales with the ratio of the substrate T to the melting point of the film, T_s/T_m (in K, not °C), known as the "homologous" or "reduced" T. Structural form also changes with the amount of additional energy being delivered to the growth surface. Three structural zones (Z1, 2, and 3) were initially identified in an evaporative deposition study which included both metals and ceramics [10], and these zones have since been observed in a wide variety of film materials deposited by all of the vapor-phase processes. A fourth "transitional" zone (ZT) between Z1 and Z2 was identified in sputter deposition [11] and has since been found prominent in other energy-enhanced processes. Occasionally, anomalous structural forms occur, in particular the whiskers, illustrated in Fig. 5.15, that have been seen in Ti, for example [12]. The occurrence of whiskers implies an extreme preference for growth along the vertical direction. The characteristic structures of the four basic zones are also illustrated in Fig. 5.15. They are described briefly below and are analyzed in more detail in the next two subsections.

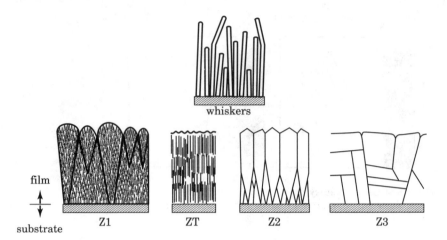

Figure 5.15 Characteristics of the four basic structural zones and of whiskers, in cross section. The ratio of substrate T to film melting T (T_s/T_m) increases in the direction Z1→ZT→Z2→Z3.

- *Z1* occurs at T_s/T_m so low that surface diffusion is negligible; that is, $\Lambda < a$ in Eq. (5.17). Z1 consists of columns typically tens of nm in diameter separated by voids a few nm across. The columns have poor crystallinity (many defects) or are amorphous. In thicker films, there becomes superimposed upon this structure an array of cones with wider voids between them. The cones terminate in domes at the surface, and the size of the domes increases with film thickness.
- *ZT* also occurs when $\Lambda < a$. It contains defected columns similar to those of Z1, but the voids and domes are absent. ZT is usually associated with energy-enhanced processes.
- *Z2* occurs at $T_s/T_m > 0.3$ or so, high enough so that surface diffusion is becoming significant. It consists of columns having tight grain boundaries between them and having a characteristic diameter which increases with T_s/T_m. Crystalline columns are less defected than in Z1 and ZT and are often facetted at the surface. The Z2 structure can also occur in amorphous films; there, the column boundaries are planes of reduced bonding rather than planes of crystallographic discontinuity.
- A transition to *Z3* occurs in certain instances at $T_s/T_m > 0.5$ or so, high enough so that considerable bulk annealing of the film is taking place during deposition. Z3 is characterized by more isotropic or equiaxed crystallite shapes. Film surfaces are often smoother at $T_s/T_m > 0.5$ for either Z2 or Z3; however, the grain boundaries can develop grooves.

The following several points about these zones need to be kept in mind.

1. All four zones cannot always be identified for a given material. In particular, Z3 is often not observed.
2. The transition from one zone to another is not always abrupt with T, and the transition T varies with deposition conditions and material. The growth mode can also change from Z3 to Z2 or from ZT to Z1 moving up through the thickness of a film.
3. The surface topography shown in Fig. 5.15 is typical, but it can vary considerably with factors such as surface-energy anisotropy and incident angle of depositing vapor.
4. Epitaxial films exhibit none of this bulk structure—at least when they are free of mosaic texture (see Sec. 5.3.4). However, they can still have highly facetted and therefore very rough surfaces when grown on planes of high surface energy, due to restructuring of the surface to minimize energy in accordance with Eq. (5.43).
5. Amorphous films exhibit bulk structure only when they are inhomogeneous, because they have no crystallographic pattern by which to define the boundaries between grains.

Homologous T is the main determinant of structure. Z1 and ZT films result from "quenched growth" processes in which thermal migration of the adsorbed material is negligible, whereas Z2 and Z3 films result from thermally activated rearrangement on or within the film, respectively. We will discuss these two regimes of structural development separately below, and we will see how the growth dynamics affect both the bulk structure and the surface topography.

5.4.1 Quenched growth

When initial bonding of the arriving vapor to the film surface is strong enough and film T is low enough, surface diffusion does not have time to occur before deposition of the next atomic layer, so that atoms become immobilized where they land. This is known as "ballistic" deposition, since the only motion of atoms which affects the deposition is the projectile direction of the arriving vapor. One might think that such deposition would result in smooth and homogeneous films, especially for vapor arriving uniformly and perpendicularly, as in high-vacuum deposition from a point source, and on very smooth substrates such as glass or chemically polished Si wafers. However, even under these conditions two destabilizing factors are always at work which generate roughness and voids. These factors are statistical roughening and self-shadowing, and we will examine them in turn below.

Statistical roughening arises because of statistical fluctuation in the vapor arrival flux. This effect can be illustrated with a simple model in which atoms of diameter a arrive perpendicularly at a steady rate but at positions randomly chosen from among the surface sites of a linear array. In this model, each atom is constrained to stick on the site it lands on even if it is on top of a pillar of atoms. The resulting surface topography after depositing an average number, $\bar{N}$, of 25 atoms per site across the array is shown in Fig. 5.16. At this point, the average column height—that is, the film thickness—is $\bar{h} = a\bar{N}$, but there is considerable variation in the heights, h, of atomic columns across the array. For large enough $\bar{N}$, as shown, this variation is mathematically described by a Gaussian distribution whose standard deviation is given by

$$\sigma = a\sqrt{\bar{N}} = \sqrt{a\bar{h}} \qquad (5.48)$$

The σ of the distribution is a measure of the roughness of the film, or the "dispersion" of h about $\bar{h}$. Note that Eq. (5.48) is analogous to Eq. (5.17) for the dispersion in the *lateral* direction that arises from surface diffusion, because they are both random processes. Statistical roughening and surface diffusion are also *competing* processes, the first increasing the film roughness and the second smoothing it out. In Z1 growth, surface diffusion is completely quenched.

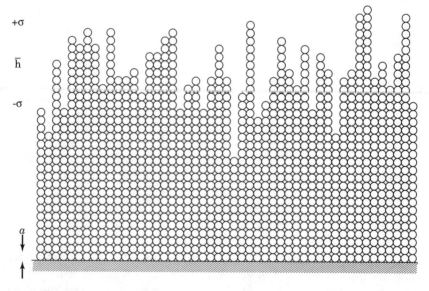

Figure 5.16 Statistical roughening in random ballistic deposition of a 25-atom-thick film. (Pascal solution courtesy of Jared Smith-Mickelson.)

5.4.1 Quenched growth

A more realistic refinement of the Fig. 5.16 model avoids the single-atom pillars by constraining neighboring pillar height differences to be ±1 atom or zero. This changes the 1D model into a 2D model, because now the columns are interacting laterally, and it changes the predicted dependence of roughness on film thickness, h. In general, roughness is found to increase as h^β, and for the 1D model of Eq. (5.48), the roughness exponent, $\beta = 1/2$. One group [13] has calculated, using the ±1 atom constraint, that $\beta = 1/3$ for the 2D case and 1/4 for the 3D case, this last corresponding to a 2D substrate. However, that work also incorporated the "solid-on-solid" assumption which excludes voids from the structure. In Z1 growth, voids are a key feature, and observations of a wide range of actual Z1 films [14] has shown that their roughness increases with $\beta = 3/4$. That is, voids increase the rate of roughening considerably. We will see below that there are many factors which influence the Z1 microstructure. Because of this and because of the geometric complexity of a voided and rough morphology, the mathematical and computational challenges of modeling such growth are formidable, and an accurate comprehensive model has yet to emerge.

In the above discussion of roughness, we have only considered the overall variation in film thickness, h, with lateral position; this is known as the "interface width," Δh, and for the 1D case it was found to be 2σ of a Gaussian distribution. A more complete description of surface topography would give Δh as a function of the lateral spatial frequency or the spatial wavelength, λ_s, of the h fluctuation, which is essentially a spatial "noise" spectrum. Topography having a given Δh over short lateral distances (short λ_s) is going to have more effect on film structure and properties than that having long λ_s, because it will have a steeper slope, $\approx 2\Delta h/\lambda_s$. We will see in the next section that surface diffusion smooths out the short-λ_s roughness.

Various methods for measuring film surface roughness after deposition are discussed in Sec. 10.1.2. For monitoring roughness *during* deposition, laser light scattering is a convenient technique that is especially sensitive to Δh for λ_s on the order of the laser wavelength. The λ_s of maximum sensitivity depends on the scattering angle being monitored [15]: it occurs at longer λ_s for angles close to the specular angle, and at shorter λ_s for angles farther away. Thus, angle-resolved scattering patterns give a more complete description of the topography [16]. Even submonolayer roughness is detectable despite the fact that the typical (Ar-ion) laser wavelength of 488 nm is 10^3 larger!

Self-shadowing is the second factor that destabilizes surface smoothness during film deposition, and it is the cause of the characteristic Z1 voided columnar structure. To understand this effect, we first make the Fig. 5.16 model more realistic by introducing the rule that arriving atoms cannot perch on top of each other, but rather settle sideways into the nearest "cradle" position in which they establish relaxed bond lengths to their nearest neighbors. This is known as "ballistic aggregation," and it is shown for 2D geometry in Fig. 5.17a. Because of the finite size of atoms ($a > 0$), aggregation can result in overhang structures such as that of Fig. 5.17b, which shadow the low areas from deposition. In effect, the area of vapor collection by the high areas increases from A_h to A_h', and that of the low areas decreases from A_l to A_l'; this is known as the "finite-size effect." Additional overhangs are produced by the atomic attraction of arriving vapor into the sidewalls of columns as shown in Fig. 5.17c. Thus, even

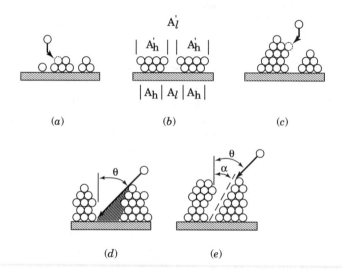

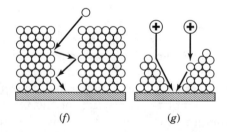

Figure 5.17 Atomistic processes in quenched-growth structure development: (a) ballistic aggregation, (b) finite-size effect, (c) sideways attraction, (d) oblique shadowing, (e) tilt effect, (f) low sticking coefficient, (g) void-filling by energetic particles due to enhanced mobility (left) and forward sputtering (right).

in perpendicular deposition, there develops a Z1 structure of columns with shadowed voids between them. One can readily see in Figs. 5.17*b* and 5.17*c* how self-shadowing increases the degree of roughening beyond what one would obtain just from statistical fluctuation in arrival rate, and this is why the roughness exponent, β, is so high for Z1 films.

When vapor is arriving obliquely at angle θ or over a range of angles, as shown in Fig. 5.17*d*, self-shadowing increases, thus increasing the T_s/T_m to which Z1 persists before surface diffusion can counteract it. Oblique incidence at a fixed *azimuthal* (rotational) angle also causes Z1 and Z2 columnar structures to tilt toward the incident direction at an angle α from the perpendicular (Fig. 5.17*e*). However, the so-called "tangent rule," which specifies that $\tan α = (1/2)\tan θ$, is not always obeyed. In addition to tilting, the columns become elliptical in cross section [17], with the long axis perpendicular to the plane of incidence except at high θ, when it can switch to parallel.

Incidence over a *range* of θ occurs when the vapor source subtends a large solid angle as viewed from the substrate, or when the process is operating in the fluid-flow regime (Kn << 1). The direction of vapor emanating even from a point source becomes randomized by collisions with background gas during fluid-flow transport to the substrate, so that the vapor can arrive with an angular distribution approaching the completely random cosine distribution. Thus, for processes such as sputter deposition and laser ablation whose operating pressure can span the range from molecular to fluid flow, raising pressure spreads the range of incident angles and increases self-shadowing, so that deposition pressure becomes an important determinant of film structure in these processes. At the even higher pressures of CVD, incidence is always random. Note that even at 10^5 Pa (1 atm), the vapor mean free path, $l ≈ 100$ nm by Eq. (2.24), is still much longer than the Z1 void diameter, so that self-shadowing persists. That is, the "fluid" never penetrates the voids. Nevertheless, there are often compensating effects in CVD that reduce self-shadowing: namely, a low sticking coefficient, S_c, for the depositing vapor, and surface diffusion of the adsorbed precursor species even at low substrate T. Low S_c allows vapor to bounce down the void walls toward the bottom as shown in Fig. 5.17*f*. Indeed, in the limit as $S_c→0$, the self-shadowing effect vanishes, but of course then there is no deposition. For $S_c > 0$, the vapor flux decreases with depth into the void channel as it progressively deposits, an effect known as "nutrient depletion," which may be thought of as partial self-shadowing. Thus, Z1 growth can occur in a CVD process as well as in PVD, even when that CVD process is producing conformal coverage on *macroscopic* topography because of fluid-flow transport on that scale. CVD coverage in channels will be discussed further in Sec. 7.3.3.

In the above discussion, we have considered the initial roughness to be due to statistical roughening of a perfectly smooth substrate. At some thickness of the growing film, self-shadowing takes over from statistical fluctuation as the dominant roughening mechanism, and at that point, void formation begins. The crossover thickness depends on all of the deposition factors mentioned above. In addition, initial roughness either on the substrate or on the film nuclei upon coalescence hastens the crossover and also increases the final roughness of the Z1 film. One might think that under the quenched Z1 growth conditions, the critical nucleus would always be only one atom and would not represent a roughening factor. However, remember that the surface-diffusion activation energy of the adsorbing vapor [E_s from Eq. (5.19)] can be lower on the substrate than on the film surface, so that the initial nucleation can involve more surface diffusion than the subsequent deposition. This usually corresponds to a nonwetting situation in which the film material bonds more readily to itself than to the substrate and thus forms 3D nuclei, which would leave a rough surface after coalescence.

As one might expect, Z1 microstructure is undesirable in a thin film unless the application specifically benefits from porosity, as in the following examples. In gas-detector applications, a film property is changed by adsorption of the gas; the amount adsorbed is proportional to surface area and therefore to porosity. High surface area is also useful in catalytic applications such as fuel-cell electrodes. Columnar porosity is useful where lateral mechanical rigidity would lead to cracking or buckling due to thermal or other stresses (see Sec. 5.6), as in ZrO_2 coatings used as thermal barriers in rocket nozzles and other high-T parts. High porosity fractions such as those occurring in whisker growth cause films of opaque materials to look black as a result of light trapping, and this is useful in some optical applications. However, for most of the applications in Table 1.1, porosity is undesirable. In optical coatings, it causes light absorption and scattering. In electronics, it decreases the conductance of metal films, increases the leakage of insulators, and causes charge trapping in semiconductors. It compromises the effectiveness of chemical-barrier films and weakens films in mechanical applications. However, it is often necessary to restrict T_s to below the Z2 regime. In such cases, Z1 can usually be transformed into the void-free ZT structure by the use of energy-enhanced deposition.

The energy-enhanced processes that can reduce void formation are those in which the energy is translational kinetic energy carried by massive particles (ions and superthermal atoms), because the void reduction mechanisms involve *momentum* transfer to the growing surface. The ions can be those of the depositing vapor itself, as in cathodic-arc and electron-beam evaporation (Chap. 8), or those of an inert gas such as Ar, as in glow-discharge plasma processes (Chap. 9). Superthermal atoms are present in supersonic beams (Sec. 4.5.4), laser ablation (Sec. 8.4), and sputter deposition (Sec. 8.5.4.3). Some of the typical energies involved are listed in Table 4.2.

Insight into the ways in which energy enhancement prevents voids has been obtained by "molecular-dynamics" computer modeling, in which all the atoms involved are followed in vibratory motion and in bond formation and breakage. In this way, four mechanisms have been identified [18]:

1. local heating due to the "thermal spike" produced upon impact, resulting in local surface diffusion
2. less trajectory curvature (Fig. 5.17c) due to the higher approach velocity
3. higher "impact mobility" of the adatoms so that they can move down into the voids
4. forward sputtering of other adatoms into the voids.

The last two effects are the dominant ones and are illustrated in Fig. 5.17g. Impact mobility amounts to a high surface diffusion rate for a few atomic distances until the excess kinetic energy upon impact becomes dissipated into the bulk. In the forward-sputtering effect, momentum transfer from the approaching particle to the adatom causes the latter to be knocked loose and scattered forward.

Figure 5.18 shows a 2D molecular-dynamics simulation of structure development versus the kinetic energy carried by atoms perpendicularly incident on a film held at $T_s = 0$ K. The lowest of the three ratios of incident energy, E_t, to the adatom potential-well depth, E_c, corresponds to thermal deposition and is dominated by ballistic aggregation, so that voids quickly develop. The effect of added E_t on void filling is quite dramatic in this simulation, as is the effect observed experimentally in energy-enhanced deposition. The highest ratio shown (1.5) would typically correspond to an E_t of 5 eV, which is characteristic of sputtered particles and of the lowest-energy ions emanating from plasmas. More recent molecular-dynamics simulations done in 3D, which uses much more computer time than 2D, have shown that the overhanging atoms predicted in 2D are actually unstable even in thermal deposition [20]. This emphasizes the importance of accurate modeling. Of course, some overhanging atoms *must* be stable in order for a Z1 structure to develop under perpendicular incidence. Collisions of energetic particles with surface atoms are discussed further in Sec. 8.5.

In practice, ZT is encouraged by depositing at low pressure so that incident particles neither become scattered into more oblique incident

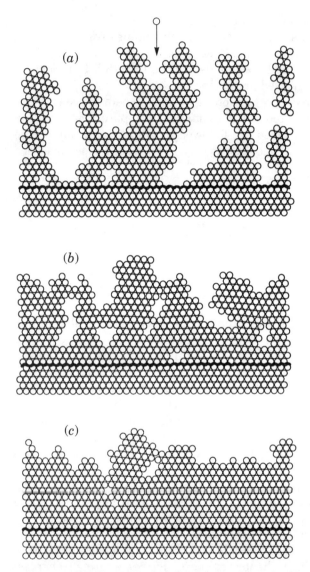

Figure 5.18 Two-dimensional molecular-dynamics simulation of the deposition of energetic atoms impinging perpendicularly onto a substrate held at 0 K. The horizontal line is the substrate interface. Normalized impinging energy E_t/E_p, as defined in text, is (a) 0.02, (b) 0.5, and (c) 1.5. (Source: Reprinted from Ref. 19 by permission.)

angles nor dissipate their kinetic energy in gas collisions. When the energy is being provided by a supplemental source of inert ions, void filling increases with ion flux and decreases with increasing deposition rate. Microstructure can also vary with *position* on substrates having convoluted topography, which is common in microcircuit fabrication, for example. Figure 5.19 illustrates schematically a sputter-deposition situation discussed by Thornton (1986), in which growth is Z1 on the top surface due to a broad angle of incidence of the vapor flux, but is ZT at the bottom of the trench due to flux collimation by the sidewalls. The sidewalls are Z1 with even higher void fraction as well as tilted columns, due to the oblique flux. Since film properties vary with microstructure, this situation is clearly undesirable. The remedy is to increase particle collimation or energy input so that the whole film becomes ZT, or to increase T_s so that it becomes Z2.

The degree of collimation depends on the solid angle that the vapor source subtends at the substrate. When one wants to coat large areas uniformly, a row of evaporation sources (Fig. 4.13) or a long rectangular-magnetron sputter source (Sec. 9.3.4) is used in conjunction with substrate transport perpendicular to the row. If, in such cases, the distance from source to substrate is made small for efficient utilization of source material, the solid angle will be large in both directions. Thus, there is a trade-off between utilization and collimation. Collimation can be improved either by increasing transport distance or, in the magnetron case, by inserting a honeycomb baffle into the transport space to trap obliquely-directed vapor. This trapping decreases mate-

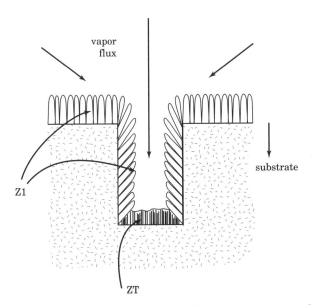

Figure 5.19 Microstructure variation with substrate geometry for sputter deposition into a trench. [Adapted from Thornton (1986).]

rial utilization by the same amount as does increasing transport distance for a given improvement in collimation, but it can also cause particulate contamination from deposits flaking off of the honeycomb.

5.4.2 Thermally activated growth

When film T is raised to where $\Lambda > a$ in Eq. (5.19), surface diffusion becomes a key factor in determining film morphology. Voids become filled by diffusing adatoms, and the film develops the characteristic Z2 columnar grain structure. Since Λ increases exponentially with T, the transition occurs over a narrow T range, and empirically it is found to occur at about 0.3 T_m for many materials, where T_m is the melting T in K. The correlation with T_m occurs because both diffusion and melting depend largely on the binding energy of atoms in the solid. Indeed, the activation energy for *bulk* solid-state diffusion has been shown to be proportional to T_m, at least for crystals of a given structure [21]. Assuming that this rule holds for surface diffusion, with a proportionality factor of B, then $E_s = BT_m$ in Eq. (5.19) and the exponential becomes proportional to T_m/T. Various factors can shift the transition T from the 0.3 T_m value, however. If background impurities are adsorbing strongly on the growing film, such as oxygen on metals, they can inhibit surface diffusion. The transition T also increases slowly with increasing deposition flux, J_r, due to the appearance of $1/\sqrt{J_r}$ in Eq. (5.19). Figure 5.20 illustrates the transition with a 2D computer simulation involving ballistic aggregation of atoms incident with negligible kinetic energy at 30° from the perpendicular, followed by surface diffusion with an activation energy of $E_s = 1$ eV. Void fraction is seen to decrease rapidly with increasing T.

A similarly rapid transition in surface *topography* occurs with the onset of surface diffusion. For example, Fig. 5.21 shows scanning-tunneling-microscope (STM) images of the surface topography of Au thermally deposited onto mica at various film temperatures, T_s. For Au, 0.3 T_m = 128° C, which corresponds to the transition between the domed topography of images a-c and the smooth terraces of d-f. In this case, the Au is growing epitaxially on its low-surface-energy (111) plane, so facetting is not observed in d-f. The increase in dome diameter with T_s in a-c probably results from an increase in the coarseness of nucleation on the mica with increasing T_s.

In cross section, the crystalline grain structure of polycrystalline Z2 films is columnar. (For epitaxial and amorphous films in this T regime, the column boundaries can vanish, as discussed earlier.) At the substrate interface, the grains of a polycrystalline film begin as nuclei and grow upwards as columns after coalescence. With increasing thickness, some columns expand at the expense of others until a limiting

5.4.2 Thermally activated growth

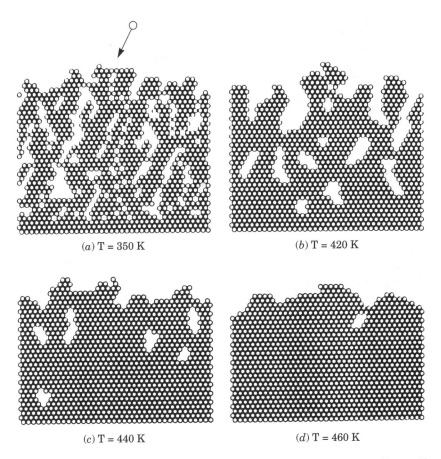

Figure 5.20 Two-dimensional computer simulation of the effect of substrate T on void filling by surface diffusion. (Source: Reprinted from Ref. 22 by permission.)

column diameter is reached, as illustrated in Fig. 5.15. The development of column size is shown in Fig. 5.22 for a 40-µm-thick columnar polycrystalline Fe film electron-beam deposited onto alumina at 80 nm/s and $T_s = 0.64 T_m$. For examination, the film was lapped through at a grazing angle to the substrate plane and then etched lightly to bring out the grain structure (see Sec. 10.1.3). Thus, the micrograph shows the grain pattern essentially parallel to the substrate as a function of thickness position, h, within the film. There is a steady increase in grain size for the first few µm of growth, after which the grain size stabilizes at about 40 µm.

The driving force for columnar grain growth in Z2 is surface-energy minimization. Computer modeling has provided insight into this process and has predicted the increase in column diameter with thickness

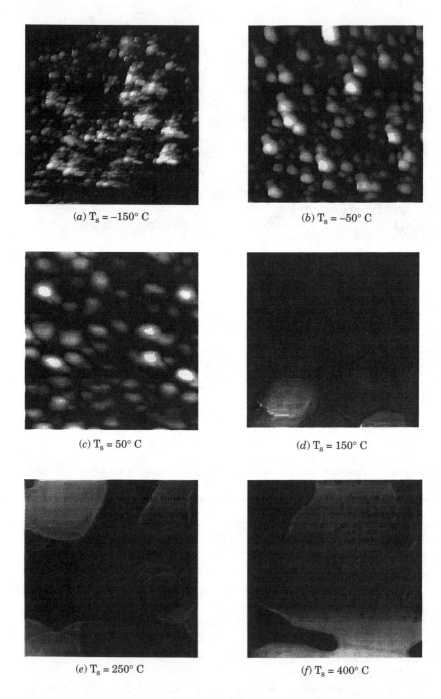

Figure 5.21 STM images of Au(111) grown on mica. Image size = 250 × 250 nm. (Source: Reprinted from Ref. 23 by permission.)

5.4.2 Thermally activated growth

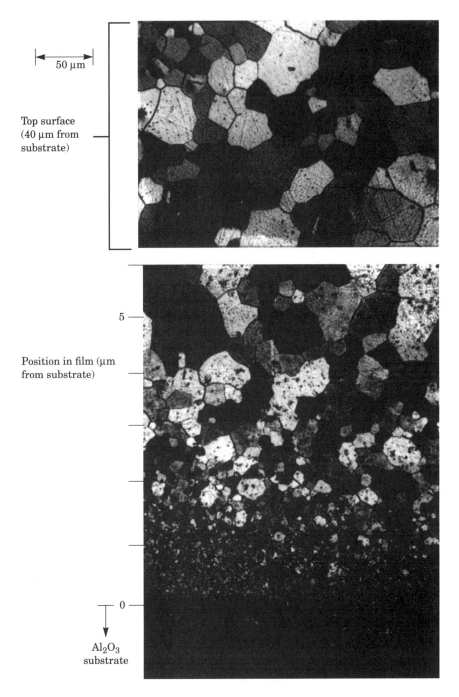

Figure 5.22 Development of columnar grain diameter with thickness in an evaporated Fe film. (Source: Author's unpublished work.)

[24], although the model cited fails to predict the experimentally observed [10] *saturation* of column diameter with increasing film thickness. The model is developed using the cross section of a growing columnar structure shown in Fig. 5.23. In the previous section, we have seen that films tend to develop roughness. In the absence of surface-energy (γ) anisotropy and for small column diameter, this roughness can be approximated as the domes shown. If surface diffusion were negligible, these domes would grow steeper with time due to self-shadowing of the cusped valleys between them, and the cusps would become the voids of the Z1 structure. In Z2, this trend is counteracted by surface diffusion, which results in net motion of adatoms toward the cusps in an attempt to satisfy the γ force balance at the cusp in Fig. 5.23. This balance is analogous to the one which determined wetting angle in Fig. 5.12. Because bonding at grain boundaries is always weaker than in the bulk of a crystal, there will always be an interfacial tension, γ_i, at the boundary between two columns. Therefore, the column surfaces do not completely flatten as a result of surface diffusion, but instead approach equilibrium dome radii such that the vertical components of γ_1 plus γ_2 balance γ_i, and thus a groove develops in the film surface along the column boundary. A similar groove develops when polished polycrystalline materials are annealed, and it is thus known as "thermal grooving." As the columns increase in diameter, they flatten on top to minimize surface area, but they retain the thermal grooves on the edges.

To satisfy the *lateral* force balance in Fig. 5.23, inspection shows that smaller-diameter columns will develop smaller radii of curvature (also see Exercise 5.10). Now, because the atoms on a surface of smaller convex radius are more exposed and less completely bonded to the bulk, they have higher γ, as we will soon show mathematically.

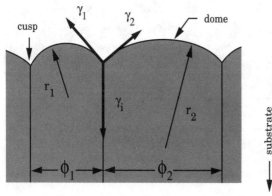

Figure 5.23 Geometry and surface energetics of columnar grain growth.

This chemical-potential driving force generates a net surface-diffusion flux from smaller columns to larger ones, causing the smaller ones to shrink and also to become more shadowed as deposition progresses, so that they eventually close out. This process continues to an extent which scales with the surface diffusion length, Λ. Now, Λ increases exponentially with T [Eq. (5.19)] or, more precisely, with T_s/T_m; and indeed, the characteristic column diameter, ϕ_c, behaves the same way [10]. Cross-sectional micrographs of metal films grown by both evaporation [12] and sputtering [11] clearly show this behavior and also show the transition from Z1 or ZT to Z2. With regard to surface *topography*, roughness features having spatial wavelength $\lambda_s < \Lambda$ become smoothed out by the surface diffusion, while those having $\lambda_s > \Lambda$ will continue to develop in accordance with the statistics discussed in the previous section.

The relationship of chemical potential to surface curvature may be derived with reference to Fig. 5.24. Here, the surface atoms of the film material are considered to be cubes of side a which have been distorted into truncated, square-based pyramids by a spherical curvature of radius r imposed on the surface. In the cross section shown, the atoms are trapezoidal, and the base exposed to the surface has area $(a + 2\delta)^2$. For $\delta \ll a$, the increase in surface area per atom versus that of a

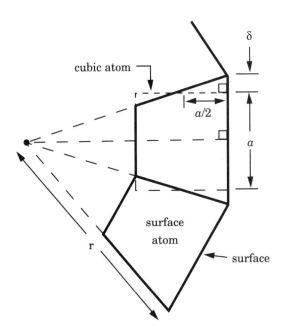

Figure 5.24 Geometry for calculating the effect of curvature on chemical potential.

flat surface (a^2) is $4a\delta$, and the corresponding increase in chemical potential is $\Delta\mu = 4a\delta\gamma_0$, where γ_0 is for a flat surface. Now, δ is related to r by similar right triangles: $(a/2 + \delta)/r \approx (a/2)/r = \delta/(a/2)$; and thus,

$$\Delta\mu = a^3\gamma_0/r = \Omega\gamma_0 K \tag{5.49}$$

where Ω = atomic volume
 $K = 1/r$ = curvature (positive is convex; negative is concave)

This is the classic expression for the curvature effect on μ.

At still higher T_s, grain-boundary migration can also occur within the bulk of the film. This is essentially bulk annealing proceeding during deposition. Figure 5.25 shows a section through the bulk of a film and parallel to the surface. A small cylindrical grain is surrounded by three larger grains. Since the curvature effect causes the μ of the atoms on the convex edge of the small grain to be larger than that of the atoms on the concave edges of the surrounding grains, atoms will transfer across the grain boundary from the small grain to the surrounding ones and eventually annihilate the small grain. Note that this process is also favorable in terms of minimization of total surface energy for the Fig. 5.25 geometry, because a decrease of dr in the radius of the small grain results in a 2πdr decrease in its boundary length with the surrounding grains, but only a 3dr increase in the total length of the surrounding three grain boundaries, so the net change in boundary length is negative. This is the same thermally-activated grain growth process that occurs in bulk metallurgy and is known as "Ostwald ripening." The velocity of grain-boundary motion is proportional to $\exp(-E_a/RT)$, where E_a is the activation energy for grain-boundary motion. Thus, columns can continue to increase in di-

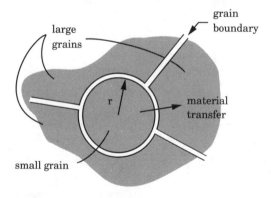

Figure 5.25 Film section parallel to the substrate showing how lateral grain growth occurs.

ameter beneath the growth surface at elevated T_s, so that their diameter approaches the film thickness to give the equiaxed grain structure characteristic of Z3.

In the above discussion of Z2, we have assumed that γ is independent of crystallographic plane. In fact, γ anisotropy is common, and it provides an additional and often dominant driving force for surface diffusion and column selection during growth. Consequently, even a film which nucleates with completely random orientation can develop its preferred crystallographic texture with increasing thickness, as those columns oriented so that they have high-γ surfaces lose material and get closed out. Moreover, a small-diameter column can be favored over a large one despite the curvature effect if the surface γ of the small column is lower; this situation is known as "abnormal" grain growth. Column selection might also be *kinetically* determined due to varying deposition rates, as pointed out in Sec. 5.3.4.

5.4.3 Amorphous films

Grain boundaries are often undesirable—even the void-free ones of the Z2 structure. In some cases, a grain boundary (gb) can be eliminated by growing epitaxially, as we will see in the next chapter, but in the more general case, the only way to eliminate them is to cause the film to deposit as an amorphous phase.

Except in the special case of "twin-plane" boundaries, grain boundaries have bonding that is weaker and less complete than within the bulk of the grains because they represent a disruption of crystalline order. This characteristic leads to a number of problems. In protective coating applications, the grain boundaries are sites for preferential chemical attack (corrosion). Their relative openness can also cause them to act as channels for the diffusion of corrosives and impurities. Thus, in diffusion-barrier applications, the grain boundaries are usually the dominant pathway for impurity diffusion, even though only a small fraction of the cross-sectional area is represented by them, because the activation energy for gb diffusion is much lower than for diffusion through the bulk of the grain. This situation is shown in the Arrhenius plot of Fig. 5.26, where diffusion through the bulk of the grain is represented by that through a single-crystal film. The steeper exponential rise of the bulk diffusion component sometimes causes it to become dominant at high T. Diffusion behavior for polycrystalline material varies considerably with grain morphology. Amorphous films are intermediate between the two in diffusion behavior: diffusion is faster than through a single crystal but can still be orders of magnitude slower than through a polycrystalline film of the same composition (Kattelus, 1988). Diffusion of the film's own atoms also occurs

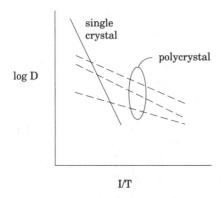

Figure 5.26 Typical diffusion behavior in a polycrystalline solid. D = diffusivity.

more rapidly along grain boundaries than through the grains, and this can cause bulk changes in film structure with time after deposition, such as hillock growth in compressively stressed films (Sec. 5.6.2) and electromigration failure in microcircuit metal lines carrying high currents. Diffusion along grain boundaries in metals appears to be inhibited by "stuffing" the boundaries (Kattelus, 1988) with a low-solubility and reactive impurity that tends to segregate to the grain boundaries and form strong bonds there (e.g., O or N in Ti). This effect may also explain the effect of Cu in inhibiting electromigration in Al microcircuit lines.

The grain boundaries cause further problems unrelated to diffusion. Mechanically-stressed films slip and fracture more easily at the boundaries because of the weak bonding. Thus, amorphous metal alloys (metallic glasses) tend to be harder and stronger than their polycrystalline counterparts. In polycrystalline semiconductors, the dangling bonds along the grain boundaries act as charge-carrier traps and thus degrade electrical behavior, as was discussed in Sec. 5.3.4. Of course, amorphous semiconductors contain dangling bonds, too. In the case of Si, most of the dangling bonds in either the amorphous or the polycrystalline phase can be terminated with H so that satisfactory electrical properties are obtained.

Because Nature likes to grow crystals, the growth of amorphous materials requires deliberately blocking this tendency, either by quenching the crystallization process (as in metastable metallic glasses) or by using a composition which does not easily crystallize. Thus, pure glassy SiO_2 crystallizes (devitrifies) as quartz; however, ordinary glass does not crystallize, because it is a solid solution of SiO_2 with various metal oxides that do not fit the quartz crystal lattice. In general, solid solutions are less likely to crystallize when their component materials have different crystal symmetries or large differences in lattice constant. *Metallic* solid solutions (alloys) usually require the above conditions as well as quenched growth ($T_s \ll T_m$) to form amorphous films, because the nondirectionality of metallic bonds makes it particularly easy for the atoms to settle into a crystalline arrangement. Likely candidates for amorphous metals are alloys having the "eutectic" (minimum T_m) composition, because low T_m is an indicator of difficulty in crystallization. *Elemental* metals crystallize even at cryogenic T! Elements with covalent bonds, which are directional, crystallize less easily. Thus, the amorphous phase of Si can be stabilized at T_s up to ≈800 K by the incorporation of ≈10 percent H to terminate dangling bonds and thus inhibit the development of tetrahedral crystalline bonding. This is done most effectively by deposition from a silane (SiH_4) plasma (more in Sec. 9.6.4.1).

5.4.4 Composites

So far, we have considered only films having uniform composition from grain to grain. Composite films are made by codepositing two or more immiscible materials which then accumulate into grains of different phases. Composites can provide properties unachievable in uniform materials. Structurally, these materials tend to be equiaxed rather than columnar, because column propagation is periodically interrupted by the nucleation of another phase.

Ceramic-metal composites ("cermets") are used as resistors, wear coatings, and optical absorbers. Resistivity is determined by the ratio of ceramic to metal; it can be varied over many orders of magnitude and controlled accurately within the high-value range between that of insulators and that of metals [25]. For example, the resistivity of the popular cermet Cr-SiO varies with composition from 10^{-4} to 10^{-2} Ω-cm. In wear-coating applications of cermets, the ceramic provides hardness while the metal provides toughness (resistance to cracking). Thus, cermets such as TiC-Mo are very hard but much less brittle than the pure ceramic material.

High optical absorption is needed in films used in solar heaters and in calorimeters for optical-power measurement ("bolometers"). This can be accomplished by light trapping in cermets, which occurs as follows. Many pure ceramic materials are transparent to visible light in single-crystal or amorphous form, but in polycrystalline form, multiple scattering of light off the grain boundaries causes them to appear white whenever there is a refractive-index (ñ) discontinuity at the grain boundaries. This happens when ñ is orientation-dependent (birefringence) or when there are voids. Thus, fused quartz (amorphous

SiO_2) is clear, and devitrification (crystallization) turns it white due to the birefringence of crystalline SiO_2. However, in cermets, each scattering event off of a metal grain is accompanied by partial absorption, so that multiple scattering results in efficient light trapping. By using a metal of low reflectivity and by optimizing metal ratio and grain size, >99 percent absorption can be obtained.

Absorptive light scattering also occurs in films consisting of whiskers (Fig. 5.15) or other highly porous structures, so such films often appear black. A very effective way to obtain a highly porous film structure is to run a PVD process at a pressure so high that homogeneous nucleation occurs during the transport step; that is, "macroparticles" condense in the gas phase. This is a variation of the gas-condensation technique of cluster formation which was discussed in Sec. 4.5.4. Homogeneous nucleation is discussed further in Secs. 7.3.2 and 9.6.2. The macroparticles settle on the substrate and agglomerate into a highly porous film. Such films are very weak mechanically, however, and cermets are far preferable from that standpoint.

5.5 Interfaces

Thin films inevitably involve interfaces, both with the bulk substrate and between layers of different film materials. The thinner the film, the larger is the fraction of material occupied by interfaces. One often wants the interfaces to be abrupt, but this is seldom the equilibrium situation. Most materials dissolve in each other to some extent, even in the solid state. This process is driven by the reduction of free energy, G, which arises from the randomization of order (configurational entropy increase) and from any heat of mixing which is released [see Eq. (4.5)]. Many elements and some compounds also *react* with each other at the interface to form new compounds having lower G of formation, $\Delta_f G$, than the starting materials. These various interfacial processes are illustrated in Fig. 5.27 and will be discussed in turn below. Interfacial reaction is particularly troublesome when eross sectionpitaxy is required, because the new compound is unlikely to match

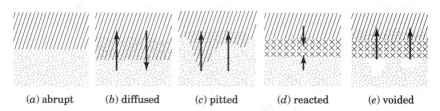

(a) abrupt   (b) diffused   (c) pitted   (d) reacted   (e) voided

Figure 5.27 Interface types. (Arrows indicate direction of diffusion.)

the other layers in crystalline symmetry or lattice constant, and it will thus disrupt the epitaxy (more in Sec. 6.3).

Both dissolution and reaction proceed by solid-state interdiffusion across the interface, and the extent to which this occurs is determined by time and T. That is, the degree of approach to equilibrium is limited by the kinetics of interdiffusion. Eq. (5.25) for the surface diffusion length also applies here to the bulk diffusion length; that is, $\Lambda = 2\sqrt{Dt}$. When reactions occur, those which involve an activation-energy barrier, E_k, have rate-limiting constants, k_k, given by Eq. (5.2). Those which have $E_k \rightarrow 0$ will have rates limited instead by diffusion of the reactants to the interface. Since D and k_k both increase exponentially with T, it is important to keep substrate temperature, T_s, as low as possible during deposition whenever interface abruptness or avoidance of interfacial compounds is important. This is one reason why films must often be deposited in the Z1 or ZT structural regime (Sec. 5.4) even when the bulk film properties of Z2 would be preferred. When high T_s is needed to activate a CVD film-forming reaction at the surface, time at T_s can be minimized by using *rapid thermal processing* (RTP), in which the substrate is heated and cooled very rapidly by maximizing thermal coupling and minimizing thermal load in accordance with the principles of Sec. 4.5.3. Another approach to the CVD problem is to use energy enhancement to activate film-forming surface reactions while keeping the bulk film T low. This energy is often supplied from a plasma (Sec. 9.3.2).

The interdiffusion rate between any two particular materials can vary considerably with their structure. In polycrystalline materials, the rate varies with grain structure because of the dominance of grain-boundary diffusion (Fig. 5.26). In epitaxial materials, diffusion occurs through the lattice either "interstitially" (that is, between atomic sites) in the case of small diffusing atoms, or substitutionally (on atomic sites) in the case of larger ones. In the latter case, the diffusion rate increases with atomic vacancy concentration, since substitutional diffusion proceeds by the diffusing atoms swapping places with neighboring vacancies. Vacancy concentration can vary considerably with film deposition conditions; in particular, it increases with ion bombardment (Sec. 8.5.3). Interdiffusion rate also increases in the presence of an electric field (electromigration) or mechanical strain.

Thus, the extent of film interaction at interfaces is hard to predict, because it is determined by rates of solid-state diffusion and reaction which are widely variable. To understand the interaction, the interface needs to be examined for the particular deposition process conditions, using the analytical techniques of Chap. 10. Nevertheless, valuable information as to what *kind* of interaction is likely to occur is available in the phase diagrams [26], which give the compositions of phases that have reached equilibrium with each other. The simplest situation consists of two elements that are completely miscible with each other and do not form any compounds. In this case, a planar interdiffused region will develop as shown in Fig. 5.27b.

Another type of binary phase diagram is represented by Al + Si in Fig. 5.28. The Al-Si interface has been studied extensively because of its importance in the metallization of integrated-circuit chips. No compounds appear between these elements, but there is a "miscibility gap"; that is, the elements have limited solubility in each other. This is indicated by the presence of two solid phases, labeled (Al) and (Si) on the phase diagram. The boundary of the (Al) phase is the solubility of Si in Al, which has a maximum value of 1.5 at.% at 577° C. No corresponding boundary is discernible on the Si-rich side, which means that Al has <0.1 at.% solubility in Si. Thus, most of the solid-state region between pure Al and pure Si consists of a two-phase mixture of Si-saturated Al and pure Si, which is labeled "(Al) + (Si)." Also shown is an intermediate composition having a minimum melting point—a "eutectic" point—at 12.2 at.% Si and 577° C. To avoid melting, T_s must never be allowed to exceed the eutectic T during deposition or subsequent processing.

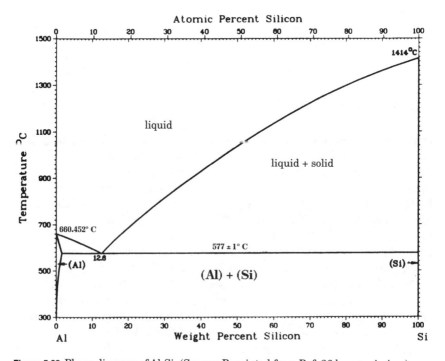

Figure 5.28 Phase diagram of Al-Si. (Source: Reprinted from Ref. 26 by permission.)

The solubility asymmetry of Al and Si means that Si will diffuse into Al, but Al will not diffuse into Si. This situation is shown in Fig. 5.27c, and it leads to the following problem. The space vacated by the out-diffusing Si becomes filled with the (Al) phase in an uneven manner, leading to the development of pits or "spikes" in the Si. When an Al film is etched off of a Si substrate, these pits are revealed quite dramatically, as seen in Fig. 5.29. Al spiking can short out shallow p/n junctions in the Si, so it is counteracted in the chip industry by inserting 100 nm or so of some film material which acts as a diffusion barrier to Al and Si, such as Ti-W alloy (Kattelus, 1988). The Al spiking of Fig. 5.29 actually resulted from breakthrough of an amorphous W-Re barrier layer due to Al reaction with the barrier. Structural considerations in the behavior of films as diffusion barriers were discussed in Sec. 5.4.3. Chemical considerations involve the degree to which the barrier film reacts with the adjoining films.

Many pairs of elements do react to form compounds rather than just interdiffusing. In addition, compound films can react at the interface to form new compounds there. In another example of semiconductor metallization problems, it is found that most metals (M) react with GaAs to form intermetallics ($M_m Ga_n$) and arsenides ($M_m As_n$). Known compounds appear as vertical lines on the binary phase diagrams, as was shown for the generic compound MY in Fig. 4.6a. When more than one compound appears on the diagram, all or only some of these may form at a film interface depending on the thermodynamic fa-

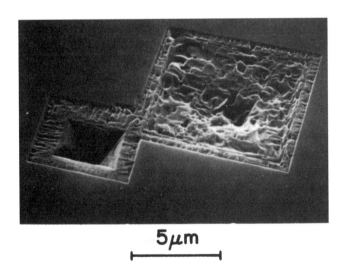

Figure 5.29 Etch pits in Si(001) substrate after removing an Al film. (Source: Reprinted from Ref. 27 by permission.)

vorability of the reaction and on kinetic factors. A reaction is favorable if the Gibbs free energy of reaction, $\Delta_r G°$, at the relevant T is negative. A favorable interfacial reaction will proceed at some rate determined by interdiffusion rate and by reaction activation energy, E_a, and this reaction rate will therefore increase exponentially with T. The $\Delta_r G°$ of a solid-state reaction that forms only pure compounds and not solid solutions depends only on the relative Gibbs free energies of formation of the reactant and product compounds from the elements, $\Delta_f G°$. Partial pressures and mole fractions are not involved. $\Delta_f G°$ is available for many compounds [28, 29]; for elements, $\Delta_f G° = 0$ by definition. Often, only the $\Delta_f G°$ at 298 K is available, but the T dependence is small. When only the enthalpy contribution to G, $\Delta_f H°$, is available, it can be used for rough estimates, because the entropy contribution, $\Delta_f S°$, is small for condensed-phase reactions. As an example of predicting whether an interfacial compound will form, consider the deposition of Si onto ZnSe, where there is the possible reaction

$$Si + 2ZnSe \rightarrow 2Zn + SiSe_2$$

However, $\Delta_f H°$ of 2ZnSe is 2×(–163 kJ/mol) while $\Delta_f H°$ of $SiSe_2$ is –29 kJ/mol, giving a $\Delta_r H°$ of +297 kJ/mol; so it is highly unlikely that the $\Delta_r G°$ of this reaction could be negative at any T. For further analysis of interfacial equilibria involving three elements, $\Delta_f G°$ data can be used to construct ternary phase diagrams [30].

The best choice for a diffusion-barrier layer is a compound with a highly negative $\Delta_f G°$ so that it will not react, and with good microstructural integrity so that it will not allow diffusion. For electronic applications such as Al/Si, the barrier must also be electrically conductive. TiN is one material that meets all three requirements fairly well, although it does form Al_3Ti at >780 K [27]. Many metal silicides, borides, and nitrides are conductive and are also significantly more stable than the metals themselves.

Solid-state interfacial reactions can only proceed by diffusion of one or more of the reacting elements through the interfacial compound layer which is being formed. Eventually, diffusion rate drops off due either to the increasing interface thickness or to the formation of an impervious layer of compound. The latter development causes the reaction to "self-limit" at the thickness when the layer becomes impervious. The oxidation of Al is a familiar example of a self-limiting interfacial reaction. Clean Al reacts spontaneously and vigorously with air to form Al_2O_3; however, once a few nm of oxide has formed, the oxide acts as a diffusion barrier to prevent further Al and O_2 from reaching each other, so the reaction stops. (Conversely, metals that form porous oxides, such as Fe in the presence of water, continue to oxidize.) Self-limiting interfacial reactions lead to the structure shown in Fig. 5.27d. Sometimes, a "sacrificial" diffusion barrier is employed whose effectiveness results from its reaction to an impervious compound with one or the other of the adjoining films. Thus, the Al/Ti/Si structure reacts to form TiAl$_3$, which inhibits Si diffusion into the Al.

One problem which can arise even with a self-limiting interfacial compound is that illustrated in Fig. 5.27e. Here, the underlying film material is diffusing through the interfacial compound, while the overlying one is not. If there is any lateral nonuniformity in the diffusion rate, as there would be with grain-boundary diffusion, material will become depleted from the fast-diffusing regions. This leads to the "Kirkendall voids" shown, which of course degrade the integrity of the structure.

When more than one compound appears on a binary phase diagram, the one that will form first at the interface between the two constituent elements can often be predicted using the concept of "effective" heat of formation [31]. First, it is assumed that upon initial interdiffusion of two elemental films A and B, the composition of the mixture will be the eutectic (minimum melting T, T_m) composition, A_xB_{1-x}, since diffusion rate is proportional to T/T_m [21]. Now, suppose that there exist two compounds, A_yB_{1-y} and A_zB_{1-z}, with heats of formation $\Delta_fH°_y$ and $\Delta_fH°_z$. The interdiffusing mixture cannot *completely* react into either compound, because the composition (x) is off-stoichiometric; so each $\Delta_fH°$ is reduced to an effective value, Δ_fH', determined by the fractional deficiency of one element. Thus, if y > x (A deficient),

$$\Delta_fH'_y = \left(\frac{x}{y}\right)\Delta_fH°_y$$

and if x > y (B deficient),

$$\Delta_fH'_y = \left(\frac{1-x}{1-y}\right)\Delta_fH°_y$$

The interfacial compound formed will be the one with the lower Δ_fH'.

5.6 Stress

Interfacial bonding of film layers to each other and to the substrate causes physical interaction as well as the chemical interaction that was discussed above. That is, the films and substrate can be held under a state of compressive or tensile stress by each other by transmitting forces across the interfaces. A film's stress affects its performance and also reveals information about the behavior of the deposition process. Stress varies widely with deposition conditions and with the physical properties of the film and substrate materials. After reviewing some basic physics of solids, we will examine first the effects of stress on film behavior and then the process factors that influence stress.

5.6.1 Physics

The basic physical behavior of solids is described by the stress-strain curve, Fig. 5.30. The force applied per unit of cross-sectional area is the stress, σ (N/m^2 or Pa), tensile being positive and compressive negative. Tensile stress along one direction, say x, causes the material to stretch along that direction by a fractional amount called the strain, $\varepsilon_x = \Delta x/x$. Up to the "yield point," the relationship is linear, and the material is said to be "elastic." The yield point is usually defined to be where the deviation from linearity reaches 0.2 percent. The slope in the elastic region is a measure of material stiffness and is called the elastic modulus or Young's modulus, Y; thus,

$$\sigma_x = Y\varepsilon_x \quad \text{or} \quad \varepsilon_x = \frac{\sigma_x}{Y} \qquad (5.50)$$

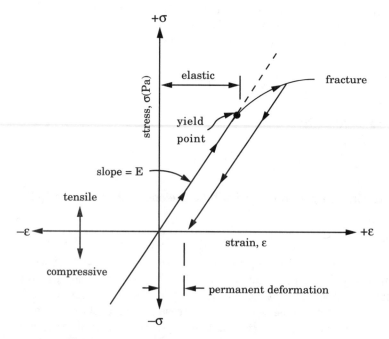

Figure 5.30 Characteristics of the stress-strain relationship.

This is Hooke's law in one dimension, and it is the same as the equation for a spring, F = kx. A typical value of Y is 10^2 GPa (10^{11} Pa) for hard materials. In the elastic region of the stress-strain curve, removing the stress returns the material to its original dimension. Applying stress higher than the yield point causes deformation which remains after removing the stress, as shown in Fig. 5.30. Finally, at the stress level called the "tensile strength," the material fractures. The curve shape is similar in the compressive quadrant. Tensile strengths range up to 1 GPa for hard materials, corresponding to a strain of about 1 percent.

The shape of the stress-strain curve for a given material depends on crystalline grain structure, T, and strain rate. When T is well below the melting point, T_m, there is usually a well defined elastic region, above which yield occurs by the "glide" of crystallographic dislocations (more in Sec. 6.6). At higher T, the material "creeps" by diffusion of atoms especially along grain boundaries, and the elastic region shrinks. For Pb, for example, it has been determined that creep becomes significant at $T/T_m > 0.4$, with T in K (Murakami, 1991). Then, the size of the elastic region depends on the *rate* of imposed stress change relative to the rate at which the stress can relax by creep.

"Uniaxial" tensile stress along x in a freestanding sample of material causes stretch along the x direction and also shrinkage along the unconstrained y and z directions, as shown in Fig. 5.31a; that is, ε_y and ε_z are negative. These three strains are related by Poisson's ratio, $\nu = -\varepsilon_y/\varepsilon_x = -\varepsilon_z/\varepsilon_x$. If there were no *volume* increase upon stressing, we would have $\varepsilon_y + \varepsilon_z + \varepsilon_x = 0$ and thus $\nu = 1/2$. This is the case for rubber, for example, and it is the upper limit of ν. However, for most materials, $\nu = 0.3 \pm 0.1$. Thin-film materials are always under *biaxial* stress, as shown in Fig. 5.31b, because they are being pulled on by the substrate in *two* dimensions. This changes the stress-strain relationship from that of Eq. (5.50). That is, σ_x causes a stretch along *x* which is σ_x/E, but σ_y causes a shrinkage along *x* which is $-\nu\sigma_y/E$, so the net strain is the sum of the two as illustrated by the following equation:

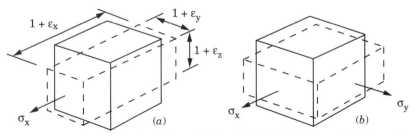

Figure 5.31 Strain resulting from (a) uniaxial and (b) biaxial stress applied to a cube of unit volume.

$$\varepsilon_x = \varepsilon_y = \varepsilon_{x,y} = \frac{(1-\nu)}{Y}\sigma_{x,y} = \frac{\sigma_{x,y}}{Y'} \qquad (5.51)$$

where Y' is sometimes known as the biaxial elastic modulus.

Thin films contain both intrinsic and extrinsic stress. Intrinsic stress is defined as that frozen in during deposition or during post-deposition treatment, and we will examine the causes of this shortly. Extrinsic stress is applied to the film by external forces, the most common one arising from differential thermal expansion with the substrate. Almost all materials expand upon heating, and the fractional linear expansion per unit ΔT is called the thermal expansion coefficient, α_T (units of K^{-1} or $°C^{-1}$). Usually, α_T decreases gradually with increasing T. For noncubic crystals or anisotropic microstructures, α_T can change with orientation as well. An extreme case is represented by materials with "layered" structures, which are discussed further at Figs. 5.7 and 6.22b. For example, pyrolytic BN has $\alpha_T = 37 \times 10^{-6}/K$ perpendicular to the layers (c direction) and $\alpha_T = 1.6 \times 10^{-6}/K$ parallel to them (a direction).

Typically, a film is deposited at elevated substrate T and then cooled to room T. If a film having a lower α_T than the substrate were *not* bonded to the substrate, the film after cool-down would be wider than the substrate by an amount δ as shown in Fig. 5.32a. The same situation would occur if the film contained intrinsic compressive stress, since this would make it want to expand, as indeed it would if it were not bonded to the substrate. However, when the film and substrate are bonded, they are constrained to the same lateral dimension. Thus, stresses develop so as to satisfy the force balance shown in Fig. 5.32b, where right-facing arrows indicate tensile σ and left-facing ones compressive σ. In the following, we will assume that $h_f \ll h_s \ll L$ as shown. For unit width in y, the force balance in the x direction, neglecting x subscripts and using Eq. (5.51), is

$$F_f = F_s \quad \text{or} \quad \sigma_f h_f = \sigma_s h_s \quad \text{or} \quad \left(\frac{Y}{1-\nu}\right)_f \varepsilon_f h_f = \left(\frac{Y}{1-\nu}\right)_s \varepsilon_s h_s \qquad (5.52)$$

Since $h_f \ll h_s$, it follows that $\varepsilon_f \gg \varepsilon_s$ (unless $Y_s \ll Y_f$). That is, essentially all of the strain appears in the film, and the film's lateral dimensions are determined entirely by those of the substrate. In a multilayer stack of films, the lateral dimensions of *all* of the films are determined by those of the substrate. It is sometimes stated in the literature that a "buffer" film was deposited underneath the desired film to reduce film stress resulting from thermal mismatch to the substrate. But this clearly cannot work unless the buffer (1) has $h_f \geq L$; (2) is capable of gross deformation without de-adhering, as is the case

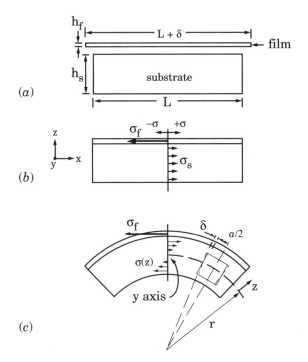

Figure 5.32 Stresses between film and substrate: (*a*) film unbonded to remove stress, (*b*) expansion component of substrate stress, and (*c*) bending component.

to some degree with indium; or (3) changes the intrinsic stress of the overlying film by affecting its grain structure.

The substrate strain is usually negligible compared to the film strain, ε_f, because $h_s \gg h_f$. Then, ε_f is given directly by the differential thermal expansion:

$$\int_{T_o}^{T_s} (\alpha_{Tf} - \alpha_{Ts})\, dT \approx (\bar{\alpha}_{Tf} - \bar{\alpha}_{Ts})(T_s - T_o) = \varepsilon_f = \left(\frac{1-\nu}{Y}\right)_f \sigma_f \quad (5.53)$$

where T_s is the deposition T, T_o is the T after cool-down, and the $\bar{\alpha}_T$ represents the average α_T over the T range. For films deposited at high T, the ε_f resulting from this thermal mismatch can approach the fracture point upon cool-down. For example, if the α_T difference is 10^{-5}/K and $T_s = 1300$ K, then $\varepsilon_f = 10^{-2}$. Note that Eq. (5.53) predicts only the *extrinsic* component of film stress. To estimate the *intrinsic* component, the total stress must be measured and the extrinsic component, calculated from Eq. (5.53), must be subtracted.

Total film stress is easily measured by the curvature it produces in the substrate. We neglected this curvature in Fig. 5.32b while examining only the *uniform* component of substrate stress, σ_s, that resulted from the force balance. Now, in Fig. 5.32c, we neglect σ_s while examining only the component of substrate stress that is due to bending. Bending results from the exertion of film force only on the front face of the substrate. This bending produces a torque or bending moment, M_f, about the y axis, which we have placed in the center of the substrate for convenience. For unit width in y,

$$M_f \, (\text{N·m}) = F_f z = \sigma_f h_f (h_s/2) \tag{5.54}$$

This moment must be balanced by the moment generated by substrate bending. The latter moment is illustrated by the stress arrows in Fig. 5.32c, where the bending stress, $\sigma(z)$, is seen to vary from tensile on the front face ($z = +h_s/2$) to compressive on the back ($z = -h_s/2$). Stress $\sigma(z)$ can be related to the substrate radius of curvature, r, by examining the trapezoidal distortion, which is illustrated to the right of the figure. In a rectangle of substrate material having variable half-height z and arbitrary half-width $a/2$, the expansion at the top edge due to the bending is δ, and the strain is $\varepsilon(z) = \delta/(a/2)$. This is the same construction as Fig. 5.24, and again we can write, by similar triangles, that $(a/2)/r = \delta/z$. Thus,

$$\varepsilon(z) = \frac{z}{r} = \left(\frac{1-\nu}{Y}\right)_s \sigma(z) \tag{5.55}$$

Integrating over the substrate thickness using this expression for $\sigma(z)$ gives the substrate bending moment:

$$M_s = \int_{-h_s/2}^{+h_s/2} \sigma(z) \cdot z \, dz = \left(\frac{Y}{1-\nu}\right)_s \cdot \frac{h_s^3}{12r} = \left(\frac{Y}{1-\nu}\right)_s I_z \cdot \frac{1}{r} \tag{5.56}$$

where I_z is the "moment of inertia" of the substrate for unit width in y. Setting this equal to M_f from Eq. (5.54) gives the desired result relating total film stress to substrate curvature:

$$\boxed{\sigma_f = -\left(\frac{Y}{1-\nu}\right)_s \frac{h_s^2}{6rh_f} = -\left(\frac{Y}{1-\nu}\right)_s \frac{Kh_s^2}{6h_f}} \tag{5.57}$$

where K (m^{-1}) is the curvature, with positive being convex on the film face and negative being concave. The same curvature occurs in the y dimension of the substrate due to bending moments along the x axis, so that the substrate is deformed into a dome. The dome may be approximated as spherical, and the curvature is independent of the size or shape of the substrate, $L_x \times L_y$, as long as $L_x, L_y \ll r$. Note that for a given σ_f, $K \propto 1/h_s^2$, so that thin substrates considerably increase sensitivity to σ_f measurement.

Very small curvatures down to 10^{-4} m^{-1} can be measured by the change in angle of reflection of a laser beam as it is scanned across the substrate. Here the beam is acting as an optical lever. Alternatively, optical interference fringes against a reference flat can be measured, in which case any variation in K over the substrate is also readily detected. The laser-beam technique can also monitor the buildup of intrinsic stress *during* deposition with the use of a ribbon-shaped substrate suspended from one end only as shown in Fig. 5.33. By the way, single-crystal Si wafers are ideal substrates to use for studying the effects of process conditions on intrinsic stress, because they are: (1) thin and well-controlled in h_s to maximize sensitivity; (2) very flat; (3) low-cost and often reusable by etching off the film; and (4) very well characterized in Y, ν, and α_T. $Y/(1 - \nu)$ is 181 GPa for Si(100) and 229 for Si(111), and α_T is 2.60×10^{-6}/K at 300 K (3.84 at 600 K and 4.38 at 1200 K). Most materials have higher α_T, although quartz and Invar (an Fe-Ni alloy) have α_T near zero.

We assumed above that σ_f is constant through the thickness of the film. In fact, σ_f can vary with z, but the curvature measurement gives only the average value across h_f. For example, we will see in Sec. 5.6.3 that intrinsic stress arises both from the growth process occurring near the surface and from any annealing which may be occurring throughout the bulk of the film during deposition at elevated T_s. Annealing will have more time to proceed during film deposition for that portion of the film deposited first. The annealing effect complicates the

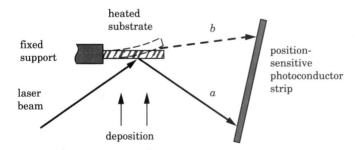

Figure 5.33 Technique for measuring intrinsic-stress build-up during deposition: (*a*) light beam before deposition and (*b*) after deposition of a compressive film.

determination of σ_f versus z using the Fig. 5.33 technique, because σ_f at a given z will continue to change after it has been measured, as deposition proceeds. Also, σ_f can vary with z whenever σ_f becomes so high that the film yields or creeps during deposition. This motion will always proceed in the direction of stress relaxation. Since it can happen more easily further from the constraining influence of a rigid substrate, it will cause σ_f to decrease with increasing z. Thus, σ_f can vary either way with z depending on conditions.

There is also *surface* stress, which can become a significant component of σ_f for very thin films [32]. Surface stress is the sum of the surface energy, γ, of Eq. (5.35) and an elastic-strain term, $d\gamma/d\varepsilon$. The latter arises from the fact that for the surface monolayer of crystalline solids, the equilibrium bond length parallel to the surface (Fig. 4.2) may be larger or smaller than that in the bulk due to charge rearrangement at the surface [33]. This means that the surface monolayer will want to expand or contract laterally, though it cannot do so because it is bonded to the bulk. The resulting strain is often *partially* relaxed by spontaneous atomic reconstruction of the surface (Sec. 6.5.3). Note that the energy stored in this strain is in *addition* to the surface energy arising from the creation of new surface by bond-breaking.

We also assumed above that $h_f \ll L$. However, when films are patterned into fine lines for integrated circuitry or are deposited over such lines, the line width becomes the relevant L, and one often has $h_f \to L$ as shown in Fig. 5.34. For example, a typical metal interconnect line on a memory chip is 1 µm wide and 1 µm thick. This is still another cause of σ_f variation with z. Recall that stress is maintained in a film by force transmitted across the interface from the constraining substrate. When $h_f \ll L$, the resulting stress in the film is parallel to the substrate as shown in Fig. 5.32b. However, the edge of a film can have no stress parallel to the substrate, because the edge is not

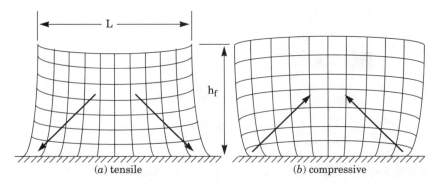

Figure 5.34 Qualitative cross-sectional strain distributions near the edges of patterned film lines that are under stress.

connected to anything. Thus, a tensile film relaxes inward as shown qualitatively in Fig. 5.34a, and also upward at the edges in accordance with Poisson's ratio. At lateral positions >h_f away from either edge, the edge effect has attenuated [34], and the film in this middle region behaves as in the $h_f \ll L$ case, with the lateral tensile stress distributed through its thickness and with a resulting compressive strain downward. The force $\sigma_f h_f$ in this middle region must be transmitted to the substrate at the edge of the film, as shown by the arrows. But at the edge, the stress is distributed only over the lower portion of the film thickness, so there is a stress *concentration* at this corner both in the tensile stress of the film and in the shear stress of the interface. When continuous films are deposited over steps in underlying layers, similar stress concentrations develop at the step corners. These concentrations can produce adherence failure.

5.6.2 Problems

Whether or not film stress causes problems in the application at hand depends on the circumstances and on the level of the stress. Actually, a small level of compressive stress can strengthen a film, because it reduces the chances of the film being put under sufficient tensile stress to cause fracturing in severe mechanical applications such as tool-bit coatings. Corrosion resistance is also improved by avoiding tensile stress. Small levels of strain of either sign can improve the properties of epitaxial structures in electronic applications, as we will see in Chap. 6. However, high stresses usually lead to problems. The upper limit to stress is the point of catastrophic failure, which is illustrated in cross section in Fig. 5.35. Tensile stress failure is characterized by cracking, which appears as a mosaic pattern when viewed from the top. The cracked film may then peel away from the substrate at the crack edges, where the stress is concentrated. Compressive stress failure is characterized by de-adherence and buckling, which from the top appears sometimes as domes or bubbles and sometimes as an undulating meander pattern looking like a mole tunnel.

At lower stress levels, other problems can still arise. The curvature induced in the substrate is unacceptable in applications where flatness is important, such as mirrors. When a film is stressed into plastic

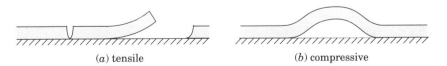

(a) tensile   (b) compressive

Figure 5.35 Catastrophic failure from film stress.

deformation, its structure degrades. In polycrystalline films, yield at grain boundaries weakens bonding there and thus aggravates the various grain-boundary problems that were discussed in Sec. 5.4.3. In epitaxial films, yield occurs by the generation and glide of dislocations within the crystal lattice (see Sec. 6.6); these defects are especially detrimental to electronic properties.

One particularly dramatic and difficult problem occurs upon compressively stressing polycrystalline films of soft metals beyond the yield point at $T/T_m > 0.4$ or so. The stress relaxes itself by transporting film material to the surface and growing "hillocks" of it there. The transport apparently occurs by diffusion of the film material along grain boundaries ("Coble creep"). Diffusion is known to be faster there than in the bulk of the grains, and in the case of Pb, the known value of the grain-boundary diffusion coefficient was found to agree with the rate of hillock growth [34]. Another clue is that hillocks tend to occur over grain boundaries. An unusually steep hillock in Al is illustrated in Fig. 5.36. Hillocks cause light scattering in optical applications and short circuits through overlying insulator films in electronic applications. The reverse problem of void formation also occurs in the same soft metals when they are stressed in *tension* beyond the yield point. Then, material is selectively transported out of certain regions of the film to relax the stress elsewhere.

The initial stress of a film after cool-down from the deposition T can often be kept below the elastic limit by minimizing deposition T and

Figure 5.36 Scanning electron micrograph of a 3 µm high hillock in a 0.7 µm thick Al film. (Source: Reprinted from Ref. 35 by permission.)

adjusting other deposition conditions, thus avoiding hillocks and other problems of the yield regime. However, when subsequent processing or use of the film involves thermal cycling, as in integrated-circuit manufacture, thermal mismatch to the substrate can increase stress beyond the yield point. Figure 5.37 is adapted from a thermal cycling study [36] between room T and 450° C of Al-1% Si and Al-2% Cu films sputter-deposited onto oxidized Si(100) at room T. It illustrates the typical hysteresis behavior that is observed in thermal cycling of films beyond their yield point. The exact shape of the curve will depend on the material, its thermal history, and the T ramp rate, but some or all of the regimes shown will be observed. This particular film contains intrinsic tensile stress after room-T deposition. Upon subsequent heating, the high α_T of the film relative to the substrate causes film stress, σ_f, to become compressive, and the slope in this elastic regime is determined by Eq. (5.53). The first decrease in slope with further T increase occurs in this film not because of yield but because of the onset of recrystallization and grain growth, which reduces the disorder frozen into the film during deposition. This amounts to a transition from a Z1 or ZT structure to a Z2 or Z3 structure (see Sec. 5.4). Since the crystallites have higher density than disordered material, these crystallization processes cause the film to want to contract. This produces a tensile stress component that partially cancels the thermal mismatch compression and thus reduces the slope. Further T increase causes the slope to reverse upon relaxation of the compressive stress due to yield. Then, at the start of the down ramp in T, elastic behavior is first observed, because the stress is now below the yield point even though the T is high. (Elastic behavior can also be observed beyond the yield point when the rate of strain change is much higher than the rate of relaxation by yield.) With further cooling, the film crosses into tension, and when the tensile stress becomes high enough, the film again yields, and the slope decreases. Finally, the slope increases back to-

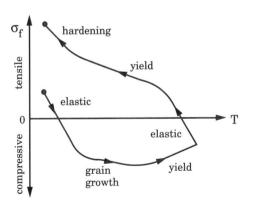

Figure 5.37 Typical hysteresis behavior during thermal cycling of metal films when α_T is higher for the film than for the substrate.

ward the elastic value near room T due to hardening of the film against yielding at the lower T. Information about how various materials yield as a function of strain and T is available in deformation-mechanism maps [37].

5.6.3 Intrinsic stress

Intrinsic stress is incorporated into the film during deposition or post-deposition treatment; that is, it is intrinsic to these processes. Here, we will focus on stress arising from the deposition process. Intrinsic stress is commonly observed and has at least three origins: chemistry, microstructure, and particle bombardment.

Chemical reactions occurring in the deposition process can produce stress whenever they continue to occur to some extent beneath the growth surface, where the film structure is beginning to become frozen. Reactions which add material to this structure produce compressive stress, and those which remove it produce tensile stress, as one would expect. For example, chemically-reactive metals such as Ti which are deposited in poor vacuums or with O_2 background gas deliberately added can develop compressive stress [38] due to oxidation proceeding beneath the surface. Conversely, plasma-deposited silicon nitride (SiN_xH_y) made using SiH_4 and NH_3 gas develops high tensile stress because the triaminosilane precursor radical, $Si(NH_2)_3$, continues to evolve NH_3 gas from beneath the growth surface as it chemically condenses toward Si_3N_4 [39] (more in Sec. 9.6.4.2). These chemical processes can also modify stress during post-deposition treatment.

The microstructure of the film and its evolution with time beneath the growth surface can produce tensile stress. In terms of the zone structure discussed in Sec. 5.4, films that are well into Z1 have little stress, because stress cannot be supported across the microvoids which separate the columns of material. However, as the film moves toward the dense ZT or Z2 structures, the microvoids collapse enough to allow atomic attraction across them. Then, tensile-strained bonds develop, and the resulting tensile stress cannot relax if the material is within its elastic limit. Additional tensile stress can develop when recrystallization of disordered material or grain growth is proceeding beneath the surface of ZT or amorphous films, due to densification as mentioned in the discussion of Fig. 5.37. At higher T/T_m well into Z2 or Z3, yield occurs more easily and partially or completely relaxes these microstructural tensile stresses. Both densification and yield can further occur during post-deposition annealing.

Bombardment of the film surface by ions or energetic neutrals can produce compressive stress both by implanting these particles into the film and by momentum transfer to surface atoms. Momentum transfer forces the surface atoms into closer proximity to each other than their relaxed bond lengths, and they become frozen in this compressed state when T is low. This is similar to the shot-peening and ball-peen-hammering processes which are used to compressively stress the surfaces of bulk metals, and it has thus been termed "ion peening" [38]. The energetic bombardment available in energy-enhanced film-deposition processes is a very effective way to counteract the tensile stress which arises from chemical or microstructural effects. In some cases, it can be controlled to just neutralize the stress.

Figure 5.38 shows the general behavior of film stress, σ_f, with process pressure, p, in sputter deposition (Thornton, 1986). Some effects of p on microstructure were discussed in Sec. 5.4.1. The transition from Z1 to ZT with decreasing p in sputtering is due both to a decrease in the spread of incident angle of depositing particles and to an increase in particle kinetic energy. This transition causes tensile stress to rise as the microvoids collapse. The height of the maximum tensile stress correlates with decreasing deposition/melting T, T_s/T_m, for many metals (Thornton, 1989), presumably because the stress is less able to anneal itself out at lower T_s/T_m. At still lower p, stress drops again due to compaction by energetic bombardment, and for some materials it becomes compressive. Compressive stress can also be obtained by using a negative bias on the substrate to increase ion-bombardment energy. Ion bombardment has the same effect in other energy-enhanced processes, and is discussed in a general context in Sec. 8.5.3.

Despite the above discussion, the origin of intrinsic stress in a given deposition situation is frequently not known, and much work remains to be done in this area to understand and control this stress.

5.7 Adhesion

Loss of film adhesion requires both high stress and weak bonding at the interface to the adjoining layer or substrate. Then, the interfacial

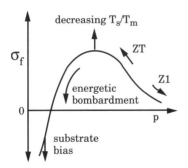

Figure 5.38 Behavior of film stress with sputter-deposition conditions.

bond will fail by either peeling or buckling as shown in Fig. 5.35, depending on the sign of the film stress. The stresses contributing to de-adherence include not only the intrinsic and thermal-mismatch stresses, but also the stress applied to the film in its application. For example, coatings on cutting tools are subjected to high local stress at the point of contact with the material being cut. Coatings on high-power laser optics are subjected to high thermal-mismatch stress during the laser pulse ("thermal shock"). Wire bonding to metal films on integrated-circuit chips generates high stress whether the thermocompression or the ultrasonic bonding method is used. In such applications, strong interfacial bonding is crucial. Conversely, if external stresses are not present in the intended application and if the initial film stress is low, weak interfacial bonding may not result in de-adherence. Also, since the shear force at the interface is proportional to film thickness in accordance with Eq. (5.52), thicker films fail more easily under a given stress than do thinner ones. Gas evolution from films can also cause de-adherence if the gas becomes trapped at the interface so that it builds up pressure there. Sputter-deposited films often contain a percent or two of Ar, and plasma-CVD films can contain tens of percent of H.

Sometimes, one *wants* to remove the film from the substrate to produce a free-standing membrane for use as, say, an x-ray window or a target in a high-energy-physics experiment, in which case interfacial bonding needs to be minimized. But more commonly, good adherence is desired. We have already examined the factors which affect interfacial stress. Here, we examine first the factors that inhibit bonding and then the process remedies used to promote bonding.

Since chemical bonding forces extend only a few tenths of nm, only one monolayer of poorly-bonded contaminant can be sufficient to prevent bonding of the depositing film material to the substrate. We saw in Sec. 3.4.2 that ordinary surfaces tend to be contaminated with water, oil, and salts. Most of this contamination can be removed by solvent degreasing followed by rinsing in deionized water, but the last monolayer or two usually remains. Even if chemical etching [40] is employed as a final step, upon exposure to the air most surfaces will readsorb a monolayer or two of water, organic vapors, and CO_2. These species will physisorb on any surface even at partial pressures well below their saturation vapor pressures whenever their bonding is stronger to that surface than to their own condensed phase, as we discussed in Sec. 5.1. In special cases such as H on Si and on GaAs (Sec. 6.3), surfaces can be chemically passivated against contaminant adsorption, but then one has a monolayer of adsorbed passivant to deal with.

The adsorbed contaminants end up sandwiched between the film material and the substrate and can block chemical bonding between the two materials. Thus, unless both the substrate and the film bond *chemically* to the adsorbate, the interfacial bonding remains weak. In some cases, chemical bonding does occur. For example, metals that form strongly bonded surface oxides can be deposited one upon the other with good adherence because the interfacial oxide forms a strong bridge of chemical bonding between the metals. Indeed, film adhesion correlates well with the free energy of formation of the film metal's oxide [41]. The metals Ti, Zr, Cr, and Al have particularly strong oxides, while Zn, Cu, and the noble metals have weak ones. Adsorbed water, organics, and CO_2 also are likely to form strong oxide and carbide bonds at the surfaces of reactive metals.

Even if the adsorbates are removed in the deposition chamber so that deposition is carried out on an atomically-clean surface, interfacial bonding will be weak if the film and substrate materials have such different bonding character (covalent, ionic, or metallic) that they do not easily bond to each other. This was discussed in Sec. 5.3.1 following Eq. (5.36) on the wetting criterion. There, the poor adherence of Au to SiO_2 was improved by inserting 10 nm or so of the "glue" layer Ti, so named because it forms both strong metallic bonds to the Au and strong covalent bonds to the oxide. Conversely, even an active metal like Ti can be peeled easily from the ionically-bonded material CaF_2. A film of CaF_2 is therefore useful as a "parting layer" in making freestanding metal membranes.

There are several ways to remove physisorbed molecules once the substrate is in the deposition chamber where it will not become recontaminated. Heating either desorbs them or activates their chemisorption, per Eq. (5.5). In the chemisorbed state, they may no longer inhibit interfacial bonding. The progress of desorption can be followed by the accompanying pressure burst or with a mass spectrometer (Sec. 3.5). Alternatively, exposure to H_2 plasma reduces many oxides and removes them as H_2O, whereas O_2 plasma oxidizes organics and removes them as CO_2 and H_2O. Bombardment with ions of >100 eV or so from a plasma or from an ion gun can remove any surface species by sputter erosion. However, a small fraction of this material is sputtered forward instead and is thus embedded beneath the surface—the so-called "knock-on" effect. Plasma operation is discussed in Chap. 9.

Energy input to the substrate surface from ions, electron beams, or UV light can desorb contaminants and can also break bonds within the surface, thereby activating the surface toward bonding to the film material. Irradiation *after* film deposition can also improve adherence as long as the film is thin enough so that the radiation penetrates to the interface [42, 43]. In energy-enhanced deposition processes such as sputtering, activation energy for interfacial bonding is carried by the depositing material itself, and this results in consistently better adhesion than is achieved using thermal-evaporative deposition of the same material.

More severe measures can be taken to further improve adherence if interface abruptness is not important. Mechanical or chemical roughening of the substrate improves adherence by increasing the bondable surface area and also by mechanically interlocking the materials on a microscopic scale. Deliberately grading the interface composition improves adherence in several possible ways: dispersion of interfacial contaminants, increase in number of bonds between the two materials, and inhibition of fracture propagation along the interface. Generally, it is not clear which mechanism dominates. Gradation can be achieved by thermal interdiffusion (Fig. 5.27b) or by "ion mixing" (Sec. 8.5.3). However, interdiffusion can *weaken* the interface if voids develop (Fig. 5.27e). Ion mixing occurs when ion bombardment in the keV range is present at the start of deposition to cause significant knock-on mixing of the film material into the substrate. When the ions are of film material itself, they mix also by their own shallow implantation. These effects typically extend over a range of a few nm. Between successive film layers, gradation can alternatively be achieved by gradually switching from the deposition of one material to the other.

One can see that the various adherence remedies each have their own problems and constraints. One should first establish that there is an adherence problem for the application at hand before taking remedial measures. The choice of remedy will usually be determined by constraints imposed by the materials involved, the interfacial structure desired, and the equipment available. Sometimes, changing the deposition conditions to reduce film stress is a better solution.

5.8 Temperature Control

It is clear by now that substrate temperature, T_s, is a very important variable in the deposition process, having profound effects on the structure and composition of films and interfaces. Since T is such a common quantity, its measurement is often treated casually. Unfortunately, T_s measurement is very difficult in the vacuum or partial-vacuum environment of most thin-film deposition processes. This is because thermal coupling to the substrate in vacuum is poor; yet one usually does not want to attach a T sensor directly to the substrate, because the substrate would become contaminated and/or needs to be moved during the process. There are noncontact techniques for measuring T, but they have shortcomings which we will examine later.

Sometimes substrates are suspended in the deposition chamber—especially when they are irregularly shaped. In this case, T control is particularly difficult unless the entire chamber is heated to T_s so that the substrate is immersed in an isothermal enclosure. Flat substrates such as glass plates or Si wafers can be placed on a heated platform. The platform T, T_h, can be measured accurately and can be controlled closely by the feedback techniques discussed in Sec. 4.5.3. It is often assumed in the literature that $T_s = T_h$ in such a situation. The fallacy of this assumption is made clear by examining the substrate-platform interface on a microscopic scale, which is done schematically in Fig. 5.39. Because most surfaces are not atomically flat, intimate atomic contact occurs only at a few points, and these points add up to a negligibly small fraction of the macroscopic interface area. Even if the substrate is clamped around the periphery to a platform which is slightly domed so that the clamping force is distributed over the whole area, the contact area will still be negligible unless one of the two surfaces is so soft that it deforms and thus conforms to the other surface. Conformable thermal-contact materials can be used at the interface, but then contamination is a concern. Vacuum grease is one such material, but its vapor pressure rises steeply with T. Ga-In eutectic alloy is liquid at room T and still has low vapor pressure at high T (see Appendix B), but it requires chemical etching to remove, and Ga also alloys with most metals. Pure In is an alternative which melts at 156° C, and there are many refractory metals with which it does not alloy [26]. However, in the majority of cases when conformable or liquid contacting layers cannot be used, contact area between substrate and platform will be negligible.

The good heat conduction provided by the solid phase can only occur where there is *atomic* contact to transfer the heat by phonon vibrations and, in the case of metals, by electrons. Therefore, for the geometry of Fig. 5.39 where atomic contact area is negligible, heat transfer under vacuum can occur only by radiation, which results in a substantial difference between T_h and T_s. When pressure is higher, gas-phase

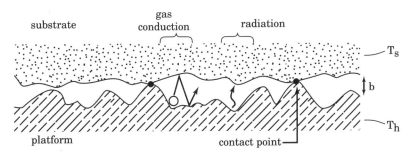

Figure 5.39 Microscopic schematic of the typical interface between a substrate and its heated platform.

heat conduction occurs in parallel to the radiation. In the following subsections, we will analyze these two mechanisms and then discuss T measurement techniques. Usually, T_s is lower than T_h because of heat loss from the substrate to the surroundings. However, if sufficient energy is arriving at the film surface during deposition, T_s will increase with deposition time and can actually become higher than T_h. This effect is especially important in energy-enhanced processes; but even in thermal evaporation, radiation is arriving from the hot evaporation source. In sputter deposition, energy is arriving both as kinetic energy of the depositing atoms and as ion bombardment.

5.8.1 Radiation

Heat transfer can occur in a vacuum only by radiation. We will analyze this process in some detail here for common substrate-heating situations in order to illustrate the large T differences which can arise and how to predict and minimize them.

The maximum amount of radiation flux which can be emitted by a surface, Φ_b, is given by the Stefan-Boltzmann blackbody radiation law [Eq. (4.19)], which is also plotted in Fig. 5.40. Real surfaces emit a fraction ε_λ of that amount, where ε_λ is the emissivity and is a function of the radiation wavelength, λ. Real surfaces also absorb only a fraction α_λ of the radiation incident on them. To allow analytical solutions to heat-transfer problems, one must make the "gray-body" assumption of constant ε_λ and α_λ over the λ involved; that is, $\varepsilon_\lambda = \varepsilon$ and $\alpha_\lambda = \alpha$. Since this is not always realistic, and since ε and α usually are not known accurately anyway, radiative-heat-transfer calculations of T_s are best backed up by calibration under actual deposition-process conditions. Nevertheless, the analytical solutions give important insight into radiative behavior as well as useful estimates of T_s. The gray-body assumption also means that $\varepsilon = \alpha$ for a given surface, a thermodynamic fact known as Kirchoff's law.

The problem with the gray-body assumption is illustrated using Fig. 5.41, which shows how *spectral* blackbody radiation, $\Phi_{b\lambda}$, is distributed in λ at 670 K. (total Φ_b is the integral of $\Phi_{b\lambda}$ over λ.) When a quartz substrate is being heated by a platform at this T_h, it absorbs the longer-λ portion of the distribution ($\alpha_\lambda \to 1$) by lattice-vibration (phonon) coupling, but it is transparent to the shorter-λ portion ($\alpha_\lambda \to 0$). If T_h is raised, the peak of the distribution shifts to shorter λ in accordance with Wien's displacement law,

$$\lambda_{peak}\ (\mu m) = 2898/T\ (K) \tag{5.58}$$

This causes the λ-averaged α of quartz and other visibly transparent or white (high-band-gap) materials to decrease with increasing T_h.

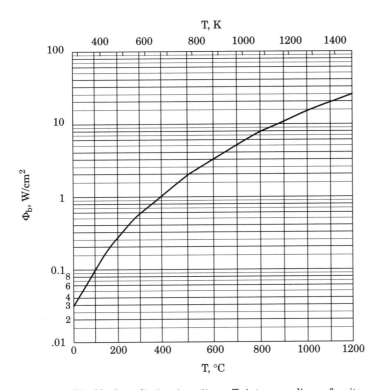

Figure 5.40 Blackbody radiation ($\varepsilon = 1$) vs. T, into a medium of unity refractive index.

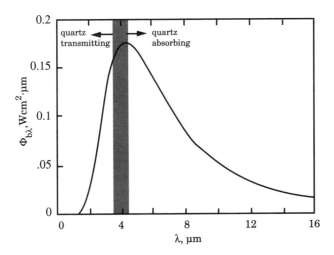

Figure 5.41 Planck's-law emissive-power function for blackbody radiation at 670 K.

The opposite situation occurs for semiconductors, whose band gaps, E_g(eV), are low enough that the fundamental absorption edge, given by λ_e (μm) ≈ $1.24/E_g$, begins to overlap the radiation λ range and cause absorption at short λ. Semiconductors are transparent only between λ_e and the much longer λ of lattice absorption, except at high T_s where thermally generated charge carriers can absorb radiation within this window. The same factors cause ε to change with λ. For example, ceramics have low ε in the visible and high ε in the infrared. The ε of metals is less dependent on λ but increases dramatically with surface oxidation, typically from 0.1 to 0.7.

To illustrate the effect of ε and α on radiative heat transfer, we will examine a few simple cases. Here, the substrate will be taken to be much wider than the heat-transfer gaps involved, so the heat flow is one-dimensional. Also, we will use ε for α, since ε = α for "gray" bodies. Consider the geometry of Fig. 5.42, where the substrate is receiving heat on face 1 and losing it to cold surroundings on face 2. A T difference, $\Delta T = T_h - T_s$, develops because the substrate is losing heat from both faces (1 and 2) but receiving it only on face 1. If, instead, the substrate were enclosed by the heater on both faces, thermodynamics would require T_s to equilibrate at T_h. Let Φ_i' denote that fraction of the radiation flux emitted from the ith surface that is not reflected back to it and readsorbed; that is, Φ_i' is the net emission. If we also

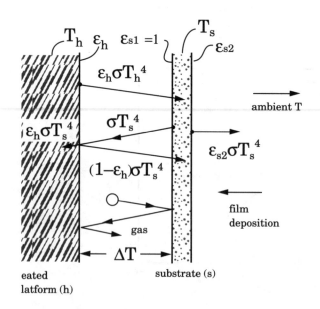

Figure 5.42 Geometry for radiative and gas-conductive heating of a substrate.

consider a gas-conductive heat flux, Φ_c, and a heat flux from the deposition process, Φ_d, we have the following heat-energy balance for the substrate in steady state:

$$\underbrace{\Phi_h' + \Phi_c + \Phi_d}_{\text{(input)}} = \underbrace{\Phi'_{s1} + \Phi'_{s2}}_{\text{(output)}} \qquad (5.59)$$

First, we will deal with vacuum, where there is no gas conduction. We will also neglect Φ_d, and we will let ε_h and ε_{s2} be < 1 and $\varepsilon_{s1} = 1$. (If all the ε values are < 1, calculations get more complicated.) Given that $\varepsilon_{s1} = \alpha_{s1} = 1$, all of the radiation emitted from surface h is absorbed by s1, and none is reflected back to h, so the net emission is just $\Phi_h' = \varepsilon_h \sigma T_h^4$ [see Eq. (4.20)]. However, part of the radiation emitted from s1 *is* reflected from h and readsorbed by s1, so Φ'_{s1} is the difference between these two fluxes:

$$\sigma T_s^4 - (1 - \varepsilon_h) \sigma T_s^4 = \varepsilon_h \sigma T_s^4$$

recalling that $(1 - \alpha_h) = (1 - \varepsilon_h)$. Thus, the heat balance of Eq. (5.59) becomes

$$\varepsilon_h \sigma T_h^4 = \varepsilon_h \sigma T_s^4 + \varepsilon_{s2} \sigma T_s^4$$

or
$$\Delta T = T_h - T_s = \left[1 - \left(\frac{\varepsilon_h}{\varepsilon_{s2} + \varepsilon_h}\right)^{1/4}\right] T_h \qquad (5.60)$$

Note here that we have assumed $T_{s2} = T_{s1}$, which is quite reasonable since heat transfer in solids is much faster than in gases or by radiation. In the limiting case where $\varepsilon_{s2} \to 0$ in Eq. (5.60), we find that $T_s \to T_h$, because the substrate is losing heat only from face 1, face 2 having been declared a perfect insulator. In the opposite limit, where $\varepsilon_{s2} = \varepsilon_{s1} = \varepsilon_s = 1$, ΔT is a function of ε_h, which is plotted as curve *a* in Fig. 5.43. There, ΔT is minimized by maximizing ε_h, as one would expect. Heater surfaces composed of ceramic have $\varepsilon_h \approx 1$, and heavily oxidized metals have $\varepsilon_h = 0.7 \pm 0.2$, while shiny metals have much lower ε_h of 0.1 ± 0.1. The minimum ΔT of curve *a*, for $\varepsilon_h = 1$, is $0.16 T_h$. For a T_h of 600° C, this ΔT would be 140°, a substantial difference indeed! It can be shown that ΔT is also $0.16 T_h$ for $\varepsilon_h = 1$ and any ε_s (see Exercise 5.14). For substrate heating from the back by radiation only, as in Fig. 5.42, ΔT can be reduced further only by reducing ε_{s2}, but of course ε_{s2} is fixed by the film being deposited. In fact, a difference in

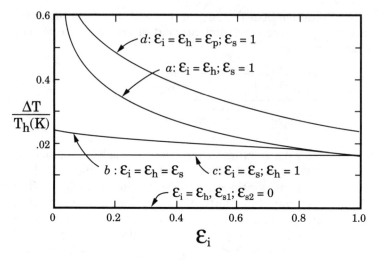

Figure 5.43 Behavior of ΔT with various ε values. Curves a, b, and c are for the geometry of Fig. 5.42, and solutions given are as follows: curve a, Eq. (5.60); curve b, Eq. (5.63); curve c, Exercise 5.15. Curve d is for the geometry of Fig. 5.45 and Eq. (5.66). Note that $\Delta T = 0$ whenever $\varepsilon_{s2} = 0$.

ε_{s2} from substrate to film will cause T_s to begin rising or falling toward a new steady-state value once deposition begins.

Consider next the case where all the ε terms are <1, and let them all be equal for simplicity. This leads to the multiple reflections shown in Fig. 5.44 between facing surfaces h and s1. Surface h emits a fraction ε of the blackbody radiation, and $(1 - \alpha)$ [$= (1 - \varepsilon)$] of this fraction is reflected from s1. A fraction α $(= \varepsilon)$ of the reflection is reabsorbed by h, and this is now a fraction $\varepsilon^2(1 - \varepsilon)$ of the blackbody radiation. Following this progression, we see that the sum of the reabsorptions is an infinite series having a closed-form solution:

$$\varepsilon^2 (1-\varepsilon) \sum_{n=0}^{\infty} (1-\varepsilon)^{2n} = \frac{\varepsilon^2 (1-\varepsilon)}{\left[1 - (1-\varepsilon)^2\right]} = f(\varepsilon) \tag{5.61}$$

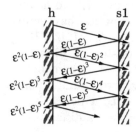

Figure 5.44 Multiple reflections between surfaces.

Thus, the net emission from h (and also from s1) is

$$\Phi_i' = [\varepsilon - f(\varepsilon)]\sigma T_i^4 = \left(\frac{\varepsilon}{2-\varepsilon}\right)\sigma T_i^4 \qquad (5.62)$$

and the heat balance for the substrate in Fig. 5.42 becomes

$$\left(\frac{\varepsilon}{2-\varepsilon}\right)\sigma T_h^4 = \left(\frac{\varepsilon}{2-\varepsilon}\right)\sigma T_s^4 + \varepsilon\sigma T_s^4 \quad \text{or} \quad \Delta T = \left[1-\left(\frac{1}{3-\varepsilon}\right)^{1/4}\right]T_h \qquad (5.63)$$

This function is plotted as curve *b* in Fig. 5.43.

Another common substrate-heating situation is illustrated in Fig. 5.45, where the substrate is held by a pallet or carrier plate that is placed in front of the heater or transported past it during film deposition. This introduces two more surfaces, p1 and p2, between the heater and the substrate, which further increases the ΔT between them. For the case where $\varepsilon_h = \varepsilon_{p1} = \varepsilon_{p2} = \varepsilon$ and $\varepsilon_{s1} = \varepsilon_{s2} = 1$, the heat balance for the pallet is essentially that of Eq. (5.62) with a second input term added for back-radiation from the substrate:

$$\left(\frac{\varepsilon}{2-\varepsilon}\right)\sigma T_h^4 + \varepsilon\sigma T_s^4 = \left(\frac{\varepsilon}{2-\varepsilon}\right)\sigma T_p^4 + \varepsilon\sigma T_p^4 \qquad (5.64)$$

The heat balance for the substrate is equivalent to that of Eq. (5.60):

$$\varepsilon\sigma T_p^4 = (1+\varepsilon)\sigma T_s^4 \qquad (5.65)$$

Combining these two heat balances to eliminate T_p yields the remarkably simple result that

$$\Delta T = \left[1-\left(\frac{\varepsilon}{3}\right)^{1/4}\right]T_h \qquad (5.66)$$

which is plotted as curve *d* in Fig. 5.43. To minimize ΔT, it is important that ε_h, ε_{p1}, and ε_{p2} all be large. The geometry of Fig. 5.45 also applies

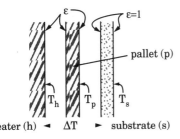

Figure 5.45 Geometry for radiative heating of a substrate held on a pallet.

to the situation where the substrate is directly facing the heater (no pallet) but is covered by a shutter on the deposition face to block deposition during vapor-source start-up. In this case, the pallet of Fig. 5.45 would become the substrate, and the substrate would become the shutter. Clearly, the substrate will be hotter when the shutter is in place, due to radiation and reflection from the shutter. When the shutter is swung away to commence deposition, the substrate will begin to re-equilibrate at a lower T (see Exercise 5.16).

Because a large ΔT always develops between the heater and the substrate whenever radiation is the only mode of heat transfer, it is best to monitor T_s directly during deposition. When this is not possible, T_s can be calibrated against T_h by using a "dummy" substrate that has a T sensor firmly attached. However, the accuracy of such a calibration will be limited by variation in the ε values with time or by variation in the heat-balance conditions such as the removal of a shutter or the input of heat from the deposition process.

5.8.2 Gas conduction

When the pressure during film deposition is more than a few Pa, gas-conductive heat transfer begins to provide a significant parallel path to radiation for thermal coupling between the heater platform and a substrate placed on it, as illustrated in Fig. 5.39. That is, the Φ_c term in the basic heat balance [Eq. (5.59)] becomes significant. From the kinetic theory of gas heat conduction (Sec. 2.8.3), we know that

$$\Phi_c = h_c \Delta T \qquad (5.67)$$

The gas heat transfer coefficient, h_c, is given by Eq. (2.32) when the gas mean free path, l, is larger that the gap, b, between the two surfaces. Then, h_c is independent of b, because the molecules are simply bouncing back and forth between the surfaces. On the other hand, when $l \ll b$, h_c is replaced by K_T/b in accordance with Eq. (2.31), where K_T is the bulk thermal conductivity of the gas.

Using Eq. (5.67) and the geometry of Fig. 5.42, and assuming for simplicity that ε_h is unity and $\varepsilon_{s1} = \varepsilon_{s2}$, we have the following heat balance on the substrate:

$$\varepsilon_s \sigma T_h^4 + h_c \Delta T = 2\varepsilon_s \sigma T_s^4 \quad \text{or} \quad \varepsilon_s \sigma \left(T_h^4 - 2T_s^4 \right) = -h_c \Delta T \qquad (5.68)$$

This can be simplified by factoring the T^4 polynomial as was done in Eq. (4.27), and by taking all the resulting T^3 terms as T_h^3 terms, which is reasonable when $\Delta T \ll T_h$. Then we have

$$\left(T_h^4 - 2T_s^4\right) \approx \left(1 + 2^{1/4} + 2^{1/2} + 2^{3/4}\right)T_h^3\left(T_h - 2^{1/4}T_s\right)$$

$$= T_h^3(6.3\Delta T - T_h) \tag{5.69}$$

or, upon inserting into Eq. (5.68),

$$\Delta T = \frac{\varepsilon_s \sigma T_h^4}{6.3\varepsilon_s \sigma T_h^3 + h_c} \quad \text{or} \quad \frac{1}{\Delta T} = \frac{6.3}{T_h} + \frac{h_c}{\varepsilon_s \sigma T_h^4} \tag{5.70}$$

Note that for $h_c \to 0$ (radiation only), $\Delta T/T_h \to 0.16$, in agreement with Fig. 5.43 for $\varepsilon_h = 1$. Conversely, for $h_c \to \infty$, $\Delta T \to 0$. For intermediate h_c, if $\varepsilon_h < 1$, the 6.3 factor in Eq. (5.70) will change, but as long as $\Delta T \ll T_h$, the linear form will remain; that is, $1/\Delta T = A + Bh_c$, where A and B are functions only of T_h and the ε values. Since by Eq. (2.32), $h_c \propto \gamma' c_v p/(\sqrt{MT})$, we can also write, assuming $T \approx T_h$, that

$$\boxed{\frac{1}{\Delta T} = A + B\frac{\gamma' c_v p}{\sqrt{MT_h}}} \tag{5.71}$$

The above formula is very useful for simplifying T_s calibration using a "dummy" substrate, which would typically have a thermocouple glued to the surface as discussed in the next subsection. When a deposition process is to be run at various conditions of p and gas composition, ΔT can vary significantly; but rather than calibrating for all process conditions, the following procedure can be used. For each T_h, the constants A and B are determined by making two measurements of T_s: one in vacuum and one at some typical process gas pressure. Then, interpolation can be used to determine ΔT for other gas compositions and pressures. If c_v is not known for a particular gas, it can be estimated by the procedures in Sec. 2.4. The thermal-accommodation factor, γ', can be taken as unity except for He, as discussed in Sec. 2.8.3. Figure 5.46 shows the Eq. (5.71) correlation for a Si substrate sitting on a heavily-oxidized metal heater at $T_h = 730$ K, using various gases. The data for the various gases fall on a straight line extrapolating to the measured ΔT under vacuum, as expected, and the effect of gas conduction on reducing ΔT is seen to be substantial at 140 Pa. The vacuum ΔT measured here corresponds to $\Delta T/T_h = 0.17$, meaning that $\varepsilon_h \approx 0.9$ from Fig. 5.43, about what is expected for a heavily oxidized metal. The He data have a slope of 0.2 times that of the other gases, which means that $\gamma' = 0.2$ for He on these particular surfaces. Al-

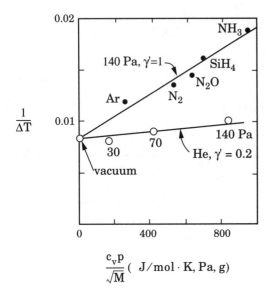

Figure 5.46 Gas-conduction correlation for $T_h = 730$ K [44].

though He is often thought of as a good heat-transfer gas because of its high atomic speed, Fig. 5.46 shows that it is actually poor because of the low γ'. Conversely, NH_3 is quite good, because its molecular weight, M, is reasonably low, and yet it has high c_v from its many molecular degrees of freedom. For process gases other than He, where $\gamma' \approx 1$, one ΔT measurement under vacuum and one at a typical process pressure are sufficient to construct the Fig. 5.46 plot from which ΔT at other process gas pressures and compositions can be interpolated.

The above discussion has assumed gas conduction with Kn > 1. However, when the gap, b, is larger or pressure, p, is higher so that Kn << 1, h_c must be replaced by K_1/b, so that the gas conduction contribution to T_s now decreases with increasing b, does *not* increase with increasing p, and does not involve γ' (see Fig. 2.9). At p approaching one atmosphere, however, "convective" gas heat transfer is added for b > 1 cm or so, due to gas circulation in T gradients (more in Sec. 7.2.3).

We have mentioned above several instances of a change in heat flux to the substrate occurring at the start of film deposition, including shutter opening, emissivity change due to the presence of the film, or process heat input. Heat flux also changes if T_h is changed between layers of a multilayer film structure. These flux changes, which we will call $\Delta\Phi$, cause a transient response in substrate T from its initial value, T_s, to a new steady-state value, T_s', as illustrated in Fig. 5.47. A similar transient response was analyzed in Sec. 4.5.3 for an evaporation crucible subjected to a change in set-point T. The heat balance given there, Eq. (4.26), can be rewritten for the present case as

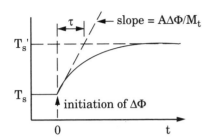

Figure 5.47 Transient response of substrate T to a step change, $\Delta\Phi$, in heat flux at time t = 0.

$$A\Delta\Phi = m_T\left(\frac{dT_s}{dt}\right)_o = c_{gs}m_s\left(\frac{dT_s}{dt}\right)_o \quad (5.72)$$

where A = substrate area
  m_T = thermal load (J/K)
  c_{gs} = heat capacity of the substrate, J/g·K

and where subscript o denotes the initial T_s slope. (When the transient of a substrate plus its carrier plate is being considered, the m_T of the carrier plate must be included too.) If the T_s slope is measured using a substrate having a T sensor attached, $\Delta\Phi$ can be found from Eq. (5.72). This procedure is especially useful for determining process heat input, Φ_d, as has been done for plasmas [45]. Here, the substrate itself is being used as a calorimeter. Equation (5.72) can also be used to estimate the time constant, τ, for T_s restabilization if $\Delta\Phi$ and the T change, $T_s' - T_s$, can be calculated from known ε values and h_c. Then, assuming exponential decay behavior,

$$\tau = \frac{T_s' - T_s}{(dT_s/dt)_o} \quad (5.73)$$

as shown in Fig. 5.47.

5.8.3 Measurement

Temperature (T) can be measured using any material property that changes with T, as most properties do. We will focus here on a few of the more common techniques. Because of the large ΔT associated with radiative heat transfer, T measurements at low pressure (where gas conduction is poor) are preferably made with the T sensor in solid contact with the object being measured. Alternatively, the emitted radiation itself can be measured with an optical or infrared pyrometer. We will first discuss contact sensors, the most common one being the thermocouple.

Thermocouples are simple devices, but their application involves several pitfalls which are easily overlooked, so we will discuss them in some detail. The device is based on the existence of an electrical potential difference, $\Delta\phi$, between any two metals in contact with each other. The chemical potential, μ, of an electron in a metal is at the Fermi level, which is lower than that of an electron in vacuum by an amount called the work function, ϕ_w. Metals differ in ϕ_w, and therefore they differ in electrical potential by $\Delta\phi = \Delta\phi_w$. Consider the thermocouple measurement circuit shown in Fig. 5.48a, which involves the common thermocouple-alloy pair Chromel and Alumel (labeled Ch and Al) plus the Cu wiring and terminals of the millivoltmeter. There are two Ch/Al junctions, a "hot" one at T_h and a "reference" one at T_r. However, *every* junction between dissimilar metals will involve a $\Delta\phi$, including the Ch/Cu junctions at T_1 and T_2. Now if the whole circuit is at the same T, then $\Delta\phi_h = -\Delta\phi_r$, since these are identical junctions in reverse; and similarly, $\Delta\phi_1 = -\Delta\phi_2$. Thus, the potentials around the circuit add up to zero, as they must to avoid generating energy out of nothing. But $\Delta\phi$ is a function of T, so if T_h is higher than T_r, a potential of $\Delta V = |\Delta\phi_h| - |\Delta\phi_r|$ is read on the meter (shown as mV). Similarly,

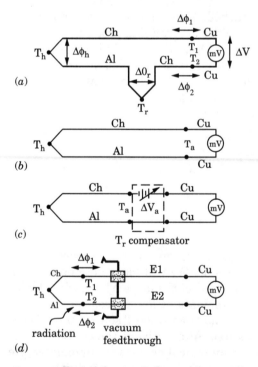

Figure 5.48 Typical thermocouple measurement circuits.

if $T_1 \neq T_2$, there will be an additional but unwanted ΔV, so these two junctions should be mounted on the same solid block.

Thermocouple-alloy pairs are formulated to both maximize and linearize their T response. For Ch/Al, $d(\Delta V)/dT \approx 0.04$ mV/K. The choice of alloy depends on process conditions. Ch/Al is resistant to oxidation. Pt/(Pt-Rh) is even more chemically resistant but is expensive. (W-Re)/(W-Re) withstands the highest T. Here, Re is alloyed into both legs to reduce brittleness, and in unequal amounts to produce a $\Delta \phi_w$.

For accurate measurements, T_r must be known, because $\Delta V \propto T_h - T_r$. Since it used to be common to use a melting ice bath for T_r, many thermocouple calibration tables are calculated for $T_r = 0°$ C. If the reference junction is omitted as in Fig. 5.48b, the *effective* reference junction becomes the terminals of the meter, since a Ch/Cu junction and an Al/Cu one in series are equivalent to a Ch/Al junction. This equivalency can be seen by imagining the Cu leg shortening until it disappears. Thus, the reference T becomes the ambient T of the meter, T_a. To use $0°$ C reference tables with readings made this way, one must first add to the meter reading the ΔV which that thermocouple would generate with a T difference of $(T_a - 0)°$ C. If the meter reading is used in a feedback loop to control T_h, omitting the reference junction will cause T_h to drift if T_a drifts. In such applications, one can insert a "reference-junction compensator" as shown in Fig. 5.48c, which electronically generates the ΔV correction signal appropriate for its own T_a. Most thermocouple monitoring meters have built-in reference-junction compensators.

Two additional junctions are present in the circuit when "extension alloys" are used, as shown in Fig. 5.48d by E1 and E2. Extension alloys avoid long runs of expensive Pt-Rh and allow W-Re to be read through a vacuum feedthrough. W wire leaks gas because it is fibrous rather than solid, so it cannot be fed through a vacuum seal. Extension alloys are formulated so that $\Delta \phi_1 = -\Delta \phi_2$ *if* $T_1 = T_2$ in Fig. 5.48d, no matter what the T. However, if $T_1 \neq T_2$, large errors can result. Thus, these junctions should be close to each other and carefully shielded from heat sources such as radiation from evaporation sources and substrates. Also, the alloys must not be reversed!

Various thermocouple attachment techniques are shown in Fig. 5.49. The thermocouple is best attached solidly to the object to be measured, and it should be of fine wire so that it does not conduct too much heat away. When the object is metal, spot-welding can be used. If each leg is separately welded to the object (a), a broken weld is readily detected as an open circuit, whereas if the thermocouple junction were welded to the metal as a unit, a broken weld might go unnoticed and give bad T_h data because of the loss in thermal contact. Contact to nonmetallic substrates can be made with braze alloy or ce-

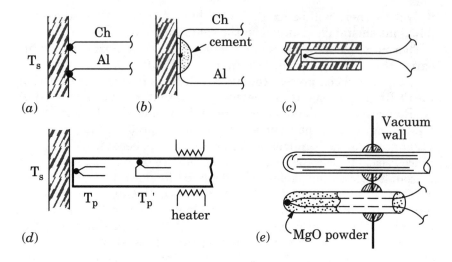

Figure 5.49 Various thermocouple attachment techniques: (a) separate spot-welding, (b) cementing, (c) burial in a well, (d) compensating probe [46], and (e) sheathed probe.

ramic cement for use at high T (b), and with polyimide at <400° C. Alternatively, the thermocouple can be buried in a deep well drilled into the substrate (c). In this case, it will reach the T of its surroundings as long as the heat lost down the wires is much less than the radiation it is receiving. Thermocouples simply pressed against a surface are often used at atmospheric pressure, but this can lead to a large error under vacuum because of the poor contact pointed out in Fig. 5.39. To cancel this error, a recently proposed compensating probe [46] uses two thermocouples, T_p and T_p', and adds heat as shown in (d) until the T drop along the probe shaft, $(T_p - T_p')$, is zero, at which point $(T_s - T_p)$ must also be zero; that is, there is no net heat flow from the substrate down the probe. For operation in electrically noisy environments such as plasma and for ease of accomplishing vacuum feedthrough, metal-sheathed thermocouples (e) are available with sheath diameters down to 0.25 mm.

There are other solid-contact T sensors which are not electrical and are therefore useful where substrate transport makes wiring awkward or where electrical interference is severe. One is an adhesive tape having dots which permanently turn black once a certain T has been reached. Another is the fluoroptic thermometer [47], which uses a dot of phosphor painted onto the substrate. The phosphor is activated by a pulse of UV light from a fiber-optic probe, and the decay time of the resulting phosphorescence is measured. Decay time decreases with increasing T in a well characterized and reproducible manner. Both of the above devices are usable only up to 600 K or so, and they also outgas somewhat under vacuum.

When *no* contact to the substrate is acceptable, T_s can be found from its radiation flux at some λ using $\Phi_\lambda = \varepsilon_\lambda \Phi_{b\lambda}$, where $\Phi_{b\lambda}$ is given by Planck's law (Fig. 5.41). "Optical pyrometers" are used for this purpose, operating in the red for >900 K and in the infrared (IR) at lower T. The lower the T, the larger an area must be sampled to gather sufficient radiation for detectability. The viewing window must be big enough so as not to occlude the area being sampled, it must be transparent to the λ band being measured, and it must be kept free of deposits. The main problem with pyrometry is uncertainty in ε_λ, which is best determined directly using a thermocouple on a "dummy" substrate to calibrate the pyrometer. However, film deposition will change ε_λ, so this should be corrected for during deposition ("adaptive calibration" [48]). Also, spurious radiation sources must be avoided, such as heater radiation being transmitted *through* the substrate [49]. Above 900 K, the area sampled can be small enough to allow observation of the mouth of a well drilled into the substrate. Such a well behaves as a blackbody and therefore has an effective ε_λ of unity, as discussed in Sec. 4.2. When a well cannot be arranged, ε_λ must be determined as above.

5.9 Conclusion

We have examined here the factors determining film structure, topography, interfacial properties, and stress. There are two principal factors: the degree of interaction of the depositing vapor with the substrate and with itself, and the amount of energy input to the deposition surface. When the energy input is thermal, care must be taken to achieve good substrate-T control.

The next four chapters examine special techniques for modifying film characteristics, including the use of single-crystal substrates to achieve epitaxy, chemical vapors to improve film conformality or to obtain selective deposition, and energy-enhanced processes to improve structure or reduce substrate T.

5.10 Exercises

5.1 A molecule has a condensation coefficient of $\alpha_c = 0.2$ on a depositing film of its own solid phase. For this molecule, what are the minimum and maximum values of the thermal accommodation coefficient, γ; trapping probability, δ; sticking coefficient, S_c; and utilization factor, η?

5.2 Assuming that the pre-exponential factor for surface diffusion is given by $k_B T/h$, (a) how low must the diffusion activation energy be to give a diffusion length, Λ, of roughly 100 nm between successive collisions with impinging vapor, for deposition of Si at 1 μm/h and $T_s = 400°$ C? (b) How much of a decrease in T_s would correspond to a ×10 decrease in Λ?

5.3 For condensation onto nonwetting substrates at 300 K using a 10^{-4} Pa overpressure, what are the classical critical-nucleus radii for the following metals, neglecting substrate contact area and γ_f anisotropy? (a) Au, with $\gamma_f = 1400$ ergs/cm^2 (dynes/cm); (b) Pb, with $\gamma_f = 560$ ergs/cm^2.

5.4 Derive Young's equation [Eq. (5.45)] using surface-energy minimization.

5.5 Derive Eqs. (5.46) and (5.47) for the critical 2D nucleus.

5.6 What are the factors that influence whether adsorbing vapor will wet a substrate?

5.7 What factors inhibit nucleation?

5.8 Name at least five ways in which 3D nucleation may, in principle, be suppressed.

5.9 Explain the factors affecting the degree of void formation in Z1 growth.

5.10 Derive the relationship between dome radius and column diameter which satisfies the surface-energy force balance of Fig. 5.23, assuming isotropic γ and negligible curvature effect on γ.

5.11 100 nm of the elastic material CaF$_2$ is deposited with an intrinsic stress of +10 kpsi onto 300° C, 0.4-mm-thick Si(100) which is then cooled to 25° C. Assume 4×10^{10} Pa for the biaxial elastic modulus of the CaF$_2$. (a) What is the level of stress (MPa) and strain in the film? (b) What are the force-balance and bending-moment contributions to the strain in the substrate at the film interface? (c) What is the radius of curvature of the substrate (in m)?

5.12 A film is being deposited with an intrinsic stress of –500 MPa onto a 0.2-mm-thick Si(111) substrate mounted as shown in Fig. 5.33. If the laser beam is hitting the surface 2 cm from the edge of the support, at what film thickness (in nm) will the deflection of the reflected beam be 0.1°?

5.13 Explain at least three ways in which substrate T can vary during film deposition when its platform T is held constant.

5.14 If $\varepsilon_h = 1$ and $\varepsilon_{s1} = \varepsilon_{s2} < 1$ in the geometry of Fig. 5.42, show that $T_s = T_h/2^{1/4}$.

5.15 Show that Eq. (5.66) follows from the heat balances in Eqs. (5.64) and (5.65).

5.16 A substrate having $\varepsilon = 1$ reaches a steady state at $T_s = 250°$ C in vacuum when placed in front of a platform having $\varepsilon = 1$ and heated to T_h and when covered on the deposition face by a shutter having $\varepsilon = 0.1$. (a) When the shutter is swung away at the start of deposition, what is the new steady state, T_s'? (b) If the substrate is a 0.4-mm-thick Si wafer ($c_g = 0.18$ cal/g, $\rho_m = 2.34$ g/cm^3), what is the time constant for T_s re-equilibration?

5.11 References

1. Feibelman, P.J. 1990. "Adsorption Energetics: First Principles Calculations of Adatom Interactions and Induced Local Lattice Relaxation." *J. Vac. Sci. Technol.* A8:2548.
2. Kang, H.C., T.A. Jachimowski, and W.H. Weinberg. 1990. "Role of Local Configurations in a Langmuir-Hinshelwood Surface Reaction: Kinetics and Compensation." *J. Chem. Phys.* 93:1418.
3. Grunthaner, P.J., F.J. Grunthaner, R.W. Fathauer, T.L. Lin, M.H. Hecht, L.D. Bell, W.J. Kaiser, F.D. Showengerdt, and J.H. Mazur. 1989. "Hydrogen-Terminated Silicon Substrates for Low-Temperature Molecular Beam Epitaxy." *Thin Solid Films* 183:197.
4. Uhrberg, R.I.G., R.D. Bringans, R.Z. Bachrach, and J.E. Northrup. 1986. "Symmetric Arsenic Dimers on the Si(100) Surface." *Phys. Rev. Lett.* 56:520.
5. Ueno, K., T. Shimada, K. Saiki, and A. Koma. 1990. "Heteroepitaxial Growth of Layered Transition Metal Chalcogenides on Sulfur-Terminated GaAs {111} Surfaces." *Appl. Phys. Lett.* 56:327.
6. Behrndt, K.H. 1966. "Phase and Order Transitions during and after Film Deposition." *J. Appl. Phys.* 37:3841.
7. Larbalestier, D.C. 1992. "High-Temperature Superconductors 1992: Bringing the Materials under Control." *MRS Bull.* August:15.
8. Givargizov, E.I. 1991. *Oriented Crystallization on Amorphous Substrates*. New York: Plenum.
9. Kanata, T., H. Takakura, H. Mizuhara, and Y. Hamakawa. 1988. "Graphoepitaxial Growth of ZnS on a Textured Natural Crystalline Surface Relief Foreign Substrate." *J. Appl. Phys.* 64:3492.
10. Movchan, B.A., and A.V. Demchishin. 1969. "Study of the Structure and Properties of Thick Vacuum Condensates of Nickel, Titanium, Tungsten, Aluminum Oxide, and Zirconium Dioxide." *Fiz. Metal. Metalloved.* 28:653.
11. Thornton, J.A. 1974. "Influence of Apparatus Geometry and Deposition Conditions on the Structure and Topography of Thick Sputtered Coatings." *J. Vac. Sci. Technol.* 11:666.
12. Bunshah, R.F., and R.S. Juntz. 1973. "Influence of Condensation Temperature on Microstructure and Tensile Properties of Ti Sheet Produced by High-Rate Physical Deposition Process." *Metallurg. Trans.* 4:21.
13. Kim, J.M., and J.M. Kosterlitz. 1989. "Growth in a Restricted Solid-on-Solid Model," *Phys. Rev. Lett.* 62:2289.
14. Messier, R., and J.E. Yehoda. 1985. "Geometry of Thin Film Morphology." *J. Appl. Phys.* 58:3739.
15. Bennett, J.M., and L. Mattsson. 1989. *Introduction to Surface Roughness and Scattering*. Washington, D.C.: Optical Society of America.
16. Pidduck, A.J., D.J. Robbins, D.B. Gasson, C. Pickering, and J.L. Glasper. 1989. "In Situ Laser Light Scattering." *J. Electrochem. Soc.* 136:3088.

17. Tait, R.N., T. Smy, and M.J. Brett. 1992. "Structural Anisotropy in Oblique Incidence Thin Metal Films." *J. Vac. Sci. Technol.* A10:1518.
18. Müller, K.-H. 1988. "Molecular Dynamics Studies of Thin Film Deposition." *J. Vac. Sci. Technol.* A6:1690.
19. Müller, K.-H. 1987. "Stress and Microstructure of Sputter-Deposited Thin Films: Molecular Dynamics Investigations." *J. Appl. Phys.* 62:1796.
20. Gilmore, C.M., and J.A. Sprague. 1991. "Molecular Dynamics Simulation of the Energetic Deposition of Ag Thin Films." *Phys. Rev. B* 44:8950.
21. Brown, A.M., and M.F. Ashby. 1980. "Correlations for Diffusion Constants." *Acta Metallurgica* 28:1085.
22. Müller, K.-H. 1987. "Models for Microstructure Evolution during Optical Thin Film Growth." *Proc. SPIE* 821:36.
23. Buchholz, S., H. Fuchs, and J.P. Rabe. 1991. "Surface Structure of Thin Metallic Films on Mica as Seen by Scanning Tunneling Microscopy, Scanning Electron Microscopy, and Low-Energy Electron Diffraction." *J. Vac. Sci. Technol.* B9:857.
24. Srolovitz, D.J., A. Mazor, and B.G. Bukiet. 1988. "Analytical and Numerical Modeling of Columnar Evolution in Thin Films." *J. Vac. Sci. Technol.* A6:2371.
25. Maissel, L. 1970. "Thin Film Resistors." Chap. 18 in *Handbook of Thin Film Technology*, ed. L.I. Maissel and R. Glang. New York: McGraw-Hill.
26. Massalski, T.B., (ed.). 1990. *Binary Alloy Phase Diagrams*, v. 1. Materials Park, Ohio: ASM International.
27. Harper, J.M.E., S.E. Hörnström, O. Thomas, A. Charai, and L. Krusin-Elbaum. 1989. "Mechanisms for Success or Failure of Diffusion Barriers between Aluminum and Silicon." *J. Vac. Sci. Technol.* A7:875.
28. Chase, M,W., et al. (eds.). 1985. *JANAF Thermochemical Tables*, 3rd ed. Washington, D.C.: American Chemical Society.
29. Wagman, D.D., et al. (eds.). 1982. "NBS Tables of Chemical Thermodynamic Properties." *J. Phys. Chem. Ref. Data* 11, suppl. no. 2. Washington, D.C.: American Chemical Society.
30. Beyers, R., K.B. Kim, and R. Sinclair. 1987. "Phase Equilibria in Metal-Gallium-Arsenic Systems: Thermodynamic Considerations for Metallization Materials." *J. Appl. Phys.* 61:2195.
31. Pretorius, R., A.M. Vredenberg, F.W. Saris, and R. de Reus. 1991. "Prediction of Phase Formation Sequence and Phase Stability in Binary Metal-Aluminum Thin-Film Systems using the Effective Heat of Formation Rule." *J. Appl. Phys.* 70:3636.
32. Andrä, W., and H. Danan. 1982. "Interface Energy as an Origin of Intrinsic Stress in Thin Films." *Phys. Stat. Sol. (a)* 70:K145.
33. Needs, R.J. 1987. "Calculations of the Surface Stress Tensor at Al(111) and (100) Surfaces." *Phys. Rev. Lett.* 58:53.
34. Murakami, M., T.-S. Kuan, and I.A. Blech. 1982. "Mechanical Properties of Thin Films on Substrates." Chap. 5 in *Treatise on Materials Science and Technology*, v. 24. New York: Academic Press.
35. Santoro, C.J. 1969. "Thermal Cycling and Surface Reconstruction in Aluminum Thin Films." *J. Electrochem. Soc.* 116:361.
36. Gardner, D.S., and P.A. Flinn. 1988. "Mechanical Stress as a Function of Temperature in Aluminum Films." *IEEE Trans. on Electron Devices* 35:2160.
37. Frost, H.J., and M.F. Ashby. 1982. *Deformation-Mechanisms Maps*. Oxford, U.K.: Pergamon Press.
38. d'Heurle, F.M. 1989. "Metallurgical Topics in Silicon Device Interconnections: Thin Film Stresses." *Internat. Mater. Reviews* 34:53.
39. Smith, D.L., A.S. Alimonda, C.-C. Chen, S.E. Ready, and B. Wacker. 1990. "Mechanism of SiN_xH_y Deposition from NH_3-SiH_4 Plasma." *J. Electrochem. Soc.* 137:614.

40. Walker, P., and W.H. Tarn. 1991. *CRC Handbook of Metal Etchants*. Boca Raton, Fla.: CRC Press (also covers many compounds).
41. Peden, C.H.F., K.B. Kidd, and N.D. Shinn. 1991. "Metal/Metal-Oxide Interfaces: A Surface Science Approach to the Study of Adhesion." *J. Vac. Sci. Technol.* A9:1518.
42. Sood, D.K., W.M. Skinner, and J.S. Williams. 1985. "Helium and Electron Beam Induced Enhancement in Adhesion of Al, Au and Pt Films on Glass." *Nucl. Instrum. and Methods in Physics Res.* B7/8:893.
43. Mitchell, I.V., G. Nyberg, and R.G. Elliman. 1984. "Enhancement of Thin Metallic Film Adhesion following Vacuum Ultraviolet Radiation." *Appl. Phys. Lett.* 45:137.
44. Smith, D.L., and A.S. Alimonda. Unpublished data.
45. Visser, R.J. 1989. "Determination of the Power and Current Densities in Argon and Oxygen Plasmas by *in situ* Temperature Measurements." *J. Vac. Sci. Technol.* A7:189.
46. Ekenstedt, M.J., and T.G. Andersson. 1991. "A Mechanical Probe for Accurate Substrate Temperature Measurements in Molecular Beam Epitaxy." *J. Vac. Sci. Technol.* B9:1605.
47. Luxtron Corporation, Mountain View, California.
48. Choi, B.I., M.I. Flik, and A.C. Anderson. 1992. "Adaptively Calibrated Pyrometry for Film Deposition Processes." In *Heat Transfer in Materials Processing*, ed. J.C. Khanpara and P. Bishop. New York: American Society of Mechanical Engineers 224:19.
49. Wright, S.L., R.F. Marks, and W.I. Wang. 1986. "Reproducible Temperature Measurement of GaAs Substrates during Molecular Beam Epitaxial Growth." *J. Vac. Sci. Technol.* B4:505.

5.12 Recommended Readings

Kattelus, H.P., and M.-A. Nicolet. 1988. "Diffusion Barriers in Semiconductor Contact Metallization." Chap. 8 in *Diffusion Phenomena in Thin Films and Microelectronic Materials*, ed. D. Gupta and P. S. Ho. Park Ridge, N.J.: Noyes Publications.

Lagally, M.G. 1993. "Atom Motion on Surfaces." *Physics Today* (November).

Lewis, B., and J.C. Anderson. 1978. *Nucleation and Growth of Thin Films*. Boston, Mass.: Academic Press.

Murakami, M. 1991. "Deformation in Thin Films by Thermal Strain." *J. Vac. Sci. Technol.* A9:2469.

Thompson, C.V. 1990. "Grain Growth in Thin Films." *Annual Rev. Mater. Sci.* 20:245.

Thornton, J.A. 1986. "The Microstructure of Sputter-Deposited Coatings." *J. Vac. Sci. Technol.* A4:3059.

Thornton, J.A., and D.W. Hoffman. 1989. "Stress-Related Effects in Thin Films." *Thin Solid Films* 171:5.

Weinberg, W.H. 1991. "Kinetics of Surface Reactions." Chap. 5 in *Dynamics of Gas-Surface Reactions*, ed. C.T. Rettner and M.N.R. Ashfold. Cambridge, U.K.: Royal Society of Chemistry.

Williams, E.D. 1994. "Surface Steps and Surface Morphology: Understanding Macroscopic Phenomena from Atomic Observations." *Surface Science* 299/300:502.

Venables, J.A., G.D.T. Spiller, and M. Hanbücken. 1984. "Nucleation and Growth of Thin Films," *Rep. Prog. Phys.*, 47:399.

Chapter 6

Epitaxy

In the previous chapter, we considered various effects of bonding across the film-substrate interface, namely, wetting, compound formation, adhesion, and stress. A new effect is added in the special case of epitaxial film growth. Epitaxy means that the crystallographic order of the film is being significantly influenced by that of the substrate as a result of some degree of matching between the two along the interface. In this chapter, we will first illustrate various symmetries of matching, and then give examples of the special thin-film structures and applications which result from having precise control over film crystallography. We will then discuss how to avoid disorder at the interface, since this disrupts epitaxy. Next, we will show how epitaxy under vacuum allows the use of molecule and electron probe beams which reveal a great deal of information about the film growth process on an atomic level, information which applies to polycrystalline growth as well. Then, we will examine the problems of controlling composition and structure. Both of these problems tend to be more difficult in the case of epitaxy because of the more stringent demands on film quality for epitaxial applications. Structure includes bulk defects such as dislocations, and it includes surface morphology (topography). We will see that both types of structure are strongly affected by small amounts of mismatch in periodicity between the film and substrate crystal lattices.

6.1 Symmetry

We saw in Fig. 5.2 that the potential energy, E_p, of a chemical bond has a sharp minimum at a particular bond length. We also know that an atom needs to have a number of bonds corresponding to its "valence" to further reduce overall E_p by filling the outer electron shell. Covalently bonded atoms have, in addition, an E_p minimum for particular angles between the bonds, whereas ionic and metallic bonds are nondirectional. These three fundamental criteria determine crystal structure, epitaxial relationships, and the atomic reconstruction of surfaces. Reconstruction will be discussed in Sec. 6.5.3.

Crystals are common in nature because their symmetry provides a bonding environment of minimum E_p for every atom, whereas in polycrystalline or amorphous material, disorder necessarily forces some atoms into bonding environments of higher E_p. Below the crystal melting point, T_m, minimum E_p usually also means minimum free energy, G; that is, it usually represents the equilibrium situation. This is because E_p is the major component of internal energy, U, and is the dominant term in Eq. (4.5) for G. The entropy term in G, which is (–TS), would be larger for a *disordered* situation, thus reducing G; but the accompanying increase in E_p usually outweighs that term. Of course, TS increases with T, and in fact T_m occurs when the TS increase on melting equals the heat of fusion absorbed, $\Delta_c H$; that is, $\Delta_c H = T_m \Delta_c S$. (Further increases in T favor the even more disordered vapor phase, as was discussed in Sec. 4.1.)

Any interruption of crystal symmetry increases E_p. Thus, surfaces and interfaces have an excess energy per unit area, γ, as discussed in Sec. 5.3.1. When a film is deposited on a single-crystal substrate, the interfacial energy, γ_i, is minimized by maximizing the density of bonds of appropriate length and angle across the interface in an attempt to merge the symmetries of the two crystals. Thus, it is energetically favorable for the film material to crystallographically align itself with the substrate so as to match the substrate's bonding symmetry and periodicity—that is, to grow epitaxially. Consequently, epitaxy occurs readily for any combination of film and substrate having some degree of matching, provided that the substrate symmetry is not masked from the film by interfacial disorder as discussed in Sec. 6.3, and provided that T is high enough to allow the depositing atoms to rearrange themselves into equilibrium positions before burial. Epitaxy of sufficient perfection for electronic-device applications is more difficult to achieve, as we will see later on. An extensive compilation of crystal structures and lattice constants of materials is given in Ref. 1.

The simplest form of epitaxy is "homoepitaxy"—the growth of a material onto itself. There, the film/substrate interface, if it is clean, vanishes into the bulk material so that the interface energy, γ_i, is zero. Conversely, "heteroepitaxy" of one material upon another generally results in $\gamma_i > 0$, and the film's preferred crystallographic orientation is often that which minimizes γ_i or, equivalently, maximizes bonding across the interface. One of the fascinating aspects of heteroepitaxy is the variety of imaginative symmetries which nature concocts to accomplish this goal, and we will examine several examples below. In practice, the only fundamental criterion for epitaxy seems to be a moderately small fractional mismatch, f, in the atomic periodicities of the two materials along the interface. We define

$$f = \frac{(a_e - a_s)}{(a_e + a_s)/2} \approx (a_e - a_s)/a_s \quad (6.1)$$

where a_e and a_s represent the atomic spacings along some direction in the film crystal and in the substrate surface, respectively. When the thermal-expansion coefficients of the two materials differ [as in Eq. (5.53)], f becomes a function of T, and then the f of interest is its value at the growth T. (However, below we will quote room-T values for convenience.) Generally, one needs f < 0.1 or so to obtain epitaxy, because for f > 0.1, so few of the interfacial bonds are well aligned that there is little reduction in γ_i. The various ways in which the interface accommodates mismatch are discussed further in Sec. 6.6. Here, we focus instead on the symmetry factor.

Much epitaxy work involves semiconducting materials for electronic applications, and most of these have either the diamond or the zinc-blende (cubic ZnS) structure. Thus, it is important to understand these two closely related structures. Their symmetry can be visualized by referring back to Fig. 5.7, which shows the array of atoms in a face-centered-cubic (fcc) unit cell of lattice constant $a_o = a = b = c$. Now imagine a second identical array of atoms superimposed on the first, and then translate it $+(1/4)a_o$ along each of the three axes. The atom which was at position 11 ends up at the center of the tetrahedron formed by atoms 8, 9, 10, and 11. In fact, it is bonded to each of them; indeed, *all* of the atoms in the structure are tetrahedrally bonded to four nearest neighbors in the same manner. This is the structure of diamond C, Si, Ge, and "gray" Sn, all of which elements have covalent bonding and a valence of four and thus fit the symmetry of this tetrahedral bonding arrangement. The zinc-blende structure has the same symmetry and is found in the III-V semiconductors such as GaAs and in the II-VI semiconductors such as ZnS. In zinc blende, the first face-centered cube contains the metal atoms, and the second contains the nonmetal atoms.

Let us now examine various ways in which crystallographic symmetry can be made to match across a heteroepitaxial interface. In the simplest and most common way, the matching atomic spacings are the crystal lattice constants, a_o, themselves. This is the case for AlAs

(a_0 = 0.562 nm at room T), GaAs (0.565 nm), and Si (0.543 nm), all of which in addition have the same zinc-blende/diamond atomic arrangement. Metals have much smaller a_0, so some match these semiconductors at $2a_0$. For Fe, for example, $2a_0$ = 0.573 nm, which is very close to the a_0 of GaAs. Thus, Fe grows epitaxially on many different planes of GaAs [2]. Figure 6.1a shows how the atomic arrangements of Fe and GaAs match up on the (001) plane. (Crystallographic notation was outlined in Sec. 5.3.1.) There is no atom shown at the center of the (001) face in the Fe unit cell, because Fe is body-centered cubic (bcc). For Cu (fcc structure), $\sqrt{2}\,a_0$ = 0.512 nm, just 6 percent less than the a_0 of Si, so Cu(001) grows on Si(001) with a 45° rotation [3]—Cu[100] direction parallel to GaAs[110]—as shown in Fig. 6.1b. However, this match works *only* for the {001} planes. CdTe (zincblende structure) has *two* epitaxial orientations [4] on GaAs(001); namely, (001) and (111). Because of the large a_0 mismatch (f = 0.15), (001)-oriented CdTe is obtained only under specific deposition conditions. CdTe(111) is obtained more easily and with higher quality because then f is only –0.7 percent for the atomic spacing shown as a in Fig. 6.1c, where the epilayer is rotated so that the direction CdTe[$\bar{2}$11] is parallel to GaAs[$\bar{1}$10] (see Exercise 6.1). This case illustrates that good epitaxy is obtainable even when only a few of the atoms in the array align, if that alignment produces low enough f.

Sometimes, crystals with very dissimilar three-dimensional structures still have similar structures along one plane, and this allows epi-

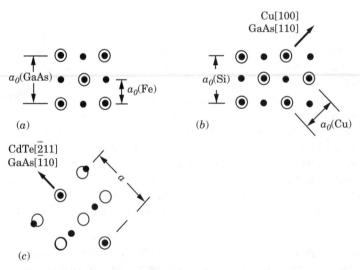

Figure 6.1 Variations on heteroepitaxial symmetry: (a) $a_0 \times 2$, (b) 45° rotation, and (c) CdTe(111) on GaAs(001). o = substrate surface atoms, • = first monolayer of epitaxy.

taxy on that one plane. This is the case for the fcc and hcp (hexagonally close-packed) lattices, which have the same close-packed atomic arrangement on (111) and (0001), respectively (see Sec. 5.3.1). This close-packed arrangement is shown for CdTe(111) in Fig. 6.1c. Actually, the only difference between fcc and hcp is the stacking sequence of these close-packed planes [1].

Some materials are "polymorphic"; that is, they have more than one possible crystal structure. Examples are C, which has the diamond and graphite phases, and Sn. Sn has the common β phase, "white" tin, which is metallic and tetragonal, and also the α phase, "gray" tin, which has the diamond structure with a_o = 0.649 nm and which is the thermodynamically stable phase only below 286 K. However, α-Sn can be grown at up to 343 K on InSb and CdTe, both of zinc-blende structure, because the a_o of α-Sn mismatches these substrates by only 0.1 percent [5]. Here, the "wrong" phase—the "pseudomorph"—has been stabilized at above 286 K by the lowering of γ_i due to interfacial bonding. At still higher T or for film thickness > 0.5 μm, growth reverts to the normal β phase. The collapse of the α phase with increasing thickness probably results from the buildup of strain energy stored in the film as a result of its 0.1 percent compression to accommodate the mismatch. Epitaxial strain energy and its relaxation mechanisms are discussed further in Sec. 6.6.

Another, recently discovered way in which epilayers can avoid misfit strain, at least in one direction, is by tilting their growth planes relative to the substrate [6]. In Fig. 6.2, the atomic spacing of the epilayer along the x axis, a_e, is slightly smaller than that of the substrate along this axis, a_s. The epilayer can correct for this misfit without strain by rotating itself about the y axis perpendicular to the page by an angle α, where cos α = a_e/a_s = (f + 1) and f is negative. This predicted relationship has been found to hold for many hcp rare earths and metals grown on bcc(211) metal substrates.

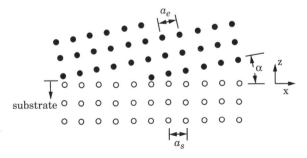

Figure 6.2 Cross-sectional view of misfit accommodation by epilayer tilt.

Above, we have considered only single-crystal substrates; but in principle, epitaxy can also occur on the individual grains of a polycrystalline substrate. Such a process would be useful in cases where the substrate needed to be polycrystalline for cost reasons. Epitaxy would improve interface quality in multilayed structures of polycrystalline films, and it would result in films with relatively few grain boundaries when grown on large-grained substrates.

The choices for good single-crystal substrates are limited. It is difficult to grow bulk crystals of low defect density which are also large enough to make useful substrate wafers. This is because defects nucleate at the slightest perturbation in melt conditions, such as vibrations, T transients, and impurity precipitation. Enormous effort has been applied to the growth of Si crystals because of their importance to the semiconductor industry, and the achievements include crystals completely free of dislocations, wafers 30 cm in diameter, reduction of unwanted impurities to below 1 ppb (10^{-9}), and very low cost. (Dislocations are discussed further in Sec. 6.6.) Si is also relatively resistant to crystallographic damage from sawing, polishing, and handling, compared to softer or less elastic materials such as metals. However, when the symmetry or a_o of Si is not appropriate, or when the film material reacts with it, another substrate may be preferred. Some of the more damage-resistant materials which are available in reasonable size, quality, and cost include Ge, GaAs, sapphire (Al_2O_3), spinel ($MgO.Al_2O_3$), quartz (SiO_2), fluorite (CaF_2), and mica.

6.2 Applications

As with other thin films, epitaxial films can provide properties or structures that are difficult or impossible to obtain in bulk materials. Indeed, many materials are easier to grow epitaxially than to grow and shape in bulk form. Compared to polycrystalline films, epitaxial films have at least four advantages, which are: elimination of grain boundaries, ability to monitor the growth by surface diffraction, control of crystallographic orientation, and potential for atomically smooth growth. The problems associated with grain boundaries were discussed in Sec. 5.4.3. Surface diffraction will be discussed in Sec. 6.4.2. Crystallographic orientation is important when the film property of interest is anisotropic. For example, the piezoelectric effect is strongest along the polar axis of a crystal, and the magnetization of Fe is easiest along {001}.

The requirements for atomically smooth growth will be discussed in Secs. 6.5.5 and 6.7. When this is achieved *and* when the growth T and time are low enough to avoid interdiffusion, multilayer epitaxial heterostructures can be grown which have compositional changes of atomic-layer abruptness, making possible some remarkable semiconductor devices which will be discussed in Sec. 6.2.1. In addition, the epitaxial growth of various materials on the basic semiconductor-device material, Si, allows the achievement of improved or expanded functionality on a single chip. Thus, GaAs on Si takes advantage of the higher switching speed of GaAs devices for microwave communications applications, and further research is underway to integrate GaAs lasers and photodetectors with Si for use as optical interconnects between computer chips. $Si_{1-x}Ge_x$ on Si can provide transistors and other devices of improved performance [7]. Epitaxial insulators such as CaF_2 on Si are being developed for electrical isolation of devices, and epitaxial metals for elimination of grain-boundary diffusion and electromigration. Epitaxy of superconductors on Si would provide zero-resistance interconnects between devices on a chip. Since most of the development effort in epitaxial technology has been devoted to the semiconductors GaAs and Si, much of the discussion in this chapter will focus on these materials. However, the same principles of epitaxial growth can be applied to other materials as well.

The atomically abrupt interfaces that are desired for most epitaxial applications are much easier to achieve using deposition processes which are *not* operating near equilibrium. This situation can be viewed in terms of Eq. (5.8), which expressed the net deposition flux, J_r, as the difference between the incorporation flux of adsorbed vapor into the film, R_r, and the re-evaporation flux of film material, J_v. As J_v increases toward R_r, the deposition process approaches equilibrium and thus $J_r \to 0$. Although finite deposition rates and good material can be obtained near (though not at) equilibrium, one problem with operating in this regime is that J_r becomes the small difference between the two much larger quantities, R_r and J_v. Thus, small fluctuations in R_r and J_v result in large fluctuations in J_r, so that film thickness and uniformity become very difficult to control. It also becomes problematic to stop deposition abruptly, because the reverse process (J_v) may not be so quickly stopped on account of the thermal mass of the substrate, in which case the surface will begin to erode away. The liquid-phase-epitaxy (LPE) process depends on near-equilibrium to regulate deposition rate, so it is useful only when structure control is not critical. There is also a vapor-phase process known as "vapor-phase epitaxy" (VPE) which operates near equilibrium. In VPE of GaAs, Ga is transported to the substrate as GaCl vapor. But the Cl-containing atmosphere can also *etch* the depositing GaAs to form GaCl and $AsCl_3$ vapor, and thus the process operates near the equilibrium between deposition and etching. Although the etching capability is useful for substrate surface cleaning, deposition-rate fluctuation and Cl contamination of the film are problematic in critical applications.

Because of the importance of atomically abrupt interfaces, we will focus in this chapter on physical and chemical vapor-deposition processes which operate far from equilibrium in the sense that $J_v \ll R_r$. This is achieved by reducing the deposition T, T_s, until J_v becomes negligible. It does not mean, however, that *all* of the steps in the deposition process are operating far from equilibrium. Recall from Chap. 5 that deposition is a series of steps: adsorption, surface diffusion, reaction, nucleation, structure development, and interdiffusion. To obtain good deposition-rate control, it is only important that either the adsorption or the reaction step be far from equilibrium. To prevent the broadening of interfaces after they are formed, it is important also that T_s be low enough so that interdiffusion is negligible during the total time of structure deposition. However, if T_s is *too* low, surface diffusion will become negligible, and structural equilibration will not occur. This is the "quenched growth" regime discussed in Sec. 5.4.1, and the crystallographic quality of epilayers is poorer in this regime than at higher T_s. Fortunately, there is often a T_s "window" within which good crystallography and sharp interfaces can both be obtained. Much of the development work in epitaxy has involved modifying processes to widen this window. We will examine this situation further in Sec. 6.5.

In addition to non-equilibrium growth, one must also have chemical compatibility and reasonably good lattice match between layers to obtain good heteroepitaxy. Chemical interactions are discussed in Sec. 6.3. The problems of lattice mismatch, f [Eq. (6.1)], are discussed in Secs. 6.6 and 6.7.

The most useful combinations of materials for semiconductor-device applications of heteroepitaxy are those in which low f can be obtained simultaneously with large band-gap (E_g) difference. Figure 6.3 shows E_g and a_0 for a wide range of cubic semiconductors. The III-V and II-VI compounds shown all have the zinc-blende cubic structure described in Sec. 6.1. The II-VI compounds can also crystallize in the hexagonally close-packed (noncubic) wurtzite structure. The IV-VI compounds have the rock-salt (NaCl) cubic structure, in which the group IV and VI atoms are both in fcc lattices, with one lattice being displaced by $(1/2)a_0$ with respect to the other along the x axis. Note that certain semiconductor combinations have quite small f and large E_g difference, such as ZnSe/AlAs/GaAs/Ge and CdTe/HgTe.

Further adjustability in E_g and a_0 is possible using solid solutions (alloys) of two or three elements or compounds having the same structure. The tie lines between materials in Fig. 6.3 show the behavior of E_g and a_0 for alloys A_xB_{1-x} formed between materials A and B. When

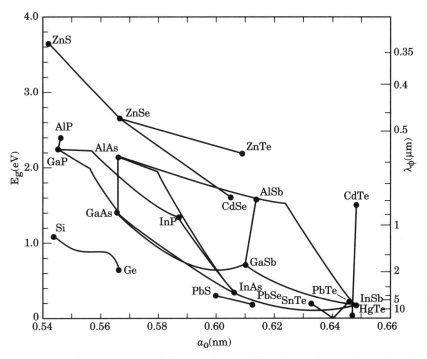

Figure 6.3 Band-gap energy vs. lattice constant at 300 K for various cubic semiconductors. (Based on an unpublished figure by H. Goronkin of Motorola Corp.; used by permission.)

there is a miscibility gap (limited solubility of A in B) in such an alloy, phase separation can be avoided by growing at a low enough T_s so that diffusion within the film is avoided, or by alternating monolayers of A and B to increase alloy ordering. The abrupt changes in slope seen for some of the alloys occur where there is a change in the conduction-band minimum from one band to another and not because of a change in crystal structure. The relationship of a_0 to x usually obeys Vegard's law; that is, linear scaling from one material to the other. The use of x as a heterostructure design variable provides continuous adjustability of a_0 or E_g. Thus, a perfect lattice match can be obtained for $(GaAs)_x(InAs)_{1-x}$ grown on the much higher-E_g semiconductor InP. Alternatively, $(AlAs)_x(GaAs)_{1-x}$, which has a low f for any x, provides over 50% adjustability in E_g. These alloys are usually written in the more compact forms $Al_xGa_{1-x}As$ or (AlGa)As. They are also commonly called ternary alloys or "ternaries" because they contain three elements, although strictly speaking they are binary alloys composed of two binary compounds. They should not be confused with ternary *compounds* such as the solar-cell semiconductor $CuInSe_2$, which exists only for the specific stoichiometric ratio indicated by its formula, rather than being a solid solution. Some "ternaries" have very small E_g, and these are useful for infrared (IR) electro-optic devices. The E_g of HgTe is zero, so $Hg_xCd_{1-x}Te$ ("mer-cad" telluride) can be tailored for photodetector applications over a wide IR range. The E_g of $Pb_xSn_{1-x}Te$ also dips to zero at about midrange in x. This "lead-salt" alloy is used for IR diode lasers tuned using x to match the vibrational absorption bands of pollutant molecules such as CO and NO, providing extremely sensitive remote detection.

The use of alloying has been further extended to the growth of "quaternaries" such as $Ga_xIn_{1-x}As_yP_{1-y}$ and $Al_xGa_yIn_{1-x-y}P$. (Note that the overall III/V ratio is always unity.) Here, a second degree of freedom (y) is obtained, so that a_0 and E_g may be adjusted independently over some range to provide lattice match as well as the desired E_g difference. These techniques of "band-gap" engineering provide great flexibility in the design of heterostructures.

6.2.1 Semiconductor devices

When atomically smooth and abrupt interfaces are achieved, some remarkable heteroepitaxial semiconductor devices become possible, as discussed in depth by Weisbuch (1991). These devices currently represent the most advanced application of epitaxy, so a few representative examples deserve special attention here. This discussion also serves as a vehicle for reviewing some important semiconductor principles which will be useful in later sections.

We first consider the modulation-doped heterojunction. The diagrams of Fig. 6.4 show the energy levels, E_e, of electrons in two semiconducting materials relative to the energy level of an electron in vacuum, E_0. Recall that valence (outer-shell) electrons in a semiconducting material can reside either in the valence band at energy E_v, where they are each part of a chemical bond in the material, or in the conduction band at the higher energy E_c, where they are free to migrate through the material. The band gap, E_g, is devoid of electrons unless it contains trap states formed by impurities or defects. By the way, an insulator is just a semiconductor with a very wide band gap, while a metal has a zero band gap and thus has migrating valence electrons.

In Fig. 6.4a, the wider-band-gap semiconductor $Al_xGa_{1-x}As$ is shown "doped" with a shallow donor impurity. Shallow donor atoms each have one more electron than they need to satisfy their bonding to neighboring atoms. Thus, that electron is very weakly bound, which gives it a shallow energy level just below the bottom of the conduction band. The extra electron easily enters the conduction band, leaving behind the immobile impurity positive ions shown and giving the ma-

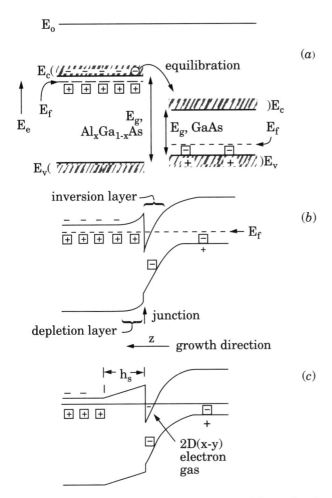

Figure 6.4 Conduction-electron behavior in a modulation-doped p-n heterojunction: (a) wide band-gap $Al_xGa_{1-x}As$ with ionized donor atoms ⊟ and separate, narrower band-gap GaAs with ionized acceptor atoms ⊞, (b) charge equilibration upon heterojunction formation, and (c) modulation-doped heterojunction with undoped "spacer" of thickness h_s.

terial some concentration of conducting electrons, say 10^{18} n/cm^3 (n denotes negative charge carriers). This is considered heavily doped material, even though the dopant is only 23 ppm of the concentration of host lattice atoms, which is 4.4×10^{22} cm^{-3}. The narrower-band-gap semiconductor GaAs is shown lightly doped with a shallow acceptor impurity, which in epitaxial material would typically be the unintentional background contaminant carbon. Acceptor atoms have one *less* electron than they need to satisfy their bonding to neighboring atoms, so they draw one from a neighbor and become the negative ions shown. The resulting "hole" in the valence band can migrate through the material by being filled with an electron from another neighbor. The unintentional hole concentration might typically be 10^{15} p/cm^3 in GaAs (p denotes positive charge carriers).

We can now look at what happens when these two semiconductors are joined in a p/n junction by growing one epitaxially upon the other. In Fig. 6.4a, the energy level, E_f, to which electrons in the $Al_xGa_{1-x}As$ are filled is close to E_c because of the n-type doping. E_f is called the Fermi level, and it is the chemical potential, μ, of the conducting electrons [defined in Eq. (4.7)]. The E_f in the GaAs is at a much lower electron energy level, E_e, close to E_v, because of the p-type doping. Thus, upon joining the two semiconductors, electrons flow in the direction of decreasing μ, from the $Al_xGa_{1-x}As$ into the GaAs. The resulting negative charge in the GaAs raises its electron μ, and the resulting positive charge in the $Al_xGa_{1-x}As$ lowers its electron μ. Equilibrium is reached when E_f (or μ) is flat across the junction, as shown in Fig. 6.4b. The transferred electrons (only one shown) are held close to the junction in an "inversion" layer of GaAs by the electrostatic attraction of the immobile positively charged dopant left behind in the $Al_xGa_{1-x}As$. The $Al_xGa_{1-x}As$ has meanwhile developed a "depletion" layer from which those electrons were transferred. If this had been a *homo*junction, say n-GaAs/p-GaAs, there would also have been hole transfer in the reverse direction. However, this is blocked in the heterojunction by the wide band gap of the $Al_xGa_{1-x}As$, which results in a barrier to hole transfer and thus maximizes the electron transfer required to reach equilibrium. Note that since decreasing energy for electrons is down in these diagrams, decreasing energy for holes is up; that is, holes "float" and electrons sink.

The key feature of this structure is the electron inversion layer, and the objective is to maximize the conductivity of this layer for application as the conducting channel in a *h*igh-*e*lectron-*m*obility field-effect *t*ransistor (HEMT). Electron conductivity in any medium is given by

$$s = \frac{1}{\rho} = \frac{j}{E} = \frac{n_e q_e u_d}{E} = n_e q_e \mu_e \quad (6.2)$$

where s = conductivity, S/cm (S = siemens = 1/Ω = conductance)
 ρ = resistivity, Ω-cm
 j = current density, A/cm^2
 E = electric field strength, V/cm
 n_e = electron concentration, cm^{-3}

6.2.1 Semiconductor devices 233

q_e = electron charge = 1.60×10^{-19} C (coulombs)
u_d = electron drift velocity, cm/s
μ_e = electron mobility, cm^2/V·s

The second equality is just Ohm's law. The third equality expresses flux in terms of concentration and velocity [see Eq. (4.35)]. The last equality defines the mobility of a charged particle as its drift velocity per unit field strength (cm/s per V/cm = cm^2/V·s). Thus, s is maximized by maximizing n_e and u_d. n_e has already been maximized in the Fig. 6.4b structure. We now consider u_d.

Electrons accelerate due to the force exerted on their charge by the field, **E**, until a steady state is reached in which this force is just balanced by the drag force of various scattering mechanisms. In the steady state, the electrons travel at a constant u_d, just as bodies falling through the atmosphere reach a terminal velocity determined by air friction. At room T, the dominant scattering centers for charge carriers traveling in a semiconductor are lattice vibrations ("phonons"), which, like all waves, have their equivalent particle behavior and thus act as scatterers. At cryogenic T, on the other hand, there are far fewer phonons, and then the dominant scattering centers become the crystal defects. If the crystal quality is very good, the dominant defects become the ionized dopant atoms themselves. This imposes the ultimate limit to the s of a homogeneously doped semiconductor. That is, at high doping levels, further increase in n_e (or n_h) just causes a compensating decrease in μ_e (or μ_h) due to the accompanying increase in the concentration of dopant-impurity scattering centers. Now we can see the advantage of the Fig. 6.4b structure, because in the layer of electrons traveling next to the interface in the GaAs, there is a very high n_e, but there are no donor ions aside from unintentional background impurities, which are orders of magnitude down in concentration at $<10^{15}$ cm^{-3}.

A further refinement of this "modulation doping" or "proximity doping" concept takes into account the finite wavelength of the conducting electrons, which is given by the deBroglie relationship,

$$\lambda_e = \frac{h}{|\mathbf{k}|} = \frac{h}{\sqrt{2m_e^* E_t}} \quad (6.3)$$

where **h** is Planck's constant, 6.63×10^{-34} J·s, and **k** is the electron momentum. (The first equality also gives the momentum of photons and thus expresses the equivalency of wave and particle nature.) The "effective" mass of the electron, m_e^*, is less than that of a free electron in vacuum, m_e, because of interaction with the solid material. The kinetic energy per electron, E_t, is, at the least, thermal: $(1/2)k_BT$ for each of the two degrees of freedom in the layer [see discussion of Eq. (2.11)]. From this, it is found that λ_e is a few tens of nm at room T (Exercise 6.2a). Now, even though the electron is confined to the GaAs side of the junction, its wave function has an exponentially decaying portion (an "evanescent tail") which extends into the $Al_xGa_{1-x}As$ by about λ_e. Therefore, the conducting electrons will "feel" ionized impurities within this distance and be scattered by them. This problem can be avoided by stopping the high p-doping at a "spacer" distance h_s from the junction as shown in Fig. 6.4c. The result is an extremely high μ_e of $>10^6$ cm^2/V·s for electrons in GaAs at cryogenic T. With such high μ_e, the electrons are behaving as a 2D gas, where they are scattered mostly by each other. They can also experience "ballistic" transport, where they undergo *no* scattering between injection into the device and collection at the other end. Clearly, atomically abrupt interfaces and monolayer precision in the thickness of h_s are needed to realize the full potential of such a device.

In another example of doping control, wavelength spreading can produce a 2D electron gas in a semiconductor of *constant* E_g if the doping is confined to a sheet significantly thinner than λ_e, as shown in Fig. 6.5. The electrons are confined to travel in a layer next to the dopant layer by the electrostatic attraction of the ionized dopant atoms, but in this travel they are scattered less than if the dopant were distributed uniformly across their path. Such a doping distribution behaves like the Dirac delta function, δ, which consists of an infinitely high, infinitesimally narrow spike at one position, z_D. The integral of such a spike along z is finite, and in the doping case it is equal to the areal dopant concentration in the layer, n_D, which is typically 4×10^{12}

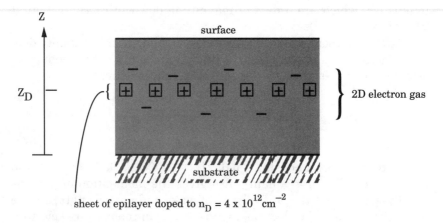

Figure 6.5 Cross section of a delta-doped structure.

cm^{-2} or about 4×10^{-3} ML (see Sec. 2.6). This type of doping is thus known as "delta" doping [8] and is also referred to as sheet or spike doping. As long as the thickness of the sheet is $<<\lambda_e$, the dopant behaves electrically like the δ function. However, if the dopant actually resides in a single monolayer, improved control can be achieved over the particular lattice sites which the dopant atoms occupy, and this improves doping efficiency (more on this in Sec. 6.5.6).

Electrons can be further confined by growing a thin layer of narrow-E_g semiconductor between layers of wider-gap material as shown by L_q in the energy level diagram of Fig. 6.6. This is the classic quantum-physics textbook problem of the "particle in a box," and today this very quantum-well structure can be engineered using heteroepitaxy. Since the electron is confined to the potential well of the narrow-E_g material, its wave function must go to zero at the boundaries (neglecting the evanescent tail). To satisfy this condition, $(1/2)\lambda_e$ must be an integer multiple, j, of the well width, L_q, as suggested by the waves drawn in Fig. 6.6. Combining this criterion with Eq. (6.3) for λ_e, we can solve for the "confinement" energy of the electron:

$$E_q = h^2 j^2 / 8 m_e^* L_q^2 \qquad (6.4)$$

This is the kinetic energy that the electron must attain to "fit" into the well, and its effect is suggested by the height above E_c at which the

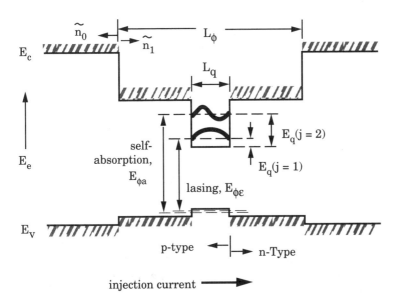

Figure 6.6 Typical quantum-well laser-diode structure.

waves are drawn in Fig. 6.6. The same effect occurs for holes confined to the valence band well, which is inverted since holes float. In this structure, the electrons are confined in only one of the translational degrees of freedom. Using special growth techniques, it is also possible to confine two of the degrees of freedom and make "quantum wires" [9].

Today's capability to design and grow quantum-well structures almost at will allows quantum effects to be exploited in practical devices. For example, quantum-well diode lasers made from III-V semiconductors are used routinely in compact-disc (CD) audio players and read-only memories (CD-ROMs). The quantum-effect advantage in lasers arises from E_q [Eq. (6.4)]. To understand this, consider that light-emitting diodes (LEDs) and diode lasers emit light having energy equal to the electron-hole radiative recombination energy, which is the energy difference between the lowest conduction-band level and the highest valence-band level, as shown by $E_{\phi e}$ in Fig. 6.6. Photon energy is related to frequency, ν_ϕ, and wavelength, λ_ϕ, by

$$E_\phi = h\nu_\phi = hc_0/\lambda_\phi \quad \text{or} \quad E_\phi(\text{eV}) = 1.24/\lambda_\phi(\mu m) \qquad (6.5)$$

with c_0 being the speed of light, 3.0×10^8 m/s. In bulk semiconductor material, some of this light is re-absorbed by excitation of electrons from the valence band back into the conduction band. This self-absorption reduces lasing efficiency. Meanwhile, the electrons (and holes) for radiative recombination are being supplied by injecting them into the conduction (and valence) bands across a forward-biased p/n junction. As injection current increases, both bands *fill* to a higher energy level for their respective carriers (up for electrons and down for holes), and this increases the self-absorption photon energy, $E_{\phi a}$, required to lift an electron into the lowest unoccupied energy level. Thus, the semiconductor becomes *transparent* to its own lasing λ. This filling effect is considerably enhanced in the quantum-well structure, as shown in Fig. 6.6, because successive energy levels have the energy E_q added. Thus, the transparency condition is achieved at lower current, which means that the threshold current at which lasing begins is reduced, and the lasing efficiency (light power out ÷ electrical power in) increases.

Lasing efficiency also increases with the intensity of the optical radiation field in the device, which is enhanced by light confinement. Laser light travels perpendicular to the paper in Fig. 6.6. It is confined in this direction by the laser die's end mirrors, which, by the way, are themselves thin-film structures producing the desired degree of destructive optical interference [see Eq. (4.43)]. Meanwhile, the light is confined laterally (left to right) by using waveguiding. Figure 6.6

shows another structure beyond the quantum well which consists of an even higher-E_g material. This produces an optical confinement region of width L_ϕ. It does so because refractive index, ñ, decreases with increasing E_g, so light trying to escape encounters a step decrease in ñ from $ñ_1$ to $ñ_0$. From Snells's law [Eq. (4.41)] and Fig. 4.18, it can be seen that at some incident angle, ϕ_1, of light on such an interface, the refracted angle ϕ_0 reaches 90°; that is, the light cannot escape. This is the angle of total internal reflection, $\phi_t = \sin^{-1}(ñ_0/ñ_1)$, and its existence means that light can be guided through the length of the laser cavity without leakage. The same consideration of being able to "fit" the optical wave into the waveguide applies here as it did for the electron in the quantum well. Since λ_ϕ is on the order of 1 µm, we thus require $L\phi \gg L_q$, and we arrive finally at the complete structure of Fig. 6.6.

The concept of making abrupt changes in epilayer composition has been further extended to make periodic structures of alternating high and low E_g called "superlattices," and ones of modulated doping called "doping superlattices." The band structure of a crystalline semiconductor arises from the periodicity of its crystal lattice, and thus the superlattice periodicity produces a superimposed band structure and associated electrical properties which can be engineered using the modulation amplitude and period of the superlattice. Interfaces also preferentially accumulate or "getter" impurities, so the many interfaces of a superlattice can be used to extract unwanted impurities from neighboring film layers. A "strained-layer superlattice" produced by lattice mismatch between the two alternating layers can be used to block the propagation of dislocation defects into overlying layers (more on this in Sec. 6.6). Magnetic/nonmagnetic metal superlattices can have greatly enhanced magnetic properties [10]. Superlattice structures can also be made with non-epitaxial films. Thus, x-ray mirrors can be made with alternating layers of materials having high and low refractive index for x-rays. For the appropriate relationship between superlattice period and optical thickness period, reflectivity is maximized [see discussion of Eq. (4.43)]. For example, a very high x-ray reflectivity of 30 percent at 21 nm was achieved for a Mo/Si superlattice mirror with 9.4-nm period [11].

6.3 Disruption

Epitaxy is particularly sensitive to degradation by impurities and defects. Moreover, complete disruption of epitaxy can occur if even a fraction of a monolayer of disordered contaminant exists on the substrate surface or accumulates on the film surface during deposition. This is because the depositing atoms need to sense the crystallographic order of the underlying material, and chemical forces extend only one or two atomic distances. An island of surface contaminant becomes the nucleus for the growth of nonepitaxial material, and this region often spreads with further deposition, as shown in Fig. 6.7, rather than being overgrown by the surrounding epilayer. Contamination can enter at any step in the thin-film process. We discussed its control in the evaporation step in Sec. 4.5.2 and in the transport step under Sec. 3.4. Substrate contamination removal for improving adhesion was addressed in Sec. 5.7. Here we focus on the additional substrate requirements that must be met to obtain epitaxy. These include crystallographic order, submonolayer surface cleanliness, and chemical inertness toward the depositing species.

Any **crystallographic disorder** at the substrate surface will be propagated into the depositing film. A few materials can be obtained as prepolished wafers with excellent surface crystallography, including those mentioned at the end of Sec. 6.1. In other cases, careful preparation is necessary to remove the disorder introduced by wafer sawing and mechanical polishing. The crystallographic damage produced by polishing-grit abrasion extends into the crystal beneath the surface scratches, to a distance of many times the grit diameter, as shown by the dislocation-line networks in Fig. 6.8a. This damaged region must be removed by chemical etching. To promote uniform etching and prevent pitting, the "chemical polishing" technique is used. In this technique, the etchant is applied to a soft, porous, flat pad which is wiped across the wafer. If the depth of etching is insufficient, some damage will remain, as shown in Fig. 6.8b, even though the surface may appear absolutely flat and smooth under careful scrutiny by Nomarski microscopy (see Sec. 10.1.2). However, these defects can be revealed by dipping the wafer in a "dislocation" etchant [12] that pref-

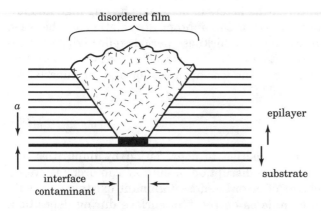

Figure 6.7 Effect of submonolayer surface contamination on epitaxy.

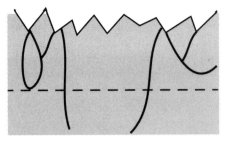
(a) Damage from abrasion extends beneath surface to level of dashed line.

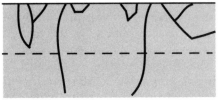
(b) Insufficient chemical polishing smooths surface but leaves subsurface damage.

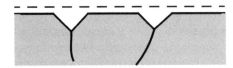
(c) Dislocation etch is applied to reveal residual bulk defects.

Figure 6.8 Crystallographic damage due to wafer sawing and mechanical polishing.

erentially attacks them and thereby decorates the surface with identifying pits and lines. The crystallographic disorder at these defects, consisting of strained and broken bonds, raises the local free energy and thereby increases reactivity toward the etchant. After sufficient chemical polishing, the only remaining defects will be those grown into the bulk crystal, as shown at the etch pits in Fig. 6.8c.

After crystallographic preparation of the substrate, **surface contamination** must be removed. In the final chemical cleaning step prior to wafer installation in the deposition chamber, one seeks to minimize residual surface contamination and also to select its composition so that it is more easily removed by the techniques available in the chamber. Many of the precleaning and in-chamber techniques were discussed in connection with adhesion promotion in Sec. 5.7; but for epitaxy, the cleanliness requirements are much more stringent. Carbon-containing molecules are generally the most difficult to remove among the common surface contaminants. They often dissociate upon heating, leaving behind graphite or carbides. Although any contaminant can be removed by sputtering under Ar^+ bombardment, this should be considered the method of last resort. Ions energetic enough to break contaminant bonds to the surface (>50 eV or so) also displace atoms of substrate material, as will be discussed in Sec. 8.5.2. This produces disorder and point defects (vacancies and interstitial atoms) near the surface, and it is not always possible to sufficiently anneal out this damage. In addition, some of the contaminant becomes displaced down into the substrate (the "knock-on" effect) rather than being removed.

For materials whose oxides can be evaporated or chemically reduced in the deposition chamber, oxidation is a useful final preparatory step because it can displace other contaminants and also passivate the surface against further adsorption. Weakly bonded oxides tend to be more volatile than their substrate materials and therefore can be removed by heating prior to deposition, as in the cases of GaAs and W. On the other hand, elements that form good ceramics and thus have very stable and refractory (nonvolatile) oxides such as Al_2O_3, SiO_2, MgO, and BeO, are not amenable to this treatment. Sometimes, nonvolatile oxides can be reduced to the metal plus H_2O in an H_2 plasma. SiO_2 can be reduced to volatile SiO by heating in an Si vapor beam. One common oxidizing solution is sulfuric/peroxide ($H_2SO_4/H_2O_2/H_2O$) in varying proportions. The peroxide should be added slowly because of the heat generated upon mixing. Ozone can also be used for oxidation and is readily generated with UV light.

The concept of displacing substrate contamination with an easily volatilized passivant has been advanced further in the case of Si with the discovery of H passivation by HF/ethanol solution [13]. Its effectiveness has since been quantified for various techniques of application [14]. The H terminates the dangling bonds at the Si surface and renders it remarkably unreactive toward O_2 and other ambient gases. Mild heating removes water weakly adsorbed on top of the H, and further heating removes the H. The T required depends on crystallographic plane.

The last cause of epitaxy disruption to be considered is reaction of the depositing species with the substrate to form **interfacial compounds** which have different symmetry and/or lattice constant (a_o) than the substrate. Some thermodynamic guidelines were given in Sec. 5.5 for predicting whether such compounds will form, based on the Gibbs free energies of formation, $\Delta_f G$. One important guideline is that interfacial compounds are more likely to form from elemental substrates and depositing species than from compound ones. For example, in the case of the Fe-GaAs interface, Fe readily grows epitaxially on GaAs (Fig. 6.1a), but GaAs on Fe does not do so [2]. Presumably, the arriving Ga and As_4 in the latter case are forming GaFe alloy and/or FeAs or $FeAs_2$, both known compounds. These reactions are inhibited in Fe epitaxy by the chemical stability of the GaAs substrate. Similarly, the alkaline-earth fluorides MF_2 (M = Mg, Ca, Ba), which evaporate molecularly, can be grown epitaxially on elemental substrates such as Si because of the stability of these vapor molecules.

Interfacial reactions can be blocked either by surface passivation or by the deposition of an epitaxial barrier layer. Passivation by adsorbed As is believed to be operative [15] in the growth of GaAs on Si. Arsenic has an appropriate bond length to Si, has a tripod-like bond configuration, and has a valence of 3 (versus 4 for Si), so it is able to fully terminate the Si surface bonds with little bond strain, resulting in very low surface energy. Arsenic passivation also prevents the formation of $SiSe_2$ in the growth of ZnSe on Si from Zn and Se_n vapors [15].

To block interfacial reactions with barrier layers, one seeks barrier materials that match the substrate crystallographically and arrive at the surface as stable compounds. For example, MgO has been successfully applied as a barrier layer in the epitaxy of the high-T_c superconductor $YBa_2Cu_3O_{7-\delta}$ (YBCO) on GaAs [16]. The growth of high-T_c superconductors such as YBCO on semiconductors is desirable for integrating high-speed superconducting devices into microcircuitry.

6.4 Growth Monitoring

Epitaxy from the vapor phase can be carried out over a wide range of ambient pressure from ultra-high vacuum (UHV) to at least one atmosphere. When pressure is high enough so that the mean free path of the source-material molecules, l, is much less than the distance, L, they must travel to reach the substrate, the process is operating in the fluid-flow regime [Kn << 1 in Eq. (2.25)]. The contributions of fluid-flow and gas-collision factors to film deposition behavior in this regime will be discussed in Chap. 7. For the present, we will focus on processes operating in the molecular-flow regime (Kn > 1). A principal advantage of this regime is that the vacuum environment allows access to the surface by particle probe beams of molecules and electrons. In Sec. 4.8, we discussed the monitoring of deposition flux by the vibrating quartz crystal and by optical techniques, both of which can be used in either flow regime. Here, we will see that the use of particle probe beams can deliver much more information about the growth process, especially when the growth is epitaxial. Indeed, most of what is known about film growth behavior on the atomic scale has been learned by monitoring epitaxy using such beams. In particular, Arthur's study of the reaction kinetics of Ga and As_4 beams with GaAs surfaces and Cho's adaptation of electron diffraction to monitoring GaAs surface structure during growth advanced the deposition technique of vacuum evaporation to the much more precisely controlled technique of molecular-beam epitaxy (MBE) [17].

The basic modulated-molecular-beam experiment is illustrated in Fig. 6.9. The nearly square waveform of the mechanically chopped incident flux is spread out somewhat by interaction of the molecules with the surface. From the shape of the resulting adsorption and desorption transients, the kinetic parameters that were discussed in Sec. 5.1 can be extracted. Although these experiments could also be done with polycrystalline films, surface behavior varies with crystallographic plane. The reaction kinetics of various deposition chemistries will be discussed in Sec. 6.5, and much of the work on GaAs has been reviewed elsewhere (Joyce, 1985). Since the technique is usually used as a research tool to investigate surface reaction kinetics rather than for routine deposition monitoring, its further discussion is beyond our present scope. The following two subsections will focus instead on the use of electron spectroscopy for surface chemical analysis and on electron diffraction for surface crystallographic analysis.

6.4.1 Electron spectroscopy

An energetic probe beam of electrons or photons can eject an electron out of an inner shell of a near-surface atom into vacuum, as shown for

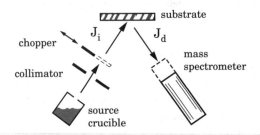

(a) beam geometry

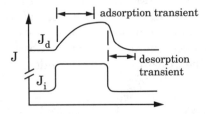

(b) typical incident and desorbing waveforms

Figure 6.9 Modulated-molecular-beam study of deposition reaction kinetics.

an atom of carbon in Fig. 6.10. The escaping electron, (2), is known as a secondary electron or a photoelectron, respectively. The inner-shell vacancy is then filled by an electron dropping down from a higher shell of that atom, (3). The potential-energy difference between the two shells is thereby released either into a photon, (4), or into the kinetic energy of an escaping outer-shell electron, (5). The electron is called an Auger ($\bar{\text{o}} \cdot \text{zh}\bar{\text{a}}'$) electron after its discoverer. These phenomena have been exploited in various analytical tools for surface elemental analysis, as summarized in Table 6.1. In all of these tools, the chemical elements can be identified by peaks in the energy spectrum of the escaping photons or electrons, since the positions of these peaks are determined by the energy levels of the electron shells in the probed atom. There are two notations used for these shells, and they have different symbols for the various quantum numbers of the atomic orbitals. In both notations, the principal quantum number, n, is listed first; it is called 1 through 7 in chemical notation and K through Q in spectroscopic notation. (Recall that the n of the outermost shell also identifies the period of an element in the periodic table.) The correspondence in notation for the other quantum numbers is listed in Table 6.2 for n = 4; the correspondence is the same for other values of n. By convention, chemical notation is used for x-ray photoelectron spectroscopy (XPS), and spectroscopic notation is used for Auger-electron spectroscopy (AES).

The four techniques listed in Table 6.1 differ considerably in the spot diameter probed (as listed) and also in the depth probed into the solid. We first discuss the two techniques that involve detection of x-ray photons. These techniques are generally used for film analysis *after* deposition (Sec. 10.2.1), but it is convenient to include them here with the electron-detection techniques. When x-rays are also used as the probe beam (x-ray fluorescence), the area probed is large, since x-rays cannot easily be focused. Also, x-rays typically penetrate >1 µm except at grazing incidence. At grazing angles of less than the critical angle, ϕ_c, total external reflection of x-rays occurs, and the probed depth becomes only

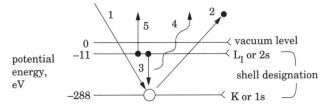

Figure 6.10 Electron and photon interactions in carbon: (1) probe beam, (2) ejected electron, (3) vacancy filling, (4) fluorescent photon, and (5) Auger electron.

TABLE 6.1 Names of Elemental-Analysis Techniques that Use X-ray or Electron Spectroscopy

Probe species (1)	Detected species	
	Photon (4)	Electron
Photon	x-ray fluorescence [10 mm]	x-ray photoelectron spectroscopy (XPS) or electron spectroscopy for chemical analysis (ESCA) (2) [100 µm]
Electron	electron microprobe [100 nm]	Auger-electron spectroscopy (AES) (5) [10 nm]

Note: Numbers in parentheses refer to Fig. 6.10; distances in brackets are approximate minimum surface diameters probed.

TABLE 6.2 Comparison of Atomic-Orbital Notations

Chemical	Spectroscopic
$4s$	N_I
$4p_{1/2}$	N_{II}
$4p_{3/2}$	N_{III}
$4d_{3/2}$	N_{IV}
$4d_{5/2}$	N_V
$4f_{5/2}$	N_{VI}
$4f_{7/2}$	N_{VII}

Note: Shells listed all have principal quantum number n = 4 = N.

10 nm or so, giving this technique the highest surface sensitivity among those in Table 6.1: $\approx 10^{10}$ atoms/cm^2. Unfortunately, it is too awkward to incorporate into film-deposition equipment. By the way, incident angle here is measured from the plane of the surface; for example, $\phi_c = 0.6°$ for a Cu(Kα) x-ray beam. This differs from the convention in light optics of measuring incidence from the perpendicular (Fig. 4.18). When the x-rays from the sample are stimulated by an electron probe beam rather than by an x-ray beam, the technique is called electron microprobe analysis and is frequently incorporated into scanning electron microscopes (SEMs). Here, the analyzed volume is determined by the teardrop-shaped volume through which the probe beam scatters before dissipating its energy. The diameter and depth of this volume are hundreds of nm, compared to the tens of nm incident beam diameter, so the electron microprobe is not considered a surface-analysis technique.

6.4.1 Electron spectroscopy

The surface-analysis techniques used for thin-film process monitoring are those that analyze escaping electrons: AES and XPS. These 100–1000 eV electrons can escape only from within a few nm of the sample surface, even though the probed depth may be much larger. The area probed is determined by the incident-electron-beam diameter in AES and by the photoelectron focusing optics in XPS. This area is much smaller in AES, as seen in Table 6.1. Both techniques have surface sensitivity of 10^{-2} to 10^{-3} ML and are therefore valuable for ascertaining the level of surface cleanliness necessary for good epitaxy. Figure 6.10 shows the electron energy levels for the element C, a ubiquitous surface contaminant. XPS typically uses the Al(Kα) x-ray line at 1487 eV as a probe. The photoelectron that this photon ejects must escape the 288-eV potential well depth (the "binding energy") of the 1s shell, and it therefore escapes with 1487 − 288 = 1199 eV of kinetic energy. In AES, a vacancy in the K shell (=1s) is filled from the L shell, and an Auger electron is ejected from L. This Auger-electron peak is designated as KLL; for higher elements, there are also LMM and MNN Auger peaks. The C(KLL) Auger electron escapes with kinetic energy equal to the K-L energy difference less the L well depth: that is, (288 − 11) − 11 = 266 eV, *independent* of the energy of the probe beam, which is a few keV. The escape energy also tells something about the chemical bonding state of the atom. The potential-well depth of an atomic inner shell shifts a few eV with bonding state, since bonding affects the electron density in the valence shell and thereby affects the screening of the inner shells. Thus, an atom stripped of valence electrons by bonding to an electronegative element such as O will have a higher binding energy. Chemical shifts such as this are easier to measure in XPS, because the peaks are sharper.

Quantitative analysis with AES and XPS requires considerable care [18], but 10–20 percent accuracy is readily achieved and is sufficient for many thin-film monitoring tasks. Besides measuring the level of substrate contamination, surface analysis can detect the surface accumulation of an unwanted depositing impurity or of a deliberate dopant impurity. The problem of dopant surface segregation is discussed in Sec. 6.5.6. One can also distinguish smooth growth from island growth by the rate at which the substrate-atom peaks attenuate, as discussed in Sec. 6.7.2. All of these data can be obtained for polycrystalline and amorphous growth as well.

A fundamental concern in any measurement is the degree to which the act of measuring is perturbing the quantity being measured. With AES and XPS, the major perturbation is the increase in chemisorption caused by the flux of energetic electrons passing through the surface. Recall from Sec. 5.1 that most vapor molecules adsorbing on a surface first enter a weakly bound precursor state; then, either they desorb or they chemisorb into a strongly bound state. Irradiation by electrons or other energetic species can supply the activation energy to increase the chemisorption rate. For example, an electron probe beam increases the fractions of adsorbing carbonaceous background gases (such as CO and CH_4) that dissociate and leave graphite on a surface rather than desorbing as molecules. Thus, the C signal increases with monitoring time for surfaces being analyzed under mediocre vacuum conditions. The degree of perturbation is much less for XPS, where both the collision cross section of the probing x-rays and the flux of escaping electrons are much less than those of the AES probe beam.

Similar perturbations to the adsorption and reaction kinetics of film-forming precursors would be expected, and this should be borne in mind when monitoring film growth with an electron beam. AES is usually used only during a growth interruption with sample transfer, because of equipment geometry problems. On the other hand, electron *diffraction* is the principal tool for continuous monitoring during epitaxial growth, as we will see in the next section. Degradation of semiconductor epilayer properties due to such monitoring has indeed been observed, so the beam (≈1 mm diameter) should be directed at an unimportant area or only used on practice runs. Even then, one might expect the beam to perturb the surface structures being probed, and to my knowledge this issue has not been addressed. Nevertheless, valuable information about surface and epitaxy behavior has been obtained by diffraction monitoring.

6.4.2 Diffraction

Both x-rays and electrons are diffracted from a crystal when their wavelength corresponds to some periodicity in the crystal lattice, and both are useful for examining thin-film crystallography. The observation of electron diffraction was among the earliest evidence for the wave nature of matter. Wavelength is related to energy by Eq. (6.5) for x-ray photons, and by Eq. (6.3) for electrons. To convert electron energy to electron-gun voltage, we note that electrons acquire kinetic energy, E_t, by acceleration in an electric field, $\mathbf{E} = -dV/dx$. The force on the electron, F, is equal to the charge ($q_e = -1.60\times10^{-19}$ C in SI units) times $\mathbf{E}$. Thus,

$$E_t = \int_x F\,dx = \int_x q_e\left(\frac{-dV}{dx}\right)dx = \int_{V_e} -q_e\,dV = -q_e V_e \qquad (6.6)$$

where V_e is the voltage drop from the electron gun's cathode (electron emitter) to the sample. Using this equation and the free-electron mass, m_e, for m_e^* in Eq. (6.3) gives

$$\lambda_e(\text{nm}) = \frac{h}{|\mathbf{k}|} = \sqrt{\frac{1.51}{V_e}} \qquad (6.7)$$

In the subsections below, we will discuss both low-energy electron diffraction (LEED), where the beam is directed perpendicular to the surface and where $V_e \sim 100$ V, and reflection high-energy electron diffraction (RHEED), where the beam grazes the surface and where $V_e \sim 10$ kV. The RHEED diffraction pattern is less sharp and more difficult to interpret, but it contains additional information on surface roughness. The RHEED pattern also can be monitored during film growth, since the electron optics are out of the way of the stream of depositing material. Since both 100-V electrons and 10-keV grazing electrons penetrate only a few atomic layers beneath the surface, electron diffraction under these conditions gives information mainly about the periodicity of the two-dimensional (2D) surface lattice, with some influence from the underlying layers. Conversely, x-ray diffraction gives information only about the bulk (3D) lattice of a thin film unless the beam is operated in the grazing-incidence mode which was discussed in the previous section.

6.4.2.1 X-rays. X-ray techniques for film analysis are generally used after deposition (Sec. 10.1.4) rather than for growth monitoring, but it is convenient to discuss them here under the topic of diffraction. The geometry of diffraction will be introduced for the case of x-ray diffraction from the 3D lattice. In Fig. 6.11a, the incident beam penetrates the lattice and scatters from each of the atoms in the 3D array. When the incident and scattering angles, θ, are equal as measured from some plane of atoms in the lattice (the x-y plane here), the path lengths of rays scattered off of any of the atoms in that plane will be the same. This results in constructive interference of those rays and produces a plane wavefront which may be thought of as having been *specularly reflected* off of that plane of atoms. The wave vectors of the incident and reflected beams, $\mathbf{k}_0$ and $\mathbf{k}_1$, describe their direction and also their wavelength, λ, by Eq. (4.44). Here, we set $|\mathbf{k}| = 1/\lambda$ rather than $2\pi/\lambda$. Now, the path-length difference between beams reflected from successive planes in the z direction is the distance 2L shown in Fig. 6.11a. For constructive interference, 2L must be an integral multiple, j, of λ; that is,

$$2L = j\lambda = 2a \sin\theta \qquad (6.8)$$

where a is the plane spacing. This is known as the Bragg condition. When "white" x-rays (broad λ range) are used, the Bragg condition is

248   Epitaxy

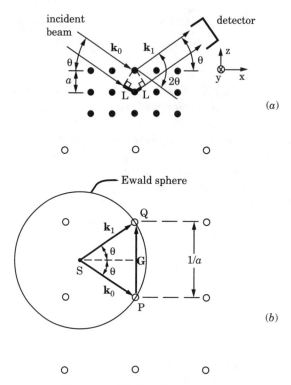

Figure 6.11 Geometry of bulk diffraction from a stack of x-y atomic planes: (a) real-space construction and (b) reciprocal-space construction.

satisfied for planes of any a, because some λ in the range will always satisfy Eq. (6.8). Then, reflected beams appear from atomic planes oriented at many angles, with beam intensity being highest for the planes having the highest areal concentration of atoms. The back-reflected beams can be recorded on photographic film placed perpendicular to the beam, giving a **"Laue" pattern,** which is useful for determining the symmetry and orientation of bulk single crystals. Other x-ray techniques are more useful for thin films, as outlined below.

In polycrystalline films, the degree of preferential orientation, or texturing, is measured using monochromatic x-rays and a **"2θ" or Bragg-Brentano scan**. There, the thin-film surface is oriented in the x-y plane of Fig. 6.11a, and θ is measured from that plane. Then, θ is scanned by rotating the sample about the y axis. Simultaneously, the x-ray detector is rotated through 2θ to keep it at the specular angle with respect to the film surface. At values of 2θ for which the atomic periodicity a perpendicular to the film surface satisfies the Bragg condition for the λ being used, a peak appears from which a can be calculated. The a value identifies the atomic plane, and the peak intensity is a *qualitative* measure of the degree of texturing; that is, intensity increases with the fraction of crystallites in the film which have that atomic plane parallel to the surface. The width of the peak, $\Delta(2\theta)$ (in radians), at half of its maximum intensity is a measure of the size of the crystal grains. This is because a larger stack of planes contributing to destructive interference at "off-Bragg" angles results in a sharper Bragg peak, as described by the Scherrer formula [19]:

$$b = 0.9\lambda/\Delta(2\theta)\cos\theta \qquad (6.9)$$

When the grains are larger than the film thickness, h, then b = h; but when they are smaller, then the grain size can be estimated from Eq. (6.9).

Other x-ray techniques are used for epitaxial films. The degree of atomic ordering can be determined by fixing the detector angle, 2θ, at the Bragg angle in Fig. 6.11a using a monochromatic beam. Then, the decrease in diffracted intensity is measured as the sample is "rocked" about the y axis. The sharper the **"rocking curve,"** the higher the crystalline perfection. Another technique is the **topograph**. There, a single diffracted beam is imaged in a microphotograph taken very close to the epilayer surface. With good beam collimation, positions on the image will correspond to positions along the surface to a resolution of a few μm. At positions where the crystal is misoriented, strained, or otherwise distorted, diffracted intensity is reduced, so the topograph reveals a variety of crystal defects. Even dislocations, which are line defects (more in Sec. 6.6), can be resolved because of the strain fields surrounding them. Finally, the **grazing-incidence** mode can be used to examine the crystallography within a few nm of the surface. There, the film surface is in the x-z plane in Fig. 6.11a, and the incident and reflected beams are almost parallel to it, so that the a of planes *perpendicular* to the film surface is being measured [20, 21].

6.4.2.2 Reciprocal space. The directions of diffracted beams may be predicted in an alternative and more versatile way than the Bragg geometry of Fig. 6.11a, and that way uses "reciprocal" space geometry. The relationship of reciprocal space to real space is illustrated in Fig. 6.12 for three cases. In all cases, a periodicity of a in real space converts to a periodicity of $1/a$ in reciprocal space. (Sometimes $2\pi/a$ is used.)

Case 1 A stack of x-y atomic planes spaced along z becomes a row of reciprocal-lattice points spaced along z. Stacks of planes in many

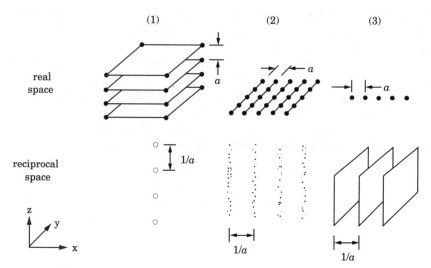

Figure 6.12 Correspondence between real and reciprocal space.

directions, as in a bulk crystal, would produce a 3D array of reciprocal-lattice points.

Case 2 Parallel rows of atoms along y and spaced along x become reciprocal-lattice "rods" along z and spaced along x. Rows along many directions in x-y would produce a 2D array of rods.

Case 3 A single row of atoms spaced along x becomes a stack of reciprocal-lattice y-z planes spaced along x.

To further examine the relationship between real and reciprocal space, observe that rows of atoms along x plus rows along y would become stacks of intersecting y-z and x-z reciprocal-lattice planes. The lines of intersection of these planes along z would form a 2D array of rods identical with the array described in Case 2. Now if rows of atoms are added along z, the corresponding reciprocal-lattice planes in x-y would intersect the array of rods, and the points of intersection would form the 3D array of reciprocal-lattice points described in Case 1.

Figures 6.11a and 6.11b compare the geometrical constructions in real space and reciprocal space for the case of Bragg diffraction from a stack of x-y planes. Figure 6.11b includes points from other planes parallel to y as well. The reciprocal-space construction operates as follows. Translate the incident-beam vector $\mathbf{k}_0$ so that it terminates on reciprocal-lattice point P as shown. Then draw a sphere having the length of that vector $(1/\lambda)$ and centered at the vector origin, S. Wherever this "Ewald sphere" intersects another reciprocal-lattice point, such as Q, there will lie the termination of a diffracted-beam vector originating at the sphere's center, such as the $\mathbf{k}_1$ shown. In vector algebra, this diffraction condition is expressed as

$$\mathbf{k}_0 + \mathbf{G} = \mathbf{k}_1 \tag{6.10}$$

where $\mathbf{G}$ is the reciprocal-lattice vector connecting the two points P and Q. From the geometry of Fig. 6.11b, it follows that

$$|\mathbf{G}| = j\left(\frac{1}{a}\right) = 2\left(\frac{1}{\lambda}\right)\sin\theta \tag{6.11}$$

which is identical to the Bragg formula, Eq. (6.8). Now the utility of the Miller indices (Sec. 5.3.1) for crystal-plane designation becomes clear, since they in fact describe orientations in reciprocal space. Moreover, any diffraction situation may now be visualized in reciprocal space. For example, when white x-rays are used to record the Laue pattern of a single crystal, the sphere has variable radius and may be visualized as expanding and contracting along S-P with its periphery fixed at P. Whenever it intersects a reciprocal lattice point, there will be a diffracted beam. In the 2θ scan, the sphere has fixed radius, and the reciprocal lattice is being rotated about P.

6.4.2.3 Low-energy electron diffraction. We now turn to the electron diffraction techniques used to monitor epitaxial film growth. We will use both the real-space and the reciprocal-space constructions to clarify the correspondence between them and also because some aspects of diffraction are more conveniently discussed in one versus the other realm, just as the electron is sometimes more conveniently thought of as a particle and sometimes as a wave. Indeed, the wave vector introduced above is also the electron momentum vector of Eq. (6.3) if one drops the $\hbar$. Figure 6.13a shows the geometry of LEED, where the beam is perpendicularly incident on a surface lattice consisting of rows of atoms along the y direction (perpendicular to the paper) and having spacing a in the x direction. The detection optics consists of a spherical phosphor screen into which the electrons are accelerated to generate phosphorescence. The screen is preceded by a spherical grid biased at just below the electron-gun voltage so that it retards any electrons that have lost energy upon being scattered from the crystal lattice. This includes, for example, electrons that have interacted with an inner shell orbital as in Fig. 6.10. These inelastically scattered electrons contain no diffraction information, since they are no longer monochromatic.

The surface-diffraction situation illustrated in Fig. 6.13a is analogous to an optical diffraction grating, and beams will occur at any angle for which the path length difference L_1 is an integral number of wavelengths; that is, whenever

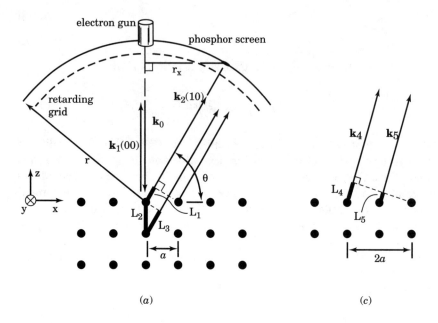
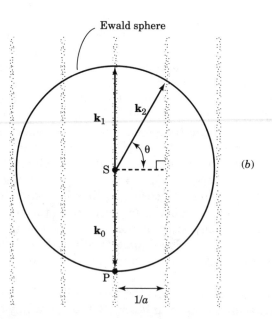

Figure 6.13 LEED geometry for perpendicular incidence: (a) real-space construction, (b) reciprocal-space construction, and (c) interference from atoms spaced at $2a$.

$$L_1 = j\lambda = a\cos\theta \tag{6.12}$$

The various beams are designated by their order in x and y as ($j_x j_y$); thus, the specularly reflected beam is the (00) beam. On a phosphor screen placed at radius r from the surface, bright spots appear at radius r_x measured in the x-y plane as shown, where

$$r_x = r\cos\theta = j\lambda r\left(\frac{1}{a}\right) \tag{6.13}$$

Equation (6.12) has been used to obtain the second equality here. In reciprocal space, the surface lattice becomes an array of rods as shown in Fig. 6.13b, with the incident-beam momentum (wave) vector terminating on a rod at point P. Wherever the Ewald sphere intersects a rod, there will lie the termination of a diffracted-beam vector, as shown for $\mathbf{k}_1$ and $\mathbf{k}_2$. Inspection shows that for this geometry,

$$\cos\theta = \frac{j(1/a)}{1/\lambda} \tag{6.14}$$

which is identical to Eq. (6.12). Remember that we are dealing with a sphere, not just the circle shown on paper, so rods are intersected in the y direction as well. The 2D pattern of spots on the screen has, in fact, precisely the same symmetry as the intersection of the reciprocal lattice with the Ewald sphere. Part of the power of the reciprocal lattice is that it allows one to visualize in this way the entire diffraction pattern at once.

As the electron beam's voltage is increased, its λ decreases in accordance with Eq. (6.7). Thus, the Ewald sphere expands, and its intersections with the rods move up in z, causing the pattern of spots on the screen to converge and new spots to enter from the periphery as the sphere reaches farther out into the array of rods. Meanwhile, there is also some elastic scattering of electrons from the second layer of atoms down from the surface, as shown in Fig. 6.13a. Contrary to the case of x-rays, this scattering is significantly attenuated relative to the surface scattering because of the high collision cross section of atoms for low-energy electrons. Still, it does cause Bragg interference with the surface scattering, so that when the path length ($L_2 + L_3$) is an integral multiple of λ, the beam is much brighter than when it is π out of phase (j + 1/2)—the "off-Bragg" condition. In terms of reciprocal space, this situation may be viewed as a partial fragmenting of the rods toward points due to a partial contribution from the z dimension. The voltages at which the Bragg peaks occur give the lattice spacing in z for the top layer of atoms. By the way, this spacing can differ by 10 percent or so from the spacing in the bulk due to bond relaxation and reconstruction of the surface lattice. LEED patterns are quite beautiful to observe. The array of spots, which elegantly replicates the symmetry of the surface lattice, expands and contracts as voltage is varied, while spots from different row spacings modulate in brightness at different voltages as they fall into and out of the Bragg condition.

LEED has been the principal tool for studying the atomic reconstruction of surfaces. This reconstruction is driven by the resulting surface-energy reduction, as discussed in Sec. 5.3.1, and it can pass through many phases of symmetry with changes in surface composition and T. These changes can also be observed with RHEED, and we will see in Sec. 6.5 how they are used to regulate surface composition during the epitaxial growth of compounds. A row of surface atoms having a periodicity of m times the unreconstructed lattice periodicity generates "extra" LEED spots of fractional order 1/m. To see how this occurs, consider Fig. 6.13c for the *un*reconstructed surface. For path length $L_4 = \lambda$, there is no beam, because the ray with vector $\mathbf{k}_4$ is exactly cancelled by the ray $\mathbf{k}_5$ which has $L_5 = (1/2)\lambda$. However, if the atoms at spacing $2a$ are raised relative to the ones between them, so that their scattering is stronger, the cancellation will be incomplete, and half-order spots will appear. These will be weaker than the primary spots but quite distinct. There are many other ways in which surface periodicity can be modulated to generate fractional-order spots. Figure 6.14a shows a typical surface array on (001) which has black atoms positioned between four nearest neighbors in the underlying layer and spaced at $2a$ in x and y. It is therefore designated as a (2×2) array, and it generates the half-order spot array shown in Fig. 6.14b. If the array also had atoms centered on each square at positions $\otimes$, it would be called a $c(2 \times 2)$ array. Alternatively, this $c(2 \times 2)$ could be described as a square array of spacing $\sqrt{2}a$ and rotated 45° about the azimuthal (z) axis, in which case it would be designated as $\sqrt{2} \times \sqrt{2}$ R45°.

In "high-resolution" LEED [22], the sharpness of the (00) beam is used as a measure of ordering and smoothness of the top few ML, by analogy to the x-ray technique of Eq. (6.9).

6.4.2.4 Reflection high-energy electron diffraction. With this background, we can now discuss RHEED, which is the principal tool for monitoring surface crystallography during epitaxial film growth. In RHEED, the incident and diffracted beams both make grazing angles to the surface, so that the electron optics are out of the way of the stream of depositing material. The transformation of a LEED pattern into a RHEED pattern may therefore be visualized in Fig. 6.13a by

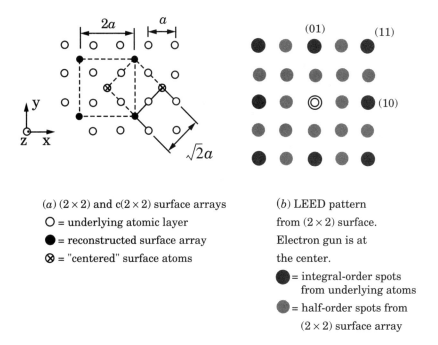

Figure 6.14 Symmetry and diffraction for a reconstructed (001) surface.

rotating the k_0 vector counterclockwise about the y axis, which causes the specularly reflected k_1 vector to rotate clockwise. In reciprocal space, the center of the Ewald sphere in Fig. 6.13b is being rotated counterclockwise about the point P at its periphery, while the tip of the k_1 vector remains on the same rod and slides down toward P as that vector rotates clockwise. The resulting geometry is shown in a perspective view in Fig. 6.15. Observe that because the reciprocal-lattice rods are now almost tangential to the Ewald sphere and because the sphere diameter is very large for a 10-keV beam, the diffraction spots are elongated into streaks on the screen at the right of Fig. 6.15a. If the sphere were a shell of zero thickness and the rods were lines of zero cross section, the pattern would still be spots. However, the sphere has finite thickness due to some λ spread and angular spread in the incident beam, and the rods have finite cross section due to disorder in the surface lattice arising from thermal vibration (phonons), atomic steps, and bond strain, as discussed further by Saloner (1986). This streaking obscures diffraction information about periodicity in the x direction, and only information on periodicity in y is obtained, by the spacing, r_y, of the streaks in the y direction. In order to observe the x periodicity, the substrate must be rotated about the z axis by an azimuthal angle, ω, of 90°.

256 Epitaxy

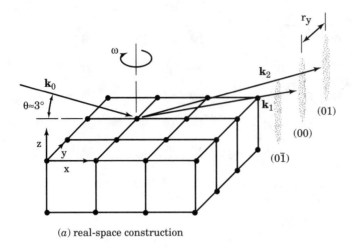

(a) real-space construction

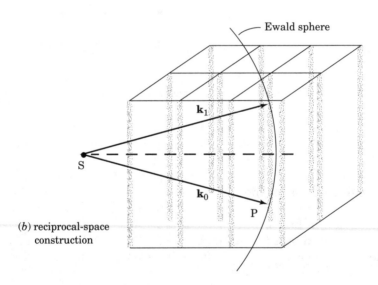

(b) reciprocal-space construction

Figure 6.15 RHEED geometry.

The occurrence of streaking may be further analyzed by the real-space construction of Fig. 6.16, which also illustrates an important effect due to surface roughening. We have so far considered only atomically flat 2D surfaces, but in real epitaxy, 3D islanding is a frequently encountered problem, as we will see in Sec. 6.7.2. A grazing electron beam can be *transmitted* through a small island, and this adds a third dimension to the diffraction geometry which is represented by the atoms 3 and 4 in Fig. 6.16. Consider first the scattering of the incident

6.4.2 Diffraction

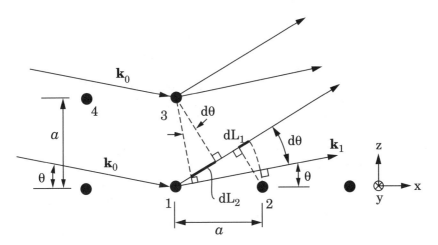

Figure 6.16 Effect of scattering-angle change on 2D and 3D diffraction.

beam off of atoms 1 and 2 in the 2D surface lattice. At the specular angle, θ, the path lengths of these two scattering events are the same, so the scattered rays are in phase, interference is constructive, and the specular spot appears on the screen. We now want to know how rapidly the scattered rays fall out of phase with each other as we move up in z on the screen, which means moving away from the specular angle by dθ. Now, the path length measured along scattered ray $\mathbf{k}_1$ is $a\cos\theta$ between atoms 1 and 2, so the change in path length, dL_1, with scattering angle is given by

$$dL_1/d\theta = d(a\cos\theta)/d\theta = -a\sin\theta \qquad (6.15)$$

which is close to zero for grazing angles of scattering. That is, the path length, and therefore the phase, changes little as we move away from the specular angle in z, and thus a streak is observed. On the other hand, for scattering off of atoms 1 and 3, the change in path length is dL_2, and by inspection of Fig. 6.16, we have

$$dL_2/d\theta = d(a\sin\theta)/d\theta = a\cos\theta \qquad (6.16)$$

which is near its maximum value for grazing angles, so that the phase changes rapidly as we move away from the specular angle in z. In terms of reciprocal space, the rods are being changed into points by the introduction of the z dimension into the sampled-atom array. Thus, islanded or otherwise roughened surfaces produce spotty RHEED patterns, and this gives a measure of the smoothness of growth to atomic-layer sensitivity.

A particularly striking RHEED phenomenon which also provides a valuable epitaxy monitoring technique involves oscillations in spot intensity as film growth proceeds. These oscillations occur only when growth is proceeding by 2D nucleation on atomic terraces as discussed in Sec. 5.3.3 and illustrated in Fig. 5.13. The various modes of epitaxial growth are discussed further in Sec. 6.7. When 2D nucleation reaches monolayer completion, the surface approaches atomic smoothness. However, when a monolayer is partially filled, there are steps of monolayer height at the edges of the 2D islands. The atoms at these steps scatter the incident electron beam much more strongly than do atoms on an atomic terrace, because the beam can impinge upon them directly, as shown in Fig. 6.17. Diffuse scattering from these steps removes electrons from the forward diffraction directions and thus attenuates the intensity of the spots. Spot intensity in fact oscillates with a period exactly equal to the monolayer growth time, allowing one to actually *count* the number of monolayers grown for precise construction of quantum-well and superlattice structures. Often the islands are anisotropic in shape, in which case the oscillation effect can be enhanced by rotating the substrate about the azimuthal angle ω (Fig. 6.15a) until the long edges of the islands are perpendicular to the incident beam. The attenuation between monolayers can be further enhanced by adjusting θ or λ so that the reflections from the top and bottom atomic-terrace surfaces are out of phase. This is the "off-Bragg" condition, where the path-length difference 2L shown in Fig. 6.17 is $(1/2)\lambda$ (see Exercise 6.7).

The quantitative analysis of RHEED oscillation behavior is much more complex than the above simple description, and it has been a subject of considerable study [23]. For example, we have considered only single-scattering (kinematic) interaction of the electron beam

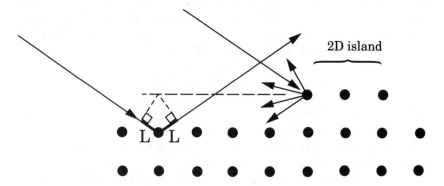

Figure 6.17 Attenuation of diffraction by 2D islands due to backscattering and off-Bragg interference.

with the lattice, whereas multiple-scattering (dynamic) events are known to significantly modify diffraction intensity. Among other effects, multiple scattering can cause a shift in the phase of the oscillations with respect to the completion of a monolayer [24], and one must therefore be cautious in using the phase of oscillation to mark the point of monolayer completion. Reflection-intensity oscillations occur with other grazing probe beams as well, including x-rays and light, so that the technique should see increased use for epitaxy in the fluid-flow regime. In MBE of KCl on KCl(001), oscillations of the reflected fraction of the KCl molecular beam were observed due to a decreased KCl trapping probability upon monolayer completion [25].

6.5 Composition Control

Film composition can vary whenever the film being deposited consists of more than one element. The control of composition is important in any alloy or compound film, not just in epitaxial films. We address it in the context of epitaxy because it is more easily studied under epitaxial conditions and because the electrical properties of epitaxial semiconducting compounds are especially sensitive to changes in composition. There are some cases in which the composition of the source vapor is identical to the desired film composition, and then composition control is not an issue. For example, ionically bonded compounds such as NaCl and CaF_2 evaporate as molecules and therefore deposit with the composition of the evaporant. More generally, however, the vapors of solid compounds are partly or fully dissociated into elemental species, as discussed in Sec. 4.4. Similarly, when the source materials are gaseous molecules, each molecular species usually carries only one of the film's constituent elements. Thus,

$$ZnCl_2(g) + H_2Se(g) \to ZnSe(c) + 2HCl(g)$$

where (g) denotes a gaseous species and (c) denotes the condensed phase of the compound being deposited.

Composition control centers on the relationship between the impingement fluxes, J_i, of the various vapor species and the resulting deposition fluxes, J_r, of the elements that constitute the film. Rarely will $J_r = J_i$ for all of the film's elements. Two basic equations for these relationships were presented in Sec. 5.1. Equation (5.7) expresses the surface reaction rate of the arriving species, R_r, in terms of its initial trapping probability, δ, and the activation energies for its reaction and desorption, E_r and E_d. Equation (5.8) states that J_r is less than R_r by the amount of re-evaporation of the reacted species. We can thus define a "sticking coefficient" for the kth elemental constituent of the film as

$$S_{ck} = \frac{J_{rk}}{v_k J_{ik}} \qquad (6.17)$$

where v_k is the number of atoms of that element in the impinging vapor species. The composition of the film will then be the ratio of the atomic impingement fluxes times the ratio of the S_c values. For a given set of vapor species, S_c is mainly a function of surface composition and substrate T, T_s, as we will see below.

6.5.1 Precipitates

Composition varies differently with J_i ratio for alloys versus compounds. Alloys can vary continuously in composition over a wide range with J_i ratio, whereas compounds have a narrow range of deviation from the stoichiometric ratio over which they can exist, as was discussed in Sec. 4.4 and illustrated in Fig. 4.6*b*. Beyond this range of deviation, a second phase forms which is richer in the excess element and is, in many cases, just the pure element. This second phase will either be vaporized or remain partly condensed as a precipitate on the film surface depending on its vapor pressure, p_v, at T_s. Clearly, the precipitate is undesirable. To avoid it in the case of an elemental second phase, one must maintain conditions such that the evaporation flux of that element from its own condensed phase, J_v, is greater than the flux at which that phase is condensing on the surface, J_c. That is, conditions must be "undersaturated." For the case of a pure material, these fluxes were given by Eqs. (4.17) and (4.18). For the present case of two (or more) impinging species reacting on the surface to form a compound film, the relevant J_i for Eq. (4.18) is the *excess* flux, J_{ie}, above the amount which is consumed in the compound-forming reaction.

To make a rough estimate of J_{ie}, it may be assumed that the arriving elements completely react with each other to form the film according to its stoichiometric formula. For example, consider this film-forming reaction typical of the II-VI semiconductors:

$$2Cd(g) + Te_2(g) \rightarrow 2CdTe(c) \qquad (6.18)$$

If the impingement flux of Cd is 3×10^{15} mc/cm$^2\cdot$s and that of Te$_2$ is 1×10^{15} mc/cm$^2\cdot$s, then only 2×10^{15} mc/cm$^2\cdot$s of the Cd can react to CdTe, so that $J_{ie} = 1\times10^{15}$ Cd/cm$^2\cdot$s. To determine whether this flux will condense, we further assume that the evaporation and condensation coefficients, α_v and α_c, in Eqs. (4.17) and (4.18) are equal. We then arrive at the condition

$$J_{vo} > J_{ie} \qquad (6.19)$$

6.5.1 Precipitates 261

for precipitate avoidance, where the upper-limit evaporation flux, J_{vo}, is related to p_v by Eq. (2.18) in accordance with the discussion following Eq. (4.16). The Eq. (6.19) condition essentially states that the excess Cd vapor is undersaturated. Elemental p_v values from Appendix B can be used to make these estimates. Since p_v increases exponentially with T, precipitate avoidance places a lower limit on T_s. Of course, if the elements were depositing in *exactly* the stoichiometric proportion, precipitates would not form at any T_s as long as the reaction to compound was proceeding rapidly enough. However, the mole-fraction boundaries of single-phase stoichiometry deviation are generally $<10^{-4}$, and it is not practical to control J_r ratios to anywhere near this degree of accuracy. There is another lower limit on T_s imposed in the case of gaseous source species by the need to activate their decomposition, as discussed in connection with Eq. (5.7).

The ultimate *upper* limit on T_s is when the film compound, or one of its elements, begins to re-evaporate faster than it is depositing. The difference between the upper and lower limits is the T_s "window" for achieving single-phase deposition. A similar lower T limit arose in considering whether a dissociatively evaporating compound would evaporate congruently—that is, in the stoichiometric ratio—or would accumulate a precipitate of the less volatile element in the source crucible (see Sec. 4.4). The behaviors of various classes of compounds described there apply here as well. That is, for metal-nonmetal compounds MY, Y is always more volatile than MY, so there is always a finite T_s window as long as Y is made to arrive in excess. Thus, for excess nonmetal arrival flux and within the T_s window, composition is self-regulating to within the single-phase region: all of the excess Y re-evaporates. This behavior applies to nonmetals from periodic table groups V, VI (the chalcogens), and VII (the halogens). For the II-VI compounds, the metals (Zn, Cd, Hg) also are more volatile than the compounds MY, so composition is self-regulating and precipitates are avoided *whichever* element is arriving in excess.

Metals other than those of group II are generally *less* volatile than their compounds, so excess arrival fluxes of these elements or the gas molecules carrying them must be avoided. This situation is illustrated for GaAs by the p_v data of Fig. 4.7. Recall from the discussion of that figure that the two curves for each element represent the Ga-rich and As-rich boundaries of the single-phase region. The minimum p_v for the more volatile element (As) over GaAs occurs at the Ga-rich boundary and is represented by the lower As_2 curve. At this boundary, the p_v of Ga is on the upper Ga curve, which coincides with that of elemental Ga. Thus, elemental Ga is more volatile than GaAs only below 640° C on the Ga-rich boundary, and only at even lower T as one moves toward the As-rich boundary. At 640° C, p_v(Ga) is only 2×10^{-5} Pa, which corresponds to much less than the arrival flux of Ga vapor required for reasonable deposition rate. Thus, it is not practical to deposit GaAs under conditions of excess Ga flux. Note also from Fig. 4.7 that because $p_v(Ga)$ decreases considerably at the As-rich boundary, the upper limit of T_s may be increased by depositing under excess As_4 flux.

Self-regulation of compound composition to within the single-phase region may also be thought of in terms of the sticking coefficients, S_c, of the impinging elements. The element whose flux is deficient will have high S_c because of its reactivity with the other element on the surface. For the element that is in excess, only that fraction of it which reacts will have high S_c, and the excess fraction will have $S_c = 0$ if that element is volatile. Thus, the overall S_c for the excess element decreases monotonically with increasing excess fraction. The compound deposition rate is governed by the J_i of the deficient element through Eq. (6.17). Although S_c for the deficient element tends to be high, it is not always unity. The most thoroughly studied case is GaAs (Joyce, 1985), where $S_c(As_4) = 1/2$ for deficient As_4 flux in the MBE T_s range of 700–900 K. Surprisingly, the unreacted arsenic desorbs as As_4 rather than as the As_2 form of arsenic vapor. This fact and reaction-kinetics data support a model which involves the interaction of *two* As_4 molecules on the surface, with half of the As reacting with Ga and the other half recombining to As_4; that is,

$$2As_4(a) + 4Ga(c) \rightarrow 4GaAs(c) + As_4(g)$$

This illustrates the potential complexity of surface-reaction kinetics in compound deposition. The detailed surface-reaction kinetics of other compound-film-forming reactions are generally not understood.

6.5.2 Alloys

In the case of alloys, composition is also governed by Eq. (6.17), although S_c is not such a strong function of composition as for compounds. Most low-p_v metals have $S_c \rightarrow 1$, so that composition is determined simply by the J_i ratio. Alloys of compounds behave similarly. III-V semiconducting films of this type were discussed in connection with Fig. 6.3. Three-element alloys of this type can have either two elements from group III or two from group V. In the former case, S_c values are near unity for the two III elements, so alloy composition is similar to the J_i ratio of these elements as long as T_s is not so high that metal re-evaporation becomes significant. For the case of two group-V elements, however, their S_c values are less than unity and are also sensitive to T_s and to the J_i ratio of the two V elements, making alloy composition control more difficult. Best control is obtained when the more reactive of the two V species is made deficient to regulate alloy composition, while the other V species is kept in excess to prevent precipitates (Joyce, 1985). This method works because the V species that is more reactive with the metal displaces the less reactive one on the surface. Surface reaction rate with the metal increases with Period for the group-V elements either in their elemental-vapor-molecule form, Y_n, or as the hydride gases, YH_3: that is, Sb > As > P > N in reactivity. In the case of N_2, reactivity is negligible without plasma activation. Passing N_2 or NH_3 gas through a plasma dissociates it and thereby activates its reaction with many metals for depositing various thin-film nitrides (see Sec. 9.3.2).

Alloys of the III-V compounds and perhaps other compounds also exhibit the fascinating phenomenon of spontaneous ordering [26, 27], whereby, for example, $Ga_{1-x}In_xAs$ arranges itself during epitaxy into alternating layers of Ga and In atoms. Presumably, this arrangement represents a state of lower Gibbs free energy than having the two metals randomly distributed on the available metal sites, because ordering results in a lower overall level of bond distortion than would random distribution of the two metals. That is, internal energy wins out over entropy in Eq. (4.5). It is also observed that the boundaries between antiphase domains (In/Ga versus Ga/In; compare Fig. 6.30) arrange themselves into regular patterns across the film surface [26].

6.5.3 Point defects and surface structure

Prevention of precipitates and regulation of alloy composition, as discussed above, represents the coarse level of composition control. Adjustment of composition within the single-phase region of deviation from the stoichiometric compound represents the fine level. Fine control is especially important to the electrical properties of semiconductors, because stoichiometry deviation is accommodated by the generation of native point defects (see Table 4.1). These defects often produce charge-carrier trap states deep within the band gap, as with the As_{Ga} antisite defect in As-rich GaAs. Point-defect concentration also determines the lattice site favored by impurity dopants and thus affects doping efficiency (more in Sec. 6.5.6). The narrow-gap lead-salt semiconductors have *shallow* point-defect states which act as dopants themselves [28]. In the case of the transparent semiconducting material indium-tin oxide (ITO), used as an electrode on flat-panel displays and other electro-optic devices, point-defect adjustment represents a compromise: metal-rich ITO has more electrical conductivity but less short-wavelength transparency.

Stoichiometry control begins with considering the vapor-solid equilibrium. The equilibrium vapor pressures, p_v, of a compound's elemental constituents depend strongly on the compound's degree of deviation from stoichiometry within the single-phase region, as was shown in Fig. 4.6c for the generic binary compound $M_{1-x}Y_x$, where x = 0.5 ± 10^{-4}, typically. The nonstoichiometric compound is, in effect, a very dilute solution of native point defects in the exactly stoichiometric compound, and the p_v of a diluent over a solution increases with diluent concentration, as anyone knows who has seen CO_2 pressure build up in a soda-pop bottle. Thus, the depositing film can be driven Y-rich by increasing the elemental incident-flux ratio, $J_i(Y)/J_i(M)$, or it can be driven M-rich by decreasing it. J_i may be thought of in terms of its "equivalent beam pressure," as expressed by p in Eq. (2.18). For $p > p_v$, conditions are supersaturated, and the film will be driven rich in that element. The ratio of point-defect concentration to p_v is not usually known, and concentration may not reach equilibrium during deposition anyway. Nevertheless, the qualitative trend is always the same, so flux ratio can be used empirically to control point-defect concentration.

When growth is epitaxial, the film's relative x position within the single-phase region is conveniently established by RHEED monitoring of surface reconstruction during deposition. The atomic reordering of crystal surfaces is driven by minimization of surface free energy, and the most energetically favorable ordering depends on surface composition, among other factors. We again use the example of GaAs, since it is the most studied compound epilayer, but the principles illustrated are quite general. Figure 6.18 shows one of the reconstructions commonly observed on As-rich GaAs(001). (Crystallographic notation was outlined in Sec. 5.3.1.) To understand the surface bonding here, recall from the description of zinc-blende symmetry in Sec. 6.1 that its atoms each have four bonds oriented in a tetrahedron. All bonds lie along $\langle 111 \rangle$ directions; so on the (001) plane, two bonds penetrate down into the crystal, and two stick up out of it and are rotated 90° about the z axis with respect to the other two. In Fig 6.18, the projections of the upward bonds from Ga (downward from As) on the x-y (001) plane are oriented along $[\bar{1}10]$. Only a few of the bonds beneath the Ga plane are shown. On top of the Ga plane at z = $(-1/4)a_o$ is an As plane from which one-fourth of the atoms are missing. To eliminate As dangling bonds, adjacent As atoms lean toward each other and cross-bond into dimers as shown. Every fourth dimer in the rows of dimers along $[\bar{1}10]$ is missing, resulting in 3/4 ML coverage by As and leaving Ga dangling bonds at the missing dimers. This arrangement has lower surface energy than does full As coverage despite the remaining dangling bonds, because surface charge neutrality is also a factor in energy minimization [29], and zinc-blende bonding is polar ($Ga^{\delta+}As^{\delta-}$).

6.5.3 Point defects and surface structure

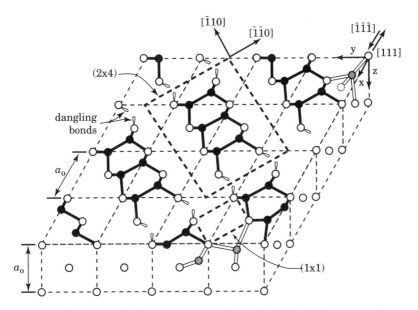

Figure 6.18 One of the As-dimer arrangements on the GaAs(001) surface. Here, dimer triplets are positioned in (2 × 4) symmetry. White circles = Ga atoms in fcc array, black circles = As atoms in dimer rows at $z = (+1/4)a_o$, and gray circles = As atoms in fcc array at $z = (-1/4)a_o$.

The 3/4-full dimer rows of this As-rich surface sometimes arrange themselves into the 2 × 4 unit cell outlined in Fig. 6.18, and sometimes into larger c(4 × 4) or c(2 × 8) unit cells [30]. (Note the comparison with the 1 × 1 unit cell of the underlying Ga plane outlined in the figure. Surface-structure notation is described in connection with Fig. 6.14.) This 2 × 4 symmetry appears in the RHEED pattern as 1/4-order streaks when the beam is directed along [110] and as 1/2-order streaks when it is directed along [$\bar{1}$10]. When this pattern is present during film deposition, the 3/4 As coverage will result in a film composition closer to the As-rich phase boundary of the single-phase region than to the Ga-rich boundary. Now, if substrate T were higher, making $p_v(As_4)$ higher, or if the $J_i(As_4)/J_i(Ga)$ flux ratio were lower, the 3/4 As coverage would not be sustainable. With decreasing As coverage, the RHEED pattern eventually changes to the 4 × 2 symmetry of the Ga-rich surface, where 4 × 2 is just 2 × 4 rotated 90° about the z axis. The rotation occurs because the Ga bonds are rotated 90° from the As bonds. In the figure, this transition amounts to shifting the stable surface down $(1/4)a_o$ from the As plane to the underlying Ga plane. This symmetry shift is seen in RHEED as a transition from 1/4-order to 1/2-order streaks or the reverse depending on the azimuthal angle (rotation about the z axis) of the electron beam. The transition provides a convenient benchmark for positioning deposition conditions within the single-phase region, and it is both more direct and more accurate than trying to control an absolute ratio of J_i values.

The symmetries described above are observed between about 720 and 870 K for GaAs MBE at 0.6 μm/h. Outside this range, other symmetries are observed, including less As-rich versions of the 2×4 having some As dimers missing [31]. One can construct a map in flux-ratio and T space which may be thought of as a surface phase diagram of RHEED symmetry [32], although it is not, strictly speaking, an *equilibrium* phase diagram when there is net deposition. Other planes of GaAs and other film materials have different surface symmetries, and these also can be used to tune deposition conditions. The RHEED pattern alone, however, does not reveal the actual atomic arrangement leading to a given diffraction symmetry, nor does it reveal all of the symmetry changes occurring within the surface-structure phase diagram. This information has instead been obtained using scanning-tunneling-microscopic (STM) observation of the actual atom positions resulting from a particular surface treatment [30].

Surface structure is important not only for controlling native point defects during deposition, but also for obtaining charge neutrality at the interface when growing one compound upon another. An excess of one or the other element relative to that needed to just satisfy the interfacial bonding can produce large electrostatic fields which disrupt smooth growth [31].

6.5.4 Gaseous sources

So far in this book, we have focused on the use of evaporating solids as vapor source material and have made only passing reference to the use of gaseous sources, although most of what has been said about deposition and epitaxy applies to any form of source species. We now introduce gaseous sources because of their utility in facilitating composition control. Here, as in Sec. 2.1, we adopt the practical distinction between a gas and a vapor rather than the strict one, the practical one defining a species as a gas if it does not condense when held above room T and below 1-atm partial pressure. We must also distinguish among vapors based on their equilibrium vapor pressure, p_v, because this determines their appropriate delivery method to the deposition reactor. Source species having $p_v < 10^2$ Pa or so at the wall T of the deposition chamber must be "physically" evaporated using heat (Chap. 4) or energy beams (Chap. 8), and they are used in physical vapor deposition (PVD) processes. PVD requires low-pressure operation and "line-of-sight" geometry from source to substrate because of wall condensation. Materials having $p_v > 10^2$ Pa or so at the wall T are used in chemical vapor deposition (CVD) processes, which are sometimes operated at 1 atm and can coat convoluted shapes or stacks of substrates as discussed in Chap. 7. Vapors with $p_v > 10^2$ Pa at *room* T can be metered and transported using the gas-flow-control equipment described in Sec. 7.1.

Most of the elements, with the exception of the alkali metals and alkaline earths (groups IA and IIA), can be converted to gases or to chemical vapors by reacting them with terminating radicals: that is, with atoms or functional groups having only one dangling bond. The resulting molecules do not chemically bond to each other and thus have high p_v. Group IA and IIA elements behave differently because they form ionically bonded molecules which *do* bond with each other—electrostatically. Commonly used gasifying radicals are H, the halogens (F, Cl, Br, I), carbonyl (CO), and H-saturated organic radicals, R, such as methyl (CH_3) and ethyl (CH_2CH_3). This gasification process is also the basic chemistry of plasma etching, which can be thought of as the reverse of CVD in this sense. Examples of gases and chemical vapors used in film deposition are AsH_3 (arsine), WF_6, $Ni(CO)_4$, and $(CH_3)_3Ga$. The last is an example of a metalorganic (or organo-metallic) compound and is used in MOCVD. When AsH_3 or PH_3 is used in molecular-beam epitaxy, it is sometimes passed through a heated tube to "crack" it into more reactive molecules [33] such as As_2.

Although early CVD was always operated at 1 atm, it can be operated at any pressure down to ultra-high vacuum (UHV). When the pressure is low enough for PVD, which means that the Knudsen number [Eq. (2.25)] is not too much less than unity, one usually has the choice of using either a physical-evaporation source or a gaseous source for a given element. The advantages of gaseous sources include: (1) uniform distribution over large areas using multiple injection orifices, (2) premixing of gases to avoid alloy compositional gradation across the substrate, (3) source replenishment that is less frequent and does not require breaking vacuum, (4) avoidance of the particle "spitting" problem often encountered with thermal sources (Sec. 4.5.1) and with energy-beam sources (Sec. 8.2), and 5) more readily achieved selective-area deposition because of higher activation energy for reaction with foreign surfaces (Sec. 7.3.3).

The main disadvantages of gaseous sources involve safety and contamination. Most of the hydrides and carbonyls are poisonous, especially arsine. The metalorganics are pyrophoric, which means that they ignite spontaneously if exposed to air. Gas safety is discussed further in Sec. 7.1.1. Regarding contamination, some elements are difficult to obtain in compounds of sufficient purity at reasonable cost. Even pure compounds may incorporate some of the gasifying radical into the depositing film. For example, the C level is higher in GaAs deposited by metal-organic CVD than by MBE. It is even higher when the CVD pressure is very low—in the molecular-flow regime. It has been proposed that at lower substrate T this contamination is due to the lack of preheating and decomposition during transport (Houng, 1992), and at higher T to the lack of H_2 overpressure [34] to drive the reaction of surface C toward CH_4. In GaAs, C is a p-type dopant, so levels of only 10^{15} cm^{-3} can be problematic in some applications.

In molecular-flow compound deposition, either the metal (M) or the nonmetal (Y), or both, or neither, may be supplied as a gas, and this has led to a proliferation of terminology whose recommended conventions (Houng, 1992) are listed in Table 6.3. In general, the M and Y_n vapors are obtained by evaporation from heated crucibles, the R_nM molecules vaporize at room T, and the H_nY are gases. Unfortunately, this categorization only includes metalorganic metal vapors and not halide ones such as WF_6. Moreover, the terminology specifies epitaxy, whereas these vapor combinations may of course be applied to polycrystalline and amorphous films as well. This just points up the complication one encounters in attempting to categorize an evolving technology, and for this reason I have tried to avoid confusion by minimizing the use of such terminology.

The common terms "cation" for M and "anion" for Y also appear in Table 6.3. Reversal of these terms can be avoided by thinking of the early vacuum-tube experiments which spawned the cathode-ray tube (CRT) and gas-plasma devices such as neon tubes. Cathode rays are electrons, so the cathode needs to be negative to emit them, and therefore it attracts positive ions—cations. Thus, the more electropositive element in a polar compound is the cation.

6.5.5 Flux modulation

In the above discussion of compound deposition, we have assumed steady impingement flux for all species at some fixed ratio which deter-

TABLE 6.3 Terminology for Vacuum Epitaxy of Compounds Using Various Types of Vapors and Gases

Cation	Anion	
	Y_n	H_nY
M	Molecular-beam epitaxy (MBE)	Gas-source MBE (GSMBE)
R_nM	Metal-organic MBE (MOMBE)	Chemical-beam epitaxy (CBE)

M = metal, Y = nonmetal, R = organic group

mines composition. During and after surface reaction of these species to form the compound, atomic rearrangement to the epitaxial crystal structure occurs if enough thermal or other energy is present. In the course of this rearrangement, various defects can become quenched in; that is, they can become buried before annealing out if substrate T, T_s, is insufficient. Also, statistical surface roughness (Fig. 5.16) develops due to limited surface-diffusion length, Λ [Eq. (5.17)], if T_s is too low. We will have more to say about roughness and defects in Sec. 6.7. Here, we focus on the use of flux modulation to reduce these problems at low T_s as well as to improve uniformity over large or convoluted areas and to provide selective-area deposition. Flux modulation involves supplying two source species to the deposition surface as alternating pulses. We will see below that both T_s and the duration of these pulses are important to film-composition control. Often, the gaseous sources of the last subsection are used for flux-modulated deposition, but we will begin with a simpler example involving elemental sources.

Consider the case of the II-VI compound semiconductor ZnS, where the Zn and S_n vapors are both more volatile than the compound. Let the substrate be held at T_s high enough so that for each vapor $J_{vo} > J_i$, where J_{vo} is the thermodynamic maximum re-evaporation flux of the condensed vapor (see Sec. 4.2), and J_i is the vapor impingement flux. This is the undersaturated condition under which neither element will condense on itself. If the Zn bonds more strongly to the substrate than it does to itself, 1 ML (one monolayer) of Zn can still be deposited in a J_i of Zn vapor; but additional Zn re-evaporates, so the deposit is *self-limiting* at 1 ML. If after 1 ML of Zn deposition, J_i is switched to S_n vapor, the same thing happens: the first ML of S deposits because of its reactivity with Zn, but additional S re-evaporates. Alternating pulses of Zn and S_n thus result in the buildup of a ZnS film, atomic layer by layer, and the process is thus known as atomic-layer epitaxy or ALE (Suntola, 1989). It was actually first applied to the deposition of *polycrystalline* ZnS using $ZnCl_2$ and H_2S. In that reaction, the Cl remains on the surface until it reacts with adsorbing H_2S in an exchange reaction to form HCl(g) and ZnS(c).

The ALE process can be operated at pressures from 1 atm to the molecular-flow regime. In addition to the II-VI compounds, various metal oxides have been deposited using MCl_n and H_2O, nitrides using NH_3, and the III-V semiconductors using metalorganics (Suntola, 1989). ALE has even been applied to an *element* (Si) using SiH_2Cl_2, which self-limits to 1 ML at 1100 K and at a low pressure of 10^{-1} Pa [35]. The Cl remaining on the surface inhibits further adsorption, but is removed as HCl by a pulse of H_2.

Self-limiting adsorption behavior has the potential to improve deposition ordering, uniformity, and area selectivity, and examples of each follow. Because the elements are being deposited not randomly but in successive layers, just as they occur in a crystal, less thermal energy is needed to help this order establish itself. For example, compound-semiconductor alloys, such as $Ga_{0.5}In_{0.5}As$, thermodynamically prefer to order themselves into alternating monolayers of GaAs and InAs, as discussed in Sec. 6.5.2. However, complete ordering is often kinetically inhibited during epitaxy [27]. ALE has the potential to bypass this kinetic barrier by artificially establishing the ordering. In another example of ordering, the placement of impurity dopants into the desired lattice sites is maximized by pulsing them at the appropriate phase of the atomic-layer deposition cycle [36].

Thickness uniformity is improved in ALE because it is now determined by the number of pulse cycles rather than by J_i as it is in continuous deposition, so J_i variation is much less important. This may be thought of as digital versus analog control of thickness. It can assist in covering substrates that are very large, odd-shaped, or stacked closely together, or that contain surface topography such as the step edges of microcircuit lines or such as particles which one wants to cover. The development of statistical roughness at low T_s is also reduced by the insensitivity of deposition thickness to statistical fluctuation in J_i. Low T_s is desirable to prevent interdiffusion of abrupt interfaces in the multilayer structures discussed in Sec. 6.2. Finally, ALE can achieve **selective-area deposition** over a wider T_s range than can continuous deposition. In either case, selectivity is obtained by masking the area where deposition is not wanted with a layer on which the precursors adsorb weakly, such as SiO_2 for ZnSe deposition. Pulsed deposition reduces the probability of the two precursors reacting to the compound film on this layer by allowing time for one to desorb before the other is introduced.

The critical factor in implementation of ALE is control of the self-limiting-adsorption condition, which is mainly determined by precursor choice and by T_s. The T_s window for deposition of a compound MY is revealed in the characteristic curve for film thickness, h, deposited per "cycle," a cycle being a pulse of M-bearing vapor followed by a pulse of Y-bearing vapor. For the II-VI compounds deposited from $M + Y_n$, the curve has the form shown in Fig. 6.19a. The flat region at 1 ML of compound per cycle is the T_s window. At T_s below the window, the metal vapor has become supersaturated so that it is not self-limiting to 1 ML. Above the window, the film is re-evaporating so that part of the 1 ML is lost between pulses. The flatness, width, and T_s position of the window depend on various process parameters such as precursor composition, J_i level, dose of each vapor per cycle ($J_i t$), and purge or pumpout time between vapor pulses. Control to within the T_s win-

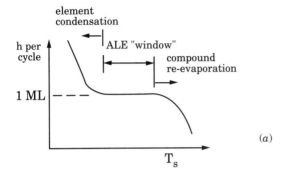

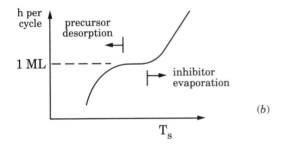

Figure 6.19 Substrate-T windows for atomic-layer-epitaxy (ALE) processes: (a) selective-adsorption mechanism and (b) adsorbate-inhibition mechanism.

dow is of course essential for a workable process, and the window boundaries must be established at the outset.

When M is nonvolatile, as is Ga, self-limiting deposition requires the use of a compound vapor. For GaAs ALE using $(CH_3)_3Ga$ and AsH_3 precursor vapors, the characteristic curve has the form shown in Fig. 6.19b, which is just the opposite of the curve described above. The mechanism here is still under investigation, but one careful study involving analysis of both the surface and the desorbing species versus T_s concluded that the degree of $(CH_3)_3Ga$ decomposition was the key factor [37]. Dissociative chemisorption of $(CH_3)_3Ga$ is self-limiting because the surface methyl groups inhibit further adsorption. But when T_s becomes so high that all of the methyl groups desorb, droplets of nonvolatile Ga can begin to accumulate on the surface as excess $(CH_3)_3Ga$ decomposes. The residence time of a CH_3 on the surface was estimated to be 1 s at 720 K and 0.1 s at 770 K. Below the window, either the $(CH_3)_3Ga$ or the AsH_3 ceases to dissociatively chemisorb; that is, $R_r \ll J_i$ in Eq. (5.7). This implies a positive chemisorption-activation energy for that precursor (E_a in Fig. 5.2). The need to dissociate the $(CH_3)_3Ga$ while retaining surface CH_3 leads to an undesirably narrow T_s window around 720 K for this process chemistry. The "adsorbate-inhibition" mechanism [37] operating here differs from the "selective-adsorption" mechanism of Fig. 6.19a in the use of the CH_3 group to inhibit adsorption rather than the use of the metal itself. Similarly, adsorbed Cl blocks deposition beyond 1 ML when GaCl is used as the Ga source in an ALE version of the traditional "hydride" CVD process [38].

Another way to implement pulsed deposition in the case of nonvolatile M is to limit the M dose to approximately 1 ML coverage. In the case of $Al_xGa_{1-x}As$, diffusion of the metals on themselves still occurs at T_s as low as 500 K, so that the M dose can migrate around and redistribute itself into a 2D layer bonded to As before the As_4 dose is initiated. This technique is known as migration-enhanced epitaxy, or MEE [39]. The persistence of RHEED oscillations for thousands of ML of deposition has confirmed that 2D, layer-by-layer growth can proceed by MEE at T_s well below that at which roughness develops in continuous growth [39].

Dose control is important in both ALE and MEE. Dose is often expressed in Langmuirs (L), where 1 L = 1 Torr-s and is roughly 1/3 of the dose which would result in 1 ML coverage if all impinging flux were to adsorb. In ALE, the required dose depends on the adsorption kinetics. Here, the simple Langmuir adsorption model sometimes applies, where it is assumed that no adsorption occurs on surface sites already occupied by the vapor being dosed, and that adsorption occurs with a fixed trapping probability δ on unoccupied surface sites (see Fig. 5.1). We further assume for the present that adsorption is irreversible; that is, re-evaporation is negligible. We then have $\delta = \zeta$, where ζ is the chemisorption-reaction probability on the bare surface [Eq. (5.7) with $k_d = 0$]. Now, the sticking coefficient, S_c, considers both the bare and the occupied surface, so it decreases exponentially toward zero as fractional surface coverage, Θ, increases, since the model specifies that $S_c = 0$ on the occupied fraction. This results in an asymptotic approach to monolayer coverage:

$$\Theta = \frac{n_s}{n_{so}} = 1 - \exp\left[\frac{-t}{(n_{so}/\zeta J_i)}\right] \quad (6.20)$$

where n_{so} is the monolayer coverage in mc/cm^2 and $(n_{so}/\zeta J_i)$ is the time constant for adsorption saturation. We have seen this familiar form of equation in several other situations of approach to steady state. The behavior of n_s and S_c with time is illustrated in Fig. 6.20. Also shown for comparison is the coverage-independent adsorption with $S_c = 1$ which occurs for nonvolatile metals. The dose, $J_i t$, required to approach monolayer coverage depends mainly on ζ, which

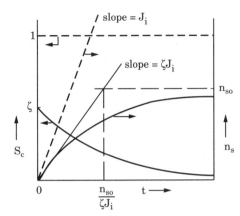

Figure 6.20 Adsorption behaviors of self-limiting precursors (solid lines) and nonvolatile metals (dashed lines); n_{so} = monolayer surface coverage.

can be <<1 and T_s-dependent for compound vapors. Thus, determination of ζ is important for process control. For ZnSe, the exponential behavior was followed by RHEED specular-spot intensity, which was higher for the Zn-covered surface than for the Se-covered surface [40]. However, Langmuir adsorption kinetics is not always obeyed. In a molecular beam study [41] of several R_3Ga compounds adsorbing on GaAs, S_c remained high with increasing Θ and then finally dropped to zero as $\Theta \to 1$, suggesting that arriving vapor could displace adsorbate sideways in order to adsorb.

One vapor cycle in ALE requires for each vapor a certain dose time to produce ≈1 ML coverage, as determined above, followed by an interval time to remove that vapor from the deposition chamber before the other vapor is introduced. For operation in the fluid-flow regime, the interval involves a purge with unreactive gas, while in the molecular-flow regime, it involves only vacuum pumping. In either case, switching times of <1 s are difficult to achieve. This results in a cycle time of >4 s and a deposition rate of ≈1/4 µm/h, which is very slow. Slow deposition not only reduces throughput, but also risks increased contamination from background gases. Slow deposition and control of process chemistry are the two main problems in ALE.

6.5.6 Doping

The final composition-control issue to be addressed involves the incorporation of deliberate impurities (dopants) into the film during deposition. We discussed the behavior of donor and acceptor dopants in semiconductors in Sec. 6.2.1, and pointed out the importance of controlling doping level and achieving abrupt steps in doping level within multilayer structures. Here, we will investigate doping control in more detail.

The choice of dopant depends both on its inherent behavior as a dopant and on the ease with which it can be incorporated during deposition. Regarding behavior, dopants can occupy either interstitial sites—between the atomic sites of the host lattice, or substitutional sites—replacing one of the lattice atoms. Many interstitial metals act as donors, but interstitials tend to have high diffusivities, so substitutional dopants are usually used instead. Substitutional dopants can be donors or acceptors. They have the desired "shallow" energy levels near the conduction-band or valence-band edge, respectively, when they have just one more or one less valence electron than they need to satisfy the lattice bonding of that site. Thus, the group-V elements like P are shallow donors in the group-IV semiconductors Si and Ge, and the group-III elements like Al are acceptors. The substitutional donor releases its extra electron to the conduction band, whereas the acceptor withdraws an electron from the valence band to complete its bonding.

In compound semiconductors, a given impurity can substitute for *either* lattice element, and we will see below how to drive it into the desired site by making that site more available at the deposition surface. Thus, group-IV dopants in III-V compounds are donors on the III sites and acceptors on the V sites; that is, they are "amphoteric" dopants. Group-II dopants like Zn in III-V compounds are acceptors on III sites, but on V sites they are deep traps, with energy levels deep within the band gap. Deep levels usually result when the impurity's valence differs by *more* than one from what is needed for that lattice site. Deep traps are undesirable unless one wants to make the film semi-insulating. Similarly, group-VI elements like Se are donors on the V sites of III-V compounds, and group-VII elements like Cl are donors on the VI sites of the II-VI compounds. *Native* point defects (Table 4.1) caused by stoichiometry deviation can also be electrically active. In the wide-gap semiconductors such as GaAs, they tend to form deep levels, such as As_{Ga} does; but in the narrow-gap semiconductors such as PbTe (see Fig. 6.3), the levels are necessarily shallow, so excess Pb is a donor, and excess Te is an acceptor. Pb is a donor because the Pb-Te bond is polar ($Pb^{\delta+}$-$Te^{\delta-}$). Thus, the Pb valence electron which would have been attracted to a neighboring Te in the exactly stoichiometric compound now sees only a Te vacancy, so that electron is released into the conduction band instead.

From among the potential film dopants discussed above, one must choose one that is also easily incorporated during epitaxial growth. Ideally, one would like every dopant atom impinging on the growth surface to be incorporated immediately into the film in an immobile and electrically active state. This occurs for Be (p-type) and Si (n-type) in GaAs deposition, for example. However, in general there are four

6.5.6 Doping

other common possibilities, all of which are undesirable. The five possibilities, illustrated at the left of Fig. 6.21, are (1) return to the vapor phase, (2) surface segregation, (3) formation of a second solid phase, (4) incorporation onto the wrong site, and (5) incorporation onto the desired site with electrical activation. We will examine each of the first four separately below.

The dopant is **returned to the vapor phase** if it either fails to bond to the lattice site or re-evaporates due to weak bonding coupled with high T_s. The bonding reaction probability is given by R_r/J_i from Eq. (5.7). An extreme example is nitrogen arriving as N_2. Because of the strong $N{\equiv}N$ bond, there is an insurmountable E_a barrier (Fig. 5.2) for dissociating this precursor, so $\zeta = 0$ and $S_c = 0$. However, N can be incorporated as an acceptor on the Se sites of ZnSe either by using NH_3 as the precursor [42] in growth from $(CH_3)_2Zn$ and H_2Se or by generating N atoms from N_2 in a microwave plasma source [43]. In the former case, E_a is being lowered, and in the latter case, it is being surmounted as shown by curve (c) in Fig. 5.2. These nitrogen results ended a long search for a successful p-dopant for this high-band-gap material and allowed blue-light-emitting ZnSe diodes and diode lasers to finally be made [43].

Even when the above kinetic barrier to dopant incorporation does not exist, a dopant B will still return to the vapor phase if its vapor pressure over the film material, p_B, is too high. Then, $J_v \rightarrow R_r$ in Eq. (5.8), so $J_r \rightarrow 0$. Eq. (4.21) expressed the p_B of a solute as $p_B = a_B x p_{vB}$, where p_{vB} is the vapor pressure of pure B, x is the dilution

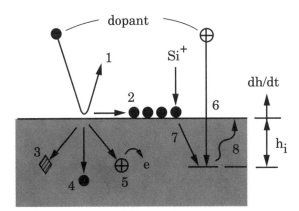

Figure 6.21 Dopant behavior during film growth: (1) re-evaporation, (2) surface segregation, (3) second-phase formation, (4) wrong lattice site, (5) electrical activation, (6) ion implantation, (7) knock-on implantation, and (8) out-diffusion.

fraction, and a_B is the "activity coefficient" which empirically accounts for the deviation of p_B from a simple Raoult's-law dilution lowering. The meaning of a_B is revealed by recalling that p_v depends exponentially on the strength of bonding into the potential well of Fig. 4.2, as given by Eq. (4.14) for a pure material evaporating from its own condensed phase. Now, in the case of a solute B, if its bond strength to the host lattice is larger than to itself, then p_B will be *lower* than that determined by dilution alone; that is, $a_B < 1$. This means that a dopant whose p_v as a pure element is very high at T_s can still incorporate if it bonds strongly to the film lattice; that is, if $a_B \ll 1$. In MBE of GaAs using the high-p_v dopants Zn, Cd, and S, it turns out [44] that $a_B \ll 1$ for S but not for Zn and Cd, so the latter two are unusable as dopants.

The above thermodynamic argument based on bonding assumes that vapor-solid equilibrium is being approached; that is, the kinetic barriers to dopant equilibration across this interface are small. Indeed, MBE incorporation levels of these and other GaAs dopants can be predicted well from doping levels measured in other deposition processes that operate much closer to equilibrium between the transport phase and the film phase, namely chloride vapor-phase epitaxy and liquid-phase epitaxy [44]. This means that the kinetic barriers indeed are small for the above MBE dopants. We will see below, however, that the introduction of a kinetic barrier by energetic-dopant burial can *increase* the incorporation of many dopants. Thus, kinetic barriers can work either for or against doping.

Even if the dopant has low enough p_v so that it does not re-evaporate, it may accumulate on the surface rather than becoming incorporated into the film, as shown by (2) in Fig. 6.21. This **surface segregation** arises from the inevitable mismatch between the relaxed bond angles and/or lengths of the dopant atom versus the lattice site. This mismatch favors the dopant's rejection during crystallization, to reduce the Gibbs free energy. The degree of rejection may be expressed in terms of an empirical segregation coefficient:

$$\kappa = \frac{n_s/a}{n} \qquad (6.21)$$

where n_s = surface concentration of dopant, atoms/cm^2
n = bulk-film concentration of dopant, atoms/cm^3
a = monolayer thickness, cm

In the literature, a often is dropped, but its inclusion has the advantages of making κ unitless and of giving $\kappa = 1$ for no segregation effect. The "zone refining" process for purifying bulk single-crystal material exploits the segregation effect to push impurities ahead of an advancing solidification front; but in the growth of multilayer structures, segregation is undesirable. Dopant *level* may not be a problem, because even for $\kappa = 10^4$, Eq. (6.21) shows that $n \approx 5\times10^{18}$ atoms/cm^3 when the surface is saturated with dopant ($\approx 10^{15}$ atoms/ML). However, when the dopant flux is turned on or off, there is a delay time while the surface re-equilibrates, and this smears out abrupt interfaces (see Exercise 6.12). Although the growth thickness for re-equilibration can be made zero in principle by pre-deposition or re-evaporation of the surface layer of dopant during growth interruptions, this complicates the process control and also risks contamination buildup during the interruptions.

Surface segregation and re-evaporation can both be dramatically reduced by using ion-enhanced processes, which will be discussed further in Sec. 8.5 and which are illustrated for the present case by (6) and (7) in Fig. 6.21. Acceleration of ionized dopant into the surface at 200 eV or so, using the ionizing source mentioned in Sec. 4.5.4, causes the dopant to become implanted ≈1 nm beneath the surface. Alternatively, thermally adsorbed dopant can be driven beneath the surface by the impact of ≈10 keV ions of depositing film material (the "knock-on" effect), as has been demonstrated for Si$^+$ from an electron beam-evaporation source [45]. In either case, if overgrowth then proceeds faster than the dopant can diffuse back to the surface, segregation and re-evaporation become kinetically inhibited, and the dopant remains frozen in the bulk. Indeed, the achievable doping level can well exceed the equilibrium solubility. Despite the "frozen" condition, there is usually enough thermal energy available to allow the dopant to settle into its substitutional lattice site and also to anneal out the shallow lattice damage caused by dissipation of the ion energy. The out-diffusion flux, J_D, can be expressed either in terms of the Fick's-law transport equation which was given earlier for surface diffusion [Eq. (5.22)], or in terms of a "diffusion velocity," u_D [see Eq. (4.35)]; thus,

$$J_D = D\left(\frac{n - n_s/a\kappa}{h_i}\right) = nu_D \qquad (6.22)$$

where D is the diffusivity (cm^2/s), h_i is the implant depth (cm) shown in Fig. 6.21, and the term $n_s/a\kappa$ is the dopant concentration just below the surface, which will be assumed in the following discussion to be negligible compared to n.

The degree of suppression of out-diffusion by overgrowth is conveniently expressed by the dimensionless ratio of growth velocity, dh/dt, to u_D. This is equivalent to the Péclet number for mass transfer used in chemical-engineering process analysis, as pointed out by Tsao (1992):

$$\text{Pe} = \frac{dh/dt}{u_D} = \frac{h_i(dh/dt)}{D} \quad (6.23)$$

When Pe >> 1, the growth velocity dominates; that is, out-diffusion is kinetically inhibited. This requires either a high growth rate or a small D, but much more adjustability is available in D because of its exponential dependence on T_s via Eq. (5.26). Even without ion enhancement, some kinetic inhibition can be achieved if T_s is low enough. For example, an abrupt increase in thermal Sb incorporation into MBE Si is observed when T_s drops below about 800 K, and the transition T increases slightly with dh/dt, as expected [46]. However, higher T_s is required for best Si quality, and then energy enhancement is called for. Thus, 200-eV Sb$^+$ can be incorporated with high efficiency [47] at 1200 K. At this T, Sb has D ≈ 10^{-17} cm^2/s in bulk single-crystal Si, so for dh/dt = 0.3 nm/s and h_i = 1 nm, Pe = 300. In fact, however, this value is unrealistically high, because D near the surface is likely to be increased by the presence of bombardment-induced excess vacancies (Sec. 8.5.2), on whose concentration the D of a substitutional impurity strongly depends. There is also evidence that ionized dopant can *drift* back to the surface in the internal electric field which results from Fermi-level pinning by surface states [8]. Because of these factors, and because of the spread and uncertainty in implant depth, quantitative determination of dopant incorporation under specific deposition conditions must be made experimentally, as reported above for Sb.

The efficiency of dopant atom *physical* incorporation can be determined by measuring J_i, n_a, and n, for which analytical techniques are discussed in Secs. 4.7, 6.4.1, and 10.2.2, respectively. Ionized-dopant flux can be determined directly by measuring current to the substrate if other ion fluxes are not interfering. However, the physically incorporated dopant may not all be electrically active if it is forming **another solid phase** as illustrated by (3) in Fig. 6.21. The *active* dopant concentration can be determined from conductivity and mobility measurements (Sec. 10.3.2) using Eq. (6.2), *if* the carriers are mostly n-type or mostly p-type rather than being "compensated" by each other. An electrically inactive second phase can form by reaction of the dopant with impurities, such as Mg + GaO → MgO in GaAs MBE [48]. At concentrations >10^{19}/cm^3 or so, inactive compounds sometimes form between the dopant and the film elements themselves.

Dopant occupying the **wrong site** in a compound semiconductor may act either as a deep trap, or as an amphoteric dopant of the wrong sign, as discussed earlier. Site preference depends both on the dopant and on the site availability at the growth surface. For the group-IV amphoteric dopants in GaAs, preference shifts from p to n with increasing elemental period number. Thus, C is always p-type and Sn is always n-type, but carrier type for Ge and Si depends on site availability. Ge is p-type under Ga-rich deposition conditions, which favor its occupying As sites, and it is n-type under As-rich conditions [28]. However, the inevitable accumulation of Ga droplets in Ga-rich growth make Ge unsuitable for p-doping of GaAs. Site availability is also determined by growth plane. For zinc-blende compounds, this effect is most pronounced on the polar {111} faces, as we will now illustrate. At the upper right in Fig. 6.18, the As atom at $z = -(1/4)a_o$ is tetrahedrally bonded to the three Ga atoms centered on the x-y, x-z, and y-z faces of the unit cell and to the corner Ga atom at $x = y = z = 0$. If this crystal were cleaved along (111), it would separate where the areal bond density is lowest—between the As and the corner Ga. Of the two new surfaces thus created, the one facing in the [111] direction consists entirely of Ga atoms (the "A" face), and the one facing in the $[\bar{1}\bar{1}\bar{1}]$ direction consists entirely of As atoms (called the "B" face). Thus, dopant atoms bonding to GaAs{111A} have As sites available, and on $\{\bar{1}\bar{1}\bar{1}B\}$ they have Ga sites. Indeed, Si is p-type on {111A} and n-type on {111B} [49]. Moving away from {111A} to the higher-order planes {j11A}, the p doping persists through $j = 3$; then for $j \geq 5$, Si is n-type. In contrast to the above behavior, the non-amphoteric dopants have less tendency to occupy the wrong site, but one should be aware of the possibility.

6.6 Lattice Mismatch

Having now dealt with avoiding precipitates and controlling point defects, we can proceed to the problem of minimizing other crystallographic defects. It is useful to think of defects in terms of their dimensionality. Point defects are zero-dimensional (0D), while precipitates or disordered regions are 3D. Planar (2D) defects, which are discussed in Secs. 5.4.3 and 6.7, include grain boundaries, twin planes, stacking faults, and antiphase-domain boundaries. Dislocations are line (1D) defects. We will see below how dislocations arise from the fractional lattice mismatch, f, at heteroepitaxial interfaces [Eq. (6.1)].

For this purpose, we consider the simple square symmetry of cubic material growing in (001) orientation on a (001)-oriented substrate, although the same principles apply to other symmetries. Figure 6.22 shows the various modes of mismatch accommodation. In the special case of perfect match (a), the lattices are naturally aligned, and the growth is therefore "commensurate" without requiring lattice strain. In (b-d), the atomic spacing of the epilayer, a_e, is larger than that of

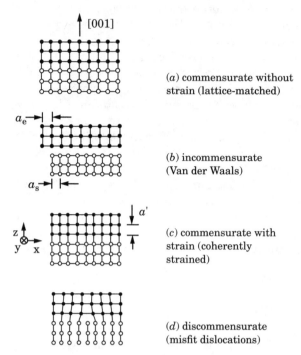

Figure 6.22 Modes of accommodating epilayer lattice (solid circles) to substrate lattice (white circles).

the substrate, a_s. In fact, f has been made quite large (0.14) here so that it may be readily observed, but it is much smaller in most heteroepitaxial systems of interest.

There are several ways in which lattice mismatch can be accommodated. In Fig. 6.22b, bonding across the interface is weak, so that the epilayer "floats" on top of the substrate and is therefore "incommensurate" with it. This mode occurs, for example, with materials having a 2D, layered structure, such as graphite and MoS_2. In such compounds, there is *no* chemical bonding perpendicular to the hexagonally close-packed and tightly bonded basal plane, so that interaction of such a film with the substrate is only by Van der Waals forces. These weak forces are often strong enough to maintain rotational alignment with the substrate and to produce a small periodic compression and expansion in the epilayer lattice, but they are not strong enough to strain the epilayer so that it fits that of the substrate. There is a small periodic distortion in a_e as the lattices fall in and out of alignment periodically across the interface, and this produces a beautiful Moiré pattern in STM images of the epilayer surface [50]. Incommensurate growth can also occur when chemical bonding is weak because of a difference in bonding character between film and substrate. Thus, PbTe sometimes grows on its low-surface-energy (001) face on BaF$_2$(111) despite a good lattice match of this substrate to PbTe(111) [51]. Chemical bonding can also be blocked by passivating the substrate surface.

In the more common situation, the epilayer is chemically bonded to the substrate, thus forming a unit called a "bicrystal." A thin epilayer with small f is likely to become strained to fit the substrate in x as shown in Fig. 6.22c, and similarly in y. This is sometimes referred to as "pseudomorphic" growth, but it really is not, because no change in crystal *structure* has occurred. It is properly termed "commensurate growth" or "coherent epitaxy." In Fig. 6.22c, it is assumed that the substrate is much thicker than the epilayer, so that the substrate is rigid and all of the strain is in the epilayer. This "coherency" strain is then just $\varepsilon_x = \varepsilon_y = \varepsilon_{x,y} = -f$, and the corresponding biaxial stress, $\sigma_{x,y}$, is given by Eq. (5.51). The biaxial stress produces a strain in z, perpendicular to the growth plane, which is given by the three-dimensional form of Hooke's law,

$$\varepsilon_z = \frac{1}{Y}(\sigma_z - \nu\sigma_x - \nu\sigma_y) = \frac{-2\nu\sigma_{x,y}}{Y} = \frac{-2\nu\varepsilon_{x,y}}{1-\nu} \qquad (6.24)$$

Here, the second equality was obtained by setting $\sigma_z = 0$ as it must be for the unconstrained direction, and the third was obtained using Eq. (5.51). In Fig. 6.22c, the epilayer is shown compressed in x and y and expanded in z in accordance with the above formula. This lattice is said to be "tetragonally" distorted, and the tetragonal strain is defined as

$$\varepsilon_T = \varepsilon_z - \varepsilon_{x,y} = -\left(\frac{1+\nu}{1-\nu}\right)\varepsilon_{x,y} \qquad (6.25)$$

X-ray diffraction measurement of the expanded atomic-plane spacing a' in z can be used with Eq. (6.24) to determine the fraction by which the epilayer lattice has compressed to fit the substrate in x and y. Electron diffraction can be used only when the change in a is larger than a few percent, because the peaks are much broader than in x-ray diffraction.

The strain energy stored per unit area in the coherently strained epilayer, U_ε, is obtained by integrating force over distance as the film is compressed toward a fit to the substrate, starting from the relaxed state shown in Fig. 6.22b. The force to maintain the compression is supplied from the rigid substrate by bonding across the interface. The integration can be done in one direction and then doubled to account for the orthogonal direction. The force, F, in the x (or y) direction, per unit width of film in y (or x), is given by Eq. (5.52), and the distance in x is the same as the strain ε_x if we use a normalized film length of $L_x = 1$. Thus,

$$U_\varepsilon \left(\frac{J}{m^2} \text{ or } \frac{N}{m} \right) = 2U_x = 2\int_0^x F_x \frac{dx}{L_x} = 2\int_0^{\varepsilon_x} \left(\frac{Y}{1-\nu}\right)\varepsilon_x h d\varepsilon_x = \left(\frac{Y}{1-\nu}\right)\varepsilon_x^2 h$$

(6.26)

The force of compression creates shear stresses in crystal planes that are not perpendicular to it, and along certain of these planes the film will "slip" to relieve stress by breaking and then reforming bonds. After slippage, there will be extra rows of substrate atoms which are not bonded to the film, such as the one shown along y in Fig. 6.22d. These features are known as misfit dislocations. Film stress is relieved by the development of a grid of such dislocations in the interface, the grid periodicity being determined by energy minimization, as we will see below. This growth mode is known as "discommensurate." In addition to the bulk film strain energy, U_ε, there is an interface energy, γ_i, as discussed in Sec. 5.3.1; but since it does not depend on h, it will be neglected below.

Usually, defects of any dimensionality (0D through 3D) are undesirable within a film unless they are introduced for a specific purpose such as doping. Films in electronic applications are particularly sensitive to degradation by defects. They disturb the lattice periodicity and thus locally alter the band structure of a semiconductor crystal, often producing charge-carrier traps or charge-recombination centers within the band gap. Defects of 1D and 2D also provide paths for electrical leakage and impurity diffusion. Thus, in heteroepitaxial growth, it is important to know what conditions have to be met to avoid the generation of misfit dislocations. Before analyzing this situation, we need to examine the nature of dislocations in more detail.

Figure 6.23 shows how the two basic types of dislocation are introduced into an otherwise perfect cubic-crystal lattice by the application of shear stress. In both cases, the dislocation lies along the y axis out of the plane of the paper, and its "core" is shown by the gray rod. Figure 6.23a is similar to Fig. 6.22d except that the shear is being applied externally instead of being the result of lattice misfit in a bicrystal. Here, the crystal has slipped along the x-y plane, and the resulting dislocation is an "edge" type. A second edge dislocation is shown starting to form at the left-hand end of the figure. There, the formation involves a switch in bonding from the two bonds labeled "1" to the one labeled "2." Application of additional shear will cause more edge dislo-

6.6 Lattice Mismatch

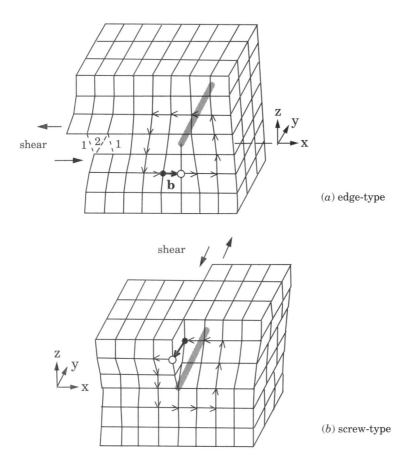

Figure 6.23 Two basic dislocations along the y axis (gray bars), their Burgers circuits starting at the white circles and ending at the black circles, and their Burgers vectors, **b** (represented by the heavy arrows).

cations to form at the left and "glide" to the right along the slip plane to take up the misfit which is being created by compression of the bottom half of the crystal and extension of the top half. The mechanical yielding of a crystal above its elastic limit (shown in Fig. 5.30) and at T below 0.4 of its melting T involves the generation and glide of a series of dislocations along slip planes. Dislocations that can glide called "glissile"; others that cannot are called "sessile." However, all edge dislocations can "climb," which is motion in the direction *perpendicular* to the slip plane—the z direction for the dislocation shown in Fig. 622d. Climb requires the diffusion of vacancy and interstitial point defects into and out of the dislocation's core and therefore operates only at high T, since diffusion increases exponentially with T [see Eq. (5.26)]. For example, under the influence of the epilayer compressive stress in

Fig. 6.22c, atoms could be squeezed out of a y-z plane and diffuse elsewhere in the crystal, leaving a missing plane of atoms as shown in the middle of Fig. 6.22d.

The edge dislocation of Fig. 6.23a has been mapped by tracing around it a counterclockwise rectangular path having an equal number of lattice units on each leg. In a perfect crystal, such a path would close on itself; but when traced around a dislocation, it does not. The distance and direction from the end of the path to its beginning is called the Burgers vector, **b**, of the dislocation, and **b** describes its strength and character in an elegant manner, by its length and by its orientation relative to the dislocation line. Note that for the edge-type dislocation, **b** lies in the slip plane and is perpendicular to the core line. Tracing a Burgers circuit around the "screw"-type dislocation of Fig. 6.23b results in a **b** that also lies in the slip plane but is *parallel* to the core line. Continuation of this Burgers circuit results in a spiral path screwing into the paper in the +y direction. Inspection shows that if a screw dislocation were to lie in the interface of a bicrystal, it would contribute nothing to the release of misfit strain, because there is no "extra" plane of atoms involved. This is the key distinction between edge and screw dislocations for epitaxy purposes—only the edge type can relieve misfit strain. A dislocation can change from screw to edge type by changing direction as it threads its way through a crystal from face to face. In doing so, **b** remains unchanged. A dislocation can never just *end* within a crystal, although it can form a closed loop within a crystal or a half loop extending in from a face and then back out the same face.

Figure 6.24 shows a "mixed" dislocation with partial edge and screw character. This dislocation is commonly observed at the interface of epilayers having an fcc, diamond, or zinc-blende crystal structure being grown on the (001) face. Only a few of the metal atoms are shown, so as to emphasize the geometry of the dislocation. In these crystal structures, this dislocation always lies along a ⟨011⟩ direction and glides along a {111} plane (for notation, see Sec. 5.3.1). The {111} planes are the glide planes because they are close-packed and also have a low density of bonds perpendicular to the plane, both of which factors facilitate gliding. For growth on (001), the glide planes are at an angle $\gamma = 55°$ to the substrate interface. This allows dislocations to glide in to the interface from the surface of the epilayer. However, this dislocation does not relieve as much stress as would a pure edge dislocation having its **b** in the plane of the interface, for two reasons. First, it has partial screw character, as measured by the angle β, which is 60°. A pure edge dislocation has $\beta = 90°$, and a pure screw dislocation has $\beta = 0°$, as seen in Fig. 6.23. Second, only that component of **b** which lies in the interface can relieve misfit strain. The elevation of **b**

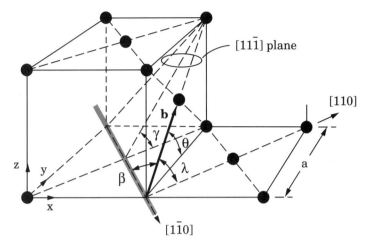

Figure 6.24 Geometry of the $(1/2)a\langle 011\rangle$ 60° dislocation in diamond-structure crystals.

out of the interface is measured by θ and is 45°. The overall efficiency of this dislocation as compared to a pure edge dislocation having **b** in the interface along [110] is measured by the cosine of the angle λ between the two Burgers vectors. Here, λ = 60°. This mixed dislocation is dominant not only because it can glide in from the surface but also because it has the shortest possible **b**, this being the distance between close-packed atoms. A short **b** means that less crystal distortion is required to nucleate the dislocation. The length of **b** is described by the lengths of its projections along the axes, which in this case are $(1/2)a$. Thus, the overall description of this dislocation in terms of its strength, core direction, and character is "$(1/2)a\langle 011\rangle 60°$," the angle referring to β. The pure edge dislocation with **b** in the interface is more common in mismatched (001) growth in the S-K or the island growth mode [52] (see Fig. 5.8), probably because the glide distance to the island edge or the climb distance to the island surface is so small.

We now return to the question of how to avoid generating dislocations in heteroepitaxial films. It is possible to do this because dislocations, while releasing the energy stored in the strained epilayer, also contain excess energy of their own. Matthews and Blakeslee pointed out that if the energy of the dislocation is greater than the energy released, it will not form [53]. This criterion, which we will derive below, has since been found to be quantitatively correct in a variety of heteroepitaxial systems.

The energy of a dislocation resides partly in the core and mainly in the strain field surrounding the core. Inspection of Fig. 6.22d shows that there is tensile strain in the lattice above the core, compressive strain below it, and shear strain all around it. Analysis of the strain field results in the following formula (Hull, 1984; p. 80) for the energy per unit length of a mixed dislocation:

$$U'_d (J/m) = F_d (N) = \frac{G_s}{4\pi}\left[\left(\frac{b\sin\beta}{1-\nu}\right)^2 + (b\cos\beta)^2\right]\ln\left(\frac{r_o}{r_i}\right)$$

$$= \frac{G_s b^2}{4\pi}\left[\frac{1-\nu\cos^2\beta}{1-\nu}\right]\ln\left(\frac{4h}{b}\right) \approx \frac{G_s b^2}{4\pi(1-\nu)}\ln\left(\frac{4h}{b}\right) \quad (6.27)$$

Here, G_s is the shear modulus of elasticity, which is the "spring constant" for shear stress just as Y is for tensile stress [Eq. (5.50)]. The two moduli are always related by

$$Y = 2G_s(1 + \nu) \quad (6.28)$$

The 4π in Eq. (6.27) comes from the radial symmetry of the strain field. The term b is the magnitude of the Burgers vector **b**; the first of the two b terms comes from the strain field of the dislocation's edge component, and the second from the screw component. The strain field extends radially outward to the edge of the crystal, so in the case of a thin film of thickness h, we set the outer radius $r_o = h$. The inner radius, r_i, is somewhat arbitrary in this continuum model, but it is usually set at (1/4)b to account for the energy in the core. The ln term may be recognized as the same one which describes the electric field about a disc-shaped contact to a thin film. The $(1 - \nu\cos^2\beta)$ term is near unity for the 60° dislocation and has been dropped in the last equality. The prime on U'_d is to distinguish this energy per unit *length* from the energy per unit *area* of Eq. (6.26). Energy/length has the units of force, and F_d is in fact the "line-tension" force of the dislocation line. It is the 1D analogy of the 2D surface tension (see Sec. 5.3.1), and it has the same effect. That is, a dislocation threading through a crystal will try to minimize its length and thus its energy by straightening itself, just as a tensioned membrane becomes flat and a droplet becomes spherical. Note also that U'_d is proportional to b^2 and thus to (strain)^2, since the lattice distortion caused by the dislocation is proportional to b. This is the same strain dependence as for the energy in Eq. (6.26). The dislocations with the smallest b are the ones most likely to form, since they have the least energy.

For two linear arrays of dislocations lying in a plane, one along each of the two axes to form a grid, the energy stored per unit area of array is

$$U_d \ (J/m^2) = 2U_d' \rho_l \tag{6.29}$$

where ρ_l (lines/m) is the linear density of the array. If the plane is the interface of a bicrystal, each of these dislocations reduces the misfit strain by $b \cos\lambda$ (see Fig. 6.24), so the strain in the epilayer after some amount of relaxation by dislocation generation is

$$\varepsilon_x = \varepsilon_y = \varepsilon_{x,y} = \rho_l b \cos\lambda - f = \rho_l b \sin\beta \cos\gamma - f \tag{6.30}$$

where the fractional lattice mismatch, f, was defined in Eq. (6.1). We now have the information we need to determine when and how many dislocations will form using the principle of energy minimization; that is, motion toward equilibrium. We have considered above only the internal-energy component, U, of the Gibbs free energy, G [Eq. (4.5)], but since a dislocation is an orderly line of defects, the entropy change, ΔS, is very small. The $p\Delta V$ term is also negligible.

Figure 6.25 shows the quadratic increase of the U_ε component of system energy U with $\varepsilon_{x,y}$, as expressed in Eq. (6.26), for two film thicknesses, h and 3h. It also shows the linear increase of the U_d component with ρ_l as expressed in Eqs. (6.27) and (6.29). The terms $\varepsilon_{x,y}$ and ρ_l are related by Eq. (6.30). The solid lines are the total energies, $U = U_\varepsilon + U_d$, versus $\varepsilon_{x,y}$ and ρ_l for the two films. The coherently strained, dislocation-free situation is shown at the left, and the completely relaxed films at the right. For the thicker film, U decreases with dislocation formation (movement to the right), so dislocations will form until the minimum U is reached as shown by the dot, pro-

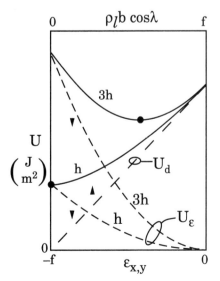

Figure 6.25 Film internal energy vs. strain in heteroepitaxy. The total energy, $U = U_\varepsilon + U_d$, is shown by the solid line, and the dots denote the equilibrium strain and dislocation density for films of thickness h and 3 h and fractional lattice mismatch f.

vided there are no kinetic barriers (which there sometimes are). Note that this film still contains some strain upon reaching equilibrium (see Exercise 6.15). If ρ_l becomes so high that the dislocation spacing is less than h, the dislocation strain fields overlap, and then the appropriate r_o in Eq. (6.27) becomes $1/2\rho_l$ rather than h. This causes U_d versus ρ_l to become sublinear toward the right and lowers the equilibrium strain.

For the thinner film in Fig. 6.25, U *increases* with dislocation formation, so the dislocations do not form, and the coherently strained epilayer is stable as shown by the dot to the right. With increasing h or f, dislocation formation becomes energetically favorable at the h or f point where $dU/d\varepsilon_{x,y} = 0$ at $\varepsilon_{x,y} = -f$; that is, when

$$-\frac{dU_\varepsilon}{d\varepsilon_{x,y}} = \frac{dU_d}{d\varepsilon_{x,y}} = \frac{dU_d}{d\rho_l}\frac{d\rho_l}{d\varepsilon_{x,y}} = \frac{1}{b\cos\lambda}\frac{dU_d}{d\varepsilon_{x,y}} \quad (6.31)$$

Here, Eq. (6.30) was used to obtain the last equality. Inserting the derivatives of Eqs. (6.26) and (6.29) into this expression, and using Eq. (6.28) to replace Y by G_s, we obtain the critical mismatch, f_c, at which dislocation formation becomes energetically favorable:

$$\boxed{f_c = \frac{1}{h}\frac{b}{8\pi(1+\nu)\cos\lambda}\ln\left(\frac{4h}{b}\right)} \quad (6.32)$$

Note that G_s has dropped out. The ln term varies slowly, so the main dependence is the inverse one, $f_c \propto 1/h$, or equivalently, $h_c \propto 1/f$. This function is shown in Fig. 6.26 for the $(1/2)a\langle 011\rangle 60°$ dislocation in Si, where $\nu = 0.22$ and $b = 0.38$ nm. It turns out that the rise in ρ_l with h is quite abrupt once h_c has been reached, so h_c can be determined accurately by observation of the grid array of dislocations at the substrate interface, using x-ray topography or using defect etching followed by Nomarski microscopy [54] (see Sec. 10.1). Equation (6.32) is found to be obeyed remarkably well for films whose strain has been allowed to relax to equilibrium by sufficient annealing during or after deposition.

Relaxation is not instantaneous, because there are activation-energy (E_a) barriers to the nucleation and growth of a dislocation which we have thus far neglected. Figure 6.25 implies that U_d increases smoothly with ε, whereas in fact it does not, because atoms are discrete. For example, Fig. 6.23*a* shows that edge dislocations nucleating at the free surface do so by the switching of a bond from position 1 to 2. The full dislocation line grows by the successive switching of bonds along the line in the y direction, and at the position where this is oc-

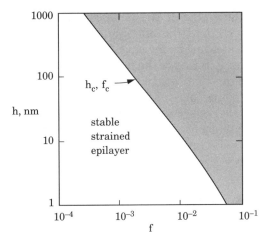

Figure 6.26 Relationship of critical thickness to lattice misfit for Si(001) epitaxy [Eq. (6.22)].

curring the line contains a "kink" which lies in the glide plane. Kinks can also nucleate in the middle and spread toward the edges. These are somewhat analogous to the kinks occurring at the edge of an atomic terrace during 2D growth (Fig. 5.13). (Steps in a dislocation line in the *climb* direction are called "jogs.") The E_a to move the kink produces the "Peierls (Pī´ - ĕrlz) lattice resistance" to dislocation glide. E_a is small in metals due to the lack of bond directionality, but it is significant in covalently bonded crystals such as semiconductors. Dislocation motion can also be "pinned" by the excess strain associated with dislocations intersecting each other [55]. As with other thermally activated processes, dislocation motion is proportional to $\exp(-E_a/k_BT)$, but here E_a also decreases with the degree of excess stress in the epilayer [56]. For 35-nm films of $Ge_{0.25}Si_{0.75}$ on Si(001), dislocation velocities have been observed [55] of 1 µm/s at 820 K and 100 µm/s at 1070 K. This means that if films are grown or annealed at sufficiently high T, full relaxation will occur, but that at lower growth T, epilayers can be frozen in a metastable, coherently strained state. Moreover, if a metastable epilayer of $<2h_c$ is overgrown with a layer of the substrate composition, it becomes stable. This is because misfit dislocations at both the top and bottom interface of the buried layer must be generated to relax its strain, so h_c is doubled from the Eq. (6.32) value.

Multilayered periodic heterostructures having epilayers of two alternating compositions (that is, superlattices) have similar h_c and f_c criteria. These have been analyzed recently using energy minimization [57]. It was early pointed out [58] that stable, dislocation-free "strained-layer superlattices" having an arbitrarily large number of periods can be grown under certain conditions. One of the epilayers must have $a_e > a_s$ [that is, f > 0 in Eq. (6.1)], and the other must have $a_e < a_s$ (f < 0). In addition, h must be less than h_c for each layer. To analyze this structure, it is useful to think in terms of the force exerted on each film to support the misfit strain: that is, the differential form of Eq. (6.26). The sign of the force is the same as that of ε_x or of f, so it is reversed when the sign of f is reversed. The net force which must be supported by bonding to the substrate is the sum of the forces due to the misfit of all the layers to the substrate. Thus, assuming the same Y and v for both epilayers, A and B, the forces cancel when

$$\sum f_i h_i = f_A h_A + f_B h_B = 0 \qquad (6.33)$$

For this condition, there is no net force, no matter how many pairs of layers h_A and h_B (periods) are grown.

Dislocations present in the substrate and intersecting its surface also enter the epilayer, and this happens at any h. These "threading" dislocations propagate up through the film and terminate at its surface as shown in Fig. 6.27a. In spite of these, it is possible to grow strained epilayers which have lower areal densities of threading dislocations than do their substrates. If an epilayer is strained *above* h_c, so that the generation of interfacial misfit dislocations is energetically favorable, the threading dislocation's intersection point with the interface becomes an easy site for such generation. This is because that

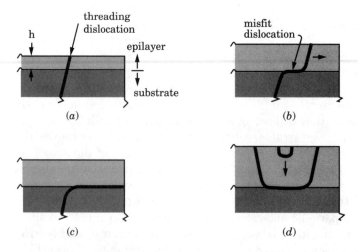

Figure 6.27 Dislocation motion in strained epilayers: (a) propagation through epilayer having h < h_c, (b) lateral motion for h > h_c, (c) termination at edge of substrate, and (d) nucleation of half-loops.

portion of the threading dislocation which lies within the film can glide laterally, laying down a misfit dislocation line along the substrate interface as it proceeds, as shown in Fig. 6.27b. Given enough time, a threading dislocation can be swept to the edge of the substrate by this mechanism, as shown in (c). In practice, however, many of the dislocations in the interface become pinned in their lateral motion by intersecting other dislocations in the grid, and this greatly limits the utility of this approach to eliminating threading dislocations from epilayers having large f.

The force-balance concept is useful in thinking about dislocation motion. Equation (6.31) for the strained-layer stability criterion may be viewed as a balance of forces. The force acting to extend the dislocation and thereby reduce film stress is $b\cos\lambda(dU_\varepsilon/d\varepsilon_{x,y})$, the $b\cos\lambda$ coming from Eq. (6.30). In general, this force causing dislocation motion under the influence of external stress is known as the Peach-Kohler force. The balancing force is the line-tension force, $dU_d/d\varepsilon_{x,y}$, which acts to minimize the dislocation's length. In Fig. 6.27b, the tension attempting to straighten the dislocation is overwhelmed by the Peach-Kohler force, causing it to extend along the interface.

Once all of the threading dislocations from the substrate have been bent into the interface, or if there are no such dislocations to begin with, this nucleation mechanism for the generation of misfit dislocations is removed, and an epilayer with $h > h_c$ is left in a metastable strained state. However, with sufficient strain energy, the activation energy for nucleation by some other mechanism is overcome, because misfit dislocations are still seen in epilayers grown on dislocation-free substrates. One commonly proposed mechanism is the nucleation of half loops at the surface, as shown in Fig. 6.27d. Such a loop can expand under the influence of the film stress and glide down toward the substrate interface along slip planes, producing two "epithreading" dislocations through the film and laying down a misfit dislocation along the interface when it arrives there. However, there is recent strong evidence for other nucleation mechanisms having lower E_a (Refs. [59], [60], and Freund, 1992), and the situation needs further study. There is also evidence that growth on small mesas increases epilayer quality [59]. Proximity of the film edge decreases the probabilities both of dislocation pinning before sweep-out and of there being a dislocation-generating defect at the surface or interface. On the other hand, the edges can act as *sources* of dislocations. The growth of defect-free heteroepitaxy with $h > h_c$ in commercially important mismatched systems such as GaAs/Si and Ge_xSi_{1-x}/Si remains a major challenge.

Strain in epilayers is not only useful for sweeping out threading dislocations. It can stabilize particular phases, including ones not found in the bulk material [61], and it can actually *improve* the performance of some electronic devices. For example, strain of either sign in the active layer of a heterojunction diode laser reduces its threshold current density below that of an unstrained layer (Weisbuch, 1991). This is because compressive strain causes a narrowing of the valence-band-energy versus momentum curve (E_v versus **k**); in other words, it decreases the effective mass of holes, m_h^*. Consequently, it takes fewer holes to fill the valence band to where the transparency condition is achieved [see discussion following Eq. (6.5)]. Tensile strain, on the other hand, by increasing m_h^* farther away from m_e^*, decreases the probability of nonradiative recombination (which involves momentum transfer), and thereby increases carrier lifetime.

In another example of a strain effect, the compressive tetragonal strain [Eq. (6.25)] of Ge_xSi_{1-x} alloy grown on Si has been shown [62], in calculations well supported by subsequent measurements, to reduce the energy gap, E_g, substantially below that of an unstrained layer of the same x, as shown in Fig. 6.28. The shaded region represents the uncertainty in the calculations. The abrupt increase in slope for the unstrained alloy above x = 0.8 is due to a change in the conduction-band minimum from the "Δ_1" to the "L_1" band. The lower E_g means

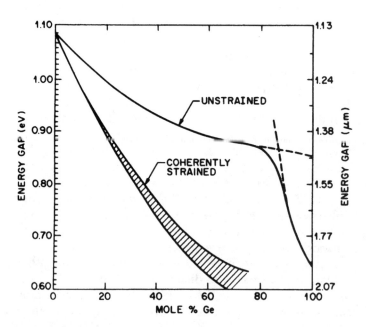

Figure 6.28 Indirect band gap of Ge_xSi_{1-x} for unstrained bulk alloy and for coherently-strained epilayer on Si(001). (Source: Reprinted from Ref. 62 by permission.)

that, to achieve a given E_g reduction, x needs to be only 40 percent as big as in an unstrained layer. This increases the latitude of design in stable heterostructures.

6.7 Surface Morphology

Up to now in the discussion of epitaxy we have tacitly assumed that the growth surface is smooth to within a monolayer or so; that is, growth is two-dimensional (2D). This is the desired growth mode both for maximizing interface abruptness and for minimizing crystallographic defects. In fact, 2D growth is often achieved. However, 3D growth is common in the first few nm of heteroepitaxy, and if conditions are unfavorable, it can persist with continued deposition. These two growth modes are not unique to epitaxial growth, and they were discussed in a more general context in Sec. 5.3. There, we found that surface energies, γ, determine which growth mode is thermodynamically favorable. When the inequality of Eq. (5.36) is satisfied, the film "wets" the substrate and grows 2D. Here, we will discuss first the behavior of 2D epitaxy, and then that of 3D. In 3D epitaxy, we will see that lattice mismatch as well as γ determine the mode of growth. At lower substrate T, T_s, where growth is "quenched," 3D growth arises from a different mechanism, namely statistical roughening.

6.7.1 Two-dimensional nucleation

Two-dimensional growth was illustrated in Fig. 5.13. There, atoms are adsorbing on an atomically smooth surface, and T_s is high enough so that $\Lambda \gg a$, where Λ is the surface-diffusion length given by Eq. (5.25), and a is the atomic spacing. Then, the adsorbing atoms behave as a 2D gas, diffusing around until they nucleate into 2D islands which eventually coalesce to become the next complete monolayer.

The atomically smooth surface is not just an idealized model. It can actually be obtained by cutting and polishing a substrate crystal on a plane that is as close as possible to a low-index crystallographic plane such as (001). Sometimes, atomic smoothness requires in addition the deposition of a "buffer layer" of 2D homoepitaxy. Figure 6.29 shows an STM photograph of a nominally (001) Si surface used to study surface diffusion during epitaxy [63]. It consists of ≈25-nm-wide terraces bounded by monolayer steps. The terraces form a downward staircase from left to right. This surface has reconstructed into dimer rows such as those illustrated in Fig. 6.18, and the rows rotate 90° with successive monolayers. The terraces with smooth outer edges have the dimer rows parallel to them and therefore have relatively low excess edge energy, β, while those with rough outer edges have undergone 2D facetting in an attempt to reduce β (see Sec. 5.3.3).

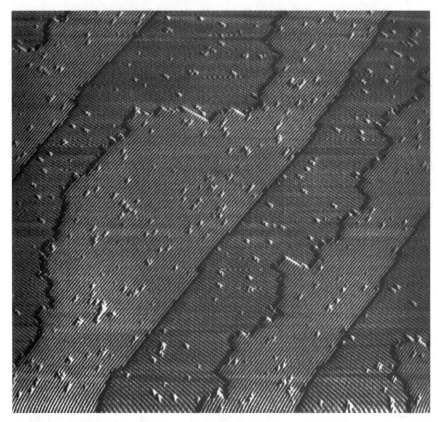

Figure 6.29 Staircase of atomic terraces on Si(001) as seen in STM. The stairs descend from top left to bottom right, and the image is 120 nm square. (Source: Previously unpublished photo courtesy of Brian S. Swartzentruber, from the laboratory of Max Lagally at the University of Wisconsin.)

On well oriented surfaces, the distances between occasional monolayer steps can much larger than Λ even when $\Lambda \gg a$. Then, 2D nucleation and coalescence occur randomly over the surface as described above. This cyclic process of island nucleation and monolayer completion can be followed easily by the accompanying oscillations in RHEED spot intensity, as discussed in Sec. 6.4.2.4. Moreover, in growing compounds by the flux-modulation techniques of Sec. 6.5.5, the flux pulses of each precursor can be "phase-locked" to the point at which RHEED shows a monolayer of the other element to be just completed. This assists the incorporation of each element into its proper lattice site and also lowers the T_s at which 2D growth can be sustained. In using phase locking, one must allow for the phase shift arising from electron-wave interference with electrons that have undergone multiple scattering [24].

6.7.1 Two-dimensional nucleation

Even with phase locking, 2D growth ultimately changes to 3D when Λ becomes smaller than a. This is the quenched-growth regime in which statistical roughening determines the surface morphology, and it will be discussed in Sec. 6.7.3. It is a kinetically inhibited process and is thus very different from the higher-T_s *equilibrium* 3D growth which is driven by free-energy minimization. The latter process requires a long Λ so that the adsorbing atoms can rearrange into 3D islands. The transition from 2D to 3D growth can be observed by a damping out of the RHEED oscillations and by an increased spottiness of the RHEED streaks due to 3D diffraction, as illustrated in Fig. 6.16.

2D nucleation can be made to occur in a more controlled manner than above by deliberately orienting the substrate surface in the *vicinity* of, but not directly on, a low-index plane: that is, on a "vicinal" plane. Then,

$$\frac{a}{L} = \tan^{-1}(\Delta\theta) \approx \Delta\theta \qquad (6.34)$$

where L is the terrace width and $\Delta\theta$ is the angle of miscut in radians. Thus, a 1° miscut gives ≈10-nm-wide terraces, and larger miscuts give narrower terraces. When $L < \Lambda$, adatoms are more likely to diffuse to the terrace edges and nucleate there than to form new nuclei within the terraces. Then, growth proceeds by the *continuous flow* of step edges across the surface rather than by cycles of nucleation and monolayer completion. This is known as "step-flow" growth. Control of nucleation in this manner can reduce crystallographic defect density, but it also removes the useful RHEED oscillations. Thus, there is a T_s *window* for RHEED oscillations, above which $\Lambda > L$ and below which statistical roughening sets in. The upper T_s boundary beyond which the oscillations damp out can be used to obtain a rough estimate of Λ versus T_s when L is known [24].

Epitaxy of compounds on elemental substrates containing single-monolayer (1-ML) steps can result in planar defects known as antiphase-domain boundaries, as illustrated in Fig. 6.30 at step a. This is because one or the other of the film's elements generally has a stronger preference for bonding to the substrate surface. Thus, for GaAs on Si or Ge, As bonds first [15]. When nuclei on adjacent terraces coalesce, the stacking of atomic layers in the growth direction is out of phase, so the bonding is distorted at the interface. These defects can be avoided when the substrate surface contains only steps of 2-ML height as shown by step d. For Si(100) miscut toward the [110] direction, these steps are thermodynamically favored because the edge energy, β, of one 2-ML step is less than twice the β of two 1-ML steps

296 Epitaxy

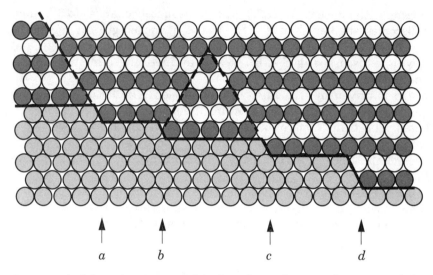

Figure 6.30 Antiphase domains created in the epitaxy of compounds on terraced elemental substrates: white circles = film cations, dark gray circles = film anion, light gray circles = substrate atoms, solid line = substrate interface, and dashed line = domain boundaries.

[29]. Even with 1-ML steps, antiphase domains can vanish with increasing h if they are overgrown with in-phase material [15], as in the case of the domain shown between steps b and c.

6.7.2 Three-dimensional rearrangement

In many heteroepitaxial situations, it is energetically favorable for the first few nm of depositing material to rearrange itself into 3D hillocks, and it will do so if there is enough thermal energy to activate the surface diffusion by which this equilibration occurs: that is, if Λ is larger than the distance between hillocks. Referring again to the three growth modes of Fig. 5.8, we see that there are two types of hillocks. In the island or Volmer-Weber growth mode, 3D islands are separated by bare substrate. This occurs when

$$\gamma_f + \gamma_i > \gamma_s \qquad (6.35)$$

where the subscripts refer to the surface energies per unit area of the film free surface, the substrate-film interface, and the substrate free surface, respectively. Then, total surface energy, $\Sigma \gamma_k A_k$ from Eq. (5.43), is minimized by islanding, which increases A_s and decreases A_f and A_i. The island growth mode is not unique to epitaxy, and it was discussed for deposition in general in Sec. 5.3.2. In epitaxy, it occurs, for example, for many semiconductors on fluorites such as CaF_2, and also for Si on Ge.

The second type of 3D-hillock growth is Stranksi-Krastanov (S-K), in which the first 1–3 ML or so grow 2D, and then hillocks begin to form. The three growth modes (2D layer, 3D island, and S-K) can be distinguished using either of the electron-spectroscopic surface-analysis techniques discussed in Sec. 6.4.1: AES or XPS. Since the electron escape depth is only a few nm, the signal from the substrate atoms, I_s, attenuates rapidly with increasing deposition dose, $J_r t$ (flux × time), but the shape of the attenuation curve varies with growth mode. Typical I_s signatures are shown in Fig. 6.31. For layer growth, I_s is uniform over the surface and attenuates exponentially with dose as expected for absorbing media in accordance with Beer's law, Eq. (4.49). For island growth, attenuation is much slower, because a portion of the substrate remains bare until the islands coalesce. Here, an approximate attenuation curve can be derived by assuming a constant pre-coalescence concentration, n*, of hemispherical nuclei which completely absorb the substrate signal. Then,

$$I_s = 1 - \Theta = 1 - \left[\frac{1}{n_{so}}\left(\frac{3\sqrt{\pi}}{2}\right)^{2/3}(n^*)^{1/3}\right](J_r t)^{2/3} \quad (6.36)$$

where Θ is the fractional surface coverage by nuclei and n_{so} is the monolayer (ML) surface concentration of atoms. A typical curve has been plotted in Fig. 6.31 by using $n_{so} = 1\times10^{15}$ atoms/cm² and $n^* = 2\times10^{11}$ nuclei/cm². The curve for S-K growth tracks that of layer growth until 3D rearrangement begins, and then there is a sharp

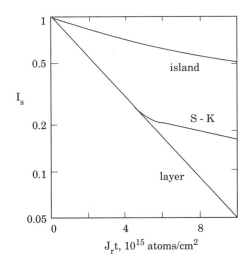

Figure 6.31 Typical signatures of the electron-spectroscopy signal from the substrate, I_s, vs. deposition dose ($J_r t$) for the three growth modes of Fig. 5.8.

transition to the curve shape for island growth. The dose at the break point tells how many ML of 2D growth preceded the transition.

There are at least three mechanisms by which S-K growth can occur, all of them involving the drive toward free-energy minimization. One is exemplified by GaAs on Si(111). There, the first ML of As bonds to the surface so as to satisfy all of the dangling bonds with very little bond distortion [15]. This produces a highly passivated surface with very low γ, so that subsequently depositing Ga and As form 3D hillocks rather than wetting it. In other words, the inequality of Eq. (6.35) has been satisfied by lowering $γ_s$ with As passivation. In the second and most often-proposed mechanism, 2D growth is energetically favorable at first, but a lattice mismatch causes the buildup of coherency-strain energy, $U_ε$, with film thickness, h, in accordance with Eq. (6.26). This term adds to the γ terms on the left side of Eq. (6.35), and if $U_ε$ becomes large enough that the inequality is satisfied, growth changes to 3D. The 3D morphology relaxes the strain by reducing substrate forces on the film and thus allowing the film material to deform toward its own lattice constant. (Thicker films relax $U_ε$ instead by dislocation generation, as discussed in Sec. 6.6.) Similar 3D-strain relaxation was examined earlier for much larger features. It was illustrated in Fig. 5.34 for the 3D features produced upon lithographic patterning of stressed films. Also, hillocks in soft metal films grow to relieve compressive stress (Sec. 5.6.2), so this may be thought of as a macroscopic (and nonepitaxial) analog of S-K growth. In a third possible S-K mechanism, some substrate orientations force the 2D film to expose a relatively high-γ crystallographic plane when this is still energetically more favorable than dewetting the substrate. However, when the film is thick enough to rearrange without dewetting, it may prefer to rearrange so as to expose facets of lower γ. The criterion here is that total surface energy summed over k different facets,

$$\sum_k γ_k A_k$$

from Eq. (5.43), be reduced by the rearrangement.

As deposition progresses beyond hillock coalescence in the 3D modes, growth often reverts to the desired 2D mode, provided that surface diffusion length is long enough to promote smoothening and provided that facetting is not energetically preferred. One reason for "buffer layer" growth on foreign substrates is to provide enough growth thickness for this process to occur. However, initial 3D growth reduces the planarity of thin heteroepitaxial layers and also introduces various crystallographic defects which tend to propagate into subsequent growth. Dislocations can be generated during the 2D→3D rearrangement process, and planar defects such as twin planes and stacking faults develop during coalescence due to mismatches at the boundaries between the strain-relaxed hillocks.

The causes and suppression of 3D growth are subjects of considerable current research. Since 3D growth is driven by free-energy minimization, it can be suppressed by freezing the kinetics of the equilibration process. For example, surface diffusion of Ge depositing on Si is frozen by preadsorption of a monolayer of passivating As on the Si (although it unfortunately dopes the Ge epilayer at the same time). The As segregates to the surface of the adsorbing Ge and thus immobilizes it by burial [64]. Although this has been termed a "surfactant" effect, the term is confusing here, because a surfactant is commonly used to assist the wetting of a film to a substrate by reducing γ_f more than it reduces γ_s, as soap does to water (see end of Sec. 5.3.1). By contrast, As passivation decreases the γ of *both* Ge and Si, and no demonstration of wetting assistance has been made. Instead, the As encapsulation is inhibiting dewetting of Ge from Si by blocking Ge surface diffusion. A surfactant *assists* thermodynamic equilibration, whereas the As here *inhibits* the kinetics of equilibration. It is always important to understand whether thermodynamics or kinetics is controlling the behavior of a process.

Freezing the kinetics of 3D-hillock formation is more commonly achieved by lowering substrate T, T_s, but too low a T_s can also freeze in crystallographic defects or cause statistical roughening. This dilemma can be reduced by the use of low-energy ion bombardment. Ions impinging with kinetic energies between about 20 and 30 eV have enough energy to displace surface atoms, but not enough energy to cause subsurface crystallographic damage (more in Sec. 8.5.1.3), so this energy range is useful for modifying epitaxial growth. In GaAs epitaxy on 653-K Si(100), for example, Ar^+ bombardment in this energy range was found to prevent the formation of 3D nuclei [65]. Evidence suggested that the bombardment was breaking up incipient 3D nuclei faster than they could reform at this low T_s.

6.7.3 Statistical roughening

The 3D roughening discussed above is driven by surface-energy minimization. At very low T_s, the kinetics of this process are frozen; that is, the surface diffusion length, Λ, is too small to allow rearrangement of depositing atoms into hillocks. Then, roughness develops by a different mechanism, namely statistical fluctuations in the deposition flux. The behavior of statistical roughening was discussed for film deposition in general in Sec. 5.4.1. It develops when Λ is insufficient to counterbalance the statistical fluctuations, and it results in the common

"Zone-1" (Z1) voided columnar film structure. For metals, the upper T_s boundary of Z1 growth scales roughly with melting point in K, according to $T_s/T_m = 0.3$.

It is important to recognize that epitaxial growth can still be obtained even when Λ is so small that statistical roughening develops. For example, both Si and Fe can be grown epitaxially at room T, where $T_s/T_m < 0.2$. This is presumably because the depositing atoms are still pulled to the most energetically favorable "cradle" position in the surface lattice by bonding forces, and this position is likely to be the epitaxial site. However, it has recently been found in Si MBE at low T_s that there is a critical *thickness* for epitaxy, h_e, above which crystalline order suddenly collapses into amorphous growth, and it is likely that this limit is being imposed by the increasing roughness of the surface [66]. The T_s dependence of h_e follows Arrhenius behavior [Eq. (5.2)]; that is, $\ln h_e \propto 1/T_s$. This suggests that the collapse is being forestalled by a thermally activated process, which is most likely a small but not negligible amount of surface diffusion, even at $T_s/T_m < 0.2$.

Similar h_e behavior has been observed in Si using ion-beam sputter deposition [67], where the depositing atoms have mean kinetic energy of several eV, compared to 0.2 eV for Si in MBE. There, h_e was found to be larger and its T_s dependence *steeper* than in MBE. The steeper slope for the energy-enhanced deposition process indicates that energetic impingement does not enhance surface diffusion, because such enhancement would *lessen* the T_s dependence. It was proposed that h_e and its T_s dependence are both increased because the energetic flux collapses incipient voids. Recall that void collapse is what causes sputter-deposited films to have the dense ZT structure rather than the Z1 structure. Void-free growth results in a lower exponent, β, for the increase of roughness with film thickness, h, where roughness is expressed by the interface width, $\Delta h \propto h^\beta$. These factors were discussed in Sec. 5.4.1. Now, assuming that epitaxy collapses when $\Delta h > \Lambda$, then h_e for the sputtered film will be larger. The h_e will also increase more steeply with T_s, since the smaller β means a larger increase in h for the same increase in Δh. These results are preliminary, however, and much more investigation is needed to understand the behavior of h_e.

6.8 Conclusion

We have seen in this chapter that the special type of thin-film deposition known as epitaxy is particularly demanding with regard to all aspects of process control. Film quality is readily degraded by small amounts of contamination, nonstoichiometry, and lattice mismatch. On the other hand, when good control is achieved, complex multilayered structures with unique properties can be fabricated with atomic-layer precision. Moreover, the precise structural and compositional nature of the epitaxial growth surface allow the use of growth monitoring techniques that give detailed information about film growth mechanisms on an atomic scale.

6.9 Exercises

6.1 For epitaxy of CdTe(111) on GaAs(001) as shown in Fig. 6.1c: (a) show by a geometrical construction similar to Fig. 5.7 that the [211] direction of CdTe does indeed lie in the (111) plane. (b) Using the a_o data of Fig. 6.3, calculate the atomic spacing a shown in Fig. 6.1c for GaAs and for CdTe.

6.2 (a) Calculate the wavelength of a thermal-energy conduction electron confined to a 2D electron gas in GaAs. (In GaAs, $m_e^* = 0.067 m_e$). (b) Show that Eq. (6.4) for the kinetic energy of confined electrons follows from Eq. (6.3). (c) How narrow must a 1D quantum well be in GaAs to raise the confined electron's kinetic energy by 0.1 eV in the ground state?

6.3 What must x be in $GaAs_xP_{1-x}$ to give $E_g = 1.8$ eV, assuming Vegard's law?

6.4 Name three causes of substrate-induced disruption of epitaxy and their remedies.

6.5 Show that the wavelength of an electron is related to the voltage drop through which it has been accelerated by

$$\lambda_e \text{ (nm)} = \sqrt{1.51/V_e}$$

6.6 Map the LEED pattern from Si(001) at 120 V and calculate the spot spacing for a screen radius of 10 cm.

6.7 What electron-beam voltage is required for off-Bragg RHEED oscillations from Si(001) at a beam incident angle of 4° from grazing?

6.8 For deposition of CdTe from Cd and Te_2 vapor at 1 μm/h using a 20 percent excess of Cd arrival flux, use the procedures of Sec. 6.5 to estimate the substrate T at which Cd precipitate will start to accumulate on the surface.

6.9 What are the advantages and disadvantages of chemical-vapor versus physical-vapor sources of depositing material?

6.10 What are the advantages and disadvantages of ALE versus epitaxy at constant vapor flux?

6.11 In depositing S-doped GaAs under conditions of 500° C substrate T, $J_r(Ga) = 1 \times 10^{15}/cm^2 \cdot s$, and J_i (S as S_2) = 1×10^{11} mc/$cm^2 \cdot s$, the doping level achieved is 1×10^{16} p/cm^3. Making reasonable assumptions for this dopant, namely no surface segregation and full electrical activation of incorporated S, what is the activity coefficient of the S under these conditions?

6.12 For Si MBE, the surface-segregation coefficient of the n-type dopant Sb under certain deposition conditions is 10^6. For 5×10^{13} Sb/cm^2 floating on the growth surface, (a) what is the bulk-film dopant concentration, n(Sb), and (b) what thickness must be grown after turning off the Sb flux before n(Sb) drops by a factor of 10? Assume no re-evaporation and full equilibration.

6.13 For a particular epilayer of CdS on Ge(100), the lattice constant a' perpendicular to the growth plane is determined by x-ray diffraction to be 0.59 nm. Assuming that a' is constant through the thickness of the film, what fraction of the coherency strain has been released by misfit-dislocation generation, and what linear density of dislocations is expected in the interface assuming that $(1/2)a_0 60°$ dislocations are dominant?

6.14 What is the difference between incommensurate and discommensurate growth?

6.15 Derive an expression for the equilibrium strain in epitaxy above the critical lattice mismatch or film thickness. If thickness is 10 percent above critical, what fraction of the biaxial misfit strain is relieved by dislocations?

6.16 For a strained-layer superlattice of equal-thickness layers of $Ga_{0.3}In_{0.7}P$ and $Ga_{1-x}In_xP$ on a GaAs substrate, determine h_c and the value of x which neutralizes shear force at the GaAs interface. Assume Vegard's law and equal thermal-expansion coefficients.

6.17 Describe three different mechanisms of 3D growth and identify their key driving forces.

6.18 In 2D epitaxy of a certain film with 0.2-nm monolayer height, the substrate T above which RHEED oscillations damp out increases from 600 to 620° C when substrate vicinality is changed from 2° to 1°. What activation energy for surface diffusion, in eV, can be derived from this information?

6.19 Show that the decrease in Auger signal from substrate atoms with increasing deposition in the island growth mode follows Eq. (6.36) for the assumptions of that equation.

6.10 References

1. Wyckoff, R.W.G. 1963. *Crystal Structures*, 2nd ed. New York: Wiley Interscience.
2. Smith, D.L. Unpublished work.

3. Chang, C.-A., J.C. Liu, and J. Angilello. 1990. "Epitaxy of (100) Cu on (100) Si by Evaporation near Room Temperature: In-plane Epitaxial Relationship and Channeling Analysis." *Appl. Phys. Lett.* 57:2239.
4. Faurie, J.P., C. Hsu, S. Sivananthan, and X. Chu. 1986. "CdTe-GaAs(100) Interface: MBE Growth, RHEED and XPS Characterization." *Surf. Sci.* 168:473.
5. Farrow, R.F.C. 1983. "The Stabilization of Metastable Phases by Epitaxy." *J. Vac. Sci. Technol.* B1:222.
6. Flynn, C.P. 1991. "Tuned Tilt of Epitaxial Crystals." *MRS Bulletin*, June:30.
7. Wang, K.L., and R.P.G. Karunasiri. 1993. "SiGe/Si Electronics and Optoelectronics." *J. Vac. Sci. Technol.* B11:1159.
8. Schubert, E.F. 1990. "Delta Doping of III-V Compound Semiconductors: Fundamentals and Device Applications." *J. Vac. Sci. Technol.* A8:2980.
9. Sundaram, M., S.A. Chalmers, P.F. Hopkins, and A.C. Gossard. 1991. "New Quantum Structures." *Science* 254:1326.
10. Falicov, L.M. 1992. "Metallic Magnetic Superlattices." *Physics Today* October:46.
11. Kim, D., H.W. Lee, J.J. Lee, J.H. Je, M. Sakurai, and M. Watanabe. 1994. "Mo-Si Multilayers as Soft X-ray Mirrors for the Wavelengths around 20 nm Region." *J. Vac. Sci. Technol.* A12:148.
12. Walker, P., and W.H. Tarn. 1991. *CRC Handbook of Metal Etchants*. Boca Raton, Fla.: CRC Press. (also covers many compounds)
13. Grunthaner, P.J., F.J. Grunthaner, R.W. Fathauer, T.L. Lin, M.H. Hecht, L.D. Bell, W. J. Kaiser, F.D. Showengerdt, and J.H. Mazur. 1989. "Hydrogen-Terminated Silicon Substrates for Low-Temperature Molecular Beam Epitaxy." *Thin Solid Films*, 183:197.
14. Fenner, D.B., D.K. Biegelsen, and R.D. Bringans. 1989. "Silicon Surface Passivation by Hydrogen Termination: A Comparative Study of Preparation Methods." *J. Appl. Phys.* 66:419.
15. Bringans, R.D. 1992. "Arsenic Passivation of Si and Ge Surfaces." *Crit. Rev. in Solid State Physics and Mater. Sci.* 17:353.
16. Fork, D.K., K. Nashimoto, and T.H. Geballe. "Epitaxial $YBa_2Cu_3O_{7-\delta}$ on GaAs(001) Using Buffer Layers." *Appl. Phys. Lett.* 60:1621.
17. Cho, A.Y., and J.R. Arthur. 1975. "Molecular Beam Epitaxy." *Progress in Solid-State Chemistry* 10:157.
18. Powell, C.J., and M.P. Seah. 1990. "Precision, Accuracy, and Uncertainty in Quantitative Surface Analysis by Auger Electron Spectroscopy and X-ray Photoelectron Spectroscopy." *J. Vac. Sci. Technol.* A8:735.
19. Cullity, B.D. 1978. *Elements of X-ray Diffraction*, 2nd ed. Reading, Mass.: Addison-Wesley, 102.
20. Segmüller, A. 1991. "Characterization of Epitaxial Thin Films by X-ray Diffraction." *J. Vac. Sci. Technol.* A9:2477.
21. Clemens, B.M., and J.A. Bain. 1992. "Stress Determination in Textured Thin Films using X-ray Diffraction." *MRS Bull.* July:46.
22. Jiang, Q., Y.-L. He, and G.-C. Wang. 1993. "Thermal Stability and Intermixing of Ultrathin Fe Films on a Au(001) Surface." *Surface Science* 295:197.
23. Cohen, P.I., P.R. Pukite, J.M. Van Hove, and C.S. Lent. 1986. "Reflection High Energy Electron Diffraction Studies of Epitaxial Growth on Semiconductor Surfaces." *J. Vac. Sci. Technol.* A4:1251.
24. Joyce, B.A. 1990. "The Evaluation of Growth Dynamics in MBE using Electron Diffraction." *J. Crystal Growth* 99:9.
25. Meyer, H.J., and B.J. Stein. 1980. "Untersuchung der Kondensation und Verdampfung von Alkalihalogenidkristallen mit Molekularstrahlmethoden." *J. Crystal Growth* 49:707.

26. Baxter, C.S., W.M. Stobbs, and J.H. Wilkie. 1991. "The Morphology of Ordered Structures in III-V Alloys: Inferences from a TEM Study." *J. Crystal Growth* 112:373.
27. Stringfellow, G.B., and G.S. Chen. 1991. "Atomic ordering in III/V Semiconductor Alloys." *J. Vac. Sci. Technol.* B9:2182.
28. Smith, D.L. 1979. "Nonstoichiometry and Carrier Concentration Control in MBE of Compound Semiconductors." *Prog. Crystal Growth and Charact.* 2:33.
29. Chadi, D.J. 1989. "Atomic Structure of Reconstructed Group IV and III-V Semiconductor Surfaces." *Ultramicroscopy* 31:1.
30. Biegelsen, D.K., R.D. Bringans, J.E. Northrup, and L.-E. Swartz. 1990. "Surface reconstructions of GaAs(100) Observed by Scanning Tunneling Microscopy." *Phys. Rev. B* 41:5701.
31. Farrell, H.H., M.C. Tamargo, and J.L. de Miguel. 1991. "Optimal GaAs(100) Substrate Terminations for Epitaxy." *Appl. Phys. Lett.* 58:355.
32. Newstead, S.M., R.A.A. Kubiak, and E.H.C. Parker. 1987. "On the Practical Applications of MBE Surface Phase Diagrams." *J. Crystal Growth* 81:49.
33. Robertson, Jr., A., and A.S. Jordan. 1993. "Equilibrium Gas-Phase Composition of Cracked AsH_3 and PH_3." *J. Vac. Sci. Technol.* B11:1041.
34. Abernathy, C.R. 1993. "Growth of III-V Materials by Metalorganic Molecular-Beam Epitaxy." *J. Vac. Sci. Technol.* A11:869.
35. Nishizawa, J., K. Aoki, S. Suzuki, and K. Kikuchi. 1990. "Silicon Molecular Layer Epitaxy." *J. Electrochem. Soc.* 137:1898.
36. Nishizawa, J., H. Abe, and T. Kurabayashi. 1989. "Doping in Molecular Layer Epitaxy." *J. Electrochem. Soc.* 136:478.
37. Creighton, J.R., K.R. Lykke, V.A. Shamamian, and B.D. Kay. 1990. "Decomposition of Trimethylgallium on the Gallium-Rich GaAs(100) Surface: Implications for Atomic Layer Epitaxy." *Appl. Phys. Lett.* 57:279.
38. Ban, V.S., G.C. Erickson, S. Mason, and G.H. Olsen. 1990. "Selective Epitaxy of III-V Compounds by Low-Temperature Hydride VPE." *J. Electrochem. Soc.* 137:2904.
39. Horikoshi, Y., M. Kawashima, and H. Yamaguchi. 1986. "Low-Temperature Growth of GaAs and AlAs-GaAs Quantum-Well Layers by Modified Molecular Beam Epitaxy." *Japan. J. Appl. Phys.* 25:L868.
40. Yao, T., and T.Takeda. 1987. "Intensity Variations of Reflection High-Energy Electron Diffraction during Atomic Layer Epitaxial Growth and Sublimation of Zn Chalcogenides." *J. Crystal Growth* 81:43.
41. Yu, M. L., U. Memmert, N.I. Buchan, and T.F. Kuech. 1991. "Surface Chemistry of CVD Reactions Studied by Molecular Beam/Surface Scattering." In *Chemical Perspectives of Microelectronic Materials II*, ed. L.V. Interrante, K.F. Jensen, L.H. Dubois, and M.E. Gross. Pittsburgh, Pa.: Materials Research Society.
42. Taike, A., M. Migita, and H. Yamamoto. 1990. "p-type Conductivity Control of ZnSe Highly Doped with Nitrogen by Metalorganic Molecular Beam Epitaxy." *Appl. Phys. Lett.* 56:1989.
43. Park, R. M., M.B. Troffer, and C.M. Rouleau. 1990. "p-type ZnSe by Nitrogen Atom Beam Doping during Molecular Beam Epitaxial Growth." *Appl. Phys. Lett.* 57:2127.
44. Heckingbottom, R., C.J. Todd, and G.J. Davies. 1980. "The Interplay of Thermodynamics and Kinetics in Molecular Beam Epitaxy of Doped Gallium Arsenide." *J. Electrochem. Soc.* 127:444.
45. Itoh, T., and H. Takai. 1989. "Partially Ionized Molecular Beam Epitaxy." Chap. 7A in *Ion Beam Assisted Film Growth*. ed. T. Itoh. Amsterdam: Elsevier.
46. Jorke, H. 1988. "Surface Segregation of Sb on Si(100) during Molecular Beam Epitaxy Growth." *Surf. Sci.* 193:569.
47. Sugiura, H. 1980. "Silicon Molecular Beam Epitaxy with Antimony Ion Doping." *J. Appl. Phys.* 51:2630.

48. Kirchner, P.D., J.M. Woodall, J.L. Freeouf, D.J. Wolford, and G.D. Pettit. 1981. "Volatile Metal Oxide Incorporation in Layers of GaAs and $Ga_{1-x}Al_xAs$ Grown by Molecular Beam Epitaxy." *J. Vac. Sci. Technol.* 19:604.
49. Wang, W.I., E.E. Mendez, T.S. Kuan, and L. Esaki. 1985. "Crystal Orientation Dependence of Silicon Doping in Molecular Beam Epitaxial AlGaAs/GaAs Heterostructures." *Appl. Phys. Lett.* 47:826.
50. Parkinson, B.A., F.S. Ohuchi, K. Ueno, and A. Koma. 1991. "Periodic Lattice Distortions as a Result of Lattice Mismatch in Epitaxial Films of Two-Dimensional Materials." *Appl. Phys. Lett.* 58:472.
51. Smith, D.L. Unpublished work.
52. Gerthsen, D., D.K. Biegelsen, F.A. Ponce, and J.C. Tramontana. 1990. "Misfit Dislocations in GaAs Heteroepitaxy on (001) Si." *J. Crystal Growth* 106:157.
53. Matthews, J.W., and A.E. Blakeslee. 1974. "Defects in Epitaxial Multilayers I: Misfit Dislocations." *J. Crystal Growth* 27:118.
54. Houghton, D.C., C.J. Gibbings, C.G. Tuppen, M.H. Lyons, and M.A.G. Halliwell. 1990. "Equilibrium Critical Thickness for $Si_{1-x}Ge_x$ Strained Layers on (100)Si." *Appl. Phys. Lett.* 56:460.
55. Hull, R., J.C. Bean, D.J. Eaglesham, J.M. Bonar, and C. Buescher. 1989. "Strain Relaxation Phenomena in Ge_xSi_{1-x}/Si Strained Structures." *Thin Solid Films* 183:117.
56. Dodson, B.W., and J.Y. Tsao. 1989. "Scaling Relations for Strained-Layer Relaxation." *Appl. Phys. Lett.* 55:1345.
57. Houghton, D.C., D.D. Perovic, J.-M. Baribeau, and G.C. Weatherly. 1990. "Misfit Strain Relaxation in Ge_xSi_{1-x}/Si Heterostructures: the Structural Stability of Buried Strained Layers and Strained-Layer Superlattices." *J. Appl. Phys.* 67:1850.
58. Matthews, J.W., and A.E. Blakeslee. 1976. "Defects in Epitaxial Multilayers III: Preparation of Almost Perfect Multilayers." *J. Crystal Growth* 32:265.
59. Fitzgerald, E.A., G.P. Watson, R.E. Proano, D.G. Ast, P.D. Kirchner, G.D. Pettit, and J.M. Woodall. 1989. "Nucleation Mechanisms and the Elimination of Misfit Dislocations at Mismatched Interfaces by Reduction in Growth Area." *J. Appl. Phys.* 65:2220.
60. Perovic, D.D., and D.C. Houghton. 1992. "Barrierless Misfit Dislocation Nucleation in SiGe/Si Strained Layer Epitaxy." In *Mechanisms of Heteroepitaxial Growth*, ed. M.F. Chisholm, B.J. Garrison, R. Hull, and L.J. Schowalter, Proceedings v. 263. Pittsburgh, Pa.: Materials Research Society.
61. Wood, D.M. 1992. "Coherent Epitaxy, Surface Effects, and Semiconductor Alloy Phase Diagrams." *J. Vac. Sci. Technol.* B10:1675.
62. People, R. 1985. "Indirect Band Gap of Coherently Strained Ge_xSi_{1-x} Bulk Alloys on (001) Silicon Substrates." *Phys. Rev. B* 32:1405.
63. Mo, Y.-W., R. Kariotis, B.S. Swartzentruber, M.B. Webb, and M.G. Lagally. 1990. "Scanning Tunneling Microscopy Study of Diffusion, Growth, and Coarsening of Si on Si(001)." *J. Vac. Sci. Technol.* A8:201.
64. Copel, M., M.C. Reuter, E. Kaxiras, and R.M. Tromp. 1989. "Surfactants in Epitaxial Growth." *Phys. Rev. Lett.* 63:632.
65. Choi, C.-H., R. Ai, and S.A. Barnett. 1991. "Suppression of Three-Dimensional Island Nucleation during GaAs Growth on Si(100)." *Phys. Rev. Lett.* 67:2826.
66. Eaglesham, D.J., H.-J. Gossmann, M. Cerullo, L.N. Pfeiffer, and K.W. West. 1991. "Limited Thickness Epitaxy of Semiconductors and Si MBE down to Room Temperature." *J. Crystal Growth* 111:833.
67. Smith, D.L., C.-C. Chen, G.B. Anderson, and S.B. Hagstrom. 1993. "Enhancement of Low-Temperature Critical Epitaxial Thickness of Si(100) with Ion Beam Sputtering." *Appl. Phys. Lett.* 62:570.

6.11 Recommended Readings

Ball, C.A.B., and J.H. Van der Merwe. 1983. "The Growth of Dislocation-Free Layers." Chap. 27 in *Dislocations in Solids*, v. 6, ed. F.R.N. Nabarro. North-Holland.

Freund, L.B. 1992. "Dislocation Mechanisms of Relaxation in Strained Epitaxial Films." *MRS Bull.* July:52.

Herman, M.A., and H. Sitter. 1989. *Molecular Beam Epitaxy: Fundamentals and Current Status*. Berlin: Springer-Verlag.

Houng, Y.-M. 1992. "Chemical Beam Epitaxy." *Crit. Rev. in Solid State and Mater. Sci.* 17:277.

Hull, D., and D.J. Bacon. 1984. *Introduction to Dislocations*, 3rd ed. Oxford, U.K.: Pergamon Press.

Joyce, B.A. 1985. "Molecular Beam Epitaxy." *Rep. Prog. Phys.* 48:1637.

Saloner, D. 1986. "Characterization of Surface Defects and Determination of Overlayer Nucleation and Growth by Surface-Sensitive Diffraction." *Appl. Surf. Sci.* 26:418.

Suntola, T. 1989. "Atomic Layer Epitaxy." *Materials Science Reports* 4:261.

Tsao, J.Y. 1992. *Materials Fundamentals of Molecular Beam Epitaxy*. Boston, Mass.: Academic Press.

Weisbuch, C., and B. Vinter. 1991. *Quantum Semiconductor Structures*. Boston, Mass.: Academic Press.

Chapter

7

Chemical Vapor Deposition

The use of chemical vapors and gases as sources of film-forming elements was introduced in Sec. 6.5.4 in the context of epitaxy in the high-vacuum regime (Kn > 1). In this chapter, we consider the use of these sources at higher pressures in the fluid-flow regime (Kn << 1), where the process is known as chemical vapor deposition (CVD). Gaseous source materials allow process operation in this regime because they do not condense (by definition) on surrounding room-T surfaces during transport to the substrate. Therefore, the line-of-sight transport geometry from source to substrate which is required in the physical vapor-deposition (PVD) processes is not required in CVD. A monolayer or two of transporting gas may *adsorb* on the room-T surfaces, but this adsorption quickly saturates. On the other hand, upon reaching the heated substrate or other hot surfaces, some fraction of the adsorbing gas reacts to form the film. This fraction is expressed by the sticking coefficient, S_c, defined by Eq. (6.17). It is sometimes referred to as the "reactive" sticking coefficient, although this term seems redundant. Often, $S_c \ll 1$ at the substrate, and this makes it possible to uniformly coat substrates having convoluted surfaces, such as microcircuit patterns, or to coat large batches of substrates on all sides at once, such as tool bits. With $S_c \ll 1$, the gas can still reach remote substrate areas despite many encounters with hot surfaces along the way. This process was illustrated in contrast to PVD behavior for the filling of a trench in Fig. 5.17*f*.

The potential for uniform coating of nonplanar substrates—that is, "conformal" coating—is a key advantage of using gaseous sources in either flow regime of Kn. Various other advantages and disadvantages were mentioned in Sec. 6.5.4. Three additional advantages apply specifically to the higher-pressure regime. One is that higher deposition rates can sometimes be obtained while still maintaining conformality. Another is that simultaneous etching of the depositing film can often be achieved by establishing sufficient partial pressure of a suitable chemical etchant. This technique can improve selectivity when one wants to deposit only on activated surface areas, or only on one crystallographic plane (anisotropic epitaxy), or only one of the possible solid phases, as in diamond versus graphitic C. The third advantage of higher process pressure is that only one stage of pumping is required ("low-pressure" or LP-CVD), or no pumping in the case of atmospheric-pressure operation (AP-CVD). However, the gas-transport problems to be discussed below are all more difficult to deal with in APCVD than in LPCVD.

The basic thin-film process sequence of gas supply, transport, and deposition from Fig. 1.1 is presented in more detail for the case of CVD in Fig. 7.1. The gas-transport step is much more complex in the fluid-flow regime than in molecular flow, and therefore much of the chapter will be devoted to examining transport behaviors—mainly, free convection, homogeneous reaction, and diffusion. The discussion of the deposition step will focus on its surface processes, since most of what

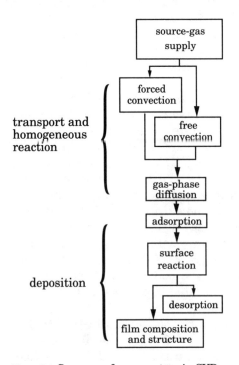

Figure 7.1 Sequence of process steps in CVD.

TABLE 7.1 Typical Overall Reactions Used in CVD

pyrolysis (thermal decomposition)	$SiH_4(g) \rightarrow Si(c) + 2H_2(g)$
	$SiH_2Cl_2(g) \rightarrow Si(c) + 2HCl(g)$
	$CH_4(g) \rightarrow C(\text{diamond or graphite}) + 2H_2(g)$
	$Ni(CO)_4(g) \rightarrow Ni(c) + 4CO(g)$
oxidation	$SiH_4(g) + 2O_2(g) \rightarrow SiO_2(c) + 2H_2O(g)$
	$3SiH_4(g) + 4NH_3(g) \rightarrow Si_3N_4(c) + 12H_2(g)$
hydrolysis	$2AlCl_3(g) + 3H_2O(g) \rightarrow Al_2O_3(c) + 6HCl(g)$
reduction	$WF_6(g) + 3H_2(g) \rightarrow W(c) + 6HF(g)$
displacement	$Ga(CH_3)_3(g) + AsH_3(g) \rightarrow GaAs(c) + 3CH_4(g)$
	$ZnCl_2(g) + H_2S(g) \rightarrow ZnS(c) + 2HCl(g)$
	$2TiCl_4(g) + 2NH_3(g) + H_2(g) \rightarrow TiN(c) + 8HCl(g)$

has previously been said about bulk film structure and interaction in Chaps. 5 and 6 also applies to CVD films. To give a flavor of the chemistry involved in CVD, Table 7.1 lists some of the commonly used *overall* reactions and their chemical types; many others are listed elsewhere [1,2]. We will see later that such overall reactions really consist of a series of reaction steps, some in the gas phase (homogeneous) and some on the surface (heterogeneous). Each reaction step has a rate determined by activation energy and process conditions, and any one of these rates can be the one that controls the film deposition rate. Alternatively, the gas-supply or transport step can be the one which controls deposition rate. Control by one or another of the steps in Fig. 7.1 has various advantages which will become clear below. It is important in CVD to determine the range of process conditions over which each step becomes the controlling one, and we will examine several techniques for doing so. The pyrolysis of silane (SiH_4) to deposit Si will often be used as an example, because it is one of the most extensively studied reactions. It is also quite complex, despite the simple overall reaction, so it illustrates all of the reaction phenomena to be discussed.

7.1 Gas Supply

Figure 7.2 illustrates typical elements of the "gas jungle" of plumbing used to supply CVD source gases and vapors to the deposition reactor. Not all of these elements will be used in a given reactor, and actual design will depend on the degree of gas hazard involved and on the operating pressures at the source and in the reactor. There are three aspects to gas-supply design: (1) protection of personnel and environ-

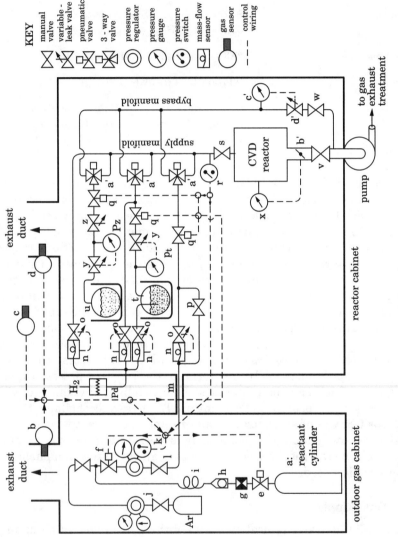

Figure 7.2 Typical elements of reactor gas-supply plumbing.

ment from the frequently hazardous properties of the reactants, (2) regulation of a steady and known flow of each gas, and (3) minimization of contamination in the process stream. We will discuss each of these aspects in turn below.

7.1.1 Safety

Gas-supply safety starts with careful design and construction to minimize the possibility of leaks or valve-operation errors. It also provides detection of hazardous conditions such as excess pressure or leaks before they become catastrophic, coupled with automatic shutdown of the gas supply. Leak prevention is also important for process purity. Plumbing should be welded wherever possible, and metal-gasketed connecting fittings should be used elsewhere. Fittings that seal directly against the tubing instead of against a welded-on flange are not recommended, because they develop leaks more easily when stressed. Valves should have bellows or diaphragm stem seals. Wherever elastomeric seals are unavoidable, the elastomer must be chemically compatible with the reactant gas.

Let us examine the safety elements in Fig. 7.2, following the gas flow path. The high-pressure gas cylinder (a) is stored in an outdoor cabinet connected to an exhaust duct, to minimize personnel exposure in the event of a leak. Gas sensors in the duct (b) as well as in the reactor-cabinet exhaust duct (c) and in the room (d) automatically shut off the pneumatically operated valves upon detecting a hazardous level. The first pneumatic valve is best installed *in* the gas cylinder at (e) if the gas supplier can do so. Otherwise, it can be mounted on a nearby panel at (f). The flow-restricting orifice (g) is mounted in the gas-cylinder valve by the supplier, and it limits the rate of gas discharge in the event of a break downstream. Discharge rate can be estimated as described in Appendix E. Further protection against a break is provided by the "excess-flow" valve (h), which slams shut when flow rate exceeds a certain level and stays shut until reset. The tubing loop (i) avoids stressing the plumbing when changing cylinders. The argon purge assembly (j) allows "rinsing" reactant gas out of the plumbing prior to cylinder change, using a series of pumpout/backfill cycles. It also minimizes intrusion of moisture during cylinder change. The argon supply must be separate for each gas so as to contain the gas in the event of backmixing. The outlet pressure gauge on the gas-pressure regulator incorporates an overpressure switch (k), which closes the pneumatic valve in the event of regulator failure. Regulator outlet pressure can creep up over time if the diaphragm's valve seat becomes contaminated or corroded. The final valve (f) isolates the gas cabinet when not in use.

Proceeding downstream from the gas cabinet, the supply line to the reactor (m) is double-contained to direct any leaking gas into the exhausted enclosures. Just after the flow-control assembly on this gas (n-o-p) as well as on the other two reactants are pneumatic valves (q) which provide hard shutoff in several instances: (1) when the gas is not being used, (2) when a leak is detected at any of the sensors, and (3) when excess downstream pressure is detected by the pressure switch at (r). This switch must be located upstream of the reactor's inlet shutoff valve (s) to prevent inadvertent backmixing of gases when (s) is closed. The gas plumbing, the low-pressure vapor sources (t and u), and the reactor are all contained in an exhausted cabinet. Downstream of the reactor, the gas-supply and bypass manifolds can be isolated from the pump and the gas exhaust system by valves (v) and (w), respectively. Pumping and exhaust treatment of hazardous gases was discussed in Sec. 3.2. Devices monitoring proper operation of the exhaust treatment system should shut the pneumatic valves (e) or (f) and also (q) in the event of malfunction.

The above discussion only illustrates some safety elements that might be desirable in a hazardous-gas supply system; it is not intended to represent a recommended system design. Actual design must follow local government and institutional codes. Hazards and recommended handling procedures for specific materials are described on the Material Safety Data Sheets (MSDSs) that the material's manufacturer is required to supply.

7.1.2 Flow control

The mass flow controller (n-o) is almost always the device used to regulate gas flow rate in CVD at any process pressure. It consists of a flow sensor (n) coupled to an electrically driven variable-leak valve (o) using a feedback loop similar to that used for T control in Sec. 4.5.3, so that it controls flow to a set point. Valve (o) does not provide reliable hard shutoff, so valve (q) is also needed. The mass flow controller is illustrated in more detail in Fig. 7.3. For high gas flows, the shunt path accommodates the excess gas to keep the sensor tube within its linear operating range of about 10 sccm. (See Sec. 2.5 for sccm defini-

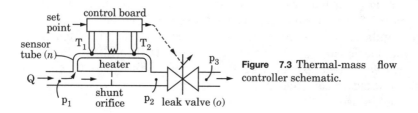

Figure 7.3 Thermal-mass flow controller schematic.

tion.) Total flow range may be changed by substituting shunt orifices. In one typical sensor-tube design, a heater establishes a T profile along the tube which is symmetric at zero flow ($T_1 = T_2$). With flow, T_1 decreases and T_2 increases in proportion to the *thermal-mass* flow rate, where the thermal mass is the number of moles times the heat capacity per *mole* at constant *pressure*, c_p. (Section 2.4 discusses the various ways of expressing gas heat capacity; it is important to use the correct one.) The T imbalance occurs because heat is taken up by the gas, carried downstream, and deposited back onto the tube. Thus,

$$T_2 - T_1 = Bc_p Q \tag{7.1}$$

where B is a proportionality constant and Q is the "mass" (molar) flow rate in sccm. This relationship relies on the gaseous thermal conduction being fast radially and slow axially. In practice, deviations from this condition lead to slight nonlinearity [3].

Manufacturers do not calibrate flow controllers with hazardous gases; they proportion the calibration from that of a standard gas such as N_2 using known constant pressures. The Q of gas A which gives the same ($T_2 - T_1$) signal as a flow $Q(N_2)$ of N_2 is found from Eq. (7.1) to be

$$Q(A) = \frac{c_p(N_2)}{c_p(A)} Q(N_2) = K_f Q(N_2) \tag{7.2}$$

where K_f is the calibration factor relative to N_2. Usually, c_p can be found in the literature. However, for accurate work it is best to calibrate the flow controller on the reactor by using the actual process gas and by spanning a range of flow rates to determine nonlinearity. This can be done easily using the gas leakup rate into a reactor closed off at the outlet. A calibrated pressure gauge such as the capacitance diaphragm gauge discussed in Sec. 3.5 is needed on the reactor, as shown at position (x) in Fig. 7.2. The enclosed volume, V_r, from the flow controller to the reactor's outlet valve (v) and including the supply manifold, must also be determined. This can be done by expansion of gas into the evacuated reactor from a flask of known volume V_0 at known pressure p_0 and use of the ideal-gas law, Eq. (2.10), whereby pV must be the same before and after the expansion. The mass flow rate is then found from the leakup rate and Eq. (2.10):

$$\frac{Q}{22{,}400} = \frac{dN_m}{dt} = \left(\frac{V_r}{RT}\right)\frac{dp}{dt} \tag{7.3}$$

where 22,400 is the molar volume, V_m (cm^3/mol), at 0° C and 1 atm, and t is in minutes for Q in sccm.

Mass flow controllers require $\approx 10^3$ Pa of pressure drop across the narrow sensor tube to drive the flow, so they are unsuitable for source vapors having very low vapor pressure, p_v. They are also unsuitable for handling vapors which decompose when moderately heated or are very reactive. In such cases, liquid-source flow-control systems [4] such as those shown at (*t*) or (*u*) in Fig. 7.2 can be used. In either case, the liquid container must be immersed in a constant-T enclosure held at somewhat below room T, to stabilize p_v and to prevent downstream condensation and resulting flow instability. Estimation of p_v was discussed in Sec. 4.1. In system (*t*), a suitable carrier gas such as H_2 or Ar bubbles through the liquid and carries the reactant vapor downstream. Carrier-gas flow rate needs to be very low so that vapor-liquid equilibrium can be sustained. Then, assuming ideal gas, the mole fraction of vapor in the stream is p_v/p_t, where p_t is the total pressure of the stream. In LPCVD or high-vacuum applications, p_t needs to be regulated by the feedback loop to leak valve (*y*). In APCVD applications requiring rapid switching of gas flow for multilayer film growth, flow toward the supply manifold may need to be increased by injecting additional carrier gas after valve (*q*) as shown. System (*u*) avoids the carrier gas but can be used only in LPCVD, because reactor pressure needs to be less than p_v to drive the flow. Flow is regulated by fixing leak valve (*z*) at a suitable conductance and then regulating the pressure upstream of it, p_z, using a pressure-control loop to leak valve (*y*). This system can be calibrated for Q versus p_z using the leakup method of Eq. (7.3). On the other hand, in system (*t*) only the carrier flow can be calibrated, and reactant-vapor flow estimation depends on knowledge of p_v.

Continuing downstream from the flow-regulating systems, pneumatically driven valves (*a'*) direct each flow stream into either the reactor supply manifold or the bypass manifold. This bypass is needed when rapid switching and restabilization of gas flow are required for multilayer films or atomic-layer epitaxy (Sec. 6.5.5). One cannot just valve off the flow at (*q*) or even at (*o*), because then pressure builds up to the upstream value (p_1 in Fig. 7.3) behind the valve and produces a flow burst when the valve is reopened. For the same reason, downstream pressure p_3 may need to be regulated using the gauge/valve control loops (*x-b'*) on the reactor outlet and (*c'-d'*) on the bypass manifold, with both manifolds held at the same pressure. Valve b' is a motor-driven, butterfly-type throttle valve as was also shown in Fig. 3.1. In LPCVD, bypass-pressure regulation is not necessary because $p_3 \ll p_1$, but reactor-pressure regulation is still needed for deposition-process control. Pressure stability upstream of the mass flow controllers is also important, even though their calibration is nominally pressure-independent by Eq. (7.1). Slow pressure drift is tolerable, but an abrupt change in p_1 of Fig. 7.2 will cause a transient flow disturbance. For example, if p_1 drops by more than $(p_1 - p_2)$, which is only $\approx 10^3$ Pa, there will actually be a temporary *backflow* through the sensor tube, which will cause the control loop to drive valve (o) wide open, resulting in a large flow burst. This transient disturbance is illustrated in Fig. 7.4. To avoid such disturbances when supply lines are shared among reactors, each mass flow controller should have its own pressure regulator.

7.1.3 Contamination

Impurities can occur in the gas or liquid as supplied by the manufacturer and can also intrude during vapor transport, just as in the case of solid source materials. Evolution ("outgassing") of impurities, especially water, from the internal surfaces of the gas-supply plumbing and control devices is a major source of contamination, and the comments of Sec. 3.4.2 on minimizing this problem in vacuum systems apply equally well here. The consequences of an impurity to the process of course depend on the film material and its application. For example, water *may* not be a serious contaminant in oxide deposition, but it is disastrous in (AlGa)As epitaxy.

The reactor must be cleaned before film deposition until the outgassing rate drops to an acceptable level. It is important to realize that this rate is not affected by the reactor operating pressure or by the ultimate vacuum level which the reactor pump can achieve. It is determined by surface conditions and is pressure-dependent only if the partial pressure of the outgassing species begins to approach its p_v, as discussed in Sec. 4.2. Thus, to avoid slowing down the cleanup rate, partial pressure should be kept low by flowing carrier gas. A long period of purging with carrier gas, or even better with a source vapor reactive with water, can clean up the plumbing just as well as can pumping to high vacuum. With sufficiently pure source materials and clean plumbing, it is possible to grow semiconductor-device-quality

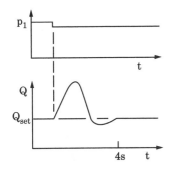

Figure 7.4 Transient flow response of a mass flow controller to a step change in upstream pressure, p_1.

epitaxial films in an APCVD reactor with no pumping at all. When it is desired to pump to high vacuum, such as for He leak checking, one must beware of contamination from backstreaming of pump oil (Sec. 3.4.1). When pumping out a gas supply line, the mass-flow-controller bypass valve (p) shown in Fig. 7.2 should be opened to increase conductance. Also note that no check valves are used in the supply plumbing, because these typically require a pressure drop of at least 0.1 atm to open. They therefore prevent both pumpout and leak checking. Backflow protection must be achieved in other ways, such as by the overpressure interlock (r-q). Despite the above procedures, there will always be some residual amount of outgassing load, Q_i. The partial pressure, p_i, of these contaminants in the reactor is minimized by using a large flow of carrier gas, Q_s, in accordance with Eq. (3.9).

Impurities coming from the source gases must be dealt with by the manufacturer. Here, the containing-cylinder material and its internal-surface preparation are as important as the chemical manufacturing process, especially with reactive gases. Carrier gases are available in very high purity, but final purification at the reactor may still be desired. For H_2, devices are available for diffusing it through hot Pd foil, which blocks all other gases. Inert gases such as Ar can be passed through "getters." These contain a reactive surface which chemisorbs reactive impurities. The Ti-sublimation type of getter can be reactivated periodically by sublimating a fresh layer of Ti. Water and O_2 can be removed by bubbling the carrier gas through Ga-In-Al alloy [5], which is liquid at room T and in which these impurities rapidly oxidize the Al. A "dew point" of $<-80°$ C is claimed for this process, the dew point being the T at which the p_v of water becomes less than its partial pressure in the gas mixture, so that dew condenses. Figure 4.4 shows that the p_v of water at $-80°$ C is 5×10^{-2} Pa, or 0.5 ppm in a 1 atm mixture. Purity of the process stream can be monitored to parts-per-billion (ppb) levels by two types of devices. Solid-state electrolytic sensors are available for specific gases such as H_2O and O_2. Mass spectrometers can detect all gases, and special ionization sources are available which operate at 1 atm, as discussed in Sec. 3.5.

The last source of impurities to be discussed is cross-contamination. When gas-stream composition is switched during the deposition of a multilayer film, gas remaining upstream from deposition of the previous layer becomes a contaminant for the next layer. To purge this remaining gas out quickly, sufficient gas velocity in the downstream direction must be maintained in all parts of the supply system downstream of the switching valves (a') in Fig. 7.2. Where "dead spots" in the flow are unavoidable, they should be made as small as possible so that diffusion can more readily clear out the lingering gas. There are unavoidable dead spots between valves (a') and the supply manifold, between the pressure switch (r) and the manifold, and between the pressure gauge (x) and the reactor. The reactor itself must be carefully designed to avoid dead spots, and this will be discussed in the next section.

7.2 Convection

Convection in a CVD reactor refers to the flow of the gaseous fluid as it moves through the reactor after being injected from the gas supply line. There is forced convection due a pressure gradient, and there can also be "free" convection due to the buoyancy of hot gas. On the other hand, when the gas gets very close to the substrate surface, flow velocity slows down due to viscous friction, and the remaining transport of reactant to the surface can only occur by *diffusion* through this relatively stationary "boundary" layer of fluid (see Fig. 7.1). Detailed knowledge of the flow pattern is needed to determine how far from the deposition surface this very important transition from convective to diffusive transport is occurring and which of these two transport steps is limiting reactant arrival rate at the surface. Convection is addressed in the subsections below, and diffusion will be addressed in Sec. 7.4. The flow pattern also determines gas residence time in the reactor and the extent of gas heating. Both of these factors influence the extent of homogeneous (gas-phase) reaction, an important aspect of CVD chemistry to be discussed in Sec. 7.3.

A complete description of fluid flow involves applying, to each point in the volume, the principles of mass, momentum, and energy conservation, along with the equation of state of the fluid. The equation of state relates pressure (p), T, and mass density (ρ_m); in CVD, the ideal-gas law, Eq. (2.10), is used. The resulting "Navier-Stokes" equations give the p, ρ_m, and velocity vector of the fluid at every point. A few analytical solutions for simple flow situations will be presented below. Some results from computer solutions of more complex flow patterns will also be presented. A full description of CVD requires adding to these equations the homogeneous reactions and the resulting variation in fluid composition through the reactor. Although some attempts at this have been reported, lack of reaction-rate data remains a major impediment, so this work will not be discussed. We will focus instead on simplifying the flow model and separately treating the various processes occurring so that simple calculations can be made. This kind of approximate analysis is easily done and is very useful in predicting which step will dominate a given CVD process and how the situation will change with process conditions.

The flow pattern of course depends on geometry, so we will consider three generic reactor designs in the following sections, as illustrated in

Fig. 7.5. There we show the substrates as flat wafers, but they may be any shape. Of the three reactors, the axisymmetric one delivers the reactant most uniformly to the substrate surface. The tube reactor has higher substrate capacity, but care must be taken to compensate for reactant depletion toward the downstream end. In these two reactors, one usually heats only the substrate region to minimize deposition elsewhere. This is conveniently done as shown in the figure, by inductively coupling radio-frequency (rf) power from the external coil to the conducting platform ("susceptor") on which the wafers lie. This arrangement is essentially a transformer which is generating a circulating current and joule (I^2R) heating in the susceptor. The third reactor is the batch type. It has the highest capacity, but there is *no* convective transport of reactant to the center surface areas of the substrates. To obtain deposition uniformity across these substrates, process conditions must be such that the reactant diffusion rate through the gap between the substrates is much faster than the deposition rate. That is, deposition rate must be limited by the reaction rate at the substrate surface. Since reaction rates increase exponentially with T [Eq. (5.16)], very good T uniformity must be established by immersing the entire reactor in a furnace. In the other reactors, the lower T uniformity obtainable with susceptor heating is acceptable as long as deposition rate is being limited *instead* by convection or diffusion.

7.2.1 Laminar flow in ducts

This simple fluid-flow model adequately describes many CVD situations. The mean fluid velocity, $\bar{u}$, needs to be low enough in CVD so that the gases will have time to diffuse to the substrate surface and react there before being swept out of the reactor. A typical value of $\bar{u}$

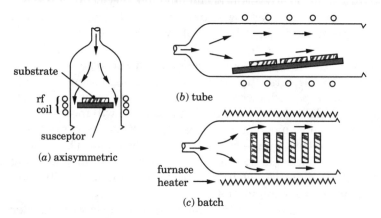

Figure 7.5 Generic reactor types.

is 4 cm/s whether the operating pressure is 10^5 Pa (1 atm) or 10^2 Pa (LPCVD). Consider a tube reactor of radius r_o = 5 cm. We can relate this u to the mass flow rate, Q (sccs), and to the volume flow rate, W (cm³/s), using the ideal-gas law, Eq. (2.10):

$$Q = W\frac{22,400}{V_m} = \pi r_o^2 \bar{u} \frac{22,400}{RT/p} \qquad (7.4)$$

Thus, Q = 310 sccs = 19 slm at 1 atm, and Q = 0.3 sccs = 18 sccm at 10^2 Pa. Most of the Q consists of carrier gas at 1 atm and of reactant at 10^2 Pa, because the Q of reactant needs to be of the same order in both cases.

At such low $\bar{u}$, the flow pattern consists of smooth layers ("lamina") of fluid moving past each other under the constraint of viscous friction. At much higher $\bar{u}$, which is not encountered in CVD, these lamina break up and the flow becomes turbulent, which means that its velocity vector changes with position and time in a chaotic manner. Turbulent flow always reverts to laminar flow in a "boundary layer" adjacent to surfaces, due to viscous drag. There is no such boundary layer in CVD, because in laminar flow, u varies smoothly across the reactor. There are *other* boundary layers in CVD which we will discuss later.

Steady-state laminar flow in cylindrical tubes is a simple one-dimensional flow situation which applies to the axisymmetric and tube reactors. The radial velocity profile, u(r), can be obtained by a force balance between pressure drop and viscous shear along the unit-length cylindrical section shown in Fig. 7.6:

$$\pi r^2 \Delta p = 2\pi r \tau = 2\pi r \eta \frac{du}{dr} \qquad (7.5)$$

Here, Eq. (2.28) relating shear stress, τ, to viscosity, η, has been used to obtain the second equality. This equation is integrated with the

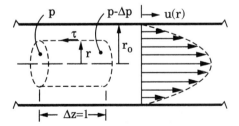

Figure 7.6 Force balance and flow-velocity profile, u(r), for steady-state laminar flow in a cylindrical tube of radius r_o.

boundary condition that $u = 0$ at the wall ($r = r_o$), which is valid for $Kn \ll 1$. We then arrive at the classic parabolic velocity profile of "Poiseuille" flow,

$$u(r) = \frac{\Delta p}{4\eta}\left(r_o^2 - r^2\right) \qquad (7.6)$$

as shown on the right-hand side of Fig. 7.6. The mean value of u is found by integrating over the cross section:

$$\bar{u} = \frac{\int_0^{r_o} u(2\pi r)\,dr}{\pi r_o^2} = \left(\frac{\Delta p}{8\eta}\right)r_o^2 \qquad (7.7)$$

For the $\bar{u}$ and r_o used above and for $\eta = 88\ \mu P = 8.8\times 10^{-6}$ Pa·s (H_2 at room T), we find that $\Delta p = 10^{-5}$ Pa/cm, a negligible value even in LPCVD. Thus, we can *neglect* total pressure variations in analyzing CVD-reactor processes, except in the special case of the thermal-transpiration pressure drop into a pressure gauge connected by a narrow tube in which $Kn > 1$ [Eq. (3.18)].

A parabolic u profile is also obtained for flow between parallel plates (Exercise 7.2). This situation would apply to a rectangular-tube reactor of large width/height ratio. For a height of y_o and for $y = 0$ at the centerline,

$$u(y) = \frac{\Delta p}{2\eta}\left(\frac{y_o^2}{4} - y^2\right) \qquad (7.8)$$

and

$$\bar{u} = \left(\frac{\Delta p}{3\eta}\right)\left(\frac{y_o}{2}\right)^2 \qquad (7.9)$$

The degree to which the reactor flow falls short of breaking into turbulence can be determined by calculating the Reynolds number, Re. Dimensionless numbers such as Re are very useful in characterizing transport situations in which the behavior can change with conditions from one mode to another. We have already encountered Kn, Pe [Eq. (6.23)], and the deposition purity ratio (Sec. 2.6). The importance of Re warrants a brief derivation here. Re is the ratio of the momentum (inertial) force to the viscous-drag force operating on the fluid. These forces respectively destabilize and stabilize the flow pattern, and when Re exceeds 1200 or so depending on geometry, turbulent instability begins. For flow in a tube, the momentum per length z of tube is $\mathbf{k} = (\pi r_0^2 z)\rho_m \bar{u}$, where ρ_m is the fluid's mass density. The force that would be required to stop this momentum within length z is $\mathbf{k}(du/dz) = \mathbf{k}\bar{u}/z = \pi r_0^2 \rho_m \bar{u}^2$. The viscous drag force on the fluid against the wall is $(2\pi r_0 z)\eta|du/dr|$, where $|du/dr|$ is evaluated at r_0 using Eqs. (7.5) and (7.7): $|du/dr| = 4\bar{u}/r_0$. Taking the ratio of these two forces and dropping the numerical factors, we have

$$\text{Re} = \frac{r_0^2 \rho_m \bar{u}}{z\eta} = \frac{L\bar{u}}{(\eta/\rho_m)} = \frac{L\bar{u}}{\nu} = \frac{L\bar{u}}{\eta}\left(\frac{pM}{RT}\right) \qquad (7.10)$$

where (r_0^2/z) has been replaced by L to represent a characteristic linear dimension of the reactor for the more general case. The "kinematic" viscosity, ν (cm^2/s, or Stokes) = η/ρ_m, may be thought of as the "momentum diffusivity" and is analogous to the mass diffusivity, D (cm^2/s). The last equality in Eq. (7.10) assumes ideal gas and shows that Re decreases with decreasing p at a given $\bar{u}$, since η is independent of p by Eq. (2.28). Using η = 88 µP = 8.8×10^{-5} g/cm·s and ρ_m = 8.2×10^{-5} g/cm^3 for H$_2$ at room T and 1 atm, so that ν = 1.07 cm^2/s, and using L = 5 cm and $\bar{u}$ = 4 cm/s, we see that the units cancel in Eq. (7.10) as they should, and that Re = 19. This is way below the onset of turbulence, and it would be even lower at lower p, so we conclude that *flow is always laminar* in CVD.

More complicated flow patterns than the parabolic one are often encountered, however. These can be caused by abrupt changes in flow path or by steep T gradients. We will consider T gradients in Sec. 7.2.3. The flow path changes first at the point of gas injection, where the supply line expands to the reactor diameter. If the expansion is gradual and u is not too high, one obtains the nearly parallel flow pattern shown in the bottom half of Fig. 7.7. However, if the expansion is too rapid, because of an abrupt diameter change or excessive u, the flow separates from the wall and recirculates in the "Hamel-flow" vortex shown in the top half of the figure [6]. Note that this is not turbulence, because the pattern does not vary with time. Such vortices are undesirable because they increase the reactant residence time, thereby lengthening gas-composition switching time and sometimes causing excessive homogeneous reaction. Their occurrence can be detected by observing the flow patterns using tracer smoke.

The flow path changes next upon encountering the susceptor, where u must drop to zero due to viscous drag along this new surface, as shown in Fig. 7.8 for the tube reactor geometry. But within only a few L lengths downstream at the low Re of CVD, the parabolic profile is re-

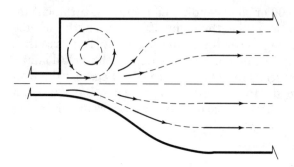

Figure 7.7 Two alternative reactor-entrance geometries and flow patterns (upper and lower halves). The upper pattern is to be avoided.

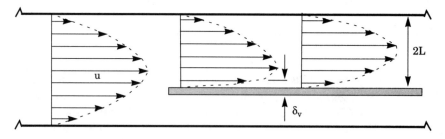

Figure 7.8 Velocity boundary layer of width δ_v forms at the susceptor leading edge in a tube reactor.

stored. The initial distortion of the profile does not affect the deposition rate significantly anyway, as we will see in Sec. 7.4. This flow situation is sometimes modeled with the susceptor acting as a wing passing through a stationary fluid at velocity $-u$. In the wing model, the position above the wing surface at which the fluid has become accelerated by 1 percent of the way toward $-u$ is defined as the edge of the velocity boundary layer, δ_v. Clearly, δ_v will expand moving downstream along the wing, but for most of the length of a typical susceptor in a CVD tube, the u profile is parabolic and does not fit this model, so the concept of δ_v has no meaning. We will see below that there *is* a meaningful δ_v in the axisymmetric flow pattern.

7.2.2 Axisymmetric flow

In this reactor geometry (Fig. 7.5a), the flow pattern is similar to one of the few two-dimensional flow situations for which an analytical solution to the Navier-Stokes equations has been found. This solution is often used in modeling CVD, but we will see below that its applica-

7.2.2 Axisymmetric flow

bility is limited. The analytical solution [7] assumes that a stream of fluid of radius r_s is approaching an infinite, planar, stationary surface at uniform velocity u_z^∞ in the $-z$ direction, as shown on the left-hand half of Fig. 7.9. At some point of approach, u_z begins to slow down, reaching a stagnation point of zero flow at $z = 0$ and $r = 0$. Meanwhile, radial velocity, u_r, begins to increase as the fluid becomes deflected by the surface. In the "potential-flow" region, far enough above the surface so that u_r is not slowed down by viscous drag against it, the functional forms

$$u_z = -2Bz \tag{7.11}$$

and

$$u_r = Br \tag{7.12}$$

are found to provide a solution [7] to the flow equations. Here, B is an unspecified constant. In this potential-flow region, the flow direction is changing while u_z remains independent of r and u_r remains independent of z, as shown in the figure. Closer to the surface, however, viscous drag causes u_r to decrease toward zero at $z = 0$, as also shown. The velocity boundary layer's edge is defined as before: it is the z value at which u_r is reduced by 1 percent from its free-stream value. With this

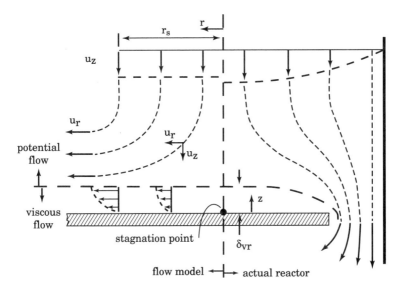

Figure 7.9 Axisymmetric flow geometry and approximate flow patterns for analytical flow model and for actual reactor.

definition and Eqs. (7.11) and (7.12), the thickness of the radial-velocity boundary layer is found to be

$$\delta_{vr} \approx 2.2 \sqrt{\frac{\nu}{B}} \qquad (7.13)$$

where ν is the kinematic viscosity [Eq. (7.10)]. Note that, according to this model, δ_{vr} is independent of r—at least for $r < r_s$. This is because two effects are cancelling each other: the increase in δ_{vr} with increasing downstream distance, r, as in Fig. 7.8, and the decrease in δ_{vr} due to the increase of u_r with r.

To estimate δ_{vr}, we need to determine B. Since for $r \geq r_s$, all of the flow has been redirected radially, it is reasonable to assume that $u_r \to u_r^\infty$ at $r = r_s$. Since kinetic energy is conserved in the potential-flow region, we also have $u_r^\infty = u_z^\infty$. Using this information in Eq. (7.12), we obtain $B = u_z^\infty/r_s$, so that

$$\delta_{vr} \approx 2.2 \sqrt{\frac{\nu r_s}{u_z^\infty}} \qquad (7.14)$$

Taking $\nu = 1.07$ for H_2 at room T and 1 atm and $u_z^\infty = 4$ cm/s as in the last section, and using $r_s = 3$ cm, we find that $\delta_{vr} = 2.0$ cm—a value almost as big as r_s itself. Even for the much higher-density gas Ar, we have $\nu = 0.13$ cm^2/s, and $\delta_{vr} = 0.70$ cm. Actual CVD conditions with a heated susceptor will cause u_z^∞ to increase as the gas expands at a fixed inlet velocity, but it will also raise ν, so that $\delta_{vr} \propto T^{1/4}$. At lower reactor pressure with the same u_z^∞, δ_{vr} will be even higher because of the higher ν ($= \eta/\rho_m$).

The fact that δ_{vr} is not much less than r_s becomes a problem for this model, because the starting assumption of an infinite planar surface is not valid in an actual CVD situation, where the susceptor radius is less than r_s as shown on the right-hand half of Fig. 7.9. The flow around the susceptor edge alters the flow along the surface for several δ_{vr} in from the edge, so that one would actually expect δ_{vr} to *decrease* with increasing r in this CVD geometry as shown in the figure, rather than to be constant as in Eq. (7.13). A second problem with δ_{vr} is that it does not directly relate to the boundary-layer edge we are really seeking, which is where the *reactant* transport changes from convective to diffusive. The δ_{vr} of Eq. (7.13) is defined by drag on u_r, whereas reactant is transported down to the surface by u_z. Thus, the z position at which u_z starts to decrease could define the edge of another velocity boundary layer, δ_{vz}. From Eq. (7.11) and our estimate of B, we obtain simply

$$\delta_{vz} = r_s/2 \qquad (7.15)$$

This, too, must be considered an oversimplification on account of edge effects, but again we see that δ_{vz} is not much less than r_s. Two main conclusions result from the above analysis of axisymmetric CVD flow with a stationary susceptor. First, the velocity boundary layers are never going to be much smaller than the susceptor radius under any reasonable flow conditions. Second, one must be careful (in any flow geometry) to select the appropriate boundary layer—the one that relates to *reactant* transport.

Rotating the susceptor disc at an angular velocity ω (rad/s) improves the above situation somewhat. The resulting flow pattern, shown in Fig. 7.10, is characterized by a boundary layer of fluid being dragged around with the disc and thrown outward by centrifugal force. This centrifugal-pumping action also sucks fluid down toward the disc along z. Thus, there are three velocity components: radial (u_r), circumferential (u_ω), and axial (u_z). Like the stationary disc, this special flow situation has an analytical solution [7, 8], given a disc radius much

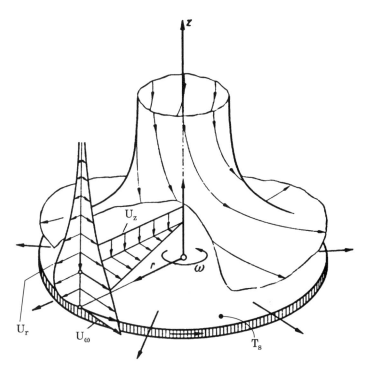

Figure 7.10 Axisymmetric flow pattern over a rotating disc. (Source: Reprinted from Ref. 7 by permission, © 1968 by McGraw-Hill Book Co.)

larger than δ_v and given no *externally* imposed u_z. Figure 7.11 shows the calculated behavior of the three velocity components versus z, all expressed in terms of the following dimensionless variables for height (ζ) and the velocities (F, G, H):

$$\zeta = z\sqrt{\frac{\omega}{\nu}} \qquad (7.16)$$

$$u_r = r\omega F(\zeta) \qquad (7.17)$$

$$u_\omega = r\omega G(\zeta) \qquad (7.18)$$

$$u_z = \sqrt{\nu\omega}\, H(\zeta) \qquad (7.19)$$

Also shown in Fig. 7.11 is the dimensionless T profile obtained when the susceptor is heated:

$$\Theta = \frac{T_z - T^\infty}{T_s - T^\infty}$$

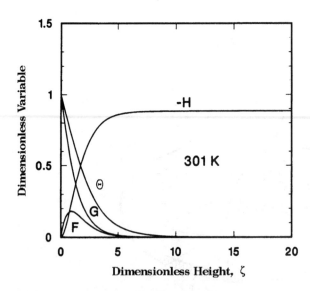

Figure 7.11 Profiles of dimensionless T (Θ) and velocity over the heated rotating disc shown in Fig. 7.10. The velocity components are radial (F), circumferential (G), and axial (H). (Source: Reprinted from Ref. 8 by permission.)

An asymptotic value of H = 0.884 for axial velocity is reached at a large height above the disc, so the corresponding u_z from Eq. (7.19) must be adjusted, by way of ω, to *match* the inlet flow velocity in the reactor tube, u_z^∞, if the calculated flow pattern is to be realized in an actual CVD reactor [8]. If ω is too low, the beneficial effect of the rotation in reducing δ_v will be lessened. If ω is too high, the excess flow being pumped will recirculate up the sidewall of the tube, producing an undesirable vortex similar to the one shown in Fig. 7.7. The onset of this flow disturbance has been observed using tracer smoke [8]. Taking ν = 1.07 for H_2 at room T and 1 atm and u_z^∞ = 4 cm/s as before, we find that the upper limit to ω is 19 rad/s or 180 rpm. Given this, we can now determine δ_{vz} by observing in Fig. 7.11 that H has dropped by 1 percent at about ζ = 5. Combining this with Eqs. (7.16) and (7.19) to eliminate ω, we have

$$\delta_{vz} \approx 4.4\nu/u_z^\infty \qquad (7.20)$$

and for the above case of H_2 at 4 cm/s, δ_{vz} = 1.2 cm. This is not much smaller than the δ_{vr} or δ_{vz} found for the stationary disc under the same conditions. However, the difference between the two δ_{vz} values will increase with r_s, since here δ_{vz} is independent of r_s; whereas, for stationary flow both δ_{vz} and δ_{vr} increase with r_s. Also, one must ask again whether δ_{vz} is the relevant boundary layer for considering reactant transport. After all, with rotation, u_r *increases* within that boundary layer (F in Fig. 7.11) due to the centrifugal action; whereas, without rotation, it decreases due to the viscous drag. This increase will assist reactant transport to the surface by reducing the z value at which diffusive transport must take over from convection. Empirically, it is often found that rotation does improve film-thickness uniformity. Another advantage of rotation is that the momentum induced thereby in the gas tends to dominate the flow pattern and thus avoid the free-convection problems to be discussed in the next section.

7.2.3 Free convection

When the susceptor is heated in either the axisymmetric or the horizontal tube reactor of Fig. 7.5, heat transfer to the adjacent gas causes the gas to expand and therefore become less dense than the cooler gas farther away. In the absence of forced downward flow, the gas above the susceptor rises because of this buoyancy, and then falls again after being recooled. This instability develops into circulating flow patterns or "roll cells" known as "free" or "natural" convection. Free convection will not develop in the isothermal batch reactor. Calculated circulation patterns (Jensen, 1989) for a horizontal tube reactor of rectangular cross section are shown in Fig. 7.12a, looking along the z-axis direction of forced convection through the tube. The direction of circulation depends on the T of the tube sidewalls. For a tube with cooled sidewalls, the flow is downward near the sidewalls and upward near the centerline; whereas, with insulated sidewalls, the pattern is reversed. The direction and velocity of this flow affect the transport rate of reactant to the substrate and therefore affect the lateral uniformity profile of the deposited film.

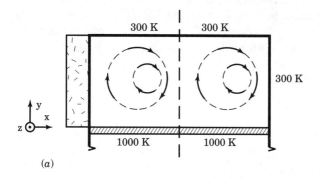

(a)

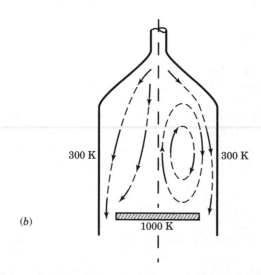

(b)

Figure 7.12 Typical free-convection roll cells: (a) horizontal rectangular-tube reactor (view along axis) with two alternate sidewall conditions, insulated (left) and cooled (right); (b) downflow axisymmetric reactor, showing two alternate flow patterns. In (b), the recirculating pattern (right) is to be avoided.

7.2.3 Free convection

The degree of free-convective flow scales with the ratio of buoyancy force to viscous force and with Re. These ratios together constitute the dimensionless Grashof number [9],

$$Gr = \frac{Ra}{Pr} = \frac{\rho_m^2 g \alpha_{TV}(T_s - T^\infty)L^3}{\eta^2} = \frac{g \alpha_{TV}(T_s - T^\infty)L^3}{\nu^2}$$

$$= \frac{gM^2 p^2 (T_s - T^\infty) L^3}{R^2 T^3 \eta^2} \qquad (7.21)$$

where g = gravitational acceleration = 980 cm/s^2
α_{TV} = dV/VdT = volumetric thermal-expansion coefficient, K^{-1}
T_s = substrate or susceptor T, K
T = mean T of circulating gas, K

Other notation is as used above. The last equality assumes the ideal-gas law, Eq. (2.10). Sometimes the Rayleigh number, Ra, is used instead of Gr to evaluate free convection. For gases, Ra ≈ Gr, because the Prandtl number, Pr, is near unity.

$$Pr = \frac{c_p}{M} \frac{\eta}{K_T} \approx 0.8 \pm 0.2$$

where c_p/M is the heat capacity per gram and K_T is the thermal conductivity. The characteristic dimension, L in Eq. (7.21), is taken as the smaller of the two dimensions in which the roll cell lies. Using ν = 1.07 for 1 atm H_2, and taking L = 5 cm, we have Gr = 1.2×10^5—high enough that free convection definitely will occur.

The critical value of Gr above which free convection becomes significant cannot be stated quantitatively, because it varies with geometry and with the amount of forced convection. For example, Fig. 7.12b shows two alternate flow patterns in an axisymmetric reactor (Jensen, 1989). In the absence of forced downward flow, u_z, the circulation pattern shown at the right typically develops for large Gr. However, there is always a finite u_z in CVD, and a large enough u_z will eliminate the circulation, producing the more parallel flow shown on the left of the figure. In addition, the flow pattern is *bistable* over some range of Gr and u_z. That is, if the circulation pattern is allowed to develop before the u_z flow is turned on, it can persist to larger u_z than if flow is started before the susceptor is heated. Reactor shape and susceptor rotation will also affect the critical value of Gr.

Circulation against the direction of u_z results in an undesirably long gas residence time in the reactor. Circulation transverse to u_z, as in Fig. 7.12a, can degrade uniformity. Inspection of Eq. (7.21) shows that reducing pressure is the most effective way of reducing Gr, since pressure is the most widely adjustable among the choices of process variables there. Low Gr is one of the main advantages of LPCVD over APCVD. In the case of the axisymmetric reactor, one can also just invert it so that the hotter gas is already at the top and there is no driving force for recirculation. However, this remedy requires supporting the substrate on its front face in a way which neither disturbs the flow pattern nor contaminates the growth surface.

Because of the difficulty of predicting flow patterns, it is best to examine them experimentally for the particular reactor at hand. This can be done easily using tracer smoke, but there are two problems. One is that smoke contaminates the reactor. The other is that in steep T gradients, the smoke patterns will be distorted from the actual flow pattern by the thermophoretic motion of particles down the T gradient, as will be discussed in Sec. 7.4.3. Alternatively, "schlieren" photography or interference holography [10] can be used to observe the patterns of optical interference fringes which result from refractive-index variation along the T gradient.

7.3 Reaction

The source gases become heated at some point during their transport to the substrate, this point depending on the type of reactor (Fig. 7.5) and the flow pattern within it. Thus, reaction often begins in the gas phase rather than occurring entirely on the substrate surface. The products of these reactions are usually more reactive with the substrate than are the source gases themselves. Excessive gas-phase reaction can produce particles of film material within the gas phase, and these settle out as powder. It is important in CVD to understand and control both the gas-phase and the surface reactions.

7.3.1 Chemical equilibrium

The simplest analysis of a reacting system assumes that all species reach chemical equilibrium with each other. Since CVD reactors are continuously producing a net change of reactant to product, they cannot be operating at equilibrium, which by definition entails no net change. Moreover, it is often necessary to operate *far* from equilibrium in order to avoid powder formation or to achieve deposition-rate uniformity over large areas, because these objectives are accomplished by flowing gas through the reactor much faster than it can react. Never-

7.3.1 Chemical equilibrium

theless, the calculation of equilibrium composition is a convenient and useful starting point for the analysis of a CVD process. At least it tells us what reactions are possible and how far they can proceed in the equilibrium *limit*. How far they actually proceed is determined by reaction rate and by gas-phase diffusion, which we will discuss later.

It was shown in Eqs. (4.4) through (4.6) that a reacting "system" operating at constant pressure, p, reaches equilibrium when its total Gibbs free energy, G, is minimized. [A system operating instead at constant V would reach equilibrium when the Helmholtz free energy, (G − pV), was minimized, but CVD always operates at constant p.] Here, G is in units of kJ for the total system, not in kJ/mol. The "system" under consideration starts as some fixed number of moles of the feed gas (supply gas) mixture, which then reacts to various gaseous and solid products as it moves through the reactor at constant p. For this mixture of reactant and product species, the total free energy is

$$G\,(kJ) = \sum_i N_{mi}\mu_i \qquad (7.22)$$

where N_{mi} is the number of moles of the ith species, and μ_i is its chemical potential. Recall from Eq. (4.7) that μ_i (kJ/mol) is the incremental Gibbs free energy per mole of i added to the mixture. We repeat here Eq. (5.28) for μ, which was derived for ideal gases.

$$\mu_i = \mu_i°(T) + RT\,\ln\frac{p_i}{p°} \qquad (7.23)$$

This gives μ_i versus the partial pressure of i, p_i, in the reactor at temperature T, relative to $\mu_i°(T)$, which is the μ_i at the standard reference pressure, $p° = 10^5$ Pa ≈ 1 atm, and at the same T. Since the μ_i of an ideal gas at a given p_i is not influenced by the presence of other species, $\mu_i°(T)$ is identical to the molar free energy of formation of that species from its elements in their standard states, $\Delta_f G_i°$ (kJ/mol). The standard state of an element is its common phase (solid, liquid, or gas) at 10^5 Pa and 298 K, and that phase is assigned G = 0 at 10^5 Pa and *all* T. Most $\Delta_f G°$ values can be found in handbooks [11,12] or in various on-line data bases [13]. Note that $\Delta_f G°$ tends to be a slowly varying and often linear function of T; some examples are shown in Fig. 7.13.

For a gas-phase species, μ_i can be found from $\Delta_f G°$ and p_i using Eq. (7.23). For condensed phases such as the depositing film, the p dependence of μ is negligible. When the film is a solid solution (an alloy) rather than a pure element or compound, the μ of each species in the solution is reduced by the mole fraction x_i to which it is diluted. For "ideal" solutions,

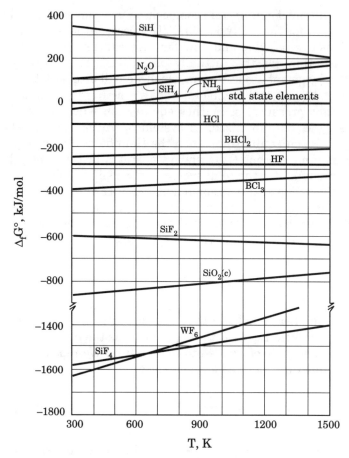

Figure 7.13 Gibbs free energies of formation for selected gaseous and solid (c) compounds at 10^5 Pa [12].

$$\mu_i = \mu_i^\circ(T) + RT \ln x_i \qquad (7.24)$$

This dilution effect is closely related to the reduction of a species' vapor pressure by dilution [Eq. (4.21)]. However, for the present, we will deal only with pure solid phases.

Once the μ_i equations for the species in a reacting system have been written, the system's equilibrium composition can be determined numerically by minimizing system G in Eq. (7.22), subject to the following constraints: non-negative N_m values, fixed total p ($= \Sigma p_i$), and conservation of the total amount of each element as it becomes distributed among the various product species (that is, conservation of mass). Programs such as SOLGASMIX [14] and BELLTHERM [15] are available for calculating the thermodynamics of complex systems that are undergoing multiple simultaneous reactions. However, to obtain more insight into equilibrium behavior, we consider instead a single generalized reaction between molecules A and B to form products C and D:

$$\nu_A A + \nu_B B \leftrightarrows \nu_C C + \nu_D D \tag{7.25}$$

Here, the ν_i values are the stoichiometric coefficients that satisfy the mass balance for all chemical elements in the reaction. The double arrow indicates that the reaction is reversible and proceeds in both directions. Some typical deposition (heterogeneous) reactions were listed in Table 7.1. Gas-phase (homogeneous) reactions can also be important, such as $SiH_4 \leftrightarrows SiH_2 + H_2$ and $BCl_3 + H_2 \leftrightarrows HBCl_2 + HCl$. (Note that all the ν_i values happen to be unity in these two reactions.) Assume for now that Eq. (7.25) is the only significant reaction of A and B, that some fraction, ξ, of A has reacted away, and that there is present an excess fraction, y, of B above the stoichiometric amount ν_B. Then, for ν_A moles of A at the start of reaction, Eq. (7.22) becomes

$$G = (1 - \xi)\nu_A \mu_A + (1 - \xi + y)\nu_B \mu_B + \xi(\nu_C \mu_C + \nu_D \mu_D) \tag{7.26}$$

and at equilibrium,

$$dG/d\xi = 0 = -\nu_A \mu_A - \nu_B \mu_B + \nu_C \mu_C + \nu_D \mu_D \tag{7.27}$$

where both ξ and y have vanished. Inserting Eq. (7.23) for μ_i gives

$$\boxed{\begin{aligned}-\Delta_r G°(T) &= -(\nu_C \mu_C° + \nu_D \mu_D° - \nu_A \mu_A° - \nu_B \mu_B°) \\ &= RT \ln \frac{(p^\nu)_C (p^\nu)_D}{(p^\nu)_A (p^\nu)_B (p°)^{\Delta \nu}} = RT \ln K\end{aligned}} \tag{7.28}$$

where $\Delta_r G°$ is the free energy of reaction calculated from $\mu°$ data at the T of interest, and K is known as the equilibrium constant of the reaction. The net change in molarity, $\Delta \nu = \nu_C + \nu_D - \nu_A - \nu_B$, drops out of Eq. (7.28) if the p_i values are given in units of 10^5 Pa ($\approx$ atm), since then the reference pressure $p° = 1$.

Equation (7.28) is the general form of the expression arrived at in Eq. (5.31), which applied to the first-order reaction of an adsorbed species into a transition state. It is the central equation of chemical equilibrium. If there are multiple simultaneous reactions occurring, their equilibria must all be satisfied simultaneously, again subject to the constraint that $p = \Sigma p_i$. This p_i sum includes nonreacting species as well, so the result will vary with dilution if $\Delta v \neq 0$ (see Exercise 7.8). The mass-balance constraint in Eq. (7.28) is contained in the v_i values, since they come from the balanced Eq. (7.25). Further discussion of equilibrium calculations is given by Smith (1980). Equation (7.28) can provide various types of information useful in CVD. It predicts the upper limit of reactant conversion to gaseous products (Exercises 7.7 and 7.8). It also tells whether deposition can occur and what solid phases are likely to form, and we give several examples of such prediction in the remainder of this section.

In integrated-circuit fabrication, the deposition of W from WF_6 vapor is sometimes used for Si contact metallization. Usually, selective deposition is desired, meaning deposition on the exposed Si areas but not on the surrounding SiO_2. The following reaction is the most favorable one on SiO_2; that is, it has the lowest $\Delta_r G°$:

$$WF_6(g) + \frac{3}{2}SiO_2(c) \leftrightarrows W(c) + \frac{3}{2}SiF_4(g) + \frac{3}{2}O_2(g) \quad (7.29)$$

where (g) and (c) denote gaseous and condensed (liquid or solid) phases. From the data of Fig. 7.13, it is found that at a typical deposition T of 700 K, $\Delta_r G° = +420$ kJ/mol for this reaction. Thus, using Eq. (7.28) with p_i in atm,

$$K = \frac{p^{3/2}(SiF_4) p^{3/2}(O_2)}{p(WF_6)} = \frac{p^3(SiF_4)}{p(WF_6)} = e^{-\Delta_r G°/RT}$$

$$= 10^{-420/2.3 \times 0.0083 \times 700} \approx 10^{-31} \quad (7.30)$$

Note that there are no p_i terms here for the condensed-phase species, in accordance with the discussion following Eq. (7.23). The above result says that even for a high WF_6 p_i of 1 atm, the product-species p_i values would have to kept below 10^{-10} atm for the W deposition reaction to reach equilibrium. For higher product p_i values, the driving force toward equilibrium is in the reverse direction, that is, etching of W in an attempt to increase the p_i of WF_6. Since keeping the product p_i values this low would require a ridiculously high flow rate of WF_6, W cannot be deposited on SiO_2 from WF_6 alone. On the other hand, the reaction of WF_6 with *elemental* Si to form W and SiF_4 has a much lower $\Delta_r G°$ of –707 kJ/mol at 700 K, and in fact this reaction occurs readily up to a limiting W thickness of ≈ 20 nm which is imposed by the need for Si to diffuse through the W film [16]. Thicker W films are deposited using the reaction

$$WF_6 + 3H_2 \leftrightarrows W(c) + 6HF(g) \tag{7.31}$$

or the reaction of WF_6 with SiH_4 gas to form W, HF, and SiF_4. Both of these reactions can be found from the data of Fig. 7.13 to be very favorable (large negative $\Delta_r G°$).

Manipulation of the p_i values which make up K can be used to drive a reaction in the desired direction. For example, C contamination in GaAs deposited from $(CH_3)_3Ga$ occurs due to CH_3 decomposition on the surface at high substrate T. Without knowing the details of the surface reaction mechanisms, one can assume an overall reaction, $C(a) + 2H_2(g) \leftrightarrows CH_4(g)$, which will be driven to the right by increasing $p(H_2)$. Indeed, the C level is found to be much lower using CVD in 1 atm of H_2 carrier gas than using chemical-beam epitaxy under vacuum (see Sec. 6.5.4).

In some depositions, more than one solid phase is possible, and then thermodynamics can be used to predict the favored phase. Thus, in depositing GaAs, liquid Ga may also be formed if one is not careful. In the GaAs-CVD T range around 1100 K, the principal volatile species produced in the decomposition of GaAs is As_2; thus, we can write

$$GaAs(c) \leftrightarrows Ga(c) + \frac{1}{2}As_2(g) \tag{7.32}$$

for which $K = \sqrt{p(As_2)/p_o}$. This means that if $p(As_2)$ falls below the value required for equilibrium ("undersaturated"), GaAs will decompose to Ga in an attempt to raise $p(As_2)$. Conversely, if $p(As_2)$ is held above that value ("supersaturated"), GaAs will remain the only condensed phase. This situation was illustrated in the p_v diagram of Fig. 4.7, where the lower As_2 line represents the phase boundary between GaAs only (above the line) and GaAs + Ga (below). This is a simple example of a "CVD phase diagram."

A more complex phase diagram is shown in Fig. 7.14, where equilibrium calculations [17] show four depositing phases appearing in different domains of substrate T and gas-feed ratio. The silicon-boride phases that are richer in B occur for higher BCl_3/SiH_4 feed ratio, as one would expect. The gas ratio within which one obtains pure SiB_3 is especially narrow. Experimentally, the SiB_3 phase was actually ob-

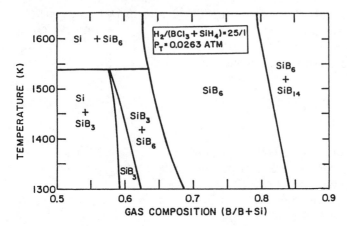

Figure 7.14 Calculated equilibrium CVD phase diagram for deposition of Si borides from BCl_3 + SiH_4. (Source: Reprinted from Ref. 17 by permission.)

tained at a much higher B/(B + Si) feed ratio of 0.75, rather than at 0.6 as shown. The likely explanation for this discrepancy [17] arises from the observation in Fig. 7.14 that the boride deposits are all richer in B than the feed gas mixture according to the equilibrium calculation. This means that if the deposition rate is being limited by the diffusive transport of gas to the surface, the gas composition at the surface will become more depleted in BCl_3 than in SiH_4 as the BCl_3 is preferentially consumed in deposition. That is, the *effective* gas ratio is lower than the feed ratio and thus corresponds more closely to the equilibrium calculation. Indeed, SiB_3 has the same B/Si ratio as the feed gas mixture at B/(B + Si) = 0.75, which is what one would expect for diffusion-limited deposition with similar diffusivities for the two gases. Here, the experimental CVD phase diagram is said to be "transport-shifted" relative to the equilibrium one. This example points up the major shortcoming of equilibrium calculations in CVD work, which is that equilibrium is seldom achieved. Diffusion-limited deposition will be discussed further in Sec. 7.4. A second impediment to the achievement of equilibrium is reaction rate, which we will discuss next.

7.3.2 Gas-phase rate

The source gases usually begin to decompose and/or react with each other as soon as they become hot enough during transport through the reactor, and before they adsorb on the surface of the depositing film. The extent of this homogeneous reaction depends both on the flow pattern and on the reaction rate, and it must be controlled in order to achieve good CVD.

7.3.2 Gas-phase rate

In Secs. 5.1 and 5.2, we developed some principles of reaction kinetics for first-order adsorption and surface-diffusion reactions. The same principles apply here in the gas phase. The rate of a first-order gas-phase reaction per unit volume may be written as

$$R_+\left(mc/cm^3 \cdot s\right) = k_+ n_A = k_+ \frac{p_A}{k_B T} \quad (7.33)$$

and that of a second-order gas-phase reaction as

$$R_+\left(mc/cm^3 \cdot s\right) = k_+ n_A n_B = k_+ \frac{p_A p_B}{(k_B T)^2} \quad (7.34)$$

Here, subscript (+) denotes the forward direction in a reversible reaction such as Eq. (7.25), and $n_{A,B}$ are the reactant concentrations in mc/cm^3. The conversion to partial pressures $p_{A,B}$ assumes the ideal-gas law [Eq. (2.10)]. Note that the units of the rate constant, k_+, will depend on the reaction order.

It is important to recognize that the order of a reaction is not just the sum of the reactant coefficients, ν_i, in the stoichiometric equation [such as Eq. (7.25)]. This is because such an equation represents an *overall* reaction, which generally involves more than one reaction step. For example, Fig. 7.15 maps the reactions believed to be important in the deposition of Si from silane by the overall reaction

$$SiH_4(g) \rightarrow Si(c) + H_2(g)$$

In general, the rate of an overall reaction is the rate of the *slowest* step in the *fastest* of various parallel reaction pathways. The fastest path-

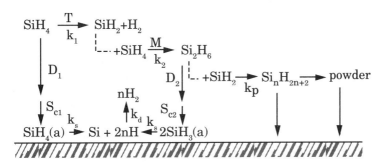

Figure 7.15 Gas-phase and surface reactions believed to be important in the thermal deposition of Si from SiH$_4$ gas: k_i = reaction rate constants, M = third body in a reactive collision, D_i = gas diffusivities, and S_{ci} = sticking coefficients.

way dominates, and the slowest step in that pathway limits its rate. The reaction order is then determined by the number of intermediate species that are reactants for that particular step. Thus, for example, reaction Eq. (7.31) for W deposition is not third-order in H_2 and first-order in WF_6 for a total of fourth-order, as implied by the v_i values, but in fact has been determined experimentally [16] to be half-order in H_2 and zeroth order in WF_6. This suggests that the rate-limiting step in this case is dissociative adsorption of H_2 on W: $(1/2)H_2(g) \rightarrow H(a)$. Fourth-order reaction implies the simultaneous collision of four reactant molecules (here, one WF_6 and three H_2 molecules) with enough energy and collision time to allow for their complete bond rearrangement into products. Clearly, this is an unlikely event, and in fact most reactions are second-order or less.

The general form of the Arrhenius expression for the rate constant, which was derived as Eq. (5.16) for surface diffusion, applies to all reactions; that is,

$$k_+ = B_+ e^{-E_a/RT} \qquad (7.35)$$

where E_a is the activation energy per mole, and where the factors constituting the pre-exponential factor B_+ depend on the type of reaction, as we will see below. Equation (5.31) for K also holds here, so that we may write

$$K = \frac{k_+}{k_-} = \frac{B_+}{B_-} e^{-(E_{a+} - E_{a-})/RT} \qquad (7.36)$$

and

$$\boxed{K = e^{-\Delta_r G°/RT} = e^{\Delta_r S°/R} e^{-\Delta_r H°/RT}} \qquad (7.37)$$

where subscript (−) denotes the reverse reaction direction. The definition of G in Eq. (4.5) has been used to obtain the last equality in Eq. (7.37), and $\Delta_r H°$ is known as the heat of reaction. Neglecting the small dependencies of $\Delta_r S°$ and $\Delta_r H°$ on T, we obtain the same T dependence of K in these two equations when $\Delta_r H° = E_{a+} - E_{a-}$. This fundamental relationship is illustrated in the familiar activation-energy diagram of Fig. 7.16. Note that a thermodynamically favorable reaction (negative $\Delta_r G°$) may have a $\Delta_r H°$ of *either* sign, depending on the entropy change, $\Delta_r S°$. Reactions having positive $\Delta_r H°$ absorb heat (they are "endothermic"), and their K increases with T. "Exothermic" reactions behave just the opposite. $\Delta_r H°$ may be calculated from H°

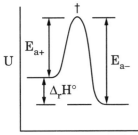

Figure 7.16 Activation-energy diagram for an exothermic reaction. The (†) at the top of the hill represents the energy level of the activated complex, and U is the internal energy of Σv_i moles in Eq. (7.25).

values found in handbooks [11, 12] or in various on-line data bases [13]. By the way, since only *relative* heat content can be measured, the H° of all species is specified to be zero at 298 K. Relating E_{a+} and E_{a-} to thermodynamics in the above way only tells us their difference, $\Delta_r H°$, unfortunately. Thus, we encounter the first difficulty of reaction kinetics: finding E_a.

The slope of an experimental Arrhenius plot, $\log_{10} R_+$ versus 1/T (K), is frequently used to calculate an E_a using Eqs. (7.33) through (7.35) or the equivalent equations for surface reactions. In thin-film work, the R_+ measured is generally the deposition rate (mc/cm²·s). The significance of such an *apparent* E_a must be viewed with some caution. Often, it is not known which reaction step is the rate-limiting one, nor even what are the elementary reaction steps. If two steps have similar rates, the E_a will be an average of the two and is likely to change with T. Often, something *other* than a reaction rate is controlling the film deposition rate, such as diffusion of reactants to the surface (Sec. 7.4) or the supply of activated species from a plasma (Sec. 9.6). In both of these cases, a very low and physically meaningless E_a will be obtained from the Arrhenius plot, and then the value of the plot becomes simply the revelation that something other than a thermally activated process is controlling the deposition rate.

Reaction kinetics is largely an empirical science, because it is not reliable to calculate either B_+ or E_a in Eq. (7.35) from first principles. Nevertheless, it is useful to examine how B_+ is likely to vary with reaction conditions, and we do so below for the cases of unimolecular and bimolecular reactions.

The **bimolecular gas-phase reaction** [Eq. (7.34)] involves the two reactant molecules colliding with enough total translational kinetic energy, ε_t, along the direction of their mutual approach to surmount the E_a "hill" of Fig. 7.16. We saw in Sec. 2.2 that the probability of a molecule having energy ε_t in a given direction is proportional to $\exp(-\varepsilon_t/k_B T)$, and this is the origin of the exponential factor in the Arrhenius equation. The top of the E_a hill represents the "activated complex" in which the bonds of the reactant molecules have become excited enough vibrationally by the collision so that those of the product molecules can begin to form. The probability of the products actually forming instead of the reactants re-forming when the activated complex decomposes depends both on the orientation of the molecules to each other upon collision (the "steric" factor) and on the entropy change of the reaction. In cases where there is only one product molecule, a third probability factor appears, because conservation of momentum prevents the activated complex from transferring its excess energy into translational kinetic energy of the products. To prevent this energetic product molecule from decomposing, its energy must instead be transferred in a collision with a nonreacting "third body" before the decomposition has time to occur. The third body can be either a nonreacting gas-phase molecule/atom or a surface. When it is a molecule or atom, the rate constant increases with total pressure as well as with the reactant concentrations n_A and n_B, so that the reaction exhibits "pseudo-third-order" kinetics. None of the above three probabilities can be predicted accurately, so they are accounted for together by a "fudge" factor, s, in the rate expression:

$$R_{II}(mc/cm^3 \cdot s) = s\,e^{-E_a/RT}\,R_c = s\,e^{-E_a/RT}\,k_c n_A n_B = k_{II} n_A n_B \tag{7.38}$$

where subscript II denotes a bimolecular reaction; R_c is the collision rate per unit volume between the two reactants, A and B; and k_c is the collision rate constant.

The upper limit of the gas-phase bimolecular reaction rate occurs when s is unity and $E_a \ll RT$. In that case, $k_{II} \to k_c$, and the reaction is said to be proceeding at "collision rate." This rate, at least, can be calculated easily from the gas kinetic theory of Chap. 2. The collision frequency (collisions/s) of some individual molecule with any other molecule is $\bar{c}/l$, where the mean speed, $\bar{c}$, and the mean free path, l, were given by Eqs. (2.3) and (2.24). This frequency times n_A is the collision rate of species A with any other molecule, per unit volume, and this rate times the fraction of B in the gas mixture, n_B/n, is the A-B collision rate, R_c (mc/cm$^{-3} \cdot$s^{-1}). Combining all of this, we have

$$R_c = \sqrt{\frac{8RT}{\pi M_r}}\left(\sqrt{2}\pi \overline{a^2} n\right) n_A \frac{n_B}{n} = \left[\sqrt{\frac{\pi RT}{M_r}}\,4\overline{a^2}\right] n_A n_B \tag{7.39}$$

where $\overline{a^2}$ is the mean-square collision diameter of the A-B pair, and $M_r = 2M_A M_B/(M_A + M_B)$ is its "reduced" mass. Comparison with Eq. (7.38) shows that the term in brackets is the k_c that we seek. For typical values of $a = 3\times10^{-8}$ cm and $\bar{c} = 4\times10^4$ cm/s (for Ar at 298 K from Fig. 2.4), we find $k_c = 1.6\times10^{-10}$ cm^3/mc·s, a very useful number to keep in mind as the upper limit of bimolecular reaction rate. When an A-B reaction is occurring at collision rate, the deficient reactant becomes half consumed within the travel time between A-B collisions. Reactions occurring at this rate may be considered to be instantaneous in CVD.

We now turn to **unimolecular gas-phase reactions** such as the silane decomposition shown as k_1 in Fig. 7.15. Although the term unimolecular implies spontaneous decomposition, the rate is governed by collisions with nonreactive neighbors, because that is how a gas molecule gains enough internal energy to dissociate—at least in a thermally controlled reaction. Having gained enough energy to reach the top of the E_a hill of Fig. 7.16, the molecule then dwells for some time, t_e, in that activated state until the internal energy, in its statistical ramblings, happens to concentrate itself into the vibrational mode(s) leading to dissociation. If t_e is less than the mean time between collisions, which is $t_c = l/\bar{c}$ (mean free path/mean speed), then the molecule is likely to dissociate before the next collision. Thus, when $t_e < t_c$, the unimolecular dissociation rate per unit volume, R_I, increases not only with concentration n_A, but also with collision frequency and, therefore, with total pressure. That is,

$$R_I \propto p n_A \qquad (7.40)$$

By comparison with Eq. (7.34), we can see that this reaction is behaving as a first-order reaction with $k_I \propto p$, or as a "pseudo-second-order" reaction. More complex molecules have longer t_e due to their larger number of internal-energy modes. Also, $t_c \propto 1/p$, so at high enough p, t_c becomes less than t_e. Then, the activated molecule is just as likely to lose energy as to gain it in the second collision, so further increase in p does *not* increase the rate at which reactant molecules achieve dissociation energy. This is the "high-pressure limit," where k_I becomes independent of p, and the unimolecular dissociation reaction exhibits true first-order kinetics.

It should now be clear why it is not possible to predict gas-phase k_i values from first principles. Similar complications arise for surface reactions, as we discovered in Sec. 5.2 by examining the simplest possible surface "reaction"—the hopping of an adsorbed molecule from one bonding site to another. Therefore, the kinetically favored pathways in a CVD reaction map such as that of Fig. 7.15 must be determined experimentally. The kinetic modeling approach to this problem is often used in CVD process studies. This involves proposing a likely reaction pathway and then fitting its k_i values to data for deposition rate versus p_i, p, and T. However, Fig. 7.15 shows that even for a simple overall CVD reaction having only one reactant and one gaseous product, there are many k_i terms, including also the sticking coefficients, S_{ci}, of the depositing precursors. Even the diffusivities, D_i, of the precursors may limit deposition rate at high p. Fitting CVD data by adjusting many k_i values is like fitting a wavy line to a polynomial: it is always possible, but there is no chemistry involved. That is to say, the achievement of a fit does not mean that one has correctly identified the reaction pathway or correctly calculated the k_i values of the individual reaction steps.

The second and more fruitful approach to the k_i values is to actually measure them individually. This difficult and tedious task has not been carried out for most CVD reactions, but it must be done to fully understand a chemical process. One way is to generate a burst of a single reactive-intermediate species by flash photolysis (light-induced dissociation) and then track its disappearance transient using some species-selective probe such as laser-induced fluorescence. For example, it has been found using this approach [18, 19] that the insertion reactions of silylene (SiH_2) into SiH_4 and Si_2H_6 shown in Fig. 7.15 by k_2 and k_p proceed at collision rates. The first reaction requires a third body (denoted by M), but k_2 was found to reach k_c by the time p reached only 130 Pa using He diluent. Significantly, these k_i values are *orders of magnitude* higher than previous estimates made by kinetic modeling. This result establishes that the gas-phase reaction rate in silane pyrolysis is governed by k_1, the unimolecular dissociation of silane. Under Si CVD conditions where k_1 was believed to be controlling deposition rate [20], rate analysis yielded $k_1 = 0.35$ s^{-1}.

A certain amount of gas-phase reaction is sometimes required to partially decompose the feed gases and thereby render them sufficiently reactive when they arrive at the film surface, as in the case of the $(CH_3)_3Ga$ and AsH_3 used in GaAs deposition. However, if the sticking coefficients, S_c, of these gas-phase products are too high, they will adsorb before they can diffuse between the stacked substrates of the batch reactor (Fig. 7.6c) or down into the crevices of rough substrates. This results in poor uniformity of film coverage, as we will see in the next section. Excessive gas-phase reaction also usually leads to gas-phase nucleation of film material as powder which can contaminate the film surface and which depletes the reactant supply ("parasitic" reaction). In Si deposition from silane, powder formation occurs by the successive insertion of SiH_2 into Si_nH_{2n+2} as shown by the polymerization rate constant k_p in Fig. 7.15. The rate of each step is second-order in concentration [Eq. (7.34)] and also increases with the dissociation rate of SiH_4 into SiH_2, which increases with both $p(SiH_4)$ and total p [Eq. (7.40)]. The resulting strong p dependence of powder formation leads to the observation of a critical p(SiH$_4$), above which it becomes noticeable. In one study [21] at T$_s$ = 1000 K and with a 1–s residence time of undiluted SiH$_4$ in the reactor, the critical p(SiH$_4$) was found to be about 130 Pa.

The extent of any gas-phase reaction occurring within a CVD reactor is proportional to residence time and rate. Residence time, t$_r$, in the *entire* reactor volume, V$_r$, is given by

$$t_r = \frac{V_r}{W} = \frac{V_r}{Q(p°/p)(T/273)} \qquad (7.41)$$

where W (cm^3/s) = total volume flow rate of gas
 Q (sccs) = total mass flow rate of gas
 p° = 1 atm
 T (K) = volume-average T in reactor

Except in isothermal reactors, only some smaller volume, V$_h$, is heated enough that reaction proceeds at a significant rate. Then, the fraction of t$_r$ which the reactant spends in that volume is V$_h$/V$_r$ when diffusional mixing is fast (low p). When mixing is slow (high p), V$_h$/V$_r$ instead roughly represents the fraction of reactant that passes through V$_h$ rather than passing around it. In either case, the fractional exposure of reactant to the "hot zone" is roughly the same. When, in addition, the fractional consumption of reactant, ξ, is small enough so that one does not need to use integration to follow the kinetics, and when the reacted composition is far enough from equilibrium so that the reverse reaction is negligible, then the amount of reactant A consumed in the controlling gas-phase reaction of rate R$_g$ is simply

$$\Delta n_A = R_g t_r (V_h/V_r) \qquad (7.42)$$

Assuming further, for simplicity, that the kinetics are first-order in A, we then obtain from Eqs. (7.33), (7.41), and (7.42) a very useful qualitative expression for fractional reactant consumption in the gas phase:

$$\xi = \frac{\Delta n_A}{n_A} = \frac{k_I V_h}{Q} \frac{p}{p°} \frac{273}{T} \qquad (7.43)$$

Note that another p factor due to residence time has been added to the one already incorporated into k$_I$.

The above two equations suggest other ways to reduce ξ besides decreasing p. One way is to decrease t$_r$ by increasing Q. However, if t$_r$ is made too small, reactant will not have time to diffuse to the surface or to encounter enough collisions there to react before being swept downstream. The other way is to decrease the hot-zone volume, V_h. The size of V_h varies considerably with reactor design and flow pattern. Clearly, the worst case is the isothermal batch reactor of Fig. 7.5c. Nevertheless, particles can be avoided in this reactor if p is sufficiently low, and the good substrate-T control of the isothermal reactor makes it useful for deposition-rate uniformity.

In cold-wall reactors, the roll cells which can occur in the flow pattern (Figs. 7.7 and 7.12) cause V_h to increase toward V_r by convectively carrying heat away from the substrate region. When these cells are avoided, V_h is instead restricted to the region of steep T gradient adjacent to the substrate surface, which is shown by Θ for the rotating-disc situation in Fig. 7.11. This steep T gradient not only minimizes V_h, but also greatly simplifies process modeling by allowing the use of the "chemical-boundary-layer" concept [22]. There, the reaction zone is approximated as a layer of uniform $T = T_s$ and of thickness δ_c against the substrate surface. The value of δ_c is adjusted so that the same amount of reaction occurs as that which would occur integrated over the actual T gradient. This concept conveniently separates the gas-phase-reaction step of the process from the convective-transport step, whereas the treatment of these two steps together amidst T gradients is a formidable problem. For the case of parabolic flow in a horizontal-tube reactor of height y_0, δ_c can be approximated by the following equation [22] when the activation energy, E_a, for the rate-limiting gas-phase reaction is >100 kJ/mol:

$$\frac{\delta_c}{y_0} = \frac{(1+\beta)(T_s/T_\infty)^{1+\beta}}{(T_s/T_\infty)^{1+\beta} - 1} \frac{RT_s}{E_a} \approx \frac{16}{E_a (\text{kJ/mol})} \quad (7.44)$$

Here, β comes from $K_T \propto T^\beta$, where K_T is the gas thermal conductivity. When heat capacity $c_v \neq f(T)$, $\beta = 0.5$ by Table 2.1. For a more typical value of $\beta = 0.7$, for $T_s = 1000$ K, and for $T_\infty = 300$ K, we obtain the last equality in Eq. (7.44). This is a useful way of making a rough estimate of hot-zone volume and therefore of the extent of gas-phase reaction.

7.3.3 Surface processes

The surface processes of CVD are adsorption of the source gases, surface diffusion, heterogeneous reaction of the adsorbates with each other and with the surface, and desorption of gaseous by-products, as shown for Si deposition in Fig. 7.15. Four quantities are used to describe the fractional consumption of reactant impinging with flux J_i on the surface, as discussed in Sec. 5.1 and Fig. 5.1. These quantities must be carefully distinguished in describing CVD, but there is some

7.3.3 Surface processes

discrepancy in the literature. Here, we define them as follows, in order of increasing amount of interaction with the surface. The **trapping probability**, δ, is the fraction of J_i that physisorbs into the precursor state instead of being immediately reflected from a clean surface, a clean surface being one whose bonding sites are all free of adsorbate. The vapor may subsequently desorb from the precursor state without reacting. The **chemisorption-reaction probability**, ζ, is the fraction of J_i that does react into a chemisorbed state with the clean surface. It may subsequently react back into a precursor state and desorb, or it may remain and become part of the film. The **sticking coefficient**, S_c, is the fraction of J_i that remains adsorbed long enough to become buried and thus permanently incorporated into the depositing film. Its value is averaged over the *whole* surface, both clean and adsorbate-occupied portions. The S_c will be less than ζ when some of the chemisorbed reactant is desorbing or when part of the surface is passivated against adsorption by being already occupied with adsorbate. Finally, the **utilization fraction**, η, is that fraction of the vapor entering the reactor that becomes incorporated into the film instead of being carried downstream and pumped away. Since there are many encounters of a vapor molecule with the surface, η can be high even when the other fractions are low.

Adsorption and reaction of a single species were considered in Sec. 5.1, where Eq. (5.7) for ζ was derived for conditions of vanishingly small and steady-state precursor surface concentration and irreversible chemisorption reaction. For that case, $\zeta \equiv S_c$. Studies on clean, hot Si under ultra-high vacuum [18, 23] approximate these conditions and have shown that the S_c of Si_2H_6 is much higher than that of SiH_4. This points up the importance of knowing the extent of gas-phase reaction (here $SiH_4 \to Si_2H_6$) in analyzing CVD kinetics. Thus, the deposition rate of Si from SiH_4 is found [24] to increase with total pressure, p, and residence time, t_r [Eq. (7.41)], at fixed $p(SiH_4)$ due to an increasing amount of gas-phase reaction with increasing p and t_r as discussed prior to Eq. (7.41).

In general, the surface concentrations of precursors and reaction by-products in CVD *cannot* be neglected as they were for Eq. (5.7). They can inhibit further adsorption or drive surface-reaction equilibrium one way or the other. Regarding by-product buildup, for example, H bonds very strongly to the Si surface and passivates it against any adsorption, as pointed out in Sec. 6.3, so that S_c is much lower on an H-covered Si surface than on a clean one. In fact, the H desorption reaction, $2H(a) \to H_2(g)$, is believed to be the rate-limiting surface step in Si CVD [18] at low pressure where the gas-phase decomposition of SiH_4 is small. A plot of deposition rate versus 1/T under these conditions then measures the E_a of H_2 desorption and gives $E_a \approx 200$ kJ/mol.

As in the gas phase, surface-reaction rates are determined by the surface concentrations of the reactants, n_s (mc/cm^2), in accordance with the rate equations, Eqs. (7.33) and (7.34) (written there for the gas phase). We now consider the steady-state n_s of a single reactant A as a function of its partial pressure, p_A, at some fixed surface T, assuming for the moment a negligible loss rate by film-forming reaction. Such a function is known as an "adsorption isotherm" and is analogous to the n-p behavior at fixed T for a gas, either ideal [Eq. (2.10)] or nonideal (Fig. 2.1). To derive a typical isotherm, we start with the Langmuir adsorption model of Sec. 6.5.5, in which adsorption is assumed to occur only on unoccupied surface sites. In Sec. 6.5.5, we also assumed a fast chemisorption reaction (that is, $\zeta \to \delta$) that was irreversible (no desorption), and this resulted in an asymptotic approach of n_s to monolayer coverage: $\Theta = n_s/n_{so} = 1$. Those assumptions were appropriate for the adsorption of a reactant upon a surface with which it bonds strongly. Here, we instead want to examine precursor reactions on less reactive surfaces, so we neglect the chemisorption reaction and add to the Langmuir model a finite desorption rate obeying first-order kinetics with a rate constant k_d. The steady-state mass balance on the surface thus becomes

$$J_i \delta (1 - \Theta) = k_d n_s = k_d n_{so} \Theta \tag{7.45}$$

which is also the limiting form of the Eq. (5.3) mass balance as the chemisorption rate constant, k_r, vanishes. From the Knudsen equation [Eq. [2.18)], we know that $J_i \propto p_A$, so adsorption on the clean surface may be thought of as a first-order reaction with $J_i \delta = k_a p_A$. Here, the $1/k_B T$ factor from Eq. (7.33) has been incorporated into k_a. Solving Eq. (7.45) for Θ and using this expression for $J_i \delta$ gives the equation for the Langmuir isotherm:

$$\boxed{\Theta = \frac{n_s}{n_{so}} = \frac{p_A}{\left(\dfrac{k_d n_{so}}{k_a}\right) + p_A}} \tag{7.46}$$

Inspection of this equation and its plot in Fig. 7.17 shows a linear region where $\Theta \propto p_A$ at low p_A, and a saturation region where Θ is independent of p_A at high p_A. These limiting regions are observed for all isotherms, not just for the Langmuir model, provided that $p_A < p_v$ so that bulk condensation does not occur. For CVD, the limiting behaviors are more important than the exact shape of the curve in between.

7.3.3 Surface processes 347

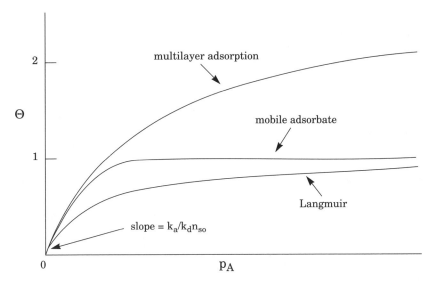

Figure 7.17 Typical adsorption isotherms: $\Theta = n_s/n_{so}$ = fractional monolayer surface coverage.

Two common variations from Langmuir behavior are also shown in Fig. 7.17. Observation of the mobile-adsorbate effect shown was cited after Eq. (6.20). The other variation involves adsorption beyond 1 ML, as embodied in the BET (Brunauer-Emmett-Teller) and other isotherms developed in the study of heterogeneous catalysis.

Now that we have established a formalism for describing adsorption and surface coverage in CVD, we will examine five effects of adsorbate concentration, n_s, on film deposition behavior, namely: pressure dependence of deposition rate, competitive adsorption, film conformality over topography, roughening, and selective deposition.

The behavior of n_s versus p_A, as embodied in the adsorption isotherm, is reflected in **film deposition rate versus p_A** in cases where the surface reaction is controlling the rate rather than another step in the Fig. 7.1 sequence. That is, the rate typically increases with p_A and then saturates at high p_A. For high δ and low k_d in Eq. (7.45) corresponding to strong adsorption, coupled with a low rate of the film-forming reaction [k_r in Eq. (5.5)], this saturation can occur at very low p_A even into the Kn > 1 regime (see Exercise 7.10). The nonlinear behavior of n_s versus p_A also points up a problem with extrapolating high-vacuum studies of surface-reaction kinetics to CVD behavior at higher pressure. High vacuum with Kn >> 1 is required for such studies, since they typically involve modulated-molecular-beam techniques (Fig. 6.9) or surface analysis by electron spectroscopy (Sec. 6.4.1). Unfortunately, n_s is much lower under high vacuum than under typical CVD conditions, and the extrapolation from one pressure regime to the other is not straightforward. In addition, the rate-limiting step may *change* with pressure. Alternatively, optical techniques of surface analysis, such as ellipsometry (Sec. 4.8.2) and infrared absorption, can be used at any pressure, and the latter technique has the advantage of being able to monitor individual adsorbates by their bond vibrational frequencies. Multiple internal reflection of the probe beam in an infrared-transparent substrate can be used to obtain sufficient infrared absorption for submonolayer sensitivity. However, a single external reflection at grazing incidence also enhances surface sensitivity [25] and is easier to implement. There is much to be done in using these techniques to understand CVD reactions under actual film-deposition conditions.

When more than one surface species is involved in a CVD reaction, **competitive adsorption** must be taken into account. For example, the deposition rate of Si is greatly reduced when a small amount of PH_3 or AsH_3 is added to the reactant feed to n-dope the Si, because the adsorption of these reactants passivates the surface against SiH_4 adsorption (see references cited in Ref. 20). The curious and troublesome phenomenon of multiple steady states of reaction behavior [26] has also been explained by competitive adsorption in the case of $TiCl_4$ and C_3H_8 reacting in the adsorbed state (Langmuir-Hinshelwood mechanism) to form TiC film and HCl gas. As the partial pressure of $TiCl_4$, $p(TiCl_4)$, is raised, deposition rate and $p(HCl)$ at first increase due to increasing $n_s(TiCl_4)$. At some level of $p(TiCl_4)$, however, deposition rate and $p(HCl)$ suddenly drop to a much lower level, because excessive $n_s(TiCl_4)$ is preventing adsorption of the other reactant, C_3H_8. When $p(TiCl_4)$ is then decreased, hysteresis is observed: that is, rate and $p(HCl)$ remain low and then finally jump up at some lower $p(TiCl_4)$. Clearly, these situations must be understood to achieve good process control.

Good **film conformality** over surface topography is one of the principal features of CVD, but it does not always occur. Conformality becomes harder to achieve when the recessed regions into which one wants the deposit to penetrate have a lateral dimension less than that needed for convective transport into them. Then, penetration can occur only by diffusive transport through the stagnant gas, as will be discussed in Sec. 7.4. This is the situation, for example, for the spaces between the stacked wafers of the batch reactor of Fig. 7.5c. When the lateral dimension of the recess is still smaller—less than the gas mean free path, diffusion no longer slows down the penetration, but still the reactant becomes progressively depleted by deposition onto the sidewalls of the recess as it penetrates deeper by molecular flow, bouncing off the sidewalls as illustrated in Fig. 5.17f. At 1 atm, molecular flow prevails in channels <100 nm wide Eq. (2.24)], and it prevails in much larger channels at LPCVD pressures. Surface diffusion (Sec. 5.2) also contributes to penetration, but conformality behavior of a wide variety of CVD films suggests that it is not a factor for penetration distances > 100 nm or so.

Conformality versus film thickness, within recesses deep enough to safely neglect surface diffusion and narrow enough to be operating in molecular flow, has been analytically modeled by assuming first-order kinetics [Eq. (7.33)] for the film-forming surface reaction and for precursor desorption, and by assuming a cosine distribution [Eq. (4.30)] for molecules scattered or desorbed from the sidewalls [27]. Results were computed for trenches of different aspect ratios (height/width), as shown in Fig. 7.18. The reactant supply becomes depleted with increasing depth into a trench because of deposition further up the sidewalls and also because of "necking" of the trench mouth, which narrows the solid angle from which deeper regions can receive deposition flux. Clearly, for good conformality one wants a low sticking coefficient, S_c, for the reactant so that most of it can bounce down to the bottom of the trench. When deposition conditions are such that the fractional surface coverage, Θ, of precursor is low, then $S_c \equiv \zeta$, where ζ is the reaction probability of a molecule impinging on bare surface as defined in Eq. (5.7). Figure 7.18a shows trench coverage computed for $\Theta = 0$ and $\zeta = 0.10$ (although the notation and terminology are different in the referenced work). The poor coverage for the deeper trenches shows that ζ needs to be still lower to obtain good coverage under conditions of low Θ, which is the "adsorption-limited" regime of deposition rate. Lower ζ can be achieved by reducing surface T. When Θ is higher, on the other hand, S_c becomes low even if ζ is high, since precursor is

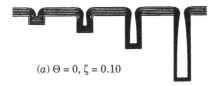

(a) $\Theta = 0, \zeta = 0.10$

(b) $\Theta = 0.01$ at top of trench, $\zeta = 1$

Figure 7.18 Calculated film-conformality profiles in trenches. (Source: Reprinted from Ref. 27 by permission.)

assumed not to adsorb on surface already covered. Figure 7.18*b* shows trench coverage for $\Theta = 0.91$ at the mouth of the trench and for $\zeta = 1$. [Θ becomes lower inside the trench due to depletion of reactant incident flux and partial pressure, p_A, in accordance with Eq. (7.46).] The figure shows that Θ needs to be still higher to obtain good coverage. This can be achieved by operating at higher p_A, further into the adsorption-saturation regime of the Eq. (7.46) isotherm. This is the "surface-reaction-limited" regime of deposition rate. However, if high p_A causes gas-phase reactions to form powder or high-S_c precursors, this remedy becomes less workable. The trench conformality problem is often encountered in integrated-circuit manufacture. It also occurs when coating the internal surface area of porous materials (see Exercise 7.15). Deposition into trenches or into cavity structures such as that of Fig. 9.35 is a good way to determine the S_c of film precursors.

Reactant depletion in recessed areas due to high S_c also aggravates **film roughening**, a topic discussed at length in Sec. 5.4.1. To summarize that discussion, substrate roughness and statistical roughening initiate the process, which is then amplified by "self-shadowing" for high S_c or by "nutrient (reactant) depletion" for low S_c. These are actually two names for the same phenomenon in different S_c ranges, that phenomenon being removal of reactant from the vapor phase by deposition. In either case, it is inherently destabilizing to surface smoothness, because the deeper the roughness features become, the more depletion there is, the slower the deposition rate is at the bottom relative to the top, and thus the faster the roughness develops. Surface diffusion is the counteracting phenomenon which stabilizes the smooth surface, and the final amplitude and scale of the roughness represent the balance between depletion and diffusion [28].

The achievement of **selective deposition** is largely determined by control of adsorbate surface coverage, n_s. The wide range of adjustability in n_s and the presence of large activation energies for the reaction of source gases together make selectivity much easier to achieve in CVD than in physical vapor deposition (PVD), where the sticking coefficients of most vapors are near unity, regardless of surface conditions. Selectivity can mean deposition on substrate areas of one composition while not on other areas, or it can mean selective deposition of one solid phase and not others, such as diamond but not graphite. (Diamond deposition uses plasma activation, which will be discussed in Chap. 9.) Area selectivity is a very important goal in integrated-circuit manufacture, as in the selective deposition of W contact metal into vias etched through SiO_2 film down to underlying Si transistors. In general, selectivity may be obtained by controlling either equilibrium or kinetics. Selective W deposition from WF_6 on Si, but not on SiO_2, was mentioned after Eq. (7.30) and is an example of equilibrium control, since there is no favorable W-deposition reaction on SiO_2. But since the reaction on Si consumes the Si, it is self-limiting, and thick deposits thus require the addition of H_2 or SiH_4 to chemically reduce the WF_6. There, deposition on SiO_2 can still be avoided, but now it is due to a *kinetic* limitation. Let us assume that the film-forming reaction is second-order in reactant surface coverage, n_s, with a rate given by $R_r = k_r n_s(WF_6) n_s(H_2)$. In the absence of chemisorption reaction with the SiO_2, these molecules will only physisorb and will thus both have low n_s because of a high k_d and possibly also a low k_a in Eq. (7.46) for surface coverage. Nevertheless, R_r is not zero, so W nuclei will eventually form after some "incubation time." Subsequent deposition on these nuclei is much faster, probably due to a lowering of the reaction's E_a by dissociative adsorption [16] of H_2 on W. Thus, selectivity is quickly lost once nuclei form. The surface W is acting here as a catalyst to activate the H_2. Nucleation behavior was discussed at length in Sec. 5.3.

Deposition on undesired areas or deposition of unwanted phases can also be avoided by raising the surface coverage of a deposition-reaction *by-product* and thus driving the reaction in the reverse direction in accordance with equilibrium behavior [Eq. (7.28)]. For example, addition of HCl product improves selectivity in the deposition of Si from SiH_2Cl_2. Presumably, the $\Delta_r G°$ is less negative on SiO_2 than on Si because of weaker bonding of Si to SiO_2 than to itself; in other words, $\Delta_f G°[Si(a)] > \Delta_f G°[Si(c)]$ because of excess surface energy, γ_i, at the $Si(a)/SiO_2$ interface. Thus, with proper partial-pressure adjustment, a balance can be achieved where deposition is occurring on the Si while etching of Si nuclei by HCl is occurring on the SiO_2. H_2 plasma can also be used [29] instead of HCl to supply active etchant. The achievement of such an equilibrium balance is more reliable than kinetic control of selective deposition, because kinetic control breaks down at nuclei formed after the incubation time or formed at spurious active surface sites such as contamination islands or scratches.

In another selectivity example, a shift in $\Delta_r G°$ with surface energy can even cause large variations in deposition rate from crystallographic plane to plane in a CVD process that is operating near equilibrium, because of variations in surface energy, γ, with surface structure. This appears to be the cause of the dramatic crystallographic selectivity of deposition rate shown in Fig. 7.19 for the GaAs "chloride" CVD process [30]. There, GaCl, $AsCl_3$, and H_2 react to deposit GaAs, and at the same time the HCl(g) by-product etches the GaAs. The {111} planes in the upper hemisphere are the Ga-rich ones, and the lower ones are As-rich. (GaAs polarity was discussed at the end of Sec. 6.5.6.) The deposition-rate anisotropy between these polar faces is more than 15/1.

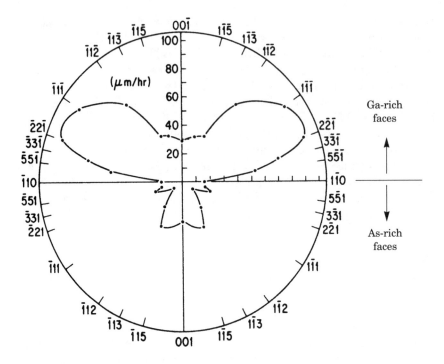

Figure 7.19 Polar diagram of GaAs deposition rate by the chloride process vs. crystallographic plane. (Source: Reprinted from Ref. 30 by permission.)

The final aspect of surface chemistry to discuss is the steep exponential T_s dependence of deposition rate which is induced when a surface reaction is the rate-controlling step of the CVD process. There are both advantages and disadvantages to this behavior. To obtain uniformity over large areas, the furnace-enclosed reactor of Fig. 7.5c is usually needed in order to obtain sufficient substrate-T control under such conditions. To compensate for the reactant depletion which occurs at substrates toward the downstream end of this reactor, an increasing T profile can be imposed along the furnace (Exercise 7.11).

In other reactors, the steep T dependence can be used to start and stop the deposition reaction abruptly by changing surface T. This feature has been exploited in a technique known as "limited-reaction" or "rapid thermal" processing, in which the substrate is rapidly heated by irradiation from a quartz-halogen lamp focussed through a quartz reactor wall. The substrate is then somewhat less rapidly cooled by thermal radiation when the desired film thickness has been reached [31]. This procedure minimizes the time at high T needed to deposit a given thickness, and it thus minimizes the amount of interdiffusion occurring between layers in a heterostructure.

A surface-T increase can also be obtained on selected areas of a substrate by directing there the focussed beam from a high-power laser, so that deposition occurs only in the irradiated areas. This is done in conjunction with conventional substrate heating to just below the deposition T, to minimize laser power and thermal shock. The laser can be programmed to scan and write arbitrary patterns of deposition for the generation of three-dimensional thin-film structures. When a short wavelength is used, there is also the possibility that photochemical activation is occurring [32] in addition to local heating; that is, direct breaking of a reactant bond due to absorption of photon energy in electron excitation. One way to identify a photo*chemical* versus a photothermal effect is to see if it still occurs at cryogenic T [33].

The above discussion of chemical reactions in CVD has presented the basic principles of equilibrium and kinetics for gas-phase and surface reactions. However, it has stopped short of attempting overall kinetic modeling, because good data on rate constants are usually unavailable. Even in the much-studied SiH_4 pyrolysis reaction, the important rates are known only for the gas phase. Nevertheless, understanding how reaction equilibria and rates vary with pressure and T is valuable in designing and optimizing CVD processes.

7.4 Diffusion

We now consider the final transport step in the CVD process sequence of Fig. 7.1: gas-phase diffusion of reactants to the substrate surface. Forced and free convection have carried the gases to the *vicinity* of the substrate, and sometimes there also has been some homogeneous reaction along the way. However, viscous friction requires that the convection velocity drop to zero at the surface whatever the flow profile, so the final transport of reactants to the surface has to occur by diffusion through a relatively stagnant boundary layer of gas.

The transition from convection to diffusion is gradual upon approaching the surface, but it is convenient to model this situation as two distinct regions: a convectively supplied reservoir of reactant, and a diffusive boundary layer between this reservoir and the surface. The surface on which film is depositing is a sink for the reactant, so the reactant concentration, n_A, just over the surface is always lower than that in the reservoir, n_∞. That concentration gradient drives the reactant diffusion flux, J_A, as shown in Fig. 7.20, and the edge of the *concentration* boundary layer in which this diffusion is occurring, δ_n, is arbitrarily but conventionally defined as the plane at which n_A has dropped by 1 percent from n_∞. We shall see that δ_n behaves very differently from δ_v, the *velocity* boundary layer, which was discussed in Sec. 7.2.2. Below, we will apply the boundary-layer model to the three reac-

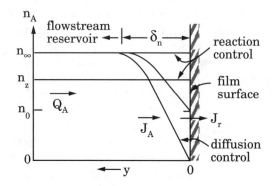

Figure 7.20 Quantities relevant to the diffusion of a single reactant, A, through the concentration boundary layer, δ_n, to the depositing-film surface at $y = 0$.

tors of Fig. 7.5. But first, we need to write the equations that describe diffusive transport.

In Sec. 2.8.1, we showed that the diffusion flux of species A in an A-B mixture is proportional to the concentration gradient by the binary diffusivity, D_{AB}, in accordance with Fick's Law, Eq. (2.27). When the total amount of gas remains constant, the countercurrent flux of B due to the displacement of B by A has to be equal and opposite to J_A, so $J_B + J_A = 0$. However, there is a complication in film deposition due to molarity changes upon reaction, such as $SiH_4(g) \rightarrow Si(c) + 2H_2(g)$. There, for every mole of SiH_4 diffusing to the surface and depositing Si, two moles of H_2 must diffuse away from the surface. This net flux of gas away from the surface, $J_B + J_A \neq 0$, is known as Stefan flow, and it sweeps A away from the surface in proportion to the mole fraction (or molecular fraction) of A in the mixture, $x_A = n_A/(n_A + n_B) = n_A/n$. Thus, the net flux of A toward the surface is the sum of the Stefan-flow and Fick's-law components:

$$J_A = x_A(J_A + J_B) - D_{AB}\frac{dn_A}{dy} \quad (7.47)$$

or, rearranging and substituting x_B for $(1 - x_A)$ and nx_A for n_A,

$$\frac{dx_A}{dy} = \frac{x_A J_B - x_B J_A}{nD_{AB}} \quad (7.48)$$

When generalized to three-dimensional diffusion of multicomponent mixtures, this becomes the Stefan-Maxwell equations [34]:

$$\nabla x_i = \sum_{j \neq i} \frac{1}{nD_{ij}} (x_i J_j - x_j J_i) \qquad (7.49)$$

When there are v_p molecules of gaseous product per molecule of reactant deposited, Eq. (7.47) becomes

$$J_A = \frac{-D \frac{dn_A}{dy}}{1 + x_A (v_p - 1)} \qquad (7.50)$$

This reduces to Fick's Law for the equimolal counterdiffusion case ($v_p = 1$) or for high dilution ($x_A \to 0$). By contrast, consider conditions leading to large Stefan flow; that is, $v_p = 2$, no dilution, and all reactant being deposited ($x_A = 0$ at $y = 0$). Then, we may take the average x_A across the x_A gradient as 1/3, so that the denominator becomes 4/3. Thus, we see that this flow effect is noticeable but not enormous, and it may be neglected for rough calculations.

7.4.1 Diffusion-limited deposition

We wish to determine the degree to which the concentration gradient required for diffusion depletes the reactant over the surface and thereby limits the film-forming surface-reaction rate. We will assume that only one reactant is diffusion-limited, and we will thus will drop some of the A subscripts from above for simplicity. Using the boundary-layer model just introduced and referring to Fig. 7.20, consider a flowstream reservoir in which the concentration of this one reactant is maintained at a level n_z by its mass flow, Q_A, into the reactor. In general, n_z will decrease with axial position, z (perpendicular to the plane of the figure), from its reactor-inlet value of n_∞ at z = 0 as it becomes depleted by the film deposition. Just over the film surface at y = 0, the reactant concentration has dropped to some lower level, n_0. For steady-state deposition and negligible film re-evaporation, the diffusion flux toward the surface, J_A atoms/cm^2·s, will be equal to the film deposition flux, J_r. Approximating dn_A/dy in Fick's law by the n_A difference across the boundary layer, δ_n, we obtain

$$J_r = J_A = -D \frac{n_z - n_0}{\delta_n} \qquad (7.51)$$

or for the fractional depletion of reactant at the surface,

$$\boxed{f_o = \frac{n_z - n_0}{n_z} = \left|\frac{J_r}{Dn_z/\delta_n}\right|}$$ (7.52)

There are two limiting cases of Eq. (7.52), and it is important to know which one applies to the CVD process being run:

1. reaction control: $f_o \to 0$, $n_0 \approx n_z$, and $J_r \ll Dn_z/\delta_n$
2. diffusion control: $f_o \to 1$, $n_0 \to 0$, and $J_r \approx Dn_z/\delta_n$

These cases are illustrated in Fig. 7.20, along with an intermediate case. The intermediate case is more difficult to analyze, but since it is also more difficult to control, it should be avoided in CVD practice anyway. Reactors in which the uniformity of reactant flow over the substrate is poor, such as the batch reactor of Fig. 7.5c, need to operate under reaction control to achieve uniformity. On the other hand, when surface-T uniformity is hard to achieve, as in the axisymmetric reactor of Fig. 7.5a, the exponential dependence of reaction rate on T means that better uniformity is obtainable under diffusion control.

One way to determine whether a process is controlled by reaction or diffusion is to estimate the quantities in Eq. (7.52). The linear deposition rate, dh/dt (nm/s), is easily measured either by thickness deposited in a given time or continuously by ellipsometry or interferometry (Sec. 4.8.2), and it is easily converted to J_r by Eq. (2.21). D often can be found in handbooks or estimated from the formula in Table 2.1. The amount by which n_z is less than the inlet value, n_∞, can be found from the value of $J_r A_r$ (mc/s) compared to Q_A (in mc/s). Here, A_r is the total deposition area *upstream* of the z position being considered; this includes all of the area heated to deposition T, not just the substrate area (see Exercise 7.12). Note that the depletion at z is less than the total fractional utilization of reactant, η, which is determined from n_z at the outlet. In the case of the axisymmetric reactor, η may be large, but the depletion in the flowstream over the substrate is still small, since that region is upstream of the deposition region. For the other reactors in Fig. 7.5, η must be kept small to ensure deposition uniformity, as we shall see below. Consequently, for present purposes we may take $n_z \approx n_\infty$.

The most uncertain quantity in Eq. (7.52) is δ_n. For the axisymmetric reactor with stationary substrate, it might be taken as δ_{vz} from Eq. (7.15), since that is where the flowstream begins to stagnate. However, if the substrate is rotating, δ_n is likely to be less than the δ_{vz} of Eq. (7.20), since the radial flow carrying fresh reactant is actually *increasing* as it approachs the substrate, as shown by curve F in Fig. 7.11. For the tube reactor (Fig. 7.5b), the solution for δ_n versus z will be presented below. For the batch reactor (Fig. 7.5c), δ_n is the entire distance from substrate edge to center, since there is negligible convective flow between substrates.

Another way to distinguish the reaction- and diffusion-controlled regimes is to measure deposition flux, J_r, versus T. The behavior of this function can be seen by considering the simple case of an irreversible first-order film-forming reaction and a surface coverage $n_s \ll$ 1 ML in the linear portion of the adsorption isotherm [Eq. (7.46)] where $n_s \propto n_0$, n_0 being the gas-phase reactant concentration just over the surface. Then,

$$J_r = k_s n_s \approx k_a n_0 \tag{7.53}$$

where the rate constant k_a here has units of cm/s. Setting this J_r equal to Eq. (7.51), solving for n_0, and substituting back into Eq. (7.53), we have

$$J_r = \frac{k_a n_z}{1 + \dfrac{k_a}{D/\delta_n}} \tag{7.54}$$

which has the following limiting values:

1. reaction control: $k_a \ll D/\delta_n$ and $J_r \to k_a n_z$
2. diffusion control: $k_a \gg D/\delta_n$ and $J_r \to D n_z/\delta_n$

Since $\log k_a \propto 1/T_s$ with a slope of $-E_a/R$ by Eq. (7.35), where T_s is the surface T in K, such an "Arrhenius plot" for a reaction-controlled process yields a steep straight line as shown for segment (1) of Fig. 7.21. As k_a becomes larger with increasing T_s, diffusion control eventually takes over in segment (2). There, the much smaller slope is determined by the relatively small combined T dependencies of D, n_z, and δ_n. Sometimes, the slope in the reaction-controlled region changes

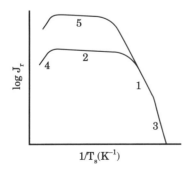

Figure 7.21 Arrhenius-plot behavior of deposition rate. Segments (1) and (3) are reaction-rate controlled with two different E_a values, whereas segments (2) and (5) are diffusion controlled at two different pressures.

with T as shown by segment (3), and this indicates a change in the rate-limiting reaction step to one with a higher E_a. In the special case of a reaction which has marginally negative $\Delta_r G°$ and which is also endothermic, the steep slope may instead represent the equilibrium constant shift toward product [30] with increasing T [Eq. (7.37)]. The slope reversal at high T in segment (4) may indicate either the onset of an equilibrium limit for an exothermic reaction, or the onset of film re-evaporation [Eq. (5.8)]. In general, one must be cautious in the interpretation of Arrhenius plots when thermodynamic and kinetic backup data are unavailable.

The effect of reactor total pressure, p, on deposition flux, J_r, depends on the extent of diffusion or reaction control and on whether it is the reactant's mole fraction, x_A ($= n_A/n = p_A/p$), or its concentration, n_A ($\propto p_A$), which is held constant as p is varied. The value of x_A can be held constant by increasing p with a fixed gas-flow ratio, using pump throttling at valve b' in Fig. 7.2. Then n_A ($=n_z$) increases, and in the reaction-controlled regime, J_r increases per Eq. (7.54). In the diffusion-controlled regime, on the other hand, the 1/p dependence of D cancels out the n_A increase and leaves J_r *independent* of p. However, it is difficult to keep x_A constant over a wide range of p, because at low p, n_A becomes too low for reasonable deposition rate, and at high p it becomes so high that homogeneous reaction to powder can occur, as discussed prior to Eq. (7.41). Therefore, it is more common to keep n_A constant by increasing reactant dilution with increasing p. In that case, J_r is independent of p in the reaction-controlled regime, but it increases with decreasing p in the diffusion-controlled regime because of the increase in D, as shown by segment (5) in Fig. 7.21. This increase in the D-limited J_r is the main reason to operate CVD at reduced p, because it allows reaction-controlled deposition at higher J_r so as to maintain uniformity with reasonable deposition rate in batch reactors.

7.4.2 Reactor models

Having now outlined diffusion behavior in general terms, we proceed to model specific reactors, namely the tube and batch types of Figs. 7.5b and c. Tube reactors can operate either in diffusion or reaction control. If f_o in Eq. (7.52) is still small even when δ_n is taken to be as large as the entire tube diameter, then reaction control applies, and n_z may be assumed constant across the tube, as shown by the horizontal line in Fig. 7.20. If f_o is *very* small ($< 10^{-2}$), then considerable *axial* diffusion will also occur, and n_z will become independent of z as well. More commonly though, axial diffusion is still small under reaction control. Then, reactant transport may be described using an Eq. (3.1) mass balance on an axially differential (dz) volume element of uniform

n_z across the reactor's cross section, as shown in Fig. 7.22a for a rectangular cross-sectional area, A_t, and for deposition occurring on area A_z per unit length in z. Since the lateral diffusional mixing is fast, the velocity profile across A_t does not matter, and we may use the "plug-flow" assumption of uniform velocity at the mean value, $\bar{u}$, as we did for the oil backstreaming problem in Eq. (3.14) and Fig. 3.4b. For first-order reaction kinetics and low surface coverage, where the reactant sticking coefficient, S_c, is independent of n_z, it can be shown (Exercise 7.14) that the reactant depletes exponentially with increasing z according to

$$\frac{n_z}{n_\infty} = \exp\left[-\left(\frac{A_z}{A_t}\right)\sqrt{\frac{T}{M}}\left(\frac{3587 S_c}{\bar{u}}\right)z\right] \quad (7.55)$$

for units of cm and s. Under these conditions, the fraction n_z/n_∞ and thus the reactant utilization fraction, η, must be kept small to obtain deposition-rate uniformity over the length of the reactor, unless a compensating $T_s(z)$ gradient is imposed to increase S_c along z. This transport situation may be thought of equivalently as a volume element of unit length in z moving through the reactor at velocity $\bar{u}$ and depositing material out at the periphery as it proceeds. Then, n_z depletes exponentially in time (and thus in z) within the volume element. This behavior is analogous to the exponential pumpout situation of Eq. (3.11).

When diffusion is too slow to maintain reactant concentration at a uniform level n_z across the tube reactor, the volume element shown in Fig. 7.22b must be used instead for the mass balance. This element is differential in both y and z, so the u profile in y must now be consid-

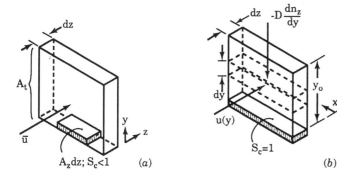

Figure 7.22 Mass-balance volume elements for determining reactant depletion in axial-flow tube reactors: (a) reaction control and (b) diffusion control.

ered. The net convective flow of reactant A into the element in the z direction, per unit width in x, is $-u(\partial n_A/\partial z)dzdy$. The net diffusive flow into the element in the y direction is

$$\left(-D\frac{\partial(\partial n_A/\partial y)}{\partial y}\right)dy\,dz$$

In steady state, n_A is constant in the volume element, so

$$u(y)\frac{\partial n_A(y,z)}{\partial z} = D\frac{\partial^2 n_A(y,z)}{\partial y^2} \qquad (7.56)$$

This illustrates the general procedure for modeling multidimensional flow in CVD reactors. In some cases, gradients in x must also be considered. Such a mass balance must be satisfied for every species in every dxdydz volume element. When homogeneous reactions are occurring, the balance must also include the rates of generation and consumption of every reacting species. Usually, solutions can only be obtained numerically. However, an analytical solution can be obtained for the Fig. 7.22b geometry when the u(y) profile is assumed to be either flat or linear [35], and these solutions are shown in Fig. 7.23 using normalized coordinates. The assumed linear u(y) profile approximately tracks the actual parabolic one where the n_A gradient is steep ($y < 0.3y_o$), as shown by the inset. The boundary conditions assumed are: uniform $n_A = n_\infty$ for z = 0, $n_A = 0$ at y = 0, and no deposition at the top of the reactor ($dn_A/dy = 0$ at $y = y_o$) or on the sides (no flux in the x direction).

Figure 7.23 provides valuable insight into diffusion-limited CVD behavior. Surprisingly, the n_A profile is only slightly affected by the choice of u(y) profile, so the simplest assumption of plug flow is satisfactory even when lateral diffusion is limiting. With that assumption, by the way, we may equivalently model the problem as a Fig. 7.22a volume element moving at $\bar{u}$ and depositing material at the bottom, as we did for the Eq. (7.55) case. This is now identical to the classic one-dimensional diffusion or heat-transfer problem involving a block of uniform semi-infinite material suddenly subjected to an n_A or T drop at y = 0, for which the solution is [36]

$$\frac{n_A}{n_\infty} = \text{erf}\frac{y}{2\sqrt{Dt}} = \frac{y}{2\sqrt{Dz/\bar{u}}} \qquad (7.57)$$

Note that the denominator of the error-function (erf) argument is equivalent to the diffusion length, Λ, from Eq. (5.25). Returning now

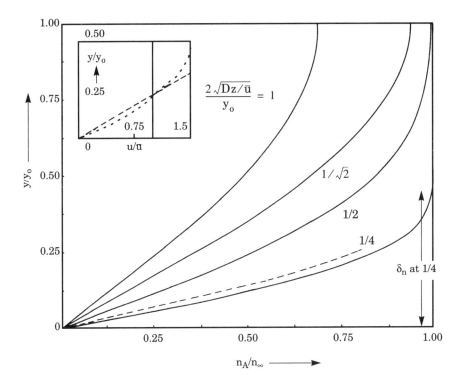

Figure 7.23 Diffusion-limited reactant concentration profiles for four successive normalized axial positions in a tube reactor: solid line = plug flow, long dashes = linear u profile, and short dashes in inset = parabolic profile. (Source: Reprinted from Ref. 35 by permission; curve labeling changed to conform to text.)

to Fig. 7.23, note that as z increases [or t in the Eq. (7.57) model], the n_A-gradient region spreads upward in y; that is, the concentration boundary layer, δ_n, expands. At a z value such that $2\sqrt{Dz/\bar{u}}/y_o = 1/2$, δ_n reaches the top wall, because that is when $n_A(y_o)$ drops by 1 percent from n_∞, this drop defining the edge of δ_n. At this point, the fractional reactant utilization averaged across y happens to be $\eta = 21$ percent. For still larger z, the semi-infinite assumption breaks down and Eq. (7.57) no longer holds, because δ_n has now spread across the reactor (throughout the reservoir). Then, n_A begins to deplete *exponentially* in z as in the case of fast lateral diffusion which we previously considered.

For smaller z, where $\delta_n < y_o$, the deposition flux, J_r, still decreases with increasing z, because the expansion of δ_n reduces the n_A gradient at the substrate, as seen in Fig. 7.23. It can be shown [35] that for the linear u(y) profile and for $n_A = 0$ at $y = 0$,

$$J_r = D\left(\frac{\partial n_A}{\partial y}\right)_{y=0} = \frac{0.89 D n_\infty}{(D y_0 z/\bar{u})^{1/3}} \quad (7.58)$$

This is slightly different from the solution that would be obtained from Eq. (7.57), due to the presence of the u profile here. Using Eq. (7.57), we would have $J_r \propto z^{-1/2}$ instead. In any case, this dropoff of film deposition rate through the tube can be counteracted by tilting the susceptor as shown in Fig. 7.5b. This constricts the flow cross section with increasing z, thus decreasing y_0 and increasing $\bar{u}$. The right amount of tilt can cancel out the increase in $(Dy_0 z/\bar{u})$ with z and thus maintain uniform deposition rate, at least until $\delta_n \to y_0$. Further downstream where $\delta_n > y_0$, or equivalently, where $\eta > 21$ percent, the exponential dropoff of n_A cannot be compensated out.

The last diffusion situation to be examined is the batch reactor of Fig. 7.5c. There, essentially no convective flow occurs between the substrates, so diffusion must carry the reactant all the way from the peripheral flowstream radially inward to the centers of the substrates. It is not possible to counteract a decrease in n_A along this path by the constriction or T-gradient techniques used above for the tube reactor, so the batch reactor must be operated in reaction control to obtain uniformity. We need to determine how high the deposition rate can be before n_A depletion causes an unacceptable drop in deposition rate at the substrate center. The exact solution to this problem involves integration over radially differential volume elements within which both diffusion and deposition are occurring. However, a considerable simplification which is nevertheless quite accurate enough for present purposes can be made by using only two volume elements and by considering the diffusion to be one-dimensional rather than radial, as shown in Fig. 7.24 by the two dashed rectangles. It is supposed that only diffusion occurs in the outer (left-hand) element, and only

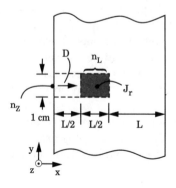

Figure 7.24 Batch-reactor model for diffusion and deposition between stacked substrates of half-width L and arbitrary y dimension; view looking down on a substrate and along the axis of the reactor tube, z.

deposition in the inner (shaded) one. The same concept of judicious separation of coupled phenomena was used to similar advantage in prior sections in the concepts of the concentration and chemical boundary layers.

In the Fig. 7.24 model, each volume element has unit size in y, a size in x of one-half the distance L from substrate edge to center, and a height in z equal to the gap, b, between substrates in the batch. For the steady-state mass balance, we set diffusive mass flow (mc/s) equal to deposition flow, noting that deposition occurs on both faces of each substrate. We omit reactant dilution to maximize diffusive flow, and use Eq. (7.50) for a typical case of $v_p = 2$. Then, using the notation of Fig. 7.24, we have

$$\frac{bD(n_z - n_L)/(1/2)L}{4/3} = LJ_r \tag{7.59}$$

For uniform deposition, the fractional reactant depletion at the center, $f_L = (n_z - n_L)/n_z$, needs to be no more than, say, 1 percent if its surface coverage is low. Then, rearranging Eq. (7.59), we have

$$J_r \leq \frac{0.015 bDn_z}{L^2} \tag{7.60}$$

Note that the upper limit of J_r is strongly dependent on substrate size, L, but is independent of pressure, p, since $n_z \propto p$ but $D \propto 1/p$. Of course, p must usually be <<1 atm when operating undiluted, to avoid powder formation. Taking typical conditions of b = 1 cm, L = 10 cm, T = 800 K, and D = 0.7 cm^2/s at room T and 1 atm, scaled to 800 K by the Table 2.1 formula, we find that $J_r \leq 5 \times 10^{15}$ atoms/cm$^2 \cdot$s or 1 nm/s, a respectable deposition rate. This is a simple procedure for roughly estimating the range of conditions suitable for operating a batch reactor.

7.4.3 Temperature gradients

Two significant perturbations to diffusion behavior occur in T gradients such as that over the substrate surface. One is "thermal diffusion," which causes the heavier and/or larger of the two molecular species in an A-B mixture to become depleted from the hot region. (In liquids but not in gases, this is known as Soret diffusion.) This behavior is described by the following equation [37], whose derivation is too involved to give here:

$$\frac{dx_A}{dy} = -\left(\frac{D_T}{D_{AB}}\right)\frac{1}{T}\frac{dT}{dy} = -\frac{k_T}{T}\frac{dT}{dy} \tag{7.61}$$

where D_T is the thermal-diffusion coefficient, and k_T is the thermal-diffusion ratio for the mixture. Measured values of k_T are ~0.2, give or take a factor of five for various mixtures [34, 37], but k_T is not easily predicted. It increases with the M differences and diameter differences of the molecules in the mixture, and decreases with increasing molecular interaction. It vanishes for M differences of <10 percent and for x_A or x_B < 0.1.

A qualitative picture of the thermal-diffusion effect can be developed using an argument similar to that used to derive the thermal-transpiration effect, Eq. (3.18). Consider the molecular flux of each species along the T gradient as it crosses some plane in space that is perpendicular to the gradient. The flux in each direction arrives at the plane from about one mean free path away toward the hot (T_h) or cold (T_c) end, and the flux is proportional to $p_i / \sqrt{M_i T}$ in accordance with the Knudsen equation [Eq. (2.19)], where i = A or B. If the initial composition is uniform along the gradient ($p_{ih} = p_{ic}$ with total p constant), then Eq. (2.19) results in a net flux, ΔJ_i, for each species toward the hot end. The ratio of these net fluxes is

$$\frac{\Delta J_A}{\Delta J_B} = \frac{J_{Ah} - J_{Ac}}{J_{Bh} - J_{Bc}} = \frac{\dfrac{p_{Ah}}{\sqrt{M_A T_h}} - \dfrac{p_{Ac}}{\sqrt{M_A T_c}}}{\dfrac{p_{Bh}}{\sqrt{M_B T_h}} - \dfrac{p_{Bc}}{\sqrt{M_B T_c}}} = \frac{p_A}{p_B}\sqrt{\frac{M_B}{M_A}} \quad (7.62)$$

If this flux ratio were equal to the p_i ratio, the situation would amount simply to an initial bulk motion of the mixture toward the hot end as in thermal transpiration. However, the flux ratio is seen to be smaller than the p_i ratio when $M_A > M_B$, so that the heavier species has an insufficient net flux and begins to segregate toward the cold end until a steady state is achieved. This segregation can significantly change reactant fluxes to the surface in CVD [38].

The second T-gradient effect is "thermophoresis," which causes small particles suspended in gas to be driven from the hot end [39], just as large molecules are. When the molecular fluxes have reached steady state and when the particle is small enough so that these fluxes can communicate with each other around it (Kn ≥ 1 based on the particle diameter), we must have equal opposing fluxes ($J_h = J_c$) along the gradient. Again, application of the Knudsen equation leads to

$$\frac{p_h}{\sqrt{T_h}} = \frac{p_c}{\sqrt{T_c}} \quad (7.63)$$

That is, p is higher against the hot face of the particle than against the cold face, and the particle is thus pushed toward the cold end. This is identical to the thermal-transpiration equation, Eq. (3.18).

Thermophoresis is a very valuable though generally unappreciated phenomenon in CVD. It is extremely effective in preventing particles homogeneously nucleated within the chemical boundary layer from settling on the film surface. They instead remain suspended within the boundary layer above the substrate and are eventually swept away in the flowstream. Scattering of laser light can be used to dramatically reveal this suspended layer. Particle rejection from the surface is so effective that I have seen quantum-well laser structures (Sec. 6.2.1) successfully grown on a routine basis under MOCVD conditions where such particles were present in abundance. Excessive homogeneous nucleation still causes other problems, however, such as parasitic consumption of source gas and contamination of the reactor with dust.

7.5 Conclusion

CVD is the most complex deposition process which we have examined, due to the many phenomena occurring simultaneously and nonuniformly in the gas phase, namely, forced convection, free convection, homogeneous reaction, and diffusion. Nevertheless, by proper reactor design and operation, it is possible to control each of these phenomena and also to separate their spatial regimes for modeling purposes. Such control provides the surface-chemistry and deposition-rate stability needed to exploit the potential advantages of working with gaseous source materials in the fluid-flow regime. The low sticking coefficient characteristic of gaseous reactants facilitates the coating of convoluted or even porous substrates, or large batches of odd shapes. It also assists selective deposition. The higher pressure of the fluid-flow regime simplifies or eliminates pumping, increases chemistry control latitude because of the higher reactant concentrations, and permits operation under diffusion control for improved film uniformity in the presence of substrate-T nonuniformity. The high T required for activation of most CVD reactions can exceed the tolerance of some substrates, but then one can turn to plasma-enhanced CVD to activate the reactions at lower T, as we will see in Sec. 9.6.

7.6 Exercises

7.1 Reactant gas is flowing into a 1500 cm^3 reactor at 200 sccm as measured on a mass flow controller calibrated for Ar. With the reactor valved off at the outlet, the rate of pressure rise is 10

Pa/min. (a) What is the actual flow rate of the reactant gas in sccm? (b) What is the c_v of the reactant gas?

7.2 For steady-state, one-dimensional laminar flow between parallel plates positioned in the x-z plane at $+y_o$ and $-y_o$, show that the fluid's velocity profile, u(y), is given by Eq. (7.8) and that the mean velocity, $\bar{u}$, is given by Eq. (7.9).

7.3 Show that the thickness of the velocity boundary layer in axisymmetric flow scales as $(T/Mp^2)^{1/4}$ for a fixed flow velocity at the room-T inlet of the reactor.

7.4 Show that the last equality in Eq. (7.21) for Gr holds assuming the ideal-gas law, and show that it is dimensionless.

7.5 Name at least six ways of suppressing recirculating roll cells in the CVD-reactor gas-flow pattern.

7.6 Show how the chemical-equilibrium equation, Eq. (7.28), is obtained from the G minimization of Eq. (7.27).

7.7 BCl_3 at a partial pressure of 10^3 Pa in 1 atm of H_2 is supplied to an APCVD reactor for the deposition of B. If the only significant gas-phase reaction produces $HBCl_2$ and HCl and reaches equilibrium, what is the fractional conversion of BCl_3? (Use the data of Fig. 7.13)

7.8 The two common forms of arsenic vapor, As_4 and As_2, are in equilibrium with each other in an Ar diluent. (a) Derive the expression for the As_2 partial pressure (in atm) in terms of K, total pressure p, and p_{Ar}. At the T for which $\Delta_r G° = 0$ and for p = 1 atm, what is the As_2/As_4 partial-pressure ratio for (b) no dilution and (c) for a feed composition of 1 at.% As_4 in Ar?

7.9 A particular deposition reaction operating at 100 Pa and 900 K in a 2-liter isothermal reactor is limited by the unimolecular dissociation, with $k_I = 0.1$ s^{-1}, of the reactant vapor, which is supplied undiluted at 100 sccm. What fraction, η, of reactant is utilized in the deposition?

7.10 A particular adsorbate, A, of M = 40 is known to have a surface residence time of 1 ms at 1000 K; in other words, the time constant for its desorption when $p_A = 0$ is 1 ms. (a) For a monolayer coverage of $n_{so} = 10^{15}$ mc/cm^2, what is the desorption rate constant in s^{-1}? (b) Assuming Langmuir adsorption kinetics, what p_A is required to sustain a steady-state surface coverage of 90 percent of a monolayer?

7.11 If the Si deposition rate from SiH_4 is being controlled by H_2 desorption in a furnace-type batch reactor, how much of a T increase is needed from the first to the last substrate along the gas flow path to maintain deposition rate uniformity if the SiH_4 utilization is 20 percent? (See data in Sec. 7.3.3.)

7.12 Si is being deposited at 10 nm/min in an isothermal reactor of 2 m² total internal-surface area. If SiH_4 is supplied at 200 sccm, what is its fractional utilization, η?

7.13 An axisymmetric cold-wall reactor operating at 1 atm in a flow of 10 sccm B_2H_6 and 5 slm Ar is depositing B onto a stationary 10-cm-diameter substrate. What is the maximum deposition rate in nm/s determined by diffusion limitations?

7.14 (a) Derive Eq. (7.55) using the stationary mass-balance volume element of Fig. 7.22a, Eq. (2.19) for the impingement rate of reactant, the ideal-gas law, and the other assumptions stated for Eq. (7.55). (b) Show that the same result is obtained using a volume element moving at the axial flow velocity. (c) For pyrolytic-graphite deposition from methane flowing at 600 sccm and 1300 Pa through a 1-cm-inside-diameter, 1400-K, isothermal tube, what is the length of tube at which the methane will be 20 percent depleted if $S_c = 10^{-4}$?

7.15 It is desired to coat the internal-surface area of porous ceramic catalyst pellets with Pt for application in crude oil hydrogenation. Assume that this is to be done by CVD from $Pt(CO)_2Cl_2$ at 800 K, and that the pores can be modeled as cylinders 20 nm in diameter and 1 mm long. Using a model similar to that of Fig. 7.24 and the molecular-flow-conductance equation [Eq. [3.6)] adjusted for T and M, determine the maximum S_c to obtain 10 percent coating uniformity from pellet surface to center.

7.16 The units of a reaction rate constant depend on the units of the reactant and the order of the reaction, and one must be careful to use consistent units. What are the SI units of k_a in Eq. (7.46)?

7.7 References

1. Pierson, H.O. 1992. *Handbook of Chemical Vapor Deposition*. Park Ridge, New Jersey: Noyes Publications.
2. Galasso, F.S. 1991. *Chemical Vapor Deposited Materials*. Boca Raton, Fla.: CRC Press.
3. Hinkle, L.D., and C.F. Mariano. 1991. "Toward Understanding the Fundamental Mechanisms and Properties of the Thermal Mass Flow Controller." *J. Vac. Sci. Technol.* A9:2043.
4. Houng, Y.-M. 1992. "Chemical Beam Epitaxy," *Crit. Rev. in Solid State and Mater. Sci.* 17:277.
5. Shealy, J.R., and J.M. Woodall. 1982. "A New Technique for Gettering Oxygen and Moisture from Gases Used in Semiconductor Processing." *Appl. Phys. Lett.* 41:88.
6. Fitzjohn, J.L., and W.L. Holstein. 1990. "Divergent Flow in Chemical Vapor Deposition Reactors." *J. Electrochem. Soc.* 137:699.
7. Schlichting, H. 1968. *Boundary-Layer Theory*, 6th ed. New York: McGraw-Hill, Chap. V.

8. Breiland, W.G., and G.H. Evans. 1991. "Design and Verification of Nearly Ideal Flow and Heat Transfer in a Rotating Disc Chemical Vapor Deposition Reactor." *J. Electrochem. Soc.* 138:1806.
9. McAdams, W.H. 1954. *Heat Transmission*, 3rd ed. New York: McGraw-Hill, Chap. 7.
10. Giling, L.J. 1982. "Gas Flow Patterns in Horizontal Epitaxial Reactor Cells Observed by Interference Holography." *J. Electrochem. Soc.* 129:634.
11. Chase, M.W., et al. (eds.) 1985. *JANAF Thermochemical Tables*, 3rd ed. Washington, D.C.: American Chemical Society.
12. Wagman, D.D., et al. (eds.). 1982. "NBS Tables of Chemical Thermodynamic Properties," *J. Phys. Chem. Ref. Data* 11, suppl. no. 2.
13. Bernard, C., and R. Madar. 1990. "Thermodynamic Analysis and Deposition of Refractory Metals." In *Proc. Chemical Vapor Deposition of Refractory Metals and Ceramics Symp.* 168:3. Pittsburgh, Pa.: Materials Research Society. These authors recommend the following on-line services: F.A.C.T., Facility for the Analysis of Chemical Thermodynamics, Ecole Polytechnique, Montreal, Québec, Canada; IVTANTHERMO, Institute of High Temperature of the Russian Academy of Sciences, Moscow; THERMODATA, Domaine Universitaire de Grenoble, BP.66, 38402 Saint Martin d'Hères cedex, France; and THERMO-CALC DATA BANK, Royal Inst. of Technol., S-10044, Stockholm, Sweden.
14. Besmann, T.M. 1977. "SOLGASMIX-PV, a Computer Program to Calculate Equilibrium Relationships in Complex Chemical Systems," ORNL/TM-5775. Oak Ridge, Tennessee: Oak Ridge National Laboratory.
15. McNevin, S.C. 1986. "Chemical Etching of GaAs and InP by Chlorine: The Thermodynamically Predicted Dependence on Cl_2 Pressure and Temperature." *J. Vac. Sci. Technol.* B4:1216.
16. Skelly, D.W., T.-M. Lu, and D.W. Woodruff. 1987. "Metallization Techniques," Chap. 3 in *VLSI Metallization*, vol. 15 of *VLSI Electronics and Microstructure Science*, ed. N.G. Einspruch, S.S. Cohen, and G. Sh. Gildenblat. Orlando, Fla.: Academic Press.
17. Spear, K.E., and R.R. Dirkx. 1990. "Predicting the Chemistry in CVD Systems." In *Proc. Chemical Vapor Deposition of Refractory Metals and Ceramics Symp.* 168:19. Pittsburgh, Pa.: Materials Research Society.
18. Jasinski, J.M., and S.M. Gates. 1991. "Silicon Chemical Vapor Deposition One Step at a Time: Fundamental Studies of Silicon Hydride Chemistry." *Accts. Chem. Res.*, 24:9.
19. Jasinski, J.M. 1994. "Gas Phase and Gas Surface Kinetics of Transient Silicon Hydride Species." In *Gas-Phase and Surface Chemistry in Electronic Materials Processing*. Pittsburgh, Pa.: Materials Research Society.
20. Holleman, J., and J.F. Verweij. 1993. "Extraction of Kinetic Parameters for the Chemical Vapor Deposition of Polycrystalline Silicon at Medium and Low Pressures." *J. Electrochem. Soc.* 140:2089.
21. Qian, Z.M., H. Michiel, A. Van Ammel, J. Nijs, and R. Mertens. 1988. "Homogeneous Gas Phase Nucleation of Silane in Low Pressure Chemical Vapor Deposition." *J. Electrochem. Soc.* 135:2378.
22. de Croon, M.H.J.M., and L.J. Giling. 1990. "Chemical Boundary Layers in CVD." *J. Electrochem. Soc.* 137:2867.
23. Buss, R. J., P. Ho, W.G. Breiland, and M.E. Coltrin. 1988. "Reactive Sticking Coefficients for Silane and Disilane on Polycrystalline Silicon." *J. Appl. Phys.* 63:2808.
24. Scott, B.A., and R.D. Estes. 1989. "Role of Gas-Phase Reactions in Silicon Chemical Vapor Deposition from Monosilane." *Appl. Phys. Lett.* 55:1005.
25. Toyoshima, Y., K. Arai, A. Matsuda, and K. Tanaka. 1990. "Real Time *in situ* Observation of the Film Growth of Hydrogenated Amorphous Silicon by Infrared Reflection Absorption Spectroscopy." *Appl. Phys. Lett.* 56:1540.

26. Haupfear, E.A., and L.D. Schmidt. 1993. "Kinetics and Multiple Steady States in the Chemical Vapor Deposition of Titanium Carbide." *J. Electrochem. Soc.* 140:1793.
27. Hsieh, J.J. 1993. "Influence of Surface-Activated Reaction Kinetics on Low-Pressure Chemical Vapor Deposition Conformality over Micro Features." *J. Vac. Sci. Technol.* A11:78.
28. Palmer, B.J., and R.G. Gordon. 1988. "Local Equilibrium Model of Morphological Instabilities in Chemical Vapor Deposition." *Thin Solid Films* 158:313.
29. Yew, T-R., and R. Reif. 1989. "Silicon Selective Epitaxial Growth at 800° C using SiH_4/H_2 assisted by H_2/Ar Plasma Sputter." *Appl. Phys. Lett.* 55:1014.
30. Shaw, D.W. 1974. "Mechanisms in Vapour Epitaxy of Semiconductors." Chap. 1 in vol. 1, *Crystal Growth: Theory and Techniques*, ed. C.H.L. Goodman. London: Plenum Press.
31. Hoyt, J.L., C.A. King, D.B. Noble, C.M. Gronet, J.F. Gibbons, M.P. Scott, S.S. Laderman, S.J. Rosner, K. Nauka, J. Turner, and T.I. Kamins. 1990. "Limited Reaction Processing: Growth of $Si_{1-x}Ge_x$/Si for Heterojunction Bipolar Transistor Applications." *Thin Solid Films* 184:93.
32. Houle, F.A. 1989. "Surface Photoprocesses in Laser-Assisted Etching and Film Growth." *J. Vac. Sci. Technol.* B7:1149.
33. Isobe, C., H.C. Cho, and J.E. Crowell. 1993. "Photochemical vs. Thermal Deposition of Group IV Semiconductors." Paper presented at annual Am. Vacuum Soc. Meeting, Chicago, November 1993.
34. Bird, R. B., W.E. Stewart, and E.N. Lightfoot. 1960. *Transport Phenomena*. New York: John Wiley & Sons, 570.
35. Van de Ven, J., G.M.J. Rutten, J.J. Raaijmakers, and L.J. Giling. 1986. "Gas Phase Depletion and Flow Dynamics in Horizontal MOCVD Reactor." *J. Crystal Growth* 76:352.
36. Crank, J. 1975. *The Mathematics of Diffusion*. Oxford, U.K.: Oxford University Press.
37. Grew, K.E., and T.L. Ibbs. 1952. *"Thermal Diffusion in Gases."* Cambridge, U.K.: Cambridge University Press.
38. van Sark, W.G.J.H.M., M.H.J.M. de Croon, G.G. Janssen, and L.J. Giling. 1990. "Analytical Models for Growth by Metal Organic Vapour Phase Epitaxy: II. Influence of Temperature Gradient." *Semicond. Sci. Technol.* 5:36.
39. Talbot, L., R.K. Cheng, R.W. Schefer, and D.R. Willis. 1980. "Thermophoresis of Particles in a Heated Boundary Layer." *J. Fluid Mech.* 101:737.

7.8 Recommended Readings

Benson, S.W. 1968. *Thermochemical Kinetics*. New York: John Wiley & Sons.

Jensen, K.F. 1989. "Transport Phenomena and Chemical Reaction Issues in OMVPE (Organometallic Vapor Phase Epitaxy) of Compound Semiconductors." *J. Crystal Growth* 98:148.

Smith, W.R. 1980. "The Computation of Chemical Equilibria in Complex Systems." *Indust. and Engineering Chem. Fundamentals* 19:1.

Stringfellow, G.B. 1989. *Organometallic Vapor-Phase Epitaxy: Theory and Practice*. Boston, Mass.: Academic Press.

Chapter 8

Energy Beams

Energy input by nonthermal means is a powerful and widely used process tool in thin-film deposition. In previous chapters, we have concentrated on the use of thermal energy alone for driving evaporation, reaction, and film structure development, with occasional reference to energy-enhancement techniques. In this chapter and the following one, we examine deposition processes in which the primary energy source is nonthermal. This energy may be directed into any of the three process steps: to vaporize the source material, to activate it during transport, or to modify film structure during deposition. The energy may be delivered by electrons, photons, or ions (meaning positive ions unless denoted negative). Table 8.1 compares the energy-beam technologies that are used for vaporization. The significance of the various characteristics tabulated will become clear later on. For now, the key feature to note is that in each of the first four processes listed, the electrons, ions, and photons, respectively, are directed at the source material in a narrow beam of at most a few mm diameter, whereas in "sputtering," the ion beam used covers a much broader area. The use of narrow beams leads to intense heating of the source material at the point of impact, so that the vaporization mechanism is thermal even though the energy input is nonthermal. Conversely, vaporization by sputtering involves direct momentum transfer from bombarding ions to the surface atoms of relatively cool source material.

There are several advantages to using energy beams for vaporization rather than the joule-heated sources of Fig. 4.8. One is that *any* material can be vaporized, no matter how refractory. In the narrow-beam processes, this is because of the very high energy density and surface T which can be achieved. In sputtering, it is because the bom-

TABLE 8.1 Energy-Beam Vaporization Methods

Text sec.	Technology	Vaporization behavior		Typical beam characteristics		Operating pressure, Pa	Evaporant typical characteristics				Requirements for insulating source (target) materials
		Incident beam species	Mechanism	Dia.	Power		Kinetic energy, E_t, eV	Fraction positive ions	Angular distrib., n in $\cos^n\theta$	Macroparticle problem	
8.2	Electron beam	10-keV electrons	Thermal	5 mm	10 kW	$<10^{-2}$	0.2	0.01 to 0.1	2–4	Small	$Y_e > 1$
8.3	Cathodic arc	20-eV ions	Thermal	10 μm	2 kW	<10	0.2 (atoms) 50 (ions)	0.7	1	Large	Not usable
8.3	Anodic arc	20-eV electrons	Thermal	5 mm	2 kW	<10	0.2 (atoms) 5 (ions)	0.2	1	Small	Not usable
8.4	Pulsed laser	4–7-eV photons	Thermal	5 mm	0.2 J in 20 ns	$<10^2$	40	0.05	10	Large	None
8.5.4	Ion-beam sputtering	1-keV ions	Momentum transfer	10 cm	1 kW	10^{-3} to 10^{-2}	5	0	0.5–2	None	Beam neutralizing filament
9.3.3	Glow-discharge sputtering	500-eV ions	Momentum transfer	20 cm	5 kW	10^{-2} to 10^{-1}	5	<<1	0.5–2	Negligible <20 Pa	rf power

barding ions have energies far exceeding chemical-bond strengths, which are a few eV (see Table 4.2). In all of these cases, one can (and must!) avoid energy input to materials enclosing the source, whereas this is not possible with conventionally heated sources. This energy control is often claimed to result in a higher-purity evaporant; however, conventional sources can be extremely clean, as evidenced by the impressive achievements of molecular-beam epitaxy (Sec. 6.2). Moreover, some of the energy involved in beam processes often arrives at other surfaces indirectly and causes contaminant desorption there. In any case, the ability to vaporize any material is an important advantage. Pulsed-laser evaporation and sputtering have a second advantage: the activated depth of source material can be only tens of nm, which results in stoichiometric (congruent) vaporization of multi-element materials. Of course, this does not guarantee a stoichiometric *deposit*, because the sticking coefficients of the vapor species can differ—but it does help. The third and principal advantage of all the Table 8.1 processes is that much of the vapor acquires energy well above the thermal energy of the source surface, and this energy can be very helpful to the deposition process. Sputtered atoms leave the surface with high energy, whereas atoms thermally evaporated by energy beams acquire most of their energy by interaction with the beam in the vapor phase. In the case of *ionized* vapor, this energy may be further increased by accelerating the ions toward the surface of the depositing film using a negative bias on the substrate.

Energy can also be directed at the deposition surface. In particular, either energetic-atom condensation or ion bombardment can result in dramatic improvement of film adherence and structure. Electron bombardment of substrates has also been used occasionally for improving film adherence, but it is much less effective and will not be discussed further here. Light-energy input to the deposition surface was already discussed briefly in Sec. 7.3.3 in relation to photoenhanced CVD. This chapter, following the progression of Table 8.1, will first examine the narrow-beam vaporization techniques and then the use of ion and atom bombardment for deposition enhancement and for sputtering.

Most of the energy-enhanced techniques involve the generation of a "plasma," which is a partially ionized gas consisting of nearly equal concentrations of positive ions and negative particles (electrons and negative ions). Plasma can be thought of as the fourth state in a progression of increasingly energetic states of matter, starting with solids with their fixed atom positions. Adding energy to this first state allows the atoms to move around each other as a liquid and then to separate completely as a gas. Adding even more energy causes gas atoms to separate into the ions and free electrons of a plasma, although there is no *abrupt* phase change as there is in going from solid to liquid to gas. Familiar plasmas include the sun, lightning, the aurora borealis, sparks, fluorescent tubes, and mercury-arc lamps. The low-pressure, diffuse plasma known as a glow discharge is especially useful in thin-film processes for sputtering, for activation of gaseous source material, and for ion bombardment of the deposition surface. Glow-discharge behavior will be treated more fully in Chap. 9.

In all of the energy-enhanced thin-film processes, the energy is coupled to atoms and molecules by electric fields—either those present in electromagnetic radiation such as lasers or microwaves, or those established between electrodes held at different voltages. Much of our discussion of the various processes will involve the details of how this coupling occurs and the consequences for thin-film properties. Controlling the level and uniformity of energy delivery is the key to being able to exploit the advantages of energetic processes without causing film damage. Acceleration of free electrons in the electric field is a key coupling mechanism, so we will precede our examination of specific Table-8.1 processes by looking at how free electrons are generated.

8.1 Electron Generation

Free electrons (e) are generated either by causing them to be emitted from surfaces or by ionizing gaseous atoms or molecules ($A \rightarrow A^+ + e$). We first discuss emission from surfaces. The electron emitter is called the *cathode*, and the electron collector which completes the circuit is the *anode*. Figure 8.1 shows voltage versus distance from cathode to anode under various conditions (*b, c, e*) to be discussed below. Also shown, to the left, is the electron energy distribution, f(E), versus electron potential energy, E, within the cathode material. That is, f(E) is the probability of finding an electron at energy E. (Note that electron potential energy increases upward here, while voltage increases downward.) In metals, f(E) has the Fermi-Dirac shape, shown here for high T and for 0 K. The mean energy is at the Fermi level, E_F, in either case. The barrier to electron emission is the difference between E_F and the energy level of a free electron in vacuum, E_o. This difference, $(E_o - E_F)$, is the work function, ϕ_W. Also illustrated in Fig. 8.1 are the three ways in which an electron can escape: it can get pushed out by thermal energy (*a*: thermionic emission), pulled out by an electric field (*e*: field emission and potential emission), or knocked out by the impact of a photon (*d*: photoemission) or energetic particle (*f*: secondary-electron emission). We will discuss each of these processes in turn.

Thermionic emission arises because f(E) becomes more diffuse about E_F with increasing T, as shown in Fig. 8.1. The electrons in the exponential tail above E_o can just *fall* out of the surface (*a*), and the

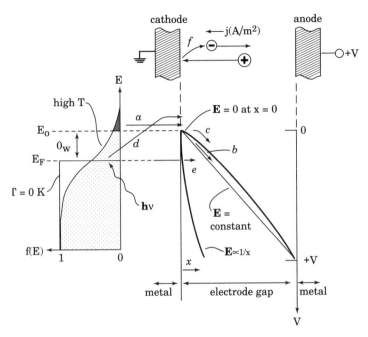

Figure 8.1 Electron emission processes: (a) thermionic emission from the high-energy tail (dark area) of the Fermi-Dirac distribution, f(E), over the surface-potential barrier, ϕ_W, (b) removal of electrons from above the surface by an applied field, **E** = constant, (c) electron removal rate limited by space charge (**E** = 0 at x = 0), (d) secondary-electron emission from the Fermi level, E_F, of a cold cathode by photon (**h**ν) impact, (e) tunneling through a potential barrier made narrow by high local **E**, and (f) secondary-electron emission by ion impact.

resulting current density of emitted electrons per unit surface area is given by the Richardson-Dushman equation,

$$j = BT^2 \exp(-\phi_W/k_B T) \tag{8.1}$$

This is analogous to the Eq. (5.16) Arrhenius rate expression derived for chemical reaction rates, which was based on a similar exponential tail in the Boltzmann distribution of gaseous molecules. The theoretical value of the constant B is 120 A/cm$^2 \cdot$K^2, but B and ϕ_W both vary with surface composition and topography. For example, native-oxide growth lowers the ϕ_W of a metal. The metal W is usually used for thermionic emitters because it can be operated hotter than any other element without evaporating, but more inert metals must be used in O$_2$ or other gases reactive with W. The properties of W-filament emitters are given in Appendix D.

Whether emitted current density can actually *reach* the Eq. (8.1) value depends on the rate at which the emitted electrons are being drawn away from the region above the surface by an applied electric field (Fig. 8.1*b*).

$$\mathbf{E} = -dV/dx \tag{8.2}$$

Note that the field vector, **E**, is positive in the direction of *conventional* (positive) current flow in accordance with Eq. (6.2) for conductivity, leading to the minus sign in Eq. (8.2). Now, if **E** is too small during thermionic emission, electrons pile up in front of the surface and produce a negative space charge in that location. The charge screens the applied field (*c*) and limits current flow. This *space-charge-limited current flow* is an important phenomenon, both in electron emission from cathodes and in ion emission from glow-discharge plasmas (Sec. 9.2), so we will derive here the Child-Langmuir law governing its behavior [Eq. (8.11)] to understand it more fully.

Several equations are needed for this purpose. First, the variation of **E** across a space-charge region of either polarity is given by one of Maxwell's equations, which is also known as the differential form of Gauss's law:

$$\nabla \cdot \mathbf{E} = n_j q_j / \varepsilon_0 \tag{8.3}$$

or for fields in the x direction only,

$$d\mathbf{E}/dx = n_j q_j / \varepsilon_0 \tag{8.4}$$

where n_j = concentration of charged particle j, m^{-3}

q_j = particle charge, C (coulombs), = -1.60×10^{-19} C for electrons

ε_0 = electrical permittivity of vacuum = 8.84×10^{-12} f/m (farads/meter) in SI units

If field strength is thought of in terms of a density of field "lines" parallel to **E**, then these equations say that the net flow of field lines through a surface enclosing a volume is equal to the charge within the volume. When Eq. (8.2) is substituted into Eq. (8.3) or (8.4) to put it in terms of voltage, the result is Poisson's equation. The second equation we need applies to electrons moving in the x direction at velocity v, where we may write the current density (flux) as

$$j = n_e q_e v \tag{8.5}$$

as in Eq. (6.2). Finally, we must have energy conservation of electrons accelerating to kinetic energy E_e through a potential drop V [see Eq. (6.6)] as follows:

$$E_e = |q_e|V = \frac{1}{2}m_e v^2 \qquad (8.6)$$

or

$$v\,(m/s) = \sqrt{\frac{2E_e}{m_e}} = \sqrt{\frac{2|q_e|V}{m_e}} \qquad (8.7)$$

where m_e = electron mass = 9.11×10^{-31} kg.

Combining Eqs. (8.5) and (8.7) to eliminate v, and substituting the resulting expression for $n_e q_e$ into Eq. (8.4), we have

$$-\frac{d\mathbf{E}}{dx} = \frac{j}{\varepsilon_0 \sqrt{2|q_e|V}} \sqrt{m_e} \qquad (8.8)$$

Multiplying both sides by dV and using Eq. (8.2) on the left, we have

$$\mathbf{E}\,d\mathbf{E} = \frac{j}{\varepsilon_0}\sqrt{\frac{m_e}{2|q_e|}} \frac{dV}{\sqrt{V}} \qquad (8.9)$$

Now, at the point in space-charge buildup when $\mathbf{E} \to 0$ at $x = 0$, the condition of space charge-limited current is reached. Integrating Eq. (8.9) with this boundary condition, taking its square root, inserting Eq. (8.2), and rearranging, we have

$$\frac{dV}{2V^{1/4}} = \sqrt{\frac{j}{\varepsilon_0}} \left(\frac{m_e}{2|q_e|}\right)^{1/4} dx \qquad (8.10)$$

Integrating this with $V = 0$ at $x = 0$ gives the classic Child-Langmuir law for one-dimensional space-charge-limited current density from a planar cathode to a planar anode which is at distance x and voltage V with respect to the cathode:

$$\boxed{j = \frac{4\varepsilon_0}{9}\sqrt{\frac{2q_e}{m_e}}\frac{V^{3/2}}{x^2} = 2.32 \times 10^{-6}\frac{V^{3/2}}{x^2}} \qquad (8.11)$$

Note that the units of x determine the units of area in j. The effect of the space-charge phenomenon is that as cathode T is increased at constant anode V, j first increases exponentially in accordance with thermionic emission [Eq. (8.1)] until the accumulating space charge causes

$\mathbf{E}|_{x=0} \to 0$, at which point j abruptly levels out at the Eq. (8.11) value. This current saturation is also shown by the dashed line in the Appendix D plot of emission versus T.

Returning to Fig. 8.1, we now address the ways other than thermionic (hot) emission in which an electron can escape from a surface; that is, the "*cold*-cathode" processes. As in the case of thermionic emission, cold-cathode emission can be space-charge-limited if anode V is insufficient. We first discuss **field emission**, which differs from the other hot- and cold-cathode processes in requiring no energy input to the cathode. It relies instead on developing a field, $\mathbf{E}$, at the surface which is so strong that the barrier becomes narrow enough for an electron to "tunnel" through by sensing the field outside via the evanescent tail of its wave function. [See discussion following Eq. (6.3).] This process is illustrated in Fig. 8.1e and is quantified by the Fowler-Nordheim equation (Lafferty, 1980; p. 26):

$$j\left(A/m^2\right) = \frac{0.054 \mathbf{E}^2}{\phi_w f_1} \exp\left(-\frac{6.83 \times 10^9 \, \phi_w^{3/2} \, f_2}{\mathbf{E}}\right) \quad (8.12)$$

where f_1 and f_2 are slowly varying, tabulated functions of $\sqrt{\mathbf{E}}/\phi_w$, and $\mathbf{E}$ is in V/m. Attaining a sufficiently high field requires sharp corners or peaks on the cathode, which have the following effect. Since like charges repel each other, electrons within the cathode preferentially accumulate at the tips of such features. This increases the charge density there, and thus the local field just outside the tip, in accordance with Eq. (8.4), since $\mathbf{E} = 0$ *within* a conducting tip. It turns out that this local field follows $\mathbf{E} \propto 1/r$, where r is the tip radius [1]. Field-emission current is difficult to control because of the exponential in Eq. (8.12), so it is best to avoid it by using smooth cathodes with rounded corners, as is done on all high-voltage equipment. In well-controlled thin-film processes, field emission is a significant source of electrons only in arc evaporation.

Photoemission and **secondary-electron emission** by impact of an energetic particle are relatively straightforward cold-cathode emission processes (Fig. 8.1*d* and *f*). For metals, if the incident particle's energy exceeds the work function, ϕ_w, electrons can be emitted. However, the yield, Y_e, in terms of number of electrons per incident particle, varies considerably from around 0.1 to 3 with particle type and energy and with surface condition and ϕ_w. Oxidation of metals increases Y_e considerably, for example. In thin-film processes, cold-cathode dominant emission is most often caused by **impact of ions**, and the dominant emission mechanism [2] depends on whether the ion velocity exceeds 10^5 m/s or so. Using the mass of Ar^+ in Eq. (8.7), one finds that this velocity corresponds to an energy of 2.1 keV, which exceeds the ion energies usually encountered in thin-film deposition. The high-energy process is called "kinetic" emission; it occurs by way of ionizing collisions beneath the surface. The low-energy process is "potential" emission, and it occurs just *before* ion impact and is similar to field emission. The intense **E** field around the ion causes an electron to tunnel out of the surface and neutralize the ion. The energy released when this electron drops into the potential well of the ion causes Auger emission of a secondary electron from either the surface or the neutralized ion [3] (see Fig. 6.10). The electron yield per ion, Y_e, depends very little on ion kinetic energy in this low-energy regime, as one might expect from the mechanism. It roughly follows the empirical formula [4]

$$Y_e \approx 0.032(0.78 E_i - 2\phi_w) \qquad (8.13)$$

As a typical example [3], Y_e for Ar^+ on Mo is 0.11. In this equation, E_i (eV) is the energy threshold for ionization of the ion's parent atom or molecule upon being struck by a photon or free electron. Thus, E_i is a measure of the potential-well depth of valence electrons in that atom, and is thus a measure of the energy released when the electron tunneling out of the surface falls into that well. By the way, the electrical potential associated with E_i, $V_i = E_i/q_e$, is known as the "ionization potential" of the atom or the "appearance potential" of the ion.

We now turn to electron generation by **gas-phase ionization**, which is the process by which plasmas are ignited and sustained. Once electrons have been emitted from the cathode, they encounter collisions with whatever gas or evaporant is present. If they have been accelerated to above the ionization potential, E_i (see above), by an applied electric field prior to collision, ionization results; that is, for gas atoms A,

$$A + e \rightarrow A^+ + 2e \qquad (8.14)$$

This reaction has generated a new electron, which also gets accelerated in the field, so that a cascade or avalanche multiplication of electrons and ions occurs, as shown in Fig. 8.2. This is the mechanism of **gas breakdown** in an electric field. Metallic-element vapors typically have E_i values of 6 to 8 eV, while other vapors have E_i values of 10 to 20 eV. (By the way, the E_i values for electron-impact ionization are very close to those for ionization by photons or other ions.) Above the E_i threshold, the ionization cross section, σ_i, depends on electron kinetic energy in the manner shown in Fig. 8.3 for various gas mole-

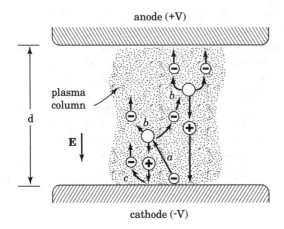

Figure 8.2 Ionization and current flow in a plasma sustained between electrodes: (a) initiating electron, (b) electron-impact ionization per Eq. (8.14), and (c) ion bombardment-induced secondary electron emission. White circles = gas atoms or molecules.

cules [5]. Note that the maximum σ_i at around 100 eV is about the same as the molecular cross section of small molecules: $\approx 4 \times 10^{-16}$ cm². At higher impact energy, σ_i *decreases* because the collision time becomes too short. Electron-impact ionization is also the operative principle of the ionization vacuum gauge (ion gauge), which was discussed in Sec. 3.5.

Gas breakdown across electrode gaps of d ~ 3 cm occurs most easily at a pressure, p, of ~30 Pa. At too low a p or d, or equivalently, too low a pd product, an electron leaving the cathode is likely to reach the anode of Fig. 8.2 before colliding with a gas molecules. At too *high* a p, on the other hand, there occur too many collisions of insufficient *energy* to result in ionization, so that breakdown becomes "quenched." The latter situation develops because with each collision, the electron loses energy by non-ionizing energy-exchange processes to be discussed further in Sec. 9.1. Ionization occurs only if the electrons are re-accelerated enough between collisions so that some eventually reach the ionizing energy, E_i. Now, the kinetic energy gained between collisions is roughly $q_e E l_e$ (eV), where $\mathbf{E}$ is the field strength and l_e is the mean free path given by Eq. (2.22). [Eq. (2.22) is approximate, because the collision cross section, σ_m, varies with the type of collision and with electron energy.] As p increases, l_e decreases, so less energy is gained. Also, for a potential drop of V between electrodes, $\mathbf{E} = V/d$, at least before the plasma ignites, so the energy gained between collisions scales inversely with d. Thus, quenching increases with the pd product for

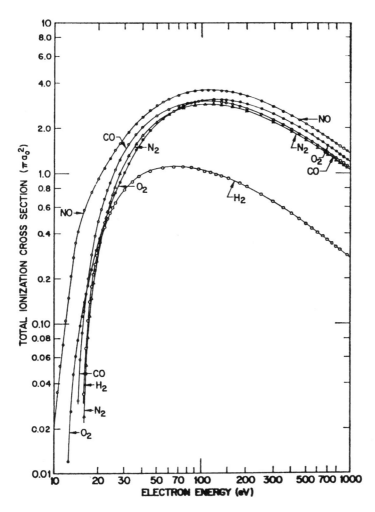

Figure 8.3 Ionization cross sections of various small molecules, normalized to the projected area of the H atom, $\pi a_o^2 = 0.88 \times 10^{-16}$ cm^2, based on its Bohr radius, a_o. (Source: Reprinted from Ref. 5 by permission.)

fixed V between electrodes. Because of the above two processes that inhibit the ionizing-collision cascade at too low and too high a value of pd, the number of ionizing collisions is a maximum at some intermediate value, and that is where the V required for breakdown is a minimum, as shown in the "Paschen curve" of Fig. 8.4.

At intermediate p, plasmas sustain themselves by gas-phase ionization and cold-cathode emission. Otherwise, they need to be supported by thermionic emission, as we will see in the next two sections. The minimum self-sustaining p can be reduced by ×30 or so by using mag-

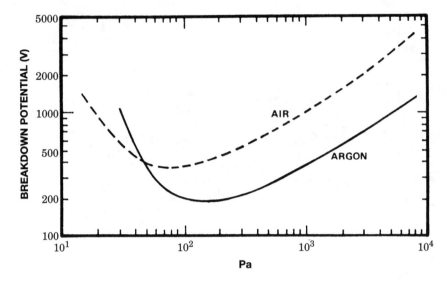

Figure 8.4 Paschen curves for gas breakdown voltage between planar electrodes at spacing d, 293 K. (Source: Reprinted from Ref. 6 by permission. Pressure units changed to Pa.)

netic fields and special techniques of rf electrical power coupling as will be discussed in Chap. 9. For now, the above concepts of electron generation are sufficient to allow us to discuss the narrow-beam vaporization methods of Table 8.1.

8.2 Electron Beams

Here, we describe the use of intense beams of high-energy electrons to evaporate source materials. Electrons thermionically emitted from a hot filament and accelerated into the source material can generate enough energy density to evaporate any material. In a typical case involving 1 A of emission accelerated through a 10-kV voltage drop, 10 kW is delivered upon impact. To avoid dissolving the filament in arriving evaporant, the filament is located out of sight of the evaporant as shown in the electron gun of Fig. 8.5, and the electron beam is pulled around to the surface by a magnetic field, **B**, shown pointing into the figure. The combined force, **F**, on an electron in electric (**E**) and magnetic fields is known as the Lorentz force and is given by

$$\mathbf{F} = \mathbf{F}_E + \mathbf{F}_B = q_e \mathbf{E} + q_e \mathbf{v} \times \mathbf{B} \qquad (8.15)$$

Using SI units, F is in newtons/m^2, q_e in coulombs, **E** in V/m, **B** in webers/m^2 = tesla (1 tesla = 10^4 gauss), and the electron velocity **v** is

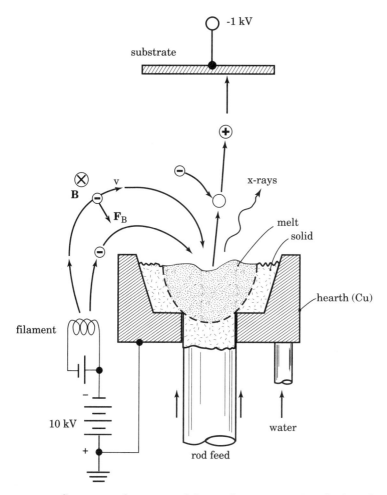

Figure 8.5 Geometry and processes of electron-beam evaporation. Circle with "x" denotes **B** field pointing *into* the paper.

in m/s. The cross-product vector, $\mathbf{F_B}$, is oriented perpendicular to both **v** and **B** as shown in Fig. 8.5, and its sign follows the right-hand-screw rule; that is, rotating **v** toward **B** drives a conventional right-hand-threaded screw in the direction of the vector. Since q_e is negative, $\mathbf{F_B}$ is in the direction shown in Fig. 8.5. The first force term in Eq. (8.15) was encountered in Eq. (6.6); it accelerates the electrons away from the filament or cathode in Fig. 8.5. The **v** so acquired causes the electrons to be deflected sideways as they cross the magnetic field lines, in accordance with the second force term. This second force is balanced by the centrifugal force of the electrons curving at radius r; that is,

$$|q_e \mathbf{v} \times \mathbf{B}| = m|\mathbf{v}|^2/r \tag{8.16}$$

Thus, the "cyclotron" or "Larmor" radius of an electron orbiting in a magnetic field is

$$r_c = m_e v_\perp / q_e |\mathbf{B}| \tag{8.17}$$

where $v_\perp$ is the component of $\mathbf{v}$ that is perpendicular to $\mathbf{B}$. For example, using SI units and numbers typical of electron-beam evaporation sources, an electron after accelerating through a 10 kV potential drop has $\mathbf{v} = 5.9 \times 10^7$ m/s by Eq. (8.7). In a field of 10^{-2} webers/m² (100 gauss), this gives $r_c = 0.034$ m = 3.4 cm. The behavior of electrons in magnetic fields will also be important in understanding magnetically confined plasmas (Chap. 9). In the electron-beam source, $\mathbf{B}$ is generated either by a permanent magnet or a solenoid coil mounted below the source. Modulation of the solenoid field can be used to steer the electron beam.

It is important to aim the beam at the center of the source material and to avoid hitting the material's container (the "hearth"). However, deposition uniformity is improved by scanning the beam around on the surface in a raster pattern using the above solenoid coil along with a second lateral steering coil. Without rastering, the evaporating material tends to be angularly distributed in a lobe-shaped pattern approximated by $\cos^n\theta$ with n = 2 to 4, rather than following a cosine distribution (n = 1; see Fig. 4.12). The lobe is apparently due to collimation arising from the depression created in the surface by rapid evaporation from the point of beam impact. This collimation occurs even with molten metals, whose viscosity is so high that flow of the melt into the depression may not be able to keep up with the evaporation rate. Even with rastering, evaporation rate and flux distribution fluctuate much more than in joule-heated crucibles. Therefore, when rate control is important, vapor flux monitoring with feedback control to the filament power should be used (see Sec. 4.7).

The source material is usually contained by a Cu "hearth" which is water-cooled to prevent its outgassing or melting. Cooling also prevents the hearth from alloying with molten source materials, because the material immediately adjacent to a cooled hearth cannot melt. Then, the source material is essentially being evaporated from a crucible formed by its own solid phase, and this makes electron-beam evaporation particularly useful for materials like Si, which alloys with everything it touches when it is molten. When there does exist a crucible material that will not contaminate the melt, it can be used as a hearth liner to reduce heat-sinking and thereby increase evaporation efficiency. Even electrically insulating source and crucible materials can be used, because usually the secondary-electron yield from beam impact is greater than unity at the high energy used, so that negative charging of the surface is avoided. The localized melting also allows the continuous feeding of a solid bar of source material up through the bottom of the hearth, so that an enormous quantity of material can be evaporated before reloading is needed, such as a 5-cm-diameter by 20-cm-long bar. The bar feed rate can be feedback-controlled based on the monitored evaporation rate, so as to keep the melt surface at a constant height. In the case of alloys, rod-feeding has the further advantage that evaporating composition has to be equal to rod composition in steady state. The transient approach to steady-state melt composition was analyzed in Sec. 4.3. The time to reach steady state can be greatly reduced by starting with a hearth charge which matches the steady-state melt composition rather than one which is the same as the rod composition.

All evaporation done with narrow, intense energy beams is subject to the "macroparticle-spitting" problem. This problem occurs even with joule-heated crucibles, as discussed in Sec. 4.5.1, but it is much more severe with energy beams. One mechanism of macroparticle ejection common to all vaporization methods and dominant for electron beams is the sudden evaporation of a nodule of a particular contaminant whose vapor pressure is much higher than that of the source material. This sudden pressure burst can knock from the surface a macroparticle of solid or liquid material typically 0.1 to 1 μm in diameter. The particle is likely to land on the depositing film, where it is most unwelcome. The velocities of such particles are high enough that they are negligibly affected by gravity over the transport distance. However, many of the particles are charged in the case of Si and probably in general, and these reportedly [7] can be deflected effectively using a transverse **E** field applied from an electrode placed off to the side at ±4 to 6 kV.

The volatile contaminant nodule which ejects the macroparticle may be an inclusion in the solid source material, a gas pocket trapped within "sintered" material (that is, hot-pressed powder), or a "slag" accumulating on the surface of molten metal by reaction of background gases with the metal or by precipitation of bulk contaminants upon melting of the metal. The macroparticle problem is minimized by using very pure source material, preferably manufactured by melting rather than by sintering. Vacuum-melted charges shaped to fit most commercial electron-beam hearths are available from materials suppliers. In the cathodic-arc-evaporation and laser-ablation methods to be discussed in the following sections, there arise other macroparticle-ejection mechanisms that are much less easily counteracted. The problem does not occur in sputtering because the energy density at the surface is much lower, although at background pressures above 20 Pa or so, particles can form in the *gas* phase (more on this in Sec. 9.6.2).

Although electron-beam evaporation is a thermal process, so that the vapor atoms leave the surface with only thermal energy of ~0.2 eV, several kinds of nonthermal energy still arrive at the film surface. One is **x-rays** generated by electron-beam impact on the source material, which is, after all, the operating principle of the x-ray tube. The x-rays generated by a 10-keV electron beam are not "hard" enough (energetic enough) to penetrate the vacuum-chamber wall or window to become a hazard to the operator, but they can be a hazard to the product. In particular, dielectric materials for electronic applications can develop undesirable charge-trapping defects or embedded charge from x-ray irradiation, and it is not always possible to anneal out this damage afterwards. A dielectric material underlying the film being deposited can also be damaged.

The second form of nonthermal energy accompanying electron-beam evaporation is a beneficial one. **Positive ions** are generated above the source by impact of the incoming beam upon the outgoing vapor, as shown in Fig. 8.5. The fraction of vapor ionized varies with beam density and vapor composition and is of the order of 0.01 to 0.1. These ions cannot get drawn into the cathode filament despite its negative bias, because the cyclotron radius of a particle increases as $\sqrt{m}$ in accordance with Eqs. (8.7) and (8.17) and is therefore hundreds of times larger for ions than for electrons. On the other hand, by biasing the substrate negatively as shown in Fig. 8.5, the ions *can* be accelerated into the depositing film. Because the resulting energy flux can be very large, it can have a profound effect on film structure even though the fractional ionization of depositing vapor is small—a fact which applies to all of the deposition techniques that involve ions. The effects of ion bombardment on film structure will be discussed in Sec. 8.5. This acceleration technique is known as "ion plating" by analogy to electroplating from liquid solution; but the term is misleading, because the ions are just modifying the deposit rather than constituting the dominant depositing species as they do in electroplating. Among the vapor-phase thin-film processes, ions are the dominant depositing species only with cathodic arcs and sometimes with high-intensity glow discharges.

Sometimes, a gaseous source species is introduced into the evaporation chamber for the purpose of forming a compound film by reaction with the evaporating species. The gaseous species will also become partially dissociated and ionized by collisions both with the electron beam and with the vapor ions if they are being accelerated toward the substrate. The ionized gas forms a plasma between the source and the substrate. This **gas activation** assists the compound-film-formation reaction and is therefore a form of "activated reactive evaporation," which will be discussed further in Sec. 9.3.2. Similar gas activation occurs in any process in which a plasma is present, and techniques other than electron-beam evaporation actually allow much higher reactant-gas pressure, because the problem of electron-beam scattering is not present.

It is important to keep in mind that in *any* of the energy-enhanced processes, the background gases in the vacuum chamber also become activated and thus become incorporated more easily into the film as contaminants. This means that the vacuum-cleanliness practices of Sec. 3.4 are even more important to observe in energy-enhanced processes than in thermal processes whenever incorporation of the background gases would degrade the film.

8.3 Arc Plasmas

The remaining energy-enhanced techniques to be discussed in this book all involve a plasma of some sort. The general concept of a plasma was introduced at the beginning of this chapter. To put the arc type of plasma into perspective, refer to Fig. 8.6, which distinguishes dc (direct-current)-discharge plasmas according to several factors; namely, the voltage drop between anode and cathode, the total current being passed between electrodes, and the current density at the cathode. It is important to know where one is operating on this map in order to control a plasma-deposition process. In dc plasmas, the current is controlled to the desired position on the map by using a current-limited power supply. The values of voltage and current shown are typical only and will vary considerably with gas composition and pressure, electrode material, and geometry. Starting in the shaded region at the left of the figure, very small amounts of current are passed even before a plasma is ignited, due to field emission and to photoemission induced by stray light. If current and voltage are allowed to increase, electron avalanching eventually occurs in the intervening gas and causes breakdown, as was discussed at Fig. 8.2. Breakdown takes the form of a spark at high pressure ($>10^4$ Pa or so) and takes the form of a much more diffuse "glow" at lower pressure. After breakdown, the plasma has been "ignited," and a continuous flow of electrons and ions between electrodes is sustained by electron-impact ionizing collisions in the gas phase, as shown in Fig. 8.2 and Eq. (8.14). In the self-sustained dc glow discharge, the electrons that initiate the avalanching are generated by secondary-electron emission induced by ion bombardment of the cathode. [Plasmas driven by *rf*

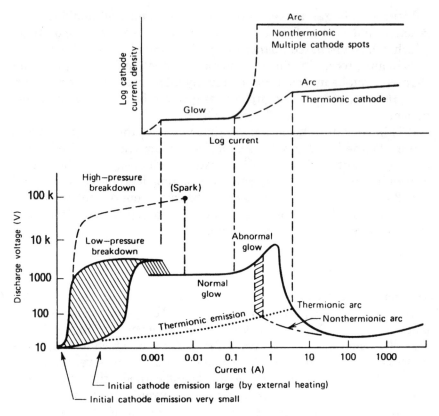

Figure 8.6 Typical current-voltage behavior of dc plasma discharges. [Source: Reprinted from Chap. 1, by J.D. Cobine, in Lafferty (1980) by permission of copyright owner, John Wiley & Sons, Inc., © 1980. Vertical axis shifted downward ×10 to correspond more closely to plasmas typical of thin-film deposition.]

(radio-frequency) power do not require this emission, as we will see in Sec. 9.4.2.2.]

As glow-discharge current is allowed to increase further in Fig. 8.6, the column of current flowing between the electrodes widens so as to keep current density, j (A/cm^2), at the cathode constant. Once the glow has filled the cathode area, the "abnormal glow" region is entered, in which further current increase is accompanied by increases in both j and V. Actually, the "abnormal" glow discharge is quite the norm in thin-film processing, because it can provide uniform j over a large area of source material or depositing-film surface. (RF power is often used instead of dc because it can be passed through insulating materials.) If plasma current is allowed to increase beyond a certain critical value in the abnormal-glow region, the plasma suddenly constricts to a much smaller area, especially at the cathode, so that j increases by *orders of magnitude*. The voltage drop developed across this "arc" plasma is also much lower because of greatly enhanced electron emission from the cathode spot; that is, plasma conductivity near the cathode is greatly increased (more on this below). In fact, enhanced electron emission at one spot on the cathode is what triggers the transition from glow to arc. The enhanced emission can be caused by field emission from a small protrusion or by the "hollow-cathode" effect at a depression (discussed at Fig. 9.6). When current is allowed to exceed 100 A or so, it tends to separate into multiple cathode spots. When electron emission from the cathode is thermionically supported by external heating, the transition to an arc is less abrupt and the arc's j is lower, as shown in Fig. 8.6, but the thermionic arc will not be considered further here. The dotted line in the figure shows what the space-charge-limited thermionic emission current would be in the *absence* of a plasma for a fixed cathode T.

Arc plasmas used to supply vapor for film deposition are nonthermionic in the sense that external heating is not applied. However, cathode j can be as high as 10^8 A/cm^2 for a typical situation of 100 A over an area 10 μm across (Lafferty, 1980; Chap. 4), and this causes intense local heating of the cathode. The power density at the cathode spot is the sum of two products: j times the ion-bombardment energy of tens of eV, and j times the voltage drop *within* the cathode material in the vicinity of the cathode spot. This voltage drop is caused by the spreading resistance from the cathode spot, and it results in joule (I^2R) heating of the cathode material. These power inputs to the cathode spot, illustrated in Fig. 8.7, heat it to thousands of degrees K, causing both thermionic emission and rapid evaporation. Evaporation from the small spot digs a pit whose sharp edges may be sites for field emission. Field emission from a hot cathode is *thermionically enhanced* from the cold value of Eq. (8.12), because at high T there are more electrons above the Fermi level, where the tunneling barrier is narrower (see Fig. 8.1). Thermionic emission and thermionically enhanced field emission are believed to dominate electron emission in arcs (Lafferty, 1980; Chap. 1). Arc plasmas differ from the dc current-flow situation in electron-beam evaporation. There, external heating supported the thermionic cathode emission, and the cathode-anode voltage drop was much higher. Insulating evaporant material could be used as the anode, because usually the secondary-electron yield, Y_e, exceeds unity due to the high voltage. Conversely, dc arcs require conducting electrodes, because the voltage drop is much too low to give $Y_e > 1$.

Arc evaporation is often carried out in vacuum, where there is no gas to break down and ignite the plasma. The plasma is instead "struck" either by sending a high-voltage pulse to the electrodes or by touching the electrodes to each other and then pulling them apart as

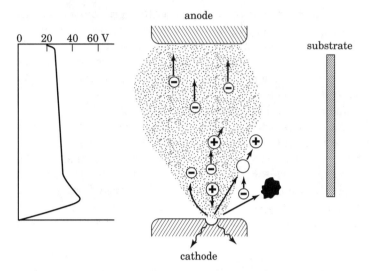

Figure 8.7 Vacuum arc behavior. Left: approximate voltage profile. Right: cathode-spot processes. Circle with plus symbol = bombarding ion, circle with minus symbol = secondary electron, white circle = vaporized cathode atom, wavy arrows = joule heating, and black mass = macroparticle.

is done in arc welding. In either case, cathode heating at a spot of local high power is apparently enough to initiate evaporation. The cathode-material's vapor itself then ionizes and sustains the plasma, which is therefore known as a *cathodic* arc. Heating of the anode also occurs because of the electron current arriving there, but the current is spread over a much larger area, as shown in Fig. 8.7. Nevertheless, if the anode is made small in diameter and is not cooled, substantial evaporation can occur from it as well. If at the same time, the cathode is made of a very refractory material such as graphite, anode vapor can be made to dominate the plasma, and then we have an *anodic* arc [8]. Evaporation for film deposition can be carried out using either a cathodic or an anodic arc. In either case, evaporation from the opposite electrode needs to be suppressed, unless both electrodes are made of the same material. Anode evaporation is effectively suppressed by cooling. Cathode evaporation cannot be suppressed completely even using graphite, so in an anodic-arc evaporation source, the cathode must be shielded from the substrate to avoid film contamination.

Anodic and cathodic arcs differ from each other in several respects as evaporation sources. The anodic arc is similar to the electron-beam source discussed in the previous section, except that fractional ionization of the vapor is higher (~20 percent) [8]. Anodic ions escape the arc with kinetic energies of ~5 eV. In the cathodic arc, fractional ionization is even higher. It increases with cathode-material melting T, and for refractory metals it is near unity [9]. The high degree of vapor ionization is in fact the principal advantage of the cathodic arc, because now *most* of the vapor arriving at the film carries enough energy to modify film structure. For many metals, the vapor ion current escaping the arc is about one-tenth of the total arc current [9]. The kinetic energy of these ions is much higher than that of anodic ions—typically 50 eV, which is much higher than can be accounted for by the 20-V total potential drop across the arc. This high energy has long baffled investigators, but the following reasonable model has been proposed recently. The intense power concentration at the cathode spot produces an extreme T and vapor pressure of cathode material, and the high electron-emission flux turns this into a plasma of very high density— that is, high electron and ion concentration. This plasma expands explosively into the vacuum [10], with the electrons accelerating much more rapidly due to their much lower mass. The resulting electron depletion from just in front of the cathode produces there a positive space charge and a positive potential hump as shown at the left of Fig. 8.7, so that **E** is *reversed* with respect to the externally applied voltage. Calculations show that plasma expansion accelerates ions *away* from the cathode in three ways under these conditions [11]. The dominant mechanism is drag on the ions produced by collisions with the faster electrons, and additional acceleration is produced by **E** and by the ion pressure gradient. The drag mechanism is analogous to the "seeding" of a light gas with a heavy one in a supersonic nozzle, which results in very high energy for the heavy gas (Sec. 4.5.4). Ionized vapor escaping either anodic or cathodic arcs may be further accelerated into the depositing film using negative bias on the substrate as in the case of electron-beam evaporation (Fig. 8.5).

Another fascinating feature of the vacuum arc is the rapid and erratic motion of the cathode spot over the surface. This motion makes the arc very difficult to probe and study. Much work was done long before thin-film applications arose, because of the importance of controlling arcing in high-power switches and lightning arrestors; yet there is still much to learn. It seems reasonable that once one spot on the cathode has become eroded away, the arc has to move to an adjacent spot, but the details of this motion have not been successfully modeled. Controlling cathode-spot motion may be a key to decreasing the problem of "macroparticle" ejection, which is the major disadvantage of the cathodic-arc process and will be discussed below. One way to control spot motion is to steer it with a magnetic field oriented parallel to the cathode surface [12]. A particularly baffling characteristic of cathode spot motion is its "retrograde" motion in a magnetic field; that is, opposite to the $\mathbf{j} \times \mathbf{B}$ direction predicted by Eqs. (8.5) and (8.15). The spot moves at typically 20 m/s, increasing [12] with **j** and **B** and decreasing with increasing background-gas pressure until it reverses direction at above 100 Pa or so (Lafferty, 1980; Chap. 4).

The ejection of cathode-metal **macroparticles** 0.1 to 1 µm in diameter makes cathodic-arc films unsuitable for applications in which low defect density is important. Consequently, the main application of the process is for wear-resistant coatings on tool bits. There, the high ionization fraction results in good adherence and compressive film stress, which improve film performance. If the macroparticles could be eliminated, the high ionization would be useful in many other applications. Macroparticles are not produced in anodic arcs, presumably due to the much lower current density at the anode [8]. But, unfortunately, the ionization fraction is much lower than in the cathodic arc.

One possible cause of macroparticle ejection is subsurface heating to a T higher than the surface T, which could cause surface material to be thrown off by the pressure of subsurface evaporation. Figure 8.7 illustrates the heat-flow situation around the cathode spot. Heat is delivered to the surface by ion bombardment and to the subsurface as joule heating from the current-spreading resistance in the metal, as discussed earlier. The subsurface heat is removed by conduction down into the bulk and up to the surface. Heat is lost at the surface partly from the heat of evaporation of the metal atoms, $\Delta_v H$, and partly from the energy absorbed by thermionic electrons escaping over the work-function (ϕ_w) barrier of Fig. 8.1. This latter "electron cooling" is essentially a heat of evaporation of the electrons. Thermal-radiation heat loss (Eq. 4.19) is much smaller than the above two factors. The solution to this heat balance for Cu is shown in dimensionless form in Fig. 8.8 for a particular value of the product of total arc current, I, and cathode-spot current density, j. The subsurface-to-surface T ratio, T/T_0, is plotted versus radial distance, r, into the cathode from a surface spot of radius r_0 for various values of the dimensionless parameter $\alpha t/r_0^2$. The thermal diffusivity appearing here,

$$\alpha \ (cm^2/s) = K_T/\rho_m c_g \qquad (8.18)$$

also appears in all transient-heat-flow equations and is analogous to the mass diffusivity, D, from Fick's law, Eq. (2.27). Using the values for Cu of thermal conductivity, K_T, mass density, ρ_m, and heat capacity per gram c_g, we have $\alpha = 0.75$ cm^2/s. The appropriate time, t, to use in $\alpha t/r_0^2$ is the dwell time of the cathode spot, which is 0.5 µs for a spot velocity of 20 m/s and for $r_0 = 10$ µm. Then, $\alpha t/r_0^2 = 0.4$, so that subsurface heating is very small. However, r_0 could become much smaller at a moment of strong field emission from a protrusion, and

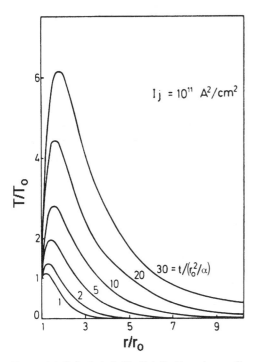

Figure 8.8 Calculated T distribution in a Cu cathodic arc. Symbols are defined in text. [Source: Reprinted from Chap. 7, by G. Ecker, in Lafferty [1980] by permission of copyright owner, John Wiley & Sons, Inc., © 1980.]

then both $\alpha t/r_o^2$ and T/T_o would become much higher. T/T_o is smaller for lower values of the Ij product than the 10^{11} A^2/cm^2 plotted (Lafferty, 1980; Fig. 7.1), but it would be expected to be higher for metals of higher electrical resistivity, because in these metals there would be more joule heating.

Another possible cause of macroparticle ejection is the force, or pressure, exerted on a molten cathode spot by the bombarding ions, which could cause metal to be splashed up from the edges of the pool. For reasonable values of ion flux and energy, this pressure can be many atmospheres (see Exercise 8.5). The observed increase in macroparticle ejection with decreasing melting T of the cathode metal [13] is consistent with this mechanism.

The macroparticles can be "filtered" out of the vapor stream by steering the atomic vapor ions through a curved duct using a crosswise magnetic field. The duct radius should be matched to the ion cyclotron radius found from Eq. (8.17). The pros and cons of various duct configurations have been discussed recently [13, 14]. In all of them, there is some loss of vapor flux and deposition rate. Moreover, merely keeping the film out of the line of sight of the cathode is not sufficient, because the macroparticles seem to be able to reflect from duct surfaces and still reach the film unless baffles are also incorporated into the duct walls to scatter the particles [14].

Arc deposition can also be operated in a **gas ambient**. One use of the gas is to scatter vapor ions as they approach the substrate so that they arrive from a wide range of angles and can thus coat complex-shaped substrates such as tool bits more evenly, at least in *cathodic*-arc deposition where ionization fraction is near unity. The ability to coat oblique or hidden surfaces is called "throwing power" in reference to the same feature of some electroplating processes. The same scattering would occur for *thermal* evaporation into a gas ambient, but there the vapor tends to condense into particles in the gas phase due to collisional cooling, whereas ions are prevented from doing so by their mutual repulsion. Gas-phase condensation was discussed in Sec. 4.5.4.

The second use of a gas ambient is for compound-film formation. The stream of electrons and ions emanating from the arc activates reactive gases enough so that full reaction of the metal vapor to stoichiometric compound films can be achieved. Thus, metal carbides, nitrides, and oxides can be formed [15] using CH_4, N_2, and O_2 at pressures of 1 to 100 Pa. TiN is a particularly useful wear-resistant coating because of its hardness, low coefficient of friction, and chemical inertness. Many other plasma processes also may be used to deposit compounds reactively from metal sources. For example, reactive glow-discharge sputtering is widely used because of its combination of high deposition energy, freedom from macroparticles, and ability to uniformly coat large areas. However, special control techniques must often be employed there to obtain stoichiometric films because of an inherent instability in the degree of cathode-surface "poisoning" by the reactive gas (more on this in Sec. 9.3.3). Arc sources do not exhibit the poisoning effect, because the evaporation *flux* (atoms/$cm^2 \cdot$s) is much higher than in sputtering due to the much higher power density, and this flux is much higher than the arrival flux of reactive gas. This behavior and the high ionization fraction are the two principal advantages of arc sources.

8.4 Pulsed Lasers

The use of a pulsed UV laser beam for vaporizing solid source material constitutes the last of the narrow-beam vaporization techniques of Table 8.1 to be discussed. High-power lasers, such as the Nd:YAG at a wavelength of λ = 1.06 μm and the CO_2 at 10.6 μm, have long been used for cutting and drilling very hard or refractory materials. Because much of the material ablates as macroparticles rather than vaporizing as atoms or molecules, "laser ablation" has received much less attention for thin-film deposition than have other vaporization techniques. However, recent work has demonstrated that very good "high-T_c" superconducting ceramic thin films can be deposited by pulsed UV-laser vaporization in an O_2 ambient (Hubler, 1992), and this has generated considerable interest in developing the technique further for film deposition. A typical high-T_c material is "1,2,3-YBCO," a mixed oxide of the approximate composition $(Y_2O_3)_{0.5}(BaO)_2(CuO)_3$. Obtaining a high superconducting-transition temperature, T_c, requires close composition control, and a key advantage of the pulsed UV beam is that it can achieve congruent evaporation of these complex materials. This advantage is shared by sputtering, and when a glow discharge is used for sputtering rather than an ion beam, one can also operate in the 50 Pa or so of O_2 ambient which is required to prevent substoichiometric O content in the YBCO. However, fast O^- ions ejected from the sputtering cathode damage the film's crystallography, and it is difficult to eliminate these ions (more on this in Sec. 9.3.3). On the other hand, sputtered films are free of macroparticles. With effort, reasonably good superconducting films have been deposited by both techniques. This is a classic example of the trade-offs involved in choosing a deposition technique, and it emphasizes the need to carefully examine all of the advantages and drawbacks of each one.

Congruent evaporation by a pulsed UV laser results from the deposition of an intense energy pulse in a shallow depth of the source material and a consequent explosive evaporation of a thin layer before it has time to disproportionate. The depth involved is either the optical absorption depth, $1/\alpha_r$, or the thermal diffusion length, Λ, whichever is larger. Recall that α_r is the optical absorption coefficient from Eq. (4.49). Taking a typical example, $\alpha_r = 10^5$/cm and $1/\alpha_r = 100$ nm for YBCO at the KrF excimer laser λ of 248 nm. For Λ, we may define a thermal diffusion length by analogy to the mass diffusion length of Eq. (5.25):

$$\Lambda = 2\sqrt{\alpha t} \qquad (8.19)$$

where α is the thermal diffusivity from Eq. (8.18). Λ is a measure of how deep into the material the depositing energy penetrates during the length of the pulse, t. Taking handbook data for BaO at room T, we find α = (4 W/cm·K)/(6.5 g/cm^3)(0.31 J/g·K) = 2.0 cm^2/s. Then, for a typical pulse length of t = 20 ns, we find Λ = 4000 nm, which is much larger than $1/\alpha_r$ for YBCO, so that thermal diffusion determines the energy deposition depth for this case. For poorer thermal conductors or more transparent materials, $1/\alpha_r$ can determine depth. To achieve the goal of shallow energy deposition, one wants both a fast pulse and a high α_r. For nonmetals, high α_r requires λ to be below the fundamental absorption edge of the material, as discussed following Eq. (5.58). Also for nonmetals, α_r increases well above its initial value when successive pulses are delivered to the same area of the surface. This is because the pulse imparts a "thermal shock" to the material, as discussed later in connection with macroparticle ejection. The shock damages the near-surface region crystallographically [16, 17] and thereby increases α_r. This phenomenon causes evaporation rate to increase with time during initial evaporation from a fresh target.

For the deposited energy to result in evaporation, there must be enough of it to first heat the material to a T of high vapor pressure, p_v, and this heat may also involve the heat of fusion ($\Delta_c H$) consumed in melting the material. Then, the heat of evaporation, $\Delta_v H$, must be supplied. Since the pulse repetition rate is very low, say 10 pulses/s, the material may be assumed to have completely cooled between pulses. A typical excimer pulse delivers 0.2 J into a 0.1-cm^2 area, corresponding to a "fluence" of 2 J/cm^2. Note that this represents a time-average power of only 2 W (J/s), 10^3 lower than in evaporation with the other Table 8.1 beams. Evaporation is possible with such low power only because the *instantaneous* power density is very high:

$$(2 \text{ J/cm}^2)/(20 \text{ ns}) = 10^8 \text{ W/cm}^2$$

which approaches that of the cathodic arc. Now, heating a BaO volume of 0.1 cm$^2 \times \Lambda$ to, say, its boiling point (p_v = 1 atm) of 2000 K requires 0.11 J, leaving 0.09 J for supplying $\Delta_c H$ and $\Delta_v H$. That is, 0.11 J is the *threshold* pulse energy for laser vaporization in this example. Since for BaO, $\Delta_c H + \Delta_v H$ = 437 kJ/mol = 1.9×10^4 J/cm^3, a thickness of 480 nm can be evaporated with this remaining 0.09 J. In practice, an evaporated depth of tens of nm is more typical [17], because much of the beam energy is either reflected from the surface or absorbed by the vapor cloud which forms over it.

Figure 8.9 shows the geometry for pulsed laser evaporation. The beam enters the vacuum chamber through a quartz window and is directed at the target of source material at an oblique angle so that the substrate can be placed directly facing the target surface. This is done because most of the evaporant is directed in a narrow lobe oriented closely perpendicular to the target, for reasons to be discussed below. Usually, the beam is scanned in a raster pattern so that the target erodes evenly. The delivery of vaporization energy in fast pulses allows two pulse-synchronized devices to be incorporated. One is a spin-

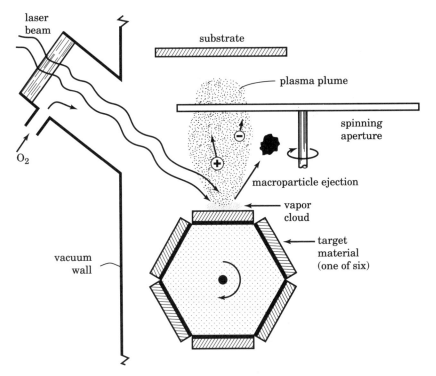

Figure 8.9 Geometry and processes of pulsed-laser vaporization.

ning polygon having a different source material on each face so that alternating-layer structures or multi-element composition control can be obtained [18]. Here, the switching time between materials is limited only by the laser pulse rate, so it can be much shorter than switching done with mechanical shutters in MBE or sputtering, or with gas valves in CVD. For applications such as superlattices, however, the close control of deposit thickness which is required can be difficult to achieve because of fluctuation in pulse energy, drifting target-surface condition, and the nonuniform deposit obtained from the narrow lobe of evaporant. The second pulse-synchronized device is a spinning disc containing an aperture synchronized to the pulses, to filter out macroparticles ejected from the target. Since the macroparticles travel much more slowly than do the vapor atoms, they can be selectively blocked by the disc. However, the hardware and geometry are awkward and not completely effective, so it would be better to avoid generating the macroparticles in the first place. Another practical problem in laser vaporization is deposition of evaporant on the beam window. When one is using a gas ambient at a pressure up in the fluid-flow regime [Kn << 1 in Eq. (2.25)], such as 100 Pa, admitting the gas at the window as shown in Fig. 8.9 will sweep approaching vapor away from it as described by Eq. (3.13) for the oil backstreaming problem.

Two evaporant components have been observed [19]. One has the $\cos\theta$ distribution of Fig. 4.12, has thermal energy, and tends to be nonstoichiometric, as in conventional thermal evaporation. The other has a sharply lobed $\cos^n\theta$ distribution, with n variously reported in the range of 4 to 14. The lobed component is closely stoichiometric and has a velocity of $\sim 10^4$ m/s [20], which is way above thermal velocity and corresponds to a particle energy of ~ 40 eV by Eq. (8.6) using typical atomic masses. This kinetic energy is a key feature of pulsed laser evaporation. It can be transported in vacuum to the film to modify its structure, or it can be used to activate reaction with a gaseous ambient such as O_2 by collisional dissociation of the gas molecules. When a gas ambient is used, the thermal vapor component can condense in the gas phase into unwanted particles (see Sec. 4.5.4), whereas the lobed component seems not to do so because of its high energy. The fraction of undesired thermal component becomes small when pulse energy is well above the threshold level discussed earlier, so that there is excess energy available to accelerate the vapor.

The mechanism of vapor acceleration has been likened to supersonic expansion from a high-pressure nozzle [20, 21], which was discussed in Sec. 4.5.4. Explosive evaporation of a layer of material by the fast pulse produces a volume of hot, high-pressure vapor calculated to be about 50 μm thick [21] and covering the much wider irradiated area of ~ 0.1 cm^2. Absorption of incoming light by the vapor raises the T and pressure still further. Most of the expansion of this vapor volume will be in the direction of steepest pressure gradient and largest cross-sectional area—that is, perpendicular to the surface—and this produces the sharp lobe. Vapor collisions during expansion convert thermal energy into directed kinetic energy as in the supersonic nozzle, resulting in the narrowed and upward-shifted velocity distribution characteristic of supersonic beams [20].

A plasma forms in the vapor plume both from thermionic electron emission from the hot surface and from photon absorption by the vapor [21]. The 5-eV energy of a 248-nm photon [from Eq. (6.5)] is not enough to ionize most vapor atoms. However, at high light intensity, additional photons can be absorbed by the same atom before the electronically excited state created by the first one has time to relax, and thus the ionization energy can be acquired by multiphoton absorption [22]. Indeed, one of the most dramatic phenomena exhibited by high-power lasers is electrical breakdown in air observed as a tiny spark suspended in space at the focal point, accompanied by a sharp "crack" noise. In pulsed laser evaporation, the plasma appears instead as the glowing plume shown in the cover photograph of this book. The gradations in color result from the decreasing level of electronic excitation in the vapor and gas atoms at positions further from the source. This plasma has two benefits. Gaseous reactants become dissociated and ionized, and this assists their reaction with the depositing vapor. Also, the ions can be accelerated into the film by application of negative bias to the substrate as in the other energy-beam techniques.

Macroparticle ejection is a serious problem in pulsed laser vaporization. It is reduced by using pure and nonporous materials as discussed in Sec. 8.2; but this does not eliminate the problem, so other ejection mechanisms must be operating. One possibility is excess subsurface heating, which we discussed also for the cathodic arc. Generation of heat throughout the light-absorption depth of $1/\alpha_r$, coupled with surface evaporative cooling, can lead to a ratio of subsurface T to surface T, $\theta = T/T_o$, which is above unity, as shown in Fig. 8.10. There, θ is plotted versus dimensionless depth, $S = \alpha_r z/B$, for various values of the parameter $B = \alpha \cdot \alpha_r \Delta_v H/V_m I$, where z is linear depth, α is the thermal diffusivity from Eq. (8.18), $\Delta_v H/V_m$ is the heat of evaporation per cm^3 (J/mol ÷ cm^3/mol), and I is the instantaneous power density, typically 10^8 W/cm^2. The plot is for $1/(1+\lambda) = 0.7$, where $\lambda = c_p T_o/\Delta_v H$. The quantity $1/(1+\lambda)$ varies little among materials, being 0.67 for Al_2O_3 and 0.82 for Cu and BaO. Now, the Fig. 8.10 solution is a steady-state one involving continuous-wave (CW) laser energy input and continuous evaporation and retreat of the surface, z being measured from the retreating surface. Since a 20-ns pulse cannot be considered steady state, Fig. 8.10 can give only a qualitative indication of T behavior during the pulse. Using the physical properties of BaO at room T and using $\alpha_r = 10^5$/cm for YBCO as before, we find B = 37, which would re-

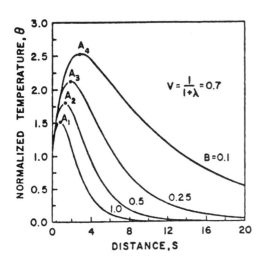

Figure 8.10 Steady-state T vs. depth in a material being laser-vaporized. Parameters are defined in text. (Source: Reprinted from Ref. 23 by permission.)

sult in a maximum T/T_0 close to unity. However, several factors could lead to substantially lower B and increased likelihood of macroparticle ejection by subsurface vapor pressure; these include lower thermal conductivity and diffusivity (α), lower optical-absorption coefficient (α_r), and higher pulse energy (I).

Another possible ejection mechanism is "thermal shock." Heating of the near-surface region causes it to want to expand in accordance with its thermal expansion coefficient, α_T (Sec. 5.6.1), and this leads to a compressive-stress transient that propagates as a shock wave through the material. This wave reflects off the surface because of the large acoustic-impedance mismatch to the vapor phase, just as light reflects off a boundary of refractive-index difference. The returning wave causes a *tensile* stress transient in the material, as shown in Fig. 8.11 for stress given in bars (= 10^5 Pa). This particular stress transient was calculated and measured for Corning CS756 glass receiving a 28-ns pulse of 0.08-J/cm^2 fluence from a ruby laser at λ = 694 nm where α_r = 65/cm for this glass. This "thermoelastic" stress is predicted [24] to scale linearly with fluence and α_r, as one would expect, so extrapolating from Fig. 8.11 to the previously discussed case of YBCO receiving 2 J/cm^2 with $\alpha_r = 10^5$/cm, we predict a peak tensile stress of 15 GPa. This is well above typical tensile strengths of 1 GPa (Sec. 5.6.1), so this appears to be a likely mechanism for macroparticle ejection.

8.5 Ion Bombardment

The impingement of energetic ions or atoms upon a solid surface produces a wide variety of effects, and there has been a vast amount of experimental and theoretical work on the subject. Many of these effects are beneficial to thin-film deposition, and those that are not can usually be avoided by controlling ion composition and keeping

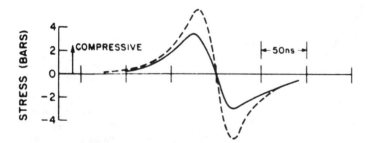

Figure 8.11 Measured (solid line) and calculated (dashed line) stress waveforms in glass resulting from a laser pulse, for conditions described in the text. Sign of stress here is opposite from conventional (see Fig. 5.30). (Source: Reprinted from Ref. 24 by permission.)

energy low. Thus, ion bombardment is one of the most important thin-film process parameters, and we will discuss it here in considerable detail. Energetic ions and atoms have similar effects on the solid, and reference to ion bombardment in this text is meant to include energetic atom or molecule bombardment. By the way, since ions can be focused into submicron-diameter beams, metal-ion beams can be used directly to deposit thin films in fine patterns. However, this is a very specialized technique and will not be discussed further here.

The effects of ion bombardment on a material differ substantially from those of electron and photon bombardment because the ion mass is of the same order as the mass of the atoms in the bombarded solid. Consider a head-on, binary, elastic collision between an impinging particle of mass m_i and energy E_i and a target particle of mass m_t which is initially at rest. The amount of kinetic energy transferred to the target particle is called the "recoil" energy and is traditionally labeled T_m for the head-on case. Conservation of energy and momentum require that

$$T_m = \frac{4 m_i m_t}{(m_i + m_t)^2} E_i = \gamma_m E_i \qquad (8.20)$$

If the impinging particle is a 10-keV electron ($m_i = 9.1 \times 10^{-31}$ kg) and the target is a Ni atom [$m_t = 59$ u (atomic mass units) $\times$ 1.67×10^{-27} kg/u], then $T_m = 0.37$ eV, which is enough to heat up the Ni atom but much less than enough to dislodge it from its lattice site. T_m is even lower for photons. At the other extreme, if $m_i = m_t$, then *all* of the energy is transferred to the target, as billiards players know. If the masses are within $\times 2$ of each other, then $\gamma_m > 0.9$.

This efficient energy transfer to individual atoms is the key feature of ion bombardment. It gives ions the ability to *move atoms around*, and all of the many effects of ion bombardment on film structure and properties arise from this one fact. When $m_i \sim m_t$, an E_i of only a few eV is enough to move a target atom from one surface site to another or to dissociate a molecular impinging ion into reactive fragments. An E_i of only a few tens of eV is enough to displace a surface atom into the bulk of the solid or to dislodge a near-surface atom into the vapor phase. The latter process is known as sputtering, which is widely used for generating source vapor by bombardment of a target at a keV or so using ion beams or glow-discharge plasmas. The sputtered particles themselves have kinetic energies of several eV, and this energy strongly affects film structure. High kinetic energy of the depositing material also occurs in many other deposition processes (Secs. 4.5.4 and 8.2–8.4), and this situation is generally known as "energetic condensation." When the depositing material comes instead from a low-energy thermal source, supplementary ion bombardment of the depositing film with a chemically inert species such as Ar^+ can be used instead to modify film structure.

In the following subsections, we consider the four categories of effects on the solid that arise from surface bombardment by ions or atoms in the energy range of a few eV to few keV; namely, surface effects, ion implantation, bulk atomic displacement, and sputtering. Between a few eV and a few tens of eV, there is an important energy *window* within which the impinging particles have enough energy to cause rearrangement of the surface but not of the bulk. Bombardment within this window is especially useful in epitaxial growth, because it can modify growth behavior without causing subsurface damage. At higher energy, and at lower energy, too, for small particles like H_2^+, the ions can penetrate the surface and become implanted, transferring energy to atoms along their path until they come to rest. The atoms receiving this energy can become displaced and mixed into the lattice, creating a variety of effects. Finally, some of the displaced atoms escape from the surface as sputtered particles.

Ion beams are usually generated in plasmas. Some plasmas have been discussed already, and glow-discharge plasmas will be discussed in Chap. 9. The type of plasma and its operating conditions are critical to the control of ion bombardment energy and flux.

8.5.1 Surface effects

8.5.1.1 Reflection. Before considering the effects of bombardment on the surface atoms, we will discuss reflection of ions from the surface back into the vapor phase, which occurs with finite probability even at high impinging energy. By Eq. (8.20), the ion retains energy $(1 - T_m)$ after a head-on collision with a surface atom. When, in addition, we have $m_i < m_t$, conservation of energy and momentum requires reversal of the direction of ion travel, so that the ion returns to the vapor phase so long as it retains enough energy to escape from the surface potential well of Fig. 5.2. In a somewhat less direct collision, the ion reflects at an oblique angle and retains somewhat more energy, as shown along with other possible ion fates in Fig. 8.12. For still less direct collisions, the ion enters the solid if it has enough energy and/or is small enough to force its way between neighboring surface atoms. If the ion's direction is aligned with a crystallographic direction along which there are channels devoid of atoms, the ion penetrates much deeper than when it is not so aligned or when the solid is amorphous. Ion channeling directions are seen clearly by eye upon examining stick-

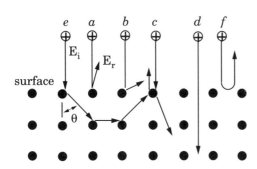

Figure 8.12 Ion fates upon impingement on a surface: (a) reflection in a head-on collision, (b) reflection in an oblique collision, (c) implantation after a grazing collision, (d) implantation with channeling, (e) reflection by a grazing collision sequence, and (f) reflection by repulsion by more than one surface atom.

and-ball crystal models. The efficiency of channeling is a useful measure of the crystallographic perfection of an epitaxial film (Sec. 10.1.4). The deeper an ion penetrates, the more likely it is that it will remain implanted rather than escaping. The binary collision dynamics of implantation are discussed in more detail in Sec. 8.5.2.2.

The maximum reflected energy, $E_r = 1 - T_m$, in a binary collision with $m_i < m_t$, can be calculated from Eq. (8.20). As an example, we consider Ar bombardment. Ar^+ is widely used for bombardment in thin-film technology because it is inert, inexpensive even in high purity, and high enough in mass ($m_i = 40$ u = 40 atomic mass units) for efficient energy transfer. Its E_r versus m_t is shown as the solid line in Fig. 8.13 for $E_i = 1$ keV. E_r increases with m_t, and it goes to zero for the *single* binary collision when $m_i \leq m_t$. However, reflection from solids is still possible when $m_i \leq m_t$, by way of multiple grazing collisions as shown in Fig. 8.12e. As an example to illustrate the mechanism, consider n sequential collisions of equal deflection angle, θ, all in the same plane and between particles of equal mass, $m = m_i = m_t$. It can be shown for this case (see Exercise 8.9) that

$$E_r = E_i \cos^{2n}(\pi/n) \qquad (8.21)$$

The cosine term approaches unity as n →∞, so that *all* of the energy, in principle, can be retained upon reflection by this mechanism. Of course, the probability of such a lineup of grazing collisions is much lower than that of a single direct hit. This reflection mechanism is analogous to the reflection of the ball from the curved basket in the game of jai-alai.

A more accurate picture of ion-solid interactions requires computer calculation. Statistics determine the probabilities of the various ion fates illustrated in Fig. 8.12 as well as the final locations of both the implanted ions and any solid atoms displaced along the way. This problem is therefore amenable to "Monte Carlo" calculation, in which

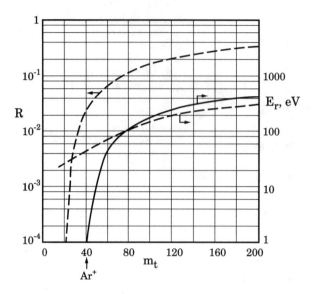

Figure 8.13 Reflected fraction (R) and energy (E_r) of Ar for Ar^+ impinging at 1 keV onto a target of atomic mass m_t. Dashed line = TRIM-code calculations [25] of R and average E_r. Solid line = E_r for head-on binary collision [Eq. (8.20)].

the trajectories of a large number of ions are followed after impact at surface positions chosen by random number. The exact shape of the ion-atom repulsive potential is important here, as we will see in the next section. The most widely used code is TRIM (TRansport of Ions in Matter; Biersack, 1987), which assumes an amorphous solid to simplify the calculations. The predictions of this code have been shown to agree well with experimental data on implantation and sputtering. It has also been used to calculate the reflected fraction, R, and energy, E_r, of ions impinging perpendicularly on the surface [25]. Results for 1-keV Ar^+ are plotted in Fig. 8.13 versus the mass of the target atoms, m_t. For $m_t > m_{Ar}$, R is quite high, and E_r is slightly lower than that predicted by Eq. (8.20) because that equation considered only head-on collisions. For $m_t < m_{Ar}$, E_r drops much less than Eq. (8.20) predicts, because of the contribution from multiple grazing collisions exemplified by Eq. (8.21). For $m_t < m_{Ar}$, R drops precipitously because of the low probability of reflection by multiple collisions.

Energetic ion reflection has important consequences when sputtering of a target is used to supply film material, because energetic ions reflected from the target bombard the depositing film, damaging it and becoming implanted in it. Ions become neutralized upon reflection, by extraction of an electron out of the surface, as was discussed in connection with Eq. (8.13), so they cannot be deflected away from the film by electric fields. When these energetic neutrals cause a damage problem, R can be kept low by choosing a sputtering gas that has $m_i > m_t$. The heavy inert gases Kr and Xe are useful here. They are very expensive, but their consumption can be minimized by using a pump that removes only the chemically active background gases, to keep the sputtering atmosphere clean. For example, a continuously depositing film of sublimated or sputtered [26] Ti pumps selectively in this way.

Further information on reflected energy has been obtained by calorimetric measurements of the fractional energy deposited into the solid [27]. These results are shown in Fig. 8.14 for noble-gas atoms impinging on Pt(111) at up to 15 eV and for their ions impinging on polycrystalline Au at above 15 eV. (Pt and Au differ by <1 percent in atomic mass.) The smooth merging of the data at 15 eV illustrates the equivalence of ion and atom bombardment with regard to their energy-deposition behavior in solids. The solid dots have been added at the fractions corresponding to γ_m in Eq. (8.20), and this tells us something

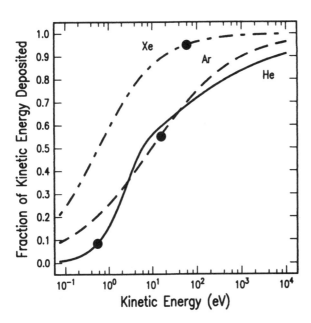

Figure 8.14 Calorimetric measurements of fractional kinetic-energy deposition from various noble-gas atoms impinging on Pt(111) at up to 15 eV, and from their ions impinging on polycrystalline Au at above 15 eV. Black dots have been added at fractions corresponding to γ_m from Eq. (8.20). (Source: Reprinted from Ref. 27 by permission.)

about the nature of the bombardment interaction. For He, energy-deposition fraction begins to exceed γ_m above a very low E_i of 0.6 eV, which means that this is enough impinging energy to cause He *penetration* into Pt. For Xe, on the other hand, energy deposition remains less than γ_m for $E_i < 60$ eV. This implies that the *effective* target mass is larger than the Au atomic mass for low E_i, making γ_m lower in Eq. (8.20). This happens for large, slow ions because their large size enables simultaneous interaction with several surface atoms and also because in a slow collision, surface-atom "recoil" displacement becomes large enough to be restricted by the atom's neighbors. Quantitative analysis of such complex multibody collisions requires "molecular-dynamics" calculations, in which the dynamic interactions of all the atoms in the vicinity of the ion are followed. The enormous computer time involved has limited such studies, but this situation is rapidly improving with computer power. Meanwhile, most work has been done using the much simpler binary-collision model, which is more accurate above the γ_m points in Fig. 8.14.

8.5.1.2 Chemical activation. Ion impact can cause chemical activation of a molecular ion or of molecules adsorbed on the surface. First of all, the ion is generally neutralized just before impact by an electron attracted from the surface into the ion's potential well, as described prior to Eq. (8.13). In the case of *molecular* ions, the energy thereby released can cause the molecule to dissociate even if its translational kinetic energy is below the dissociation threshold of a few eV [28]. At higher translational energies, some energy is transferred to the surface upon impact, and some is transferred to the vibrational modes of the molecule's bonds, causing further dissociation [28]. This means that, for molecular ions, the collision is *inelastic*, and Eq. (8.20) is not strictly obeyed. If the dissociated fragments (free radicals) are reactive with other surface species, compounds will form. Thus, molecular-ion bombardment is a useful way of activating a reaction to form a compound film, such as TiN from Ti and N_2^+ or SiO_2 from Si and O_2^+.

We note in passing that if the ion's atoms retain enough translational energy after dissociation, they will also become implanted. The starting translational energies of the atoms to use in calculating implant depth are determined by assuming that the velocities of all the ion's atoms remain unchanged by dissociation. Consequently, the incident translational energy, E_i, of a molecule AB becomes distributed to its dissociated atoms in proportion to mass; that is,

$$E_A = (E_i - E_b)m_A/m_{AB}$$

where E_b is the dissociation energy. In general, atoms of different masses will not have the same implant depth for the same starting velocity, as we will see in Sec. 8.5.2.2.

Bombarding ions can also dissociate *adsorbed* molecules and thereby activate them chemically. Ion-activated reaction of adsorbates with a surface to form volatile products is a key mechanism in the plasma etching of integrated-circuit patterns. Ion activation of adsorbed CVD precursors is an important mechanism in film deposition downstream of high-flux, broad-area ion sources (Sec. 9.6.3). It also allows the deposition of submicron dots or lines of thin film by using a focused ion beam. Such film "writing" is useful for microcircuit and photomask repair and has finer resolution than the laser-induced CVD discussed at the end of Sec. 7.3.3: that is, ~0.1 μm versus ~1 μm. The observation that a single ion can cause the reaction of many adsorbed molecules means that their dissociation occurs not primarily by direct impact, but instead by transfer of ion energy into the substrate and back up to the surface atoms *around* the impact point through the "collision cascade" [29], which will be discussed further in Sec. 8.5.2. Deposition rate of Au was found to correlate well with the amount of energy returned to the surface by the cascade. When the precursor is a metalorganic, however, C content of the film can be tens of percent [29]. One should also keep in mind that in any energy-enhanced film-deposition process, dissociation of adsorbed background impurities such as H_2O and CO also is occurring, so that such processes are more sensitive to background contamination than are thermal deposition processes.

8.5.1.3 Atomic displacement. We now examine the ways in which bombarding ions can move the surface atoms around, as illustrated in Fig. 8.15. Some of the displacements shown also involve the subsurface, and they will be discussed in subsequent sections. For now, we will concentrate on displacements which involve only the surface. These dominate in the energy window between a few eV and a few tens of eV. They still occur at higher energies, but subsurface effects then become more important.

The adsorbed impurity atom a shown in Fig. 8.15 has received enough vibrational energy from an ion impact to break its bond to the surface and desorb. This energy may also be delivered indirectly through the collision cascade, as discussed in the previous section. Ion-induced desorption makes low-energy inert-gas ion bombardment a valuable technique for cleaning substrates prior to film deposition. At higher energies, adsorbates are also removed by sputtering, which involves upward momentum transfer through the collision cascade as

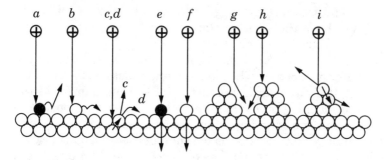

Figure 8.15 Effects of bombarding ions on surface atoms: (*a*) adsorbate removal or chemical activation, (*b*) lateral displacement, (*c*) surface vacancy created by sputtering, (*d*) surface vacancy created by displacement to an adatom position, (*e*) knock-on implantation of an impurity atom, (*f*) knock-on implantation of a film atom, (*g*) crevice filling by ion-enhanced surface mobility, (*h*) crevice filling by forward sputtering, and (*i*) breakup of 3D nucleus. Note that processes *c* through *f* also involve the subsurface.

shown in *c*. However, other subsurface effects occurring at higher energies degrade single-crystal substrates and, moreover, the impurity itself can become implanted by forward recoil as shown in *e*. Thus, energy control is important in bombardment-cleaning. Cleaning efficiency also decreases with increasing adsorption bond strength. Finally, for adsorbed molecules whose atoms are reactive with the substrate, bombardment-induced desorption proceeds in competition with reaction of the molecule's atoms with the substrate as the molecule becomes dissociated by the energy received. Thus, with metal substrates other than the noble ones, it is difficult to clean away the common contaminants water, CO_2, and hydrocarbons, because of the competing formation of oxides and carbides.

The remaining surface-displacement effects to be discussed involve atoms of the bulk solid rather than adsorbates. First of all, energetic condensation of film material at above a few eV of vapor-atom impact energy has a dramatic and important effect on the bulk microstructure of the film. The same effect can also be achieved by adding a sufficient flux of inert-ion bombardment to depositing material being supplied from low-energy, thermal sources. In either case, the depositing energy collapses the voids of the "Zone 1" microcolumnar film structure as they develop, and it thus changes the structure to the densified "Zone T" structure. This transition was discussed in Sec. 5.4.1 and was illustrated by molecular-dynamics modeling results in Fig. 5.18. As the voids progressively collapse with increasing impingement energy, film stress first becomes more tensile due to interatomic attraction across the shrinking voids, and then it becomes less tensile as full material density is approached. At still higher energy, where implantation can occur, stress becomes compressive due to the embedding of atoms into the lattice at neighbor distances less than their relaxed distance. This widely observed stress behavior with increasing energy was shown in Fig. 5.38 of Sec. 5.6.3 for the case of sputter deposition, where first the impinging energy of the sputtered atoms was increased by decreasing ambient pressure to lessen gas scattering. Then, impinging energy was increased still further by adding negative substrate bias to accelerate Ar^+ ions into the film. Mechanisms of void collapse were discussed at the end of Sec. 5.4.1, and two are illustrated again here in Fig. 8.15. That is, enhanced surface mobility of the depositing ions or atoms enables them to diffuse down into the void valleys (*g*), and overhanging atoms are forward-sputtered into the valleys by ion impact (*h*).

Low-energy ion bombardment can also planarize surface topography in heteroepitaxial growth in cases where there is a thermodynamically driven tendency to form three-dimensional (3D) rather than 2D nuclei. The case of GaAs MBE growth on Si(100) was discussed in Sec. 6.7.2. Study of that case using supplementary Ar^+ bombardment from a glow-discharge plasma source indicated that the ions were destabilizing and breaking up incipient 3D nuclei [30], as shown in Fig. 8.15*i*. For the similar case of InAs on Si(100), x-ray rocking-curve measurements of epilayer crystallographic perfection versus ion energy [31] showed a clear maximum in perfection at ≈30 eV. Lower energy was not enough to suppress 3D growth, and higher energy caused subsurface damage. The actual optimum energy may be shifted from that reported, because it is difficult to accurately determine the electrical bias of a plasma ion source. Nevertheless, the existence of the energy window was clearly demonstrated. Determination of damage threshold will be discussed further in Sec. 8.5.2.1. Ion *flux* is important as well as energy. In the above work, ion flux was always larger than the film-atom deposition flux, and this is probably a good specification to use in designing such processes.

It is often proposed that ion bombardment causes enhanced surface diffusion, as shown in Fig. 8.15*b*, with the implication that resulting diffusion lengths can be orders of magnitude larger than interatomic distances, as they can be in thermal surface diffusion. However, there is an important distinction between thermal and ion-induced adatom motion. Thermal adatom motion continues indefinitely, or in thin-film deposition at least for a time limited only by burial under the next depositing layer, so that the diffusion length increases with monolayer growth time per Eq. (5.25). However, ion-induced adatom motion stops as soon as the excess energy imparted to the adatom by the ion becomes dissipated through coupling into the surface. For instance, in a 3D molecular-dynamics simulation of 10-eV Ag atoms depositing onto

Ag(111) at 300 K, it was found that the atoms moved an average of only 0.44 nm between their arrival and their settling into final lattice sites [32]. The main effect of the energetic Ag deposition was surface smoothening by the displacement of atoms off of roughness peaks as shown in Fig. 8.15*i*. This smoothening was also seen in the GaAs/Si work discussed above [30], where the authors pointed out that enhanced *diffusion* would have *assisted* the kinetics of 3D-nucleus equilibration, in contrast to what they observed. In another example, namely the study of low-T Si sputter-epitaxy discussed in Sec. 6.7.3, the critical epitaxial thickness was found to be even more steeply dependent on T than reported for thermal deposition, whereas if ion-induced surface diffusion had been dominant, there would have been much *less* T dependence. Indeed, it appears that all reported surface effects of ion bombardment on film structure development can be explained by short-range motions of only a few interatomic distances. This rapid dissipation of the excess energy is to be expected based on the steep corrugation of the surface potential with lateral distance as shown in Fig. 5.4*a*. In effect, the traversing adatom is undergoing a collision with every surface atom it passes. Under these circumstances, one would expect long diffusion lengths only in the case of a close-packed surface plane of small atoms and very weakly bound adatoms, say perhaps Au on graphite.

The last surface-displacement effect to be discussed here involves native point defects: surface vacancies and adatoms. These can be produced by sputtering, for example, as illustrated in Fig. 8.15*d*. This is not strictly a surface effect, since the sputtering process necessarily involves the subsurface and requires ion energy above the sputtering threshold of a few tens of eV. Nevertheless, if energy is not excessive, and if substrate T is sufficient, the ion-induced bulk defects can be annealed out faster than they are created, as a result of their thermal mobility. Those point defects that are created on or diffuse to the surface are much more mobile than the bulk point defects. On Pt(111), for example, surface vacancies were found to coalesce into vacancy islands [33] at above ~700 K, which is still only one-third of the melting point. In another example—sputter epitaxy of TiN on MgO(100)—negative substrate bias of up to 500 V was used to accelerate plasma ions into the depositing film [34]. At a surface T of 920 K, 300 V of bias produced a minimum in TiN dislocation-loop density which was ×30 less than that obtained without bias. At lower T, loop density instead increased monotonically with bias. The improvement at 920 K was attributed to the ability of surface point defects to diffuse to the locations of incipient dislocation loops and annihilate them. Here, ion bombardment is increasing the *population* of surface point defects, but it is the surface T that is making them mobile. This is the surface analog of radiation-enhanced bulk diffusion (more in Sec. 8.5.2). In another effect, ion-induced surface defects on the substrate can increase film nucleation density by acting as energetic sites for nucleation. Much more work is needed, however, to fully understand the effects of defects induced by ion bombardment during film growth.

8.5.2 Subsurface effects

When the ion-bombardment energy window for surface effects is exceeded at a few tens of eV, particle penetration into the bulk material begins, producing a variety of effects which can be either beneficial or damaging to a depositing film. Some of these effects have been mentioned already. Figure 8.12c and d showed direct implantation of ions into the bulk. Implantation can also occur for atoms to which the ion has transferred energy in accordance with Eq. (8.20). When the recoil energy of these atoms is sufficient to penetrate the surface, they are called "knock-on" atoms. Figures 8.15e and f illustrate knock-on (also called recoil) implantation of adsorbed impurity atoms and of bulk-material atoms, respectively. As penetration of the ion and of the knock-on atoms proceeds, additional knock-on atoms continue to be generated in the bulk until the energies of all particles have dropped below the threshold for atomic displacement. This knock-on sequence is called the collision cascade, and the four basic types of displacement it produces are illustrated in Fig. 8.16. Displacement of a bulk atom from its lattice site produces either a vacancy plus an interstitial—two complementary native point defects known as a "Frenkel pair" (a), or it produces an interstitial only, with the incoming particle replacing the displaced one (b). If the displaced atom acquires enough upward-directed energy to escape from the surface upon arriving there, it becomes a "sputtered" atom (c). When a bulk displacement occurs near

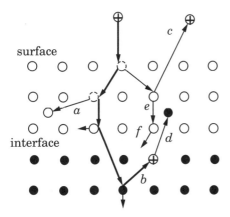

Figure 8.16 Types of atomic displacements produced in the collision cascade. The heavy line is the path of the impinging ion. (a) Frenkel-pair creation, (b) replacement collision, (c) sputtering, and (d) ion mixing. All displaced atoms shown are primary knock-ons except atoms (e) and (f), which are secondary and tertiary knock-ons.

a buried interface with the result that an atom of one material is displaced into the other, "ion mixing" of the interface has occurred. Note that upward (d) as well as downward mixing can occur. Indeed, sputtering may be thought of as upward mixing of target material into the vapor phase. The above events have been illustrated here using the "primary" knock-on atoms—those directly displaced by the ion. The same events can also be produced by sufficiently energetic knock-on atoms such as the secondary knock-on atom shown in (e).

A small part of the energy in the collision cascade is transferred to the elevated potential energy of the vacancies and displaced atoms and to the kinetic energy of sputtered atoms. However, most of the energy is instead transferred to vibrational motion in atoms which are *not* displaced out of their lattice sites. This vibrational energy propagates outward into the lattice as phonons, becoming heat as it becomes shared randomly among the atoms in the vicinity of the cascade. If the energy density of the cascade is so high that its energy becomes shared randomly *within* the cascade region before it has time to propagate outward, the cascade becomes a "thermal spike" [35]. This occurs only for large ions, which lose their energy within a short penetration depth, and for energies above a few keV. Most collision cascades encountered in thin-film deposition are not thermal spikes, and they behave differently in terms of the kinetics of energy transfer.

The vacancies and interstitials produced by the collision cascade constitute ion-induced radiation damage, which is usually undesirable in thin films. In particular, it degrades the electronic properties of semiconductors by creating charge-carrier traps. Also, since diffusion in solids occurs by way of vacancies and interstitials, the damage causes "radiation-enhanced" diffusion of impurities and diffusional intermixing of interfaces. Increasing the film-deposition T causes an exponential rise in the diffusion rates of all of the above point defects in accordance with Eq. (5.26), which applies to both surface and bulk diffusion. Some of these defects diffuse to the surface, but others become buried by the deposition or diffuse deeper into the bulk. Native point defects can also condense beneath the surface to form dislocations [36], although complementary (Frenkel-pair) defects instead annihilate each other. The final film structure results from a balance between the defect generation rate caused by the energy-deposition flux and the defect loss rate caused by annihilation in the bulk and at the surface.

Below, we will quantify as much as possible the threshold for the onset of subsurface effects, and we will develop the theory of implantation depth which determines the extent of the collision cascade. Then in Sec. 8.5.3, we will discuss in more detail the specific effects on film properties that result from ion bombardment above the subsurface threshold. Finally, Sec. 8.5.4 examines sputtering and its use both for source-material vaporization and for modification of the depositing film.

8.5.2.1 Energy thresholds. Two ion-bombardment-energy thresholds, E_o, are of concern here: that for penetration of the ions beneath the surface, and that for Frenkel-pair creation. There is also a sputtering threshold, which we will discuss in Sec. 8.5.4. The penetration threshold depends primarily on the "size" of the ion relative to the interatomic channels available to it at the surface. We will see in the next section that the effective sizes of two colliding atoms decrease somewhat with increasing collision energy, but the main factor influencing penetration threshold is the increase in ion size with atomic number, Z. Figure 8.17 shows two measures of the energy threshold, E_o, for penetration of noble-gas ions of atomic number, Z_i, into heavy metals. One [37] is the threshold for detectable gas *trapping* in polycrystalline W, "detectable" being a fraction of $\approx 10^{-5}$ here. The other threshold,

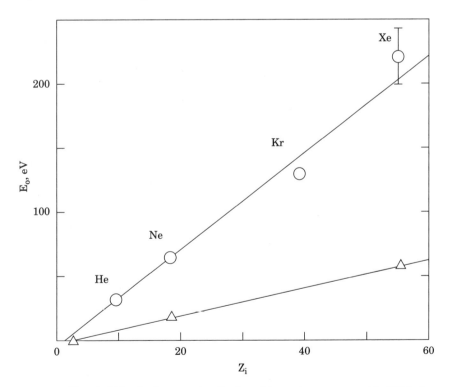

Figure 8.17 Threshold impinging energies, E_o, for noble-gas ion *entrapment* in W (circles, from Ref. 37) and *penetration* into Au or Pt (triangles, from Fig. 8.14).

taken from Fig. 8.14, is the point at which energy deposition into Au or Pt exceeds the T_m of Eq. (8.20), indicating ion *penetration*. In both cases, E_0 is seen to increase linearly with Z_i and to approach zero for the smallest ions. From this behavior, it can be inferred that He^+ and H_2^+ penetration will certainly occur if these ions are present in any enhanced-energy process. Thus, small ions must be avoided when their entrapment in the film would be undesirable. The large difference in E_0 between the two sets of data in Fig. 8.17 cannot be due to differences among the three metals involved, since they all have about the same Z and atomic density. Rather, it is due to the difference in what is being measured. The higher E_0 values represent entrapment, while the lower ones represent energy deposition by penetration, not necessarily accompanied by entrapment. During film deposition, the amount of entrapment of penetrating unreactive ions will decrease with increasing film T and with decreasing deposition rate, both of which assist outdiffusion of the ions. Entrapment also increases when bombardment is along crystallographic channeling directions, as shown in Fig. 8.12d.

The ion-energy threshold for subsurface atomic displacement with Frenkel-pair creation behaves differently from that for ion penetration. Specifically, small, light ions can penetrate without causing any subsurface displacements, while large, massive ions can create subsurface vacancies by way of knock-on atoms generated at the surface, without the ions themselves ever penetrating. The principal determinant of the displacement threshold is the fraction of energy transferred to an individual subsurface atom, γ_m in Eq. (8.20). Thus, the threshold is at a minimum when $m_i = m_t$ and increases for either lighter or heavier ions. From the above considerations, we reach the important conclusion that if it is desired to move surface atoms around without causing either ion penetration or subsurface displacements, the widest bombardment-energy window will be obtained by using a very heavy ion, such as Xe^+. Another threshold determinant is the bond energy with which the atom is anchored in its lattice site. This energy is roughly proportional to the heat of sublimation, $\Delta_s H$, which measures the energy required to break the bonds of a surface atom to the lattice. The use of $\Delta_s H$ is convenient because it is readily obtained from thermodynamic-data tables as the difference between the heats of formation of the gaseous and solid phases at the T of interest: $\Delta_s H = \Delta_f H(g) - \Delta_f H(c)$.

One common way to measure displacement thresholds is to use electron bombardment. Electrons are so light that they must impinge at hundreds of keV to cause a displacement, as discussed following Eq. (8.20). Electrons this energetic penetrate many μm, producing homogeneous bulk damage that can be measured, for example, by the re-

8.5.2 Subsurface effects

sulting resistivity increase in metals [38]. From the threshold for resistivity increase and Eq. (8.20), one can calculate the minimum energy, E_f, that a lattice atom must acquire to create a Frenkel pair. The corresponding impinging-*ion* energy threshold for displacement is then found similarly from $E_o = E_f/\gamma_m$. Figure 8.18 shows E_f data for many elements plotted versus $\Delta_s H$. The data for the metallic elements were obtained as described above [38]. The group-IV-element data were quoted in Ref. 39, except for one of the graphite points, which was obtained by detecting the onset of distortion in the KLL Auger spectral line of graphite after Ne$^+$ bombardment [40]. Note that, overall, there is a rough correlation between E_f and $\Delta_s H$, whereby $E_f \approx 5\Delta_s H$ within a factor of two for the 30 elements shown. The scat-

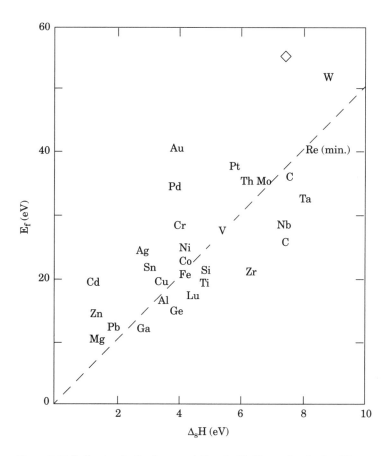

Figure 8.18 Bulk-atomic-displacement threshold, E_f, vs. heat of sublimation, $\Delta_s H$. Metal-element data from Ref. 38, group-IV-element data as quoted in Ref. 39, except upper point for graphite (C) from Ref. 40. ◇ = diamond. Slope of dashed line = 5.

ter in the data suggests that other factors besides $\Delta_s H$ determine E_f, likely ones being the "openness" and the elastic modulus (stiffness) of the lattice into which the displaced atom must wedge itself.

How is this high a threshold energy accounted for? Taking Si as an example, we have E_f = 22 eV from Fig. 8.18. The vacancy and self-interstitial formation energies calculated for Si are ≈ 4 eV each [41, 42], and the activation energy for interstitial migration is ≈ 0.2 eV [41]. The remaining 14 eV is consumed in displacing the interstitial far enough away from the vacancy so that the Frenkel pair becomes stable rather than spontaneously recombining due to attraction of the interstitial back into the potential well of the vacancy, an attraction which results from the localized strain fields surrounding both point defects. Of course, the stable point defects may still diffuse thermally and annihilate each other given enough time. Molecular-dynamics simulations of 10-eV Si atoms [42] and Ag atoms [32] impinging on their own lattices have shown that no stable vacancies are created at that impact energy, which is consistent with the 22-eV threshold of Fig. 8.18.

8.5.2.2 Penetration depth. The depth to which an ion penetrates a solid before losing all of its directed energy and coming to rest is of great interest in high-energy physics and in ion implantation of dopants into integrated circuits, as well as in thin-film deposition. It is also by far the best understood of all aspects of ion bombardment in thin-film deposition. We will therefore examine it in detail here for the insight it gives into the nature of ion-solid interaction. When the penetration depth, or "range," of the ion is more than a few atomic distances, it can be estimated to within about 10 percent using LSS (Lindhard, Scharff, and Schiott) theory, which proceeds as follows (Ziegler, 1985). The rate of ion energy (E) loss with depth is proportional to atom concentration, n(atoms/nm^3), and to the "stopping power" of those atoms for the ion; that is,

$$dE/dx \text{ (eV/nm)} = n(S_n + S_e) \quad (8.22)$$

Here, S_n (eV·nm^2/atom) is the nuclear stopping power, which arises from collisions with the atoms (nuclear repulsion), and S_e is the electronic stopping power, which arises from drag on the ion by the electron cloud within the solid. Below a few keV, which is the energy range of interest for film deposition, S_n is much larger than S_e, while at higher E, S_e becomes dominant. The central problem, then, is to estimate S_n, and to do this we must describe the ion-atom binary collision in detail.

8.5.2 Subsurface effects

We neglect the shallow potential well representing the chemical-attraction portion of the interatomic potential function which was shown in Fig. 4.2, and consider only the repulsive portion. *If* both atoms were stripped of all electrons, the mutual repulsive force, F_+, would be the force on one nucleus (say the impinging ion) due to the immersion of its charge, q_i, in the electric field, $\mathbf{E}$, radiating from the other nucleus (say the target atom); that is, $F_+ = q_i E_t$, as in Eq. (8.15). The strength of $\mathbf{E}_t$ is proportional to q_t and inversely proportional to the surface area of the sphere over which $\mathbf{E}_t$ is distributed at the interatomic distance, r; thus,

$$F_+ = q_i \mathbf{E}_t = \frac{q_i q_t}{4\pi r^2 \varepsilon_0} = \frac{9 \times 10^9 q_e^2 Z_i Z_t}{r^2} \qquad (8.23)$$

which is just Coulomb's law.

In SI units (newtons, coulombs, meters), the proportionality factor ε_0 (the permittivity of free space) is $1/(36\pi \times 10^9)$ $C^2/N \cdot m^2$, q_e is the proton (or electron) charge (1.60×10^{-19} C), and the Z values are the atomic numbers of the nuclei. Now, the potential energy, E_+, of an impinging ion at distance r from its target ion is the integral of the repulsive force as the ion is brought in from infinity to r. If Eq. (8.23) were obeyed (all electrons stripped), this energy would be

$$E_+(eV) = \int_\infty^r -F_+ dr = \frac{9 \times 10^{18} q_e Z_i Z_t}{r(nm)} \qquad (8.24)$$

Here, we have switched from the SI energy units of $C \cdot V$ to the more convenient units of eV (by dividing by q_e) and nm. This "Coulomb potential" is *screened*, however, by the electrons orbiting both atoms, and this fact is accounted for by a dimensionless screening-function correction factor, $\Phi_u(r)$, when calculating the actual interatomic potential energy,

$$E_p = E_+ \Phi_u(r) \qquad (8.25)$$

It has been found empirically that hundreds of calculated and measured interatomic potentials are well correlated by the screening-function formula given in Fig. 8.22. There, x is the "reduced" radius,

$$x = r/a_u \qquad (8.26)$$

where a_u is the "screening radius," given by

$$a_u \text{ (nm)} = 0.0468/(Z_i^{0.23} + Z_t^{0.23}) \tag{8.27}$$

Figure 8.19 shows the resulting correlation and compares it to other potential functions. Figure 8.20 shows the behavior of E_p for the typical cases of Ar^+ and H^+ colliding with Ni and compares it with the Coulomb potential. These curves more accurately describe what is meant by the "size" of an ion, and they show the degree by which H^+ is smaller than Ar^+. By contrast to this screened Coulomb collision, a hard-sphere (billiard-ball) collision would have an E_p rising from zero to infinity at the mean sphere diameter. The screened Coulomb collision is the basic model for describing the interaction of energetic ions or atoms with each other.

It is instructive to calculate a typical trajectory resulting from the screened repulsive potential. To do this, we first calculate the mutual repulsive force,

$$F = -\frac{dE_p}{dr} = -\Phi_u \frac{dE_+}{dr} - E_+ \frac{d\Phi_u}{dr} \tag{8.28}$$

Then, the deceleration and deflection of the ion and the acceleration of the target atom can be calculated from $F = m(dv/dt)$. The results are

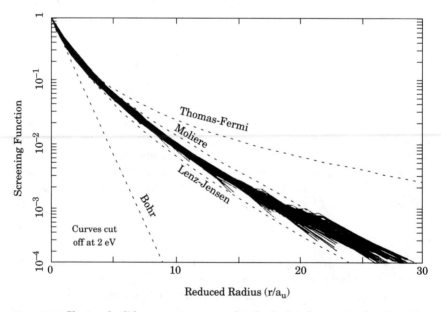

Figure 8.19 Clustered solid curves are measured and calculated screening functions, Φ_u, for hundreds of interatomic potentials between atoms of various elements. Their mean is represented by the formula given in Fig. 8.22. [Source: Reprinted from Ziegler (1985) by permission of Macmillan Publishing Co., copyright © 1985 by Pergamon Press.]

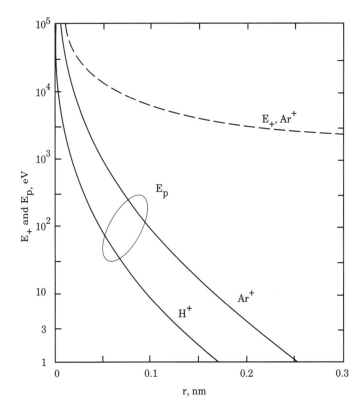

Figure 8.20 Interatomic potential-energy curves for ions colliding with Ni. E_+ is the Coulomb potential, and E_p is the screened potential of Eq. (8.25). The hard-sphere potential would be a vertical line at the mean sphere diameter.

shown in Fig. 8.21 for a Ni target atom, positioned at the coordinate origin, being struck by an Ar ion impinging at an offset, or "impact parameter," p, of 0.05 nm. The calculation has been done for three incident energies, E_i, at the same p. For each case, the calculation was stopped when F dropped to 1 eV, and the ion and atom positions at that point are shown. Two important features of the collision reveal themselves here. One is that at E_i = 30 keV, target-atom displacement is negligible, while at 30 eV, it is about half of the interatomic distance. Thus, the binary-collision model is very accurate at the higher energy but much less so at the lower one, where the restoring force of lattice stiffness becomes a significant correction. The more complex molecular-dynamics calculations are required to account for this lattice interaction. The second feature shown is the decrease in deflection angle with increasing E_i. This arises from the "softness" of E_p relative

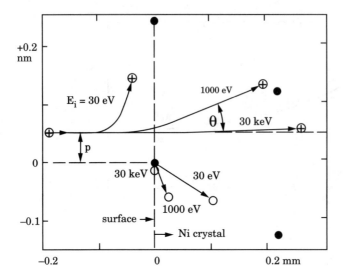

Figure 8.21 Trajectories of impinging Ar ion (circle with plus symbol) and target Ni surface atom (white circle) in their binary collision at three impinging energies, E_i, for a fixed impact parameter of p = 0.05 nm. Positions of neighboring Ni atoms in a (111) plane are shown for reference, but their interaction potentials with the ion are neglected. (Pascal solution courtesy of Jared Smith-Mickelson.)

to the hard-sphere case. A faster ion penetrates more closely toward the target nucleus before being deflected, so that the target nucleus appears smaller and the collision therefore becomes more glancing (less direct) for the same value of p. By contrast, hard-sphere collisions would produce the same deflection angle for all E_i.

For purposes of calculating the nuclear stopping power, S_n, it is not necessary to plot the entire trajectory but only to know the final deflection angle, θ, and final ion energy as a function of p. This is because conservation of energy and momentum make the amount of energy transferred in the collision, $T(\theta)$, a unique function of θ, regardless of the shape of E_p, as shown in Exercise 8.9a. The relationship of θ to p does depend on E_p, however, as described in Biersack (1987). Finally, S_n is found by integrating $T(\theta)$, or equivalently $T(E_i, p)$, over all p:

$$S_n(E_i) = \int_0^{p_{max}} T(E_i, p)\, 2\pi p\, dp \qquad (8.29)$$

When S_n and E_i are converted to the reduced stopping power $S_n(\varepsilon)$ and reduced energy (ε) of LSS theory as defined below, the $S_n(\varepsilon)$ for all

8.5.2 Subsurface effects

pairs of atoms fall on the same universal curve, which is shown in Fig. 8.22. The reduced energy is defined as

$$\varepsilon = \frac{a_u}{r_o} = \frac{a_u m_t E_i}{9 \times 10^9 q_e^2 Z_i Z_t (m_i + m_t)} = \frac{0.0325 M_t E_i \,(\text{eV})}{Z_i Z_t \left(Z_i^{0.23} + Z_t^{0.23} \right)(M_i + M_t)} \quad (8.30)$$

Here, the second equality is in SI units, whereas the third has E_i in eV. The term a_u is the screening radius from Eq. (8.27), and r_o is the distance of closest approach in a head-on, unscreened (Coulomb) collision, which can be found from conservation of energy and momentum as outlined in Exercise 8.13. The reduced nuclear stopping power is defined as

$$S_n(\varepsilon) = \frac{\varepsilon}{\pi a_u^2 T_m} S_n(E_i) \quad (8.31)$$

where T_m was defined in Eq. (8.20). Here, the denominator is essentially a stopping power, consisting of a characteristic collision cross

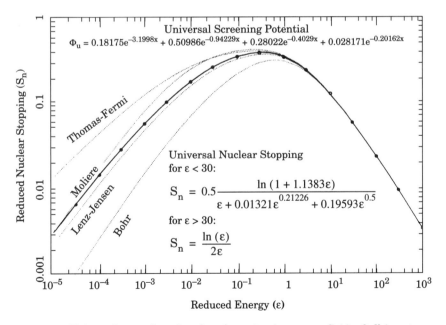

Figure 8.22 Universal curve for reduced nuclear stopping power, $S_n(\varepsilon)$, of all ion-atom pairs vs. energy based on LSS theory. [Source: Reprinted from Ziegler (1985) with permission of Macmillan Publishing Co., copyright © 1985 by Pergamon Press.]

section, πa_u^2, times the head-on energy transfer, T_m. *If* the actual effective collision cross section did not decrease with increasing E_i (hard-sphere case), then $S_n(E_i)$ would be proportional to this denominator, so we would have $S_n(\varepsilon) \propto \varepsilon$ and a slope of unity in Fig. 8.22. The falloff from unity slope arises from the increasing proximity of the collision with increasing E_i, as shown in Fig. 8.21. For $\varepsilon > 1$ or so, the electron screen has been fully penetrated, and the slope reverses. At still higher ε in the Rutherford-scattering regime of MeV, most collisions result in very little deflection, so that the energy transferred is determined more by the amount of time that the colliding atoms spend exchanging energy than by E_i. Thus, $S_n(\varepsilon) \propto 1/\varepsilon$ at high ε.

To give a feeling for the magnitudes of the above quantities in a typical film-deposition situation, we calculate S_n for 1-keV Ar^+ bombarding Ni. From the Z and M values, we find $\gamma_m = 0.96$, $a_u = 0.0114$ nm, and $\varepsilon = 9.36 \times 10^{-3}$, so that S_n (eV·nm^2/atom) = $41.9 S_n(\varepsilon) = 7.1$ nm^2·eV from Fig. 8.22. Then, using n = 91.3 atoms/nm^3 for crystalline Ni, we have dE/dx = 650 eV/nm for the nuclear-stopping component of Eq. (8.22). Thus, we are in the positive-slope regime of Fig. 8.22 for film-deposition work, except for very-low-Z cases such as H-Be, where $S_n(\varepsilon)$ peaks at $E_i = 97$ eV. Note that for $E_i \sim 1$ keV or less, dE/dx is so high that most energy is lost within a few atomic distances, except for the lightest ions or for channeling directions.

The final task is to calculate the penetration depth, or range, of the ion. In doing so, the smaller but non-negligible electronic stopping component, S_e, must also be accounted for, as described in Ziegler (1985). S_e continues to rise monotonically with ε rather than falling off as S_n does, so that it eventually becomes dominant for $\varepsilon \gg 1$. For the $\varepsilon < 1$ regime of interest here, $S_e/S_n \approx 1/4$ for Si^+ in Si and Ge^+ in Ge, for example (Davies, 1992). One way of calculating ion range is to use these stopping data in an analytical expression involving the Boltzmann transport equation. This equation deals with systems of randomly colliding particles characterized by some velocity distribution function, $f(\mathbf{v})$. Given an initial $f(\mathbf{v})$ which is localized in space and/or time, such as that represented by an impinging ion, the transport equation describes the space and time transient of $f(\mathbf{v})$ relaxation. At the lowest E_i (1 keV) for which range calculations are plotted in Ziegler (1985), the ranges of most ion-atom combinations are 2.6±1 nm; and at 2 keV, they are 4.1±2 nm. By contrast, the range of H^+ is much larger, both because H^+ is small and because its energy transfer per collision is low per Eq. (8.20); for example, the H^+ ranges at 1 and 2 keV are 20 and 36 nm into Si, and 7 and 14 nm into Ni. There is always some spread, or "straggling," of ion range about these mean values due to the statistical nature of the process, and this can also be calculated.

At <1 keV, ion range begins to approach one interatomic distance, so that the concepts of a statistical distribution of particles and of the continuum property S_n begin to break down. Then, it is more appropriate to follow a large number of *individual* ion trajectories via Monte-Carlo techniques as described in Sec. 8.5.1.1, where each impact parameter, p, is selected randomly in Monte-Carlo fashion, and then the shape of the Eq. (8.25) interatomic potential is used to calculate T_m and θ. Above 1 keV, both the analytical and the Monte-Carlo approach give identical results for range. In both approaches, an isotropic, amorphous solid is generally assumed for simplicity, so that channeling is not treated. However, one version of the TRIM Monte-Carlo program does include crystalline structure.

Note that the penetration depth of *knock-on* atoms such as those shown in Fig. 8.15*e* and *f* can be calculated in the same two ways. First, the maximum initial knock-on energy, T_m, is found from Eq. (8.20), and then the knock-on atom is treated as the impinging particle. Of course, collisions having p > 0 will transfer energy less than T_m to the knock-on atom. Also, the effective cross section for energy transfer near to the T_m value decreases with increasing ion energy, as shown in Fig. 8.21. Thus, T_m gives only the *maximum* range, and many of the knock-ons will penetrate much less.

8.5.3 Bulk-film modification

At this point, we have some understanding of two important subsurface phenomena caused by ion bombardment; namely, implantation and atomic displacement. We now examine the changes in film properties that these phenomena produce. As before, we use the term "ion" to refer also to energetic atoms and molecules bombarding the surface.

Ion implantation into a depositing film can be either unwanted, as with sputtering gas or background gas, or deliberate, as with dopants and compound-film constituents. **Implantation of inert gas**, such as Ar from the plasma used in sputter-deposition, is generally undesirable. It distorts the crystalline lattice, imparts compressive stress, increases resistivity, and can diffuse out when the film is later heated, producing gas bubbles at the interface when an overlying film has been deposited. The amount of implanted gas can reach several at.%, but it can be kept much smaller by selecting $m_i > m_t$ to minimize ion reflection from the sputtering target (Fig. 8.13). It can be further reduced by increasing substrate temperature, T_s, to activate out-diffusion during deposition. The quantity of gas remaining implanted is a balance between its burial rate by depositing film material and its out-diffusion rate. A similar balance is involved when implantation is used to increase the **incorporation of a desired dopant** which would prefer to segregate to the surface or evaporate, as was discussed in Sec. 6.5.6. Consequently, fractional ion incorporation increases with increasing implant depth and deposition rate and with decreasing T_s. By the way, a dopant may be implanted either directly as an impinging ion or by knock-on implantation using another ion following the dopant's thermal adsorption (Fig. 8.15e). The out-diffusion rates of both inert gases and dopants are likely to be much higher than predicted from the bulk diffusivity, D, at T_s, because D is "radiation-enhanced" by the presence of bombardment-induced excess vacancies in the subsurface region. On the other hand, when reactive ions that form stable compounds with the film become implanted, they do not diffuse out. This presents the risk of increased contamination from common background gases whenever an energy-enhanced deposition process is used. Conversely, **reactive implantation** can be used deliberately to form compound films, such as TiN from Ti and N_2^+, and it can also increase the content of that element in the film beyond what would otherwise be achievable, thus allowing nonequilibrium stoichiometry adjustment.

Regardless of whether a bombarding ion becomes permanently implanted into the film, the collision cascade which accompanies its nuclear stopping results in **displacements of film atoms** from their lattice sites whenever the recoil energy of those atoms exceeds the displacement threshold, E_f, shown in Fig. 8.18. These displacements increase the local concentration of vacancy and self-interstitial point defects, and they also cause intermixing of the two materials in the vicinity of an interface. Molecular-dynamics simulation [43] of collision cascades perpetrated by 5-keV primary-knock-on atoms in Cu and Ni showed the resulting damage to consist of an excess of vacancies at the core of the cascade, surrounded by a shell containing excess interstitials. This is what one would expect from the fact that atoms are being knocked *out* of the core by the collisions. At the high particle energy of this study, a thermal spike of ≈4000 K was produced at the core, and the core consequently melted. The final rms displacement of Cu atoms was estimated at 23 nm and that of Ni at 12 nm. Lower-energy cascades are less likely to produce thermal spikes, and the displacements are smaller, but the structure consisting of vacancies surrounded by interstitials is expected to be a general feature. Below, we will discuss first some ways of estimating the amount of atomic displacement and then the effects of displacement on film properties.

A great deal of work on radiation damage to materials by energetic particles was done in the development of nuclear reactors [44], and much of this also applies to the relatively low-energy ions involved in thin-film deposition. For ion impingement energies E_i *well above* the

8.5.3 Bulk-film modification

displacement threshold ($E_i \gg E_f$), the number of Frenkel pairs produced per ion may be roughly estimated as follows. First, there will be some number of primary-knock-on atoms (PKs) of energy E_γ generated in successive binary collisions as the ion is slowed down in the material. For a PK, in turn, to generate an *additional* displacement, it must have at least energy E_f to displace a second atom plus another amount E_f to carry itself out of the new vacancy so created; that is, $E_\gamma \geq 2E_f$. For $E_f < E_\gamma < 2E_f$, on the other hand, the PK instead undergoes a replacement collision, filling the new vacancy and producing no additional displacements, as shown for an ion in Fig. 8.16b. And for $E_\gamma < E_f$, the PK simply comes to rest in an interstitial site. But at high energy, $E_\gamma \gg 2E_f$, the number of Frenkel pairs created by each PK is given by $E_\gamma/2E_f$, which is the Kinchen-Pease formula [44]. Now suppose for simplicity that some constant fraction γ of the ion impingement energy, E_i, is transferred to each PK; that is, $E_\gamma = \gamma E_i$. This fraction will depend on the atomic-mass ratio, m_i/m_t in accordance with Eq. (8.20). Then for $\gamma E_i \gg E_f$, there will be a maximum of $(1/\gamma)$ PKs generated, and the total number of displacements per ion will be given by

$$N_f = \left(\frac{\gamma E_i}{2E_f}\right)\left(\frac{1}{\gamma}\right) = \frac{E_i}{2E_f}$$

which is independent of m_i/m_t. With the addition of a correction multiplier of about 0.8 from transport theory and with allowance for energy loss to electronic stopping [45], we have the following modified Kinchen-Pease formula for the total number of Frenkel pairs, N_f, or vacancies, N_v, or interstitials, N_i, generated per ion when $\gamma E_i \gg E_f$:

$$\boxed{N_f = N_v = N_i = \frac{0.8 \mathrm{x} E_i}{2E_f}} \qquad (8.32)$$

The fraction, x, of ion energy deposited into atomic motion instead of into electron interaction may be estimated from the respective stopping powers: $x = S_n/(S_n + S_e)$. Experimental results for Si and Ge show that $x \approx 0.8$ when the reduced energy, ε, is less than unity (Davies, 1992).

For lower E_i, the Monte-Carlo approach (Sec. 8.5.1.1) is more appropriate for calculating N_v and N_i. Figure 8.23 shows TRIM-code calculations of the vacancy distribution versus depth resulting from H^+ and Si^+ bombardment of Si. The number of vacancies produced per unit depth by an H^+ is only 2 percent of that produced by an Si^+, but the

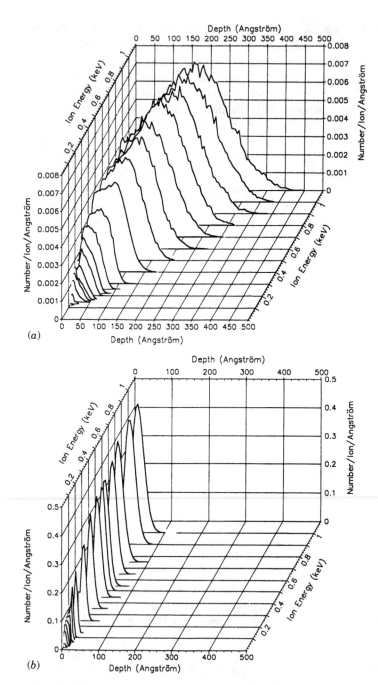

Figure 8.23 TRIM-code calculation of vacancy depth distributions resulting from ion impingement at various energies: (a) H^+ and (b), Si^+. (Source: Reprinted from Ref. 46 by permission.)

8.5.3 Bulk-film modification

vacancies are distributed 5 times deeper. These two differences arise from the much lower stopping power of H^+. Note also that the ion-energy threshold for vacancy production by H^+ is about 120 eV. This, when multiplied by $\gamma_m = 0.133$ for H-Si from Eq. (8.20), implies a Si-displacement threshold, E_f, of 16 eV, which is the same as that observed in Fig. 8.23b for vacancy production by Si^+, where $\gamma_m = 1$. This is therefore presumably the E_f which was specified to calculate Fig. 8.23, though it is somewhat below the 22 eV reported for Si in Fig. 8.18. The accuracy of displacement calculations is always limited by the accuracy to which E_f is known, whether Monte-Carlo or Eq. (8.32) is used.

To illustrate the effects of energy and mass on vacancy production, we calculate total N_v from Fig. 8.23 by integrating the depth distribution assuming a symmetrical distribution extending out of the surface. The resulting placement of some vacancies above the surface is satisfactory for the comparison we wish to make with Eq. (8.32), which also neglects the surface. The above integral is plotted in Fig. 8.24 and shows that indeed $N_v \propto E_i$, as predicted by Eq. (8.32). However, N_v is much less for H^+ than for Si^+ – 1.1 versus 18 at 1 keV, whereas no mass dependence is predicted by Eq. (8.32). Solving Eq. (8.32) at 1 keV

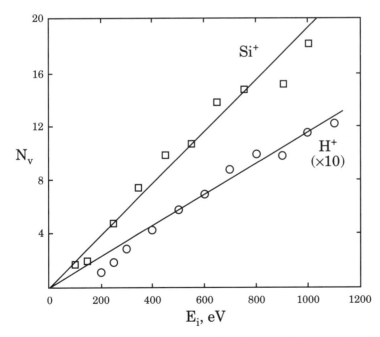

Figure 8.24 Total number of vacancies produced per ion, N_v, vs. bombardment energy, E_i, estimated from the data of Fig. 8.23.

with x = 0.8 and E_f = 16 eV, we find N_v = 19.9, which agrees well with the TRIM result of 18 for Si^+ ions (Fig. 8.24) using the same E_f. On the other hand, agreement is poor for H^+ because $\gamma_m E_i$ (=133 eV) is much closer to threshold, so that many of the H^+ collisions with atoms within the film transfer insufficient energy to displace an atom. Remember that Eq. (8.32) assumes that *all* of E_i is absorbed by displaced atoms. This example demonstrates how vacancy production drops off when $\gamma_m E_i$ becomes as small as E_f, and how a small ion/target mass ratio reduces the fraction of E_i that goes into vacancy production rather than heat. Another effect apparent in Fig. 8.23 is that as E_i continues to decrease toward E_f, the vacancy distribution shifts more toward the surface. The interstitials (not plotted) shift this way too, but they will always be distributed somewhat deeper than the vacancies because of their net forward displacement by the bombardment. Thus, there will be some E_i below which *only* interstitials are being generated, and this may be responsible for the cases of film density greater than bulk-crystal density which have been reported [47].

The excess vacancies resulting from the collision cascade are responsible for **radiation-enhanced diffusion** in the near-surface region of materials being bombarded by ions. They are also likely to modify the **structural evolution** of depositing films, as did the surface vacancies discussed in Sec. 8.5.1.3. The interstitials, on the other hand, produce intermixing and compression of thin films. **Ion mixing** of two film materials across an interface was illustrated in Fig. 8.16*d*. The depth of the mixed region can be roughly estimated as the implantation range of a knock-on atom launched from the interface at the maximum energy, T_m in Eq. (8.20), which it can receive from the bombarding ion's remaining energy at that depth. For example, under 1-keV bombardment by ions having γ_m near unity, the first few layers of deposit will become mixed ~3 nm into the substrate in accordance with the typical range quoted at the end of the previous section. Adsorbed impurities become similarly mixed, at least that fraction of them which is not sputtered away instead. Both the mixing of the film-substrate interface and the dispersion of intervening adsorbates greatly assist **film adhesion**, a topic discussed more broadly in Sec. 5.7. For E_i exceeding a few keV, where thermal spikes begin to appear, collision-cascade mixing is further enhanced by the chemical-potential driving force [48] which is momentarily activated by the high local T— that is, the Gibbs free energy of mixing. Finally, interstitials often have high diffusivities, and then they can diffuse from the near-surface region deep into the bulk at elevated T. When such contamination is undesirable or when atomically abrupt interfaces are required, as in some epitaxial-semiconductor applications, interstitial generation and mixing must be avoided by keeping $\gamma_m E_i < E_f$.

8.5.3 Bulk-film modification

Compressive film stress is the second principal effect of interstitial generation by ion bombardment. This process is known as "ion peening," as discussed in Sec. 5.6.3. Specifically, the wedging of interstitials between lattice sites forces surrounding atoms outward into closer proximity to their neighbors, causing the whole lattice to want to expand. However, a film bonded to a substrate cannot expand laterally (in x and y), so it instead develops a biaxial compressive stress, $\sigma_{x,y}$, which is proportional to the lateral lattice distortion or biaxial strain, $\varepsilon_{x,y}$, by the factor $Y/(1-\nu)$ as given in Eq. (5.51), where Y and ν are the Young's modulus of elasticity (stiffness) and Poisson's ratio of the film material. Assuming that $\varepsilon_{x,y}$ is proportional to the fraction of atoms placed into interstitial sites by the ion bombardment accompanying film deposition, we have

$$\sigma_{x,y} = B\left(\frac{Y}{1-\nu}\right)\frac{J_i N_i}{J_r} = B\left(\frac{Y}{1-\nu}\right)\frac{J_i N_i}{n_m N_A (dh/dt)} \quad (8.33)$$

where B = proportionality constant
 N_i = number of interstitials produced per bombarding ion
 J_i = ion flux, ions/cm^2·s
 J_r = deposition flux, atoms/cm^2·s
 n_m = molar concentration, mol/cm^3
 N_A = Avogadro's number = 6.02×10^{23} atoms/mol
 dh/dt = film deposition rate, cm/s

The J_i in the above equation includes both energetic film material and other energetic species such as Ar$^+$. In laser ablation and the cathodic arc, most of the depositing ions and atoms that constitute J_r have sufficient E_i to generate interstitials, so long as ambient pressure is low enough to prevent their thermalization by scattering during transport. In sputter-deposition, on the other hand, only a fraction of J_r in the high-energy tail of the sputtered-particle energy distribution has this much energy, and this fraction is roughly proportional to the displacement threshold of the target material being sputtered, as will be discussed in the next section. However, high-energy neutralized sputtering-gas ions reflected from the target also impinge upon the film. Their flux increases dramatically when $m_i < m_t$, as seen in Fig. 8.13, so they probably dominate the film-compression effect when $m_i < m_t$. In addition, when electronegative elements such as O are present in sputtering, negative ions can be formed at the target surface and accelerated into and across the plasma to bombard the film with E_i approaching the target bias potential of ~500 eV. In any of the energy-enhanced deposition processes, one can also accelerate ions into the film by applying a negative bias to the substrate, thereby increasing N_i by Eq. (8.32). For insulating films or substrates, "rf bias" can be used whenever a plasma is present, as will be discussed in Sec. 9.4.3. In electron-beam and arc evaporation, the ions accelerated by substrate bias are those of the film material, whereas in sputtering, they are those of the sputtering gas (usually Ar^+). J_i can also be supplied from an auxiliary ion gun, as is often done in conjunction with thermal evaporation. Whenever J_i is not the film material itself, its effectiveness at compressing the film can be enhanced by decreasing the deposition rate, as seen in Eq. (8.33).

N_i is the most difficult factor to estimate in Eq. (8.33). TRIM calculation is preferred over Eq. (8.32), although the two methods were found to agree fairly well for Si^+/Si in Fig. 8.24. However, when E_i is still lower, as in the case of sputtered atoms, the binary-collision premise inherent in both methods is no longer valid, and molecular-dynamics calculations are needed. Finally, total N_i must include not only the displaced film atoms, but also that fraction of implanted impurities such as Ar which remains trapped in the film.

The increase in compressive stress with material stiffness, Y, is the third important factor in Eq. (8.33). In one study of films deposited at fixed rate (dh/dt) by ion-beam sputtering [49], it was found that $\sigma_{x,y} \propto Y/(1-\nu)n_m$ for a wide range of metals and compounds. This correlation is surprising, since it implies constant N_iJ_i over this range of materials. The same study also quoted various other data showing $\sigma_{x,y} \propto \sqrt{E_i}$, whereas Eq. (8.32) and Fig. 8.24 predict $N_i \propto E_i$ and thus $\sigma_{x,y} \propto E_i$. Studies of film compression by particle bombardment are further complicated by other factors that affect the stress besides those in Eq. (8.33). For example, compound-forming reactions increase it (Sec. 5.6.3), and plastic-flow relaxation decreases it once the yield point is reached [50]. To summarize the situation on compressive-stress prediction, there is a large body of data showing that compression increases with J_i/J_r, E_i, and Y, as predicted by Eqs. (8.32) and (8.33). However, the functional form of the relationship and the quantification of compression have not yet been adequately modeled. Nevertheless, the process trends outlined here can be used to increase film compression in order to counteract tensile stress, to increase film density, or to improve those film properties that benefit from moderate compression, as was more fully discussed in Sec 5.6.3. It has also been suggested that the stored energy of compression enhances preferred orientation in mechanically anisotropic materials by providing a thermodynamic driving force toward crystallite orientations that have the more compliant (lower-Y) directions parallel to the growth plane [51]. Of course, many other factors also influence preferred orientation (see Sec. 5.3.4). If any film is *too* compressive, substrate bowing (Fig. 5.32c), delamination (Fig. 5.35b), or other problems can arise.

Ion bombardment can also cause the formation of **metastable phases** of the film material, which would not be seen in thermal deposition, such as $(III\text{-}V)_{1-x}(IV_2)_x$ alloys [52], cubic rather than hexagonal BN [53], and amorphous phases. Regarding the latter, it has long been known that many crystalline materials subjected to large doses of high-energy radiation, not necessarily ions, convert to an amorphous "metamict" phase [44]. Recently, electron spectroscopy of Si and Ge crystal surfaces subjected to low-energy Ar^+ bombardment (100 to 5000 eV) showed amorphization above a critical density of deposited energy amounting to 12 eV per atom for Si and 14 for Ge [54]. In another study, thermally evaporated Ge was subjected to Ar^+ bombardment during film deposition [47], and film density, ρ_m, was inferred from ellipsometric measurements. At first, ρ_m increased toward the value for crystalline Ge with increasing $J_i E_i/J_r$ [symbols from Eqs. (8.32) and (8.33)], as would be expected from the collapse of "Zone-1" microvoids. Then, ρ_m underwent a sudden drop of about 7 percent at $J_i E_i/J_r = 18$ eV, *independent* of ion energy over the range 15 to 600 eV. This is about the energy needed to displace every depositing Ge atom (Fig. 8.18), suggesting a phase change from microcrystalline to amorphous with an accompanying decrease in ρ_m, but more study is needed to understand what is happening here.

Ion-induced radiation damage to the *substrate* can also be important. Si(100) substrates predamaged by 100-keV P^+ or As^+ bombardment and then etched to expose the buried damage layer show **enhanced nucleation density** in diamond deposition [55]. Nucleation is a major problem in diamond deposition, as was illustrated in Fig. 5.3.

Despite all of the important subsurface effects discussed above that involve implantation and displacement, most of the energy lost by the bombarding ion as it is stopped within the film ends up as heat. This heat input is often far greater than the heat of condensation of the film material, and it can cause severe substrate-T control problems when the substrate's thermal contact to its T-controlled platform is poor, which it usually is (see Sec. 5.8). Thus, when only the platform T is reported in energy-enhanced film-deposition work, one must be aware that substrate T might have been hundreds of degrees higher (see Exercise 8.15).

8.5.4 Sputtering

Sputtering is the final subsurface effect of ion bombardment to be discussed. The sputter-erosion of solid material by positive-ion bombardment is widely used as a source of vapor for thin-film deposition because of the following unique combination of advantages over other techniques: any material can be volatilized by sputtering, compounds are volatilized stoichiometrically, and the film deposition rate can be made uniform over very large areas. Furthermore, the kinetic-energy distribution of sputtered atoms falls largely within the energy window for displacing surface atoms on the depositing film without causing subsurface damage, as discussed at the beginning of Sec. 8.5. In addition to the use of sputtering to supply vapor, partial sputter-erosion of the film as it is being deposited ("resputtering") can be used to improve film coverage over rough topography and to enhance preferential crystallographic orientation of the deposit, although it sometimes produces too much compressive stress. A small amount of "intrinsic" resputtering always occurs, largely due to the impingement of fast neutrals reflected from the target material. Sputtering can also be used to clean the substrate prior to deposition ("backsputtering"), but this should be done in a separate chamber or with a shutter blocking the target surface, so that the target does not become contaminated. Below, we will discuss first the equipment used for sputtering and then the mechanism of sputtering.

8.5.4.1 Equipment. The two common setups for sputter deposition are shown in Fig. 8.25. Their operating characteristics were listed in Table 8.1. In both, inert-gas ions are generated in a glow-discharge plasma. The references made below to glow-discharge terminology and operation will be very brief, since the subject is treated more fully in Chap. 9. In both setups, the ions are accelerated into a "target" consisting of the material to be sputtered. The target must be well bonded to a water-cooled Cu backing plate to dissipate the heat created by bombardment and thus to prevent outgassing or melting of the target material. Since sputtered material emanates in roughly a cosine distribution (Fig. 4.12), the substrate is best placed directly opposite the target as shown. In the ion-gun setup of Fig. 8.25a, ions are extracted from the plasma and accelerated toward the target by the positive bias of the gun, impinging at an angle θ which may be varied. In the glow-discharge setup of (b), ions are accelerated out of the plasma into the target by the voltage drop between the two (the "cathode fall") established by the target bias, so the ions always impinge perpendicularly on the target. By also biasing the *substrate* negatively, ions may be accelerated out of the same plasma into the substrate to enhance resputtering. When the substrate is electrically insulating, rf rather than dc bias must be used. When the target is insulating, rf power must be used to bias it in the glow-discharge configuration, while in the ion-gun configuration, an electron-emitting filament (Appendix D) must be used instead to neutralize the ion beam. Confinement of the

8.5.4 Sputtering

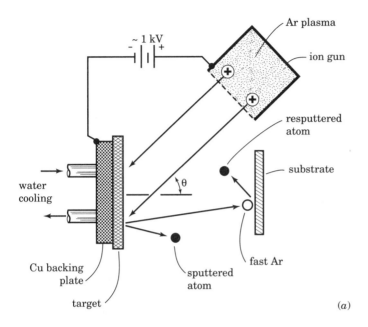

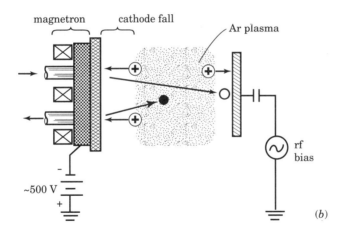

Figure 8.25 Typical setups for film deposition sputtering: (a) ion beam and (b) glow discharge.

plasma near the target is improved by the magnetic field of the "magnetron" sputter source shown, although glow-discharge sputtering can also be done in the "diode" mode without magnets. In either the magnetron or the diode, the ion flux and energy at the target are interrelated by the electrical impedance of the plasma, since the target is directly coupled to it. Conversely, in the ion-gun configuration, the ion flux is determined by the intensity of the plasma in the gun, while the ion energy is separately controlled by the bias of the gun assembly. This separate control of flux and energy, coupled with the adjustability in ion impingement angle, make the ion gun more useful for process research on sputtering, whereas the directly coupled glow discharge is simpler and higher in sputtering rate for routine film deposition.

The process pressure is an important consideration in sputtering. This is the pressure p_2 in the generalized vacuum system of Fig. 3.3, and we will refer to this system below. In the ion-beam setup of Fig. 8.25a, p_2 must be low enough to ensure a Knudsen number Kn > 1 so that the beam does not become scattered during transport: typically, $p_2 < 10^{-2}$ Pa. Conversely, p_2 in the glow-discharge setup must be *high* enough to ensure enough electron collisions to sustain the plasma, in accordance with the discussion accompanying Fig. 8.4. With the magnetron, minimum p_2 is about 30× lower than with the diode—about 0.1 Pa versus 3 Pa—because of magnetic confinement of the electrons, but p_2 can be even lower with the ion gun. Operating at lower p_2 decreases the partial pressure of outgassing contaminants, p_i, in accordance with Eq. (3.9) for throttled pumping, if Ar mass flow, Q_s, and outgassing rate, Q_i, are taken as fixed. To minimize p_i, Q_s should be set at the maximum throughput that the vacuum pump can handle, and then the throttle should be opened to reach the minimum p_2 that will sustain the plasma. There is a perception that ion-beam sputtering is cleaner than glow-discharge sputtering because p_2 is lower. However, Q_i is often *higher* in ion-beam sputtering because much more contaminating surface area is exposed to plasma in the gun assembly, and because a hot filament is often operating. Also, the film contamination fraction increases with the *ratio* of contaminant to film-material impingement flux; that is, inversely with the "purity ratio" of Sec. 2.6. Film deposition flux is much lower in ion-beam sputtering, because the ion flux to the target is lower, and because the substrate cannot be situated as close to the target.

Film purity is further assisted by a clean target and clean process gas. The inert gases are available in extremely high purity. A new target, or one that has been exposed to air, should be presputtered to clean the surface before films are deposited. The guidelines for target selection are the same as those outlined for evaporants in Sec. 4.5.2, and the following additional guidelines apply. Whenever possible, one should avoid target material made by "hot-pressing" or "sintering" of powders." These processes are used to form target shapes from materials that are difficult to cast from a melt or difficult to machine into shape, and sometimes there is no alternative. However, such materials inevitably contain internal surface area and voids which increase the outgassing of adsorbed impurities and sometimes also result in ejection of poorly bonded macroparticles which break loose upon being uncovered by the sputter-erosion. The hot isostatic pressing (HIP) process for powdered material operates at very high pressure and thus minimizes void fraction. Finally, in the case of cast alloy targets, fine grain size is preferred so as to minimize segregation of constituent elements into nodules during ion bombardment, because these are another source of macroparticle ejection.

8.5.4.2 Yield. We now turn to the task of describing the sputtering mechanism quantitatively. That is, how do ion-beam and target properties determine ion-energy threshold and sputtered-particle yield and energy? Here, we consider the sputtering of *atoms*, although sputtered material can also include clusters of two or three atoms [56]. Regarding **energy threshold**, consider that volatilization by sputtering occurs when an atom in the solid receives enough outward-directed kinetic energy to overcome its chemical binding energy to the surface, E_b. It has been experimentally determined that the minimum kinetic energy required is about equal to the heat of sublimation, $\Delta_s H$. This seems reasonable, although one might actually expect a higher minimum energy based on the fact that sublimation occurs preferentially from the least tightly bound atoms such as those at the outside corners of kinks (see Fig. 5.13), whereas sputtering occurs at random points of ion impact across the surface. Recall for comparison that E_f, the minimum energy for atomic displacement *within* the solid, is roughly 5 times $\Delta_s H$ by Fig. 8.18. The factor of 5 difference between the atom kinetic energies required for these two processes comes partly from the larger binding energy in the bulk than on the surface and partly from the extra energy required to stabilize the Frenkel pair against spontaneous recombination in the bulk. The corresponding *ion*-energy threshold, E_0, for sputtering depends on the ion and target-atom masses, which we will designate as M_i and M_t (rather than m_i and m_t) to denote atomic mass units (u). Knowledge of E_0 is especially important when one wants to *avoid* sputtering, for example from the film or from a potentially contaminating deposition-chamber surface in the operation of an energy-enhanced process.

We distinguish two basic sputtering mechanisms for threshold calculations. Outward-directed energy can be delivered to the sputtering atom either directly from a reflected ion as in Fig. 8.15c or indirectly through the collision cascade as in Fig. 8.16c. For $M_i > M_t$, only the latter mechanism applies, because in that case the ions cannot be reflected in a binary collision [Eq. (8.20)]. The reflected-ion mechanism has the lower threshold, being more direct, and this threshold will be minimum when both the ion and the sputtered atom are travelling nearly perpendicular to the surface. Then, by Eq. (8.20), the energy retained by the ion following reflection in a head-on collision within the solid is $(1 - \gamma_m)E_i$, and the fraction of this energy transferred to the sputtered atom in a head-on collision is γ_m. Thus, the ion-energy threshold for sputtering when $M_i < M_t$ is

$$E_o = \Delta_s H/(1 - \gamma_m)\gamma_m \qquad (8.34)$$

which agrees with experimental data to within a factor of two [57] for $M_i < 0.3M_t$. The E_o of Eq. (8.34) decreases with increasing M_i and reaches a minimum of $4\Delta_s H$ at $\gamma_m = 0.5$ and $M_i = 0.17M_t$, which is only *coincidentally* close to the E_f threshold of Fig. 8.18. The E_o for higher M_i/M_t is not so easily predicted, but experimental data for $M_i > 0.3M_t$ fit the empirical formula [57]

$$E_o = 8\Delta_s H(M_i/M_t)^{2/5} \qquad (8.35)$$

The next quantity of interest is the **sputtering yield**, Y_s atoms per ion. For now, we will consider ions impinging perpendicularly on the surface. The measured behavior of Y_s versus E_i just above threshold follows the characteristic curve shown in Fig. 8.26 with normalized coordinates. The curves for each ion-target combination have been displaced into alignment vertically by a yield multiplication factor, f_s, to give a normalized yield, Y_s/f_s, where f_s has been arbitrarily set at unity for H^+–Au. For $M_i/M_t < 1$, f_s fits the empirical formula [57]

$$f_s = 0.75 M_t \gamma_m^{5/3} \qquad (8.36)$$

When using sputtering as a source of vapor for film deposition, one wants to know Y_s further above threshold, where sputtering is more efficient. Figure 8.27 shows typical Y_s behavior for the case of a Ni target. Considerable experimental Y_s data on other elemental targets is given in the literature [58, 59] and in Appendix C. In Fig. 8.27, measured Y_s is seen to be in good agreement with Monte-Carlo calculations using the TRIM code (Biersack, 1987). Recall that the Monte-Carlo method averages many ion histories as the ions undergo binary collisions within the target. The fraction of ions reflected (Fig. 8.13) and the bulk-displacement distributions (Fig. 8.23) can also be obtained. For displacement and sputtering calculations, the trajectories of both the ion and all of the knock-on atoms in the collision cascade must be followed until their energies become insignificant. This procedure uses Eq. (8.25) for the collision interaction potential and also takes into account the electronic stopping, S_e, of Eq. (8.22). Whenever a knock-on atom reaches the surface in its trajectory, it is counted as a sputtered

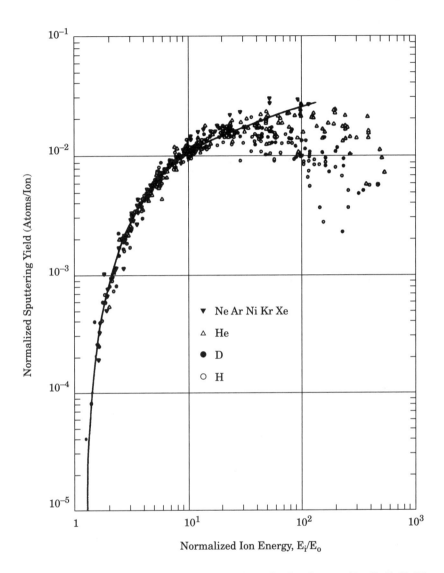

Figure 8.26 Sputtering yield, Y_s, near threshold, E_o, for the elements Be, C, Al, Si, Ti, V, Fe, Ni, Zr, Mo, Ta, W, and Au, normalized as Y_s as described in the text. (Source: Reprinted from Ref. 57 by permission.)

atom if its component of kinetic energy perpendicular to the surface is larger than $\Delta_s H$. Atoms of lower energy are instead "refracted" back into the solid just as light is refracted back into a solid medium at the surface (Fig. 4.18). Regarding behavior near threshold in Fig. 8.27, the lowest incident-ion energies for which Y_s is plotted are consistently higher by about ×2 for H, D, and He than the thresholds calculated

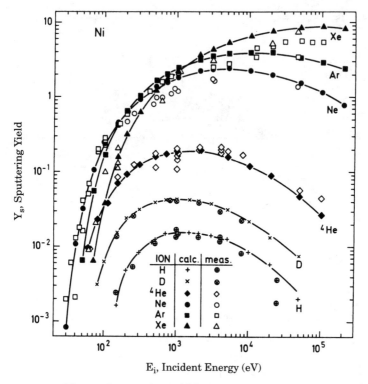

Figure 8.27 Measured sputtering yield for Ni and TRIM-code calculation for various ions impinging perpendicularly. [Source: Reprinted from Ziegler (1985) by permission of Macmillan Publishing Co., copyright © 1985 by Pergamon Press.]

from Eq. (8.34). This is because the Fig. 8.27 data do not extend to low enough Y_s to actually be observing the threshold, as is apparent by comparison with Fig. 8.26.

Y_s can also be determined analytically using the Boltzmann transport equation to describe the collision cascade. Recall that this equation was also used to describe ion stopping in the range calculation of Sec. 8.5.2.2. This derivation of Y_s is complex, but it results in the following surprisingly simple and highly successful formula [60]:

$$Y_s = \frac{4.2\alpha S_n}{\Delta_s H} \quad (8.37)$$

As above, this Y_s is for ions impinging perpendicularly. Here, the nuclear stopping power, $S_n(E_i)$, is in $eV \cdot nm^2/atom$ and may be evaluated from Fig. 8.22 and Eq. (8.31). The factor α has the theoretical dependence on M_t/M_i shown by the dashed line in Fig. 8.28. Agree-

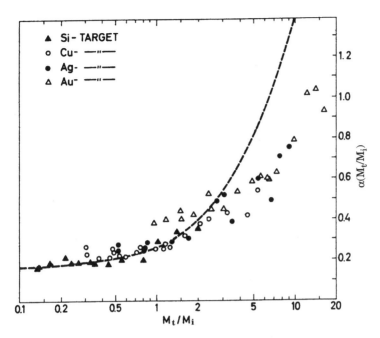

Figure 8.28 The factor α of Eq. (8.37). Dashed line was calculated from transport theory [60], and data points from measured sputtering yields for the targets indicated and ions of various masses, M_i. (Source: Reprinted from Ref. 58 by permission.)

ment with α values found from experimental Y_s data is seen to be very close except at high M_t/M_i. The rise in α for $M_t/M_i > 1$ is due to the onset of the more direct and therefore more efficient reflected-ion mechanism (Fig. 8.15c). The scaling of Y_s with S_n predicted by Eq. (8.37) may be observed by comparing the shapes of the Y_s and S_n curves versus ion energy in Figs. (8.27) and (8.22). This scaling results from the fact that when the ion transfers its energy to knock-on atoms within a shallower depth, those atoms have a greater chance of reaching the surface before they lose too much energy to escape. Finally, the inverse dependence of Y_s on $\Delta_s H$ may be understood by analogy to the reasoning behind Eq. (8.32) for the number of displacements produced by an ion. That is, higher $\Delta_s H$ means that fewer of the knock-on atoms reaching the surface will have enough energy to escape. The agreement of Eq. (8.37) with experimental Y_s is generally very good when incident energy, E_i, is well above threshold. Taking the case of Ar^+–Ni at 1000 eV as an example, we find $S_n = 7.1$ eV·nm^2/atom from Fig. 8.22 and Eq. (8.31). This gives $Y_s = 2.0$ from Eq. (8.37), which falls right on the experimental data of Fig. 8.27.

To estimate **deposition rate** from a sputter source, one needs to know E_i, Y_s, the ion current to the target (I_i), the target-substrate geometry, and the gas pressure. In ion-beam sputtering, one can measure E_i and I_i directly. In glow-discharge sputtering, one measures the total power and voltage drop of the plasma: P and V. From these, I_i is roughly determined by assuming that all of this power is dissipated in ion bombardment of the target and that the ions are accelerated to maximum energy given the voltage drop available (that is, no scattering), in which case I_i = P/V. These are approximate assumptions, as discussed further in Secs. 9.3.1 and 9.4.4, but they are suitable for rough estimates. Sputtered material then leaves the target at a rate

$$Q_s \text{ (atoms/s)} = Y_s I_i / q_e \qquad (8.38)$$

where q_e is the electron (ion) charge, 1.60×10^{-19} C. Material leaving a particular point on the target is angularly distributed in roughly the cosine distribution which was shown in Fig. 4.12 for thermal evaporation. In sputtering, there can be deviations from this distribution characterizable as $\cos^n\theta$. Usually, $0.5 < n < 2$, n being affected mainly by crystallography and ion energy [61]. However, because Q_s is distributed over a much larger area of the target than is typical of thermal evaporation, this distribution and its variability are of little concern, and uniform deposition over large substrate areas is easily obtained with sputtering. When gas pressure is high enough that scattering and thermalization of the sputtered particles occurs, as in glow-discharge sputtering without a magnetron, some of the particles will diffuse back to the target rather than proceeding to the substrate, thus reducing deposition rate. This and other pressure effects are discussed further in Sec. 9.3.3. Because so many factors determine sputter-deposition rate, determining it to better than a factor of two for particular process conditions requires direct measurement. Rate will then be reproducible to within a few percent in successive depositions at the same power and gas pressure. Reproducibility and control of deposition rate are easier in sputtering than in evaporation, because Q_s depends linearly on power in Eq. (8.38), whereas vapor pressure depends exponentially on the T of an evaporation source [Eq. (4.14)].

When there is more than one element present at the surface being sputtered, Y_s determination becomes more complicated. We consider here first adsorbates and then bulk constituents. **Adsorbates** are present when sputter-cleaning a surface or when carrying out reactive sputtering of a metal target in a reactive gas for compound-film deposition (more in Sec. 9.3.3). In both cases, the rate of adsorbate removal is expected to be inversely proportional to the adsorption energy (Fig. 5.2) by analogy to Eq. (8.37). Meanwhile, the Y_s of the *bulk* material will be reduced, both because of the stopping power of the adsorbate for the ions and because most of the sputtered atoms come from the top monolayer.

Regarding the sputtering of a **compound or alloy** target, the Q_s of Eq. (8.38) for each element must first be multiplied by its fractional surface coverage, Θ, to give the sputtering rate for that element: ΘQ_s. On a fresh target, Θ may be the same as the bulk atomic fraction. However, the ΘQ_s ratio between elements will generally be different than the Θ ratio itself, because the Y_s values are different for each element. Thus, as sputtering proceeds, the volume of target material within an ion-mixing depth of the surface becomes depleted in the high-Y_s element until a steady state is reached in which the ΘQ_s ratio *is* equal to the bulk atomic ratio. This is the same as the wire-feed alloy-evaporation problem of Eq. (4.25), except that here the volume of mixed material is orders of magnitude smaller, and therefore the time to reach steady state is negligible. Conversely, in the case of a film being subjected to significant *resputtering* while it is being deposited, depletion of the high-Y_s element will not be able to occur at the film surface because of the continual deposition of fresh material from the target and at the target bulk ratio. Thus, the film will always end up being depleted in the high-Y_s element when compared to the composition of the target material.

Bonding differences among the constituents of compound or alloy targets affect Y_s in various ways. The near-surface composition of a sputtered target or film may be shifted by segregation of a weakly bonded element to the surface due to radiation-enhanced diffusion [62]. Also, the Y_s of an element in a compound or alloy may be very different from the pure-element value because of differences in bonding strength as represented by $\Delta_s H$ in Eq. (8.37). Finally, elements strongly bonded to each other can sputter as diatomic molecules rather than as atoms.

It was mentioned above that adsorbates can reduce the Y_s of the bulk material. However, when adsorbate reaction with the bulk is activated by ion bombardment and when the reaction product is volatile, Y_s increases, and the process is called **"chemical sputtering"** [63]; for example, $SF_6(a) + Si(c) \to SiF_4(g)$. In a related process called "chemically enhanced physical sputtering" [63], the adsorbate reacts with the bulk to form a molecule having higher Y_s than the bulk, such as $F(a) + Si(c) \to SiF_n(a)$. These processes are encountered more in ion-activated plasma etching than in thin-film deposition, but it is important to be aware of them.

Crystal orientation also affects Y_s, so that resputtering of the film surface during deposition of a polycrystalline film can result in a preferred crystallographic orientation, or texture, which is different from the one determined thermodynamically and discussed in Sec. 5.4.2. In some cases, it has been found that the preferred plane parallel to the surface corresponds to having ion channeling directions perpendicular to the surface [64, 65]. This seems reasonable, because channeling directions have lower ion-stopping power, S_n, which reduces the energy transfer per unit volume of crystallite and reduces Y_s in Eq. (8.37). Film crystallites nucleating with other orientations will be preferentially made amorphous by the higher energy transfer, or they will be resputtered away. In other cases [66–68], however, it has been found that the preferred plane is the *close-packed* one, which does not have a channeling direction perpendicular to it. This also seems reasonable, since close-packed planes provide more complete bonding for surface atoms and therefore have higher $\Delta_s H$. Whether S_n or $\Delta_s H$ dominates the Y_s change with orientation is likely to depend on various factors such as crystal structure, ion mass, and ion energy.

We will discuss one final factor affecting sputtering yield: the ion-beam **incident angle**, θ, of Fig. 8.25a. The Y_s referred to above has always been for perpendicular impingement, and now we wish to relate that Y_s to $Y_s(\theta)$. From Eq. (8.37), we learned that shallower collision cascades increase Y_s because the displaced atoms can reach the surface more easily. Thus, we would expect to find $Y_s\theta \propto S_n/\cos\theta$. However, at very grazing angles approaching 90°, the *fraction* of ions penetrating the surface is expected to drop abruptly toward zero as more of them become deflected away from the surface by their first collision with a surface atom, and also as they ultimately have an insufficient perpendicular component of energy to penetrate the surface at all (see Fig. 8.17). The experimental $Y_s(\theta)$ data shown in Fig. 8.29 support these expectations, although $Y_s(\theta)/Y_s$ does not follow $1/\cos\theta$ exactly. Note that the height of the peak in $Y_s(\theta)/Y_s$ at around 60° varies considerably among the target metals shown. For ion incidence at off-perpendicular angles, the angular distribution of the sputtered material tends to be lobed near the specular-reflection angle of the ion [69] rather than following a cosine distribution, because of the higher probability of forward-collisional events compared to events involving momentum reversal. The peak size of this lobe increases with decreasing ion energy [61, 69], because at lower energy the collision cascade contains fewer events and therefore does not become as randomized directionally.

The increase in $Y_s(\theta)$ with θ and the lobe effect both have an important application in the improvement of film coverage over steep surface topography and in the **planarization** of surface roughness. Rough surfaces are always difficult to cover evenly with materials of high melting point, T_m, when they are deposited in a process of high sticking coefficient and at low substrate temperature, T_s. At $T_s/T_m <$ 0.3 or so, the process is in the "quenched-growth" regime of Sec. 5.4.1, where surface diffusion is negligible. In this regime, depositing material preferentially accumulates on outside corners of surface steps as

8.5.4 Sputtering

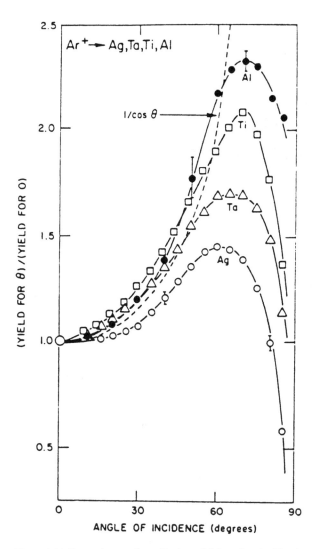

Figure 8.29 Dependence of sputtering yield on ion incident angle, θ, for 1.05-keV Ar⁺. (Source: Reprinted from Ref. 69 by permission.)

shown in Fig. 8.30, because these areas see a larger solid angle of the incident vapor than do the sidewalls and inside corners of the steps. The accumulating film increases this "self-shadowing" of the inside corners, resulting in very poor sidewall coverage. This effect greatly reduces the current-carrying capacity and reliability of metal lines deposited over steps in integrated circuitry, for example. Partial resputtering of the film during deposition dramatically improves step

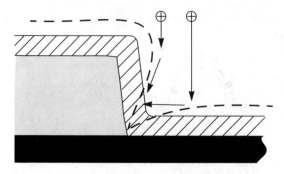

Figure 8.30 Improvement of film coverage over steep steps by resputtering during deposition. Thickness profiles are shown without (dashed line) and with (area with hash marks) resputtering.

coverage due to preferential forward-sputtering of material from the outside corner into the inside corner as shown in Fig. 8.30. To a lesser extent, material resputtered from the lower flat areas also improves sidewall coverage. When films are deposited over grooves or depressions whose width is not much larger than their height (high aspect ratio), sufficient resputtering can result in complete filling of these low areas and effective planarization of the surface, as shown for Al deposition in Fig. 8.31. The bombardment energy required to produce this much resputtering does deposit considerable heat, so that substrate overheating can be a problem.

The self-shadowing effect is more severe in sputtering than in thermal evaporation, especially when large targets are used at close substrate spacing, because of the resulting broad range of angles from which the sputtered atoms approach the substrate. This can even produce the microcolumnar voids of the generally undesirable "Zone-1" film structure despite the high kinetic energy of the sputtered particles, which tends to collapse such voids. The sputtered flux can be **collimated** by inserting "honeycomb" shielding in front of the target to trap the more obliquely directed atoms [70]. However, a large penalty is paid in deposition rate, and flaking of deposits from the shields can contaminate the film.

8.5.4.3 Energy. The high translational kinetic energy, E_t, of sputtered particles has been cited several times above as being a principal advantage of this film-deposition process, because of its beneficial effects on film adherence and structure. Application of transport theory to the collision cascade results in the Thompson distribution [71] of E_t, which is usually written as

8.5.4 Sputtering

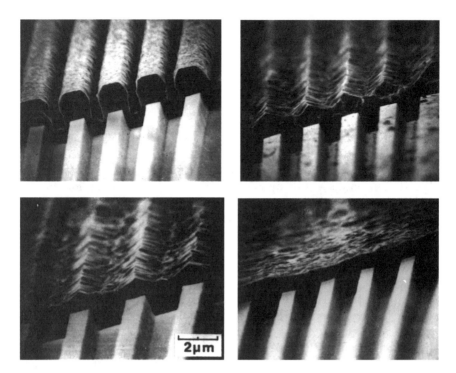

Figure 8.31 Cross-sectional scanning electron micrographs of sputtered Al coverage over a grooved substrate for various resputtering rates given as a percentage of the deposition rate: top left, 0%; top right, 40%; bottom left, 50%; bottom right, 70%. (Source: Reprinted from Ref. 68 by permission.)

$$\frac{dN}{dE_t} \propto \frac{E_t}{(E_t + E_b)^3} \qquad (8.39)$$

Here, dN/dE_t is the number of particles within a differential energy interval, and E_b is the surface binding energy of target atoms. Normalization of this distribution gives the probability of a sputtered particle having kinetic energy within a differential E_t interval:

$$\frac{dN/dE_t}{N} = \frac{2E_b E_t}{(E_t + E_b)^3} \qquad (8.40)$$

Here, as in the case of the sputtering threshold-energy calculations of the previous section, it is usually assumed that $E_b \approx \Delta_s H$. Fitting of Eq. (8.40) to measured E_t distributions [72] shows that this assumption is accurate to within 20 percent in many cases, and to within a factor of two in general.

The key features of Eq. (8.40) are a very broad maximum in probability centered at $(1/2)\, E_b \approx (1/2)\, \Delta_s H$, and a $1/E_t^2$ dependence for $E_t \gg E_b$. In practice, there is also a high-energy cutoff in the distribution corresponding to T_m, which is the maximum energy transferable from ion to sputtered atom by Eq. (8.20). All of these features are also seen in Monte-Carlo calculations of sputtering using the TRIM code (Biersack, 1987), although in that case the high-energy behavior is closer to $1/E_t^{3/2}$. Surprisingly, the maximum energy transfer, T_m, does not appear in Eq. (8.40), meaning that sputtered-particle energy is predicted to be *independent* of incident ion energy. Experimentally, however, the E_t distribution does shift up slightly in E_t with increasing ion energy [73]. For example, it has a maximum at 2 eV for Fe sputtered with 500-eV H^+ and 3 eV with 2500-eV H^+. By the way, the deviation of the Eq. (8.40) distribution from the Maxwellian distribution of Eq. (2.1) is one of the main pieces of evidence that sputtering is caused by collision cascades rather than by thermal spikes.

It is most convenient that the bulk of the Eq. (8.40) distribution falls within the desirable energy window for activating surface processes such as reaction and densification without producing bulk atomic displacements in the film (see Exercise 8.19), and this is one of the principal advantages of the sputter-deposition process. However, the high-energy tail of the distribution will cause *some* displacements and also some resputtering of the depositing film. This component of resputtering amounts to a few percent of the deposition for many elements [74]. For $M_t > M_i$, the number of displacements and the percent resputtering of the film both increase sharply due to additional energy input from neutralized gas ions reflected from the target [25], as illustrated in Fig. 8.25a. These neutrals have significant reflection only for $M_t > M_i$, as plotted in Fig. 8.13. Resputtering caused either by the high-energy tail of the Eq. (8.40) distribution or by the reflected neutrals is known collectively as "intrinsic" resputtering, because it is inherent to the process. Additional resputtering may be achieved if desired by applying negative bias to the substrate as discussed earlier.

8.6 Conclusion

Energy beams of electrons, photons, or ions can be used to volatilize even the most refractory source material for supplying vapor to a thin-film deposition process. Using laser pulses or ion-sputtering, compounds are more easily volatilized in the stoichiometric ratio than using the other energy-beam processes or thermal evaporation. The vapor produced by energy beams always contains energetic particles in the form of ions or fast neutral atoms, although the particle mix and energy vary considerably with the type of beam used and the operating conditions. These energetic particles impinging on the depositing film have dramatic and usually beneficial effects on film structure and composition because of their ability to displace surface atoms upon impact. The energy impinging on the film can be further increased by accelerating ions into the film. Particle bombardment may displace only surface atoms or may also displace atoms in the bulk, depending on whether the threshold energy for bulk displacement is exceeded, and therefore the control of particle energy is important in achieving the desired film properties. In semiconductor epitaxy, for example, bulk displacement degrades film electronic properties, while in metallurgical applications, it increases adherence and compressive stress. The theory of binary atomic collisions is well developed and can be used to accurately predict the amount and depth of atomic displacement.

The presence of a plasma is a general characteristic of all energy-beam techniques. In the next chapter, we will examine the behavior of the widely used glow-discharge type of plasma in more detail. Control of the glow discharge is the key to controlling ion impingement energy and flux in many processes.

8.7 Exercises

8.1 Verify the derivation given for the Child-Langmuir law by writing down all of the steps.

8.2 A 0.040-inch diameter, 3-cm long W filament is operating at 2200 K and ground potential in vacuum, 10 cm away from a collecting anode. What bias voltage is required on the anode to just avoid space-charge-limited current flow, assuming that the filament can be considered a planar emitter?

8.3 For electrons traversing an electrode gap of 10 cm and a potential difference of 1 kV, what is the lowest pressure of O_2 at which 20 percent of the electrons will cause ionization? (Neglect the energy dependence of ionization cross section.)

8.4 Make a table listing the advantages and potential disadvantages of using nonthermal energy in a thin-film process—at the source, during vapor transport, and on the depositing film.

8.5 Calculate the pressure exerted on a cathode by ion bombardment from an arc plasma, assuming Ti^+ ions arriving at 20 eV with a flux of 10^8 A/cm^2. The ion velocity from Eq. (8.7) may be used in Eq. (2.8), except that the ions should be assumed to stop at the surface rather than to bounce off it.

8.6 BaO is being evaporated by a pulsed laser. Assuming that the pulse energy is used entirely to heat and evaporate the material and to accelerate it to 40 eV, what fraction of the energy is used in the acceleration?

8.7 Why can't films be deposited by laser ablation at 1 atm?

8.8 Show that conservation of energy and momentum result in Eq. (8.20).

8.9 (a) By conservation of energy and of the lateral and longitudinal components of momentum, calculate the energy retained by an ion in a grazing collision with a stationary atom of equal mass, where the deflection angle is θ, as shown in Fig. 8.12e. (b) Show that the retroreflected energy after n sequential collisions at angle θ in the same plane, with $n\theta = \pi$, is given by Eq. (8.21).

8.10 For Si deposition at 10 μm/h with supplementary Ar^+ bombardment, what ion-current density (mA/cm^2) is needed to produce a 1:1 Ar^+:Si flux ratio?

8.11 Distinguish four types of subsurface atomic-displacement collisions.

8.12 For H^+ impinging on Ni at energy E_i, estimate the E_i threshold for generating Frenkel pairs (a) just below the surface, and (b) at a depth of 20 nm. For simplicity, assume an energy-independent stopping power.

8.13 Consider an atomic nucleus of mass m_i, atomic number Z_i, and initial energy E_i, colliding head-on with a second nucleus, m_t, Z_t, initially at rest. Using conservation of momentum and of kinetic plus potential energy, derive the following expression for the distance of closest approach in SI units [as in Eq. (8.30)]:

$$r_o = 9 \times 10^9 \, q_e \, 2 Z_i Z_t (m_i + m_t)/E_i m_t$$

Note that at r_o, the velocities of the two nuclei must be the same.

8.14 What are the three sources of energetic-ion/atom bombardment of the film during sputter deposition, and what are the factors that influence their respective fluxes and energies?

8.15 Sputtered Mo atoms having an average kinetic energy of 3 eV are depositing at 1×10^{16} atoms/$cm^2 \cdot$s onto one face of a 1-mm thick glass sheet suspended in vacuum. For the glass, $c_g = 0.84$ W·s/g·°C and, $\rho_m = 2.76$ g/cm^3; for the Mo, the heat of sublimation is $\Delta_s H = 659$ kJ/mol, and c_g may be neglected. Assume that heat sources other than the sputtered Mo are negligible. (a) What is the initial rate of glass-T increase from room T in °C/s? (b) What fraction of this rate is due to $\Delta_s H$? (c) What is the steady-state glass T assuming only radiative heat loss at an emissivity of 0.1 for the Mo surface and 0.9 for the glass surface?

8.16 Show that the ion-energy threshold for sputtering given by Eq. (8.34) has a minimum value of $4\Delta_s H$ at $M_i = 0.17 M_t$.

8.17 Using the analytical (not the Monte-Carlo) approach, estimate the sputtering yield, Y_s, for Xe^+ bombarding Pb at 500 eV.

8.18 What is the difference between resputtering, backsputtering, presputtering, and forward-sputtering?

8.19 Using Eq. (8.40), and also taking into account the high-energy cutoff, determine what fraction of the Ge that is being sputtered by 500-eV Ar^+ falls above the energy threshold for subsurface atomic displacement in the depositing film.

8.8 References

1. Feynman, R.P., R.B. Leighton, and M. Sands. 1964. *The Feynman Lectures on Physics*, vol. 2. Reading, MA: Addison-Wesley, 6–13.
2. Zalm, P.C., and L.J. Beckers. 1985. "Ion-Induced Secondary Electron Emission from Copper and Zinc." *Surface Science* 152:135.
3. Hagstrum, H.D. 1977. "Low Energy De-excitation and Neutralization Processes Near Surfaces." In *Inelastic Ion-Surface Collisions*, ed. N.H. Tolk, et al. New York: Academic Press.
4. Baragiola, R.A., E.V. Alonso, J. Ferron, and A. Oliva-Florio. 1979. "Ion-Induced Electron Emission from Clean Metals." *Surface Science* 90:240.
5. Rapp, D., and P. Englander-Golden. 1965. "Total Cross Sections for Ionization and Attachment in Gases by Electron Impact. I: Positive Ionization," *J. Chem. Phys.* 43:1464.
6. Thornton, J.A. 1982. "Plasmas in Deposition Processes." Chap. 2 in R.F. Bunshah et al., *Deposition Technologies for Films and Coatings*. Park Ridge, NJ: Noyes Publications, 40.
7. Tatsumi, T., H. Hirayama, and N. Aizaki. 1989. "Si Particle Density Reduction in Si Molecular Beam Epitaxy using a Deflection Electrode." *Appl. Phys. Lett.* 54:629.
8. Meassick, S., C. Chan, and R. Allen. 1992. "Thin Film Deposition Techniques Using the Anodic Vacuum Arc." *Surface and Coatings Technology* 54,55:343.
9. Kimblin, C.W. 1973. "Erosion and Ionization in the Cathode Spot Regions of Vacuum Arcs." *J. Appl. Phys.* 44:3074.
10. Anders, A., S. Anders, A. Förster, and I.G. Brown. 1992. "Pressure Ionization: Its Role in Metal Vapour Vacuum Arc Plasmas and Ion Sources." *Plasma Sources Sci. Technol.* 1:263.
11. Hantzsche, E. 1992. "A Hydrodynamic Model of Vacuum Arc Plasmas." *IEEE Trans. on Plasma Science* 20:34.
12. Swift, P.D., D.R. McKenzie, and I.S. Falconer. 1989. "Cathode Spot Phenomena in Titanium Vacuum Arcs." *J. Appl. Phys.* 66:505.
13. Martin, P.J., R.P. Netterfield, A. Bendavid, and T.J. Kinder. 1992. "The Deposition of Thin Films by Filtered Arc Evaporation." *Surface and Coatings Technology* 54/55:136.
14. Falabella, S., and D.M. Sanders. 1992. "Comparison of Two Filtered Cathodic Arc Sources." *J. Vac. Sci. Technol.* A10:394.
15. Randhawa, H., and P.C. Johnson. 1987. "Cathodic Arc Plasma Deposition Processes and Their Applications." *Surface and Coatings Technology* 31:303.
16. Webb, R.L., L.C. Jensen, S.C. Langford, and J.T. Dickinson. 1993. "Interactions of Wide Band-Gap Single Crystals with 248 nm Excimer Laser Radiation." *J. Appl. Phys.* 74:2323, 2338.
17. Brannon, J. 1993. *Excimer Laser Ablation and Etching*. New York: American Vacuum Society.

18. Boyce, J.B., et al. 1989. "In Situ Growth of Superconducting YBa$_2$Cu$_3$O$_y$ Films by Pulsed Laser Deposition." *Proc. SPIE Symp. on Processing of Films for High-T$_c$ Superconducting Electronics* 1187:136.
19. Venkatesan, T. 1988. "Observation of Two Distinct Components during Pulsed Laser Deposition of High T$_c$ Superconducting Films." *Appl. Phys. Lett.* 52:1193.
20. Zheng, J.P., Z.Q. Huang, D.T. Shaw, and H.S. Kwok. 1989. "Generation of High-Energy Atomic Beams in Laser-Superconducting Target Interactions." *Appl. Phys. Lett.* 54:280.
21. Singh, R.K., O.W. Holland, and J. Narayan. 1990. "Theoretical Model for Deposition of Superconducting Thin Films using Pulsed Laser Evaporation Technique." *J. Appl. Phys.* 68:233.
22. Ready, J.F. 1971. *Effects of High-Power Laser Radiation*, chap. 5. Orlando, Fla.: Academic Press.
23. Gagliano, F.P., and U.C. Paek. 1974. "Observation of Laser-Induced Explosion of Solid Materials and Correlation with Theory." *Applied Optics* 13:274.
24. Bushnell, J.C., and D.J. McCloskey. 1968. "Thermoelastic Stress Production in Solids." *J. Appl. Phys.* 39:5541.
25. Eckstein, W., and J.P. Biersack. 1986. "Reflection of Heavy Ions." *Z. Phys. B* 63:471.
26. Window, B. 1993. "Removing the Energetic Neutral Problem in Sputtering." *J. Vac. Sci. Technol.* A11:1522.
27. Winters, H.F., H. Coufal, C.T. Rettner, and D.S. Bethune. 1990. "Energy Transfer from Rare Gases to Surfaces: Collisions with Gold and Platinum in the Range 1-4000 eV." *Phys. Rev. B* 41:6240.
28. Martin, J.S., J.N. Greeley, J.R. Morris, B.T. Feranchak, and D.C. Jacobs. 1994. "Scattering State-Selected NO$^+$ on GaAs(110): the Effect of Translational and Vibrational Energy on NO$^-$ and O$^-$ Product Formation." *J. Chem. Phys.* 100:6791.
29. Ro, J.S., C.V. Thompson, and J. Melngailis. 1994. "Mechanism of Ion Beam Induced Deposition of Gold." *J. Vac. Sci. Technol.* B12:73.
30. Choi, C.-H., R. Ai, and S.A. Barnett. 1991. "Suppression of Three-Dimensional Island Nucleation during GaAs Growth on Si(100)." *Phys. Rev. Lett.* 67:2826.
31. Choi, C.-H., and S.A. Barnett. 1989. "Nucleation and Epitaxial Growth of InAs on Si(100) by Ion-Assisted Deposition." *Appl. Phys. Lett.* 55:2319.
32. Gilmore, C.M., and J.A. Sprague. 1991. "Molecular Dynamics Simulation of the Energetic Deposition of Ag Thin Films." *Phys. Rev. B* 44:8950.
33. Poelsema, B., R. Kunkel, L.K. Verheij, and G. Comsa. 1990. "Mechanisms for Annealing of Ion Bombardment Induced Defects on Pt(111)." *Phys. Rev. B* 41:11609.
34. Hultman, L., S.A. Barnett, J.-E. Sundgren, and J.E. Greene. 1988. "Growth of Epitaxial TiN Films Deposited on MgO(100) by Reactive Magnetron Sputtering: the Role of Low-Energy Ion Irradiation during Deposition." *J. Crystal Growth* 92:639.
35. Børgesen, P., R.E. Wistrom, H.H. Johnson, and D.A. Lilienfeld. 1989. "The Influence of Hydrogen on Ion Beam Mixing of Multilayer Films." *J. Mater. Res.* 4:821.
36. Kimerling, L.C. 1991. "Defect Engineering." *MRS Bull.* December: 42.
37. Kornelsen, E.V. 1964. "The Ionic Entrapment and Thermal Desorption of Inert Gases in Tungsten for Kinetic Energies of 40 eV to 5 keV." *Can. J. Physics* 42:364.
38. Vajda, P. 1977. "Anisotropy of Electron Radiation Damage in Metal Crystals." *Rev. Mod. Phys.* 49:481.
39. Brice, D.K., J.Y. Tsao, and S.T. Picraux. 1989. "Partitioning of Ion-Induced Surface and Bulk Displacements." *Nuclear Instr. and Methods in Physics Res.* B44:68.
40. Steffen, H.J., D. Marton, and J.W. Rabalais. 1992. "Defect Formation in Graphite during Low Energy Ne$^+$ Bombardment." *Nuclear Instr. and Methods in Physics Res.* B67:308.

41. Bar-Yam, Y., and J.D. Joannopoulos. 1984. "Intrinsic Defects in Si: Formation and Migration Energies." In *Proc. 17th Internat. Conf. on the Physics of Semiconductors.* New York: Springer-Verlag.
42. Kitabatake, M., P. Fons, and J.E. Greene. 1991. "Molecular Dynamics and Quasidynamics Simulations of the Annealing of Bulk and Near-Surface Interstitials Formed in Molecular Beam Epitaxial Si due to Low-Energy Particle Bombardment during Deposition." *J. Vac. Sci. Technol.* A9:91.
43. Diaz de la Rubia, T., R.S. Averback, and H. Hsieh. 1989. "Molecular Dynamics Simulation of Displacement Cascades in Cu and Ni: Thermal Spike Behavior." *J. Mater. Res.* 4:579.
44. Kinchin, G.H., and R.S. Pease. 1955. "The Displacement of Atoms in Solids by Radiation." *Rep. on Prog. in Physics* 18:1.
45. Sigmund, P. 1969. "On the Number of Atoms Displaced by Implanted Ions or Energetic Recoil Atoms." *Appl. Phys. Lett.* 14:114.
46. Veprek, S, F.-A. Sarrott, S. Rambert, and E. Taglauer. 1989. "Surface Hydrogen Content and Passivation of Silicon Deposited by Plasma Induced Chemical Vapor Deposition from Silane and the Implications for the Reaction Mechanism." *J. Vac. Sci. Technol.* A7:2614.
47. Messier, R., J.E. Yehoda, and L.J. Pilione. 1990. "Ion-Surface Interactions: General Understandings." Chap. 19 in *Handbook of Plasma Processing Technology*, ed. S.M. Rossnagel, J.J. Cuomo, and W.D. Westwood. Park Ridge, N.J.: Noyes Publications.
48. Cheng, Y.-T., A.A. Dow, and B.M. Clemens. 1989. "Influence of Ion Mixing on the Depth Resolution of Sputter Depth Profiling." *J. Vac. Sci. Technol.* A7:1641.
49. Windischmann, H. 1987. "An Intrinsic Stress Scaling Law for Polycrystalline Thin Films Prepared by Ion Beam Sputtering." *J. Appl. Phys.* 62:1800.
50. Window, B., F. Sharples, and N. Savvides. 1988. "Plastic Flow in Ion-Assisted Deposition of Refractory Metals." *J. Vac. Sci. Technol.* A6:2333.
51. McKenzie, D.R. 1993. "Generation and Applications of Compressive Stress Induced by Low Energy Ion Beam Bombardment." *J. Vac. Sci. Technol.* B11:1928.
52. Mei, D. H., Y.-W. Kim, D. Lubben, I.M. Robertson, and J.E. Greene. 1989. "Growth of Single-Crystal Metastable $(GaAs)_{1-x}(Si_2)_x$ Alloys on GaAs and $(GaAs)_{1-x}(Si_2)_x$/GaAs Strained-Layer Superlattices." *Appl. Phys. Lett.* 55:2649.
53. Wada, T., and N. Yamashita. 1992. "Formation of cBN Films by Ion Beam Assisted Deposition." *J. Vac. Sci. Technol.* A10:515.
54. Bock, W., H. Gnaser, and H. Oechsner. 1993. "Modification of Crystalline Semiconductor Surfaces by Low-Energy Ar^+ Bombardment: Si(111) and Ge(100)." *Surf. Sci.* 282:333.
55. Lin, S.J., S.L. Lee, and J. Hwang. 1992. "Selective Deposition of Diamond Films on Ion-Implanted Si(100) by Microwave Plasma Chemical Vapor Deposition." *J. Electrochem. Soc.* 139:3255.
56. Franzreb, K., A. Wucher, and H. Oechsner. 1992. "Formation of Neutral and Positively Charged Clusters (Ag_n and Ag_n^+; $n \leq 4$) during Sputtering of Silver." *Surf. Sci. Lett.* 279:L225.
57. Bohdansky, J., J. Roth, and H.L. Bay. 1980. "An Analytical Formula and Important Parameters for Low-Energy Ion Sputtering." *J. Appl. Phys.* 51:2861.
58. Anderson, H.H., and H.L. Bay. 1981. "Sputtering Yield Measurements." Chap. 4 in *Sputtering by Particle Bombardment I*, ed. R. Behrisch. Berlin: Springer-Verlag.
59. Matsunami, N., et al. 1984. "Energy Dependence of the Ion-Induced Sputtering Yields of Monatomic Solids." *Atomic Data and Nuclear Data Tables* 31:1.
60. Sigmund, P. 1981. "Sputtering by Ion Bombardment: Theoretical Concepts." Chap. 2 in *Sputtering by Particle Bombardment I*, ed. R. Behrisch. Berlin: Springer-Verlag.
61. Valeri, S. 1993. "Auger Electron Emission by Ion Impact on Solid Surfaces." *Surf. Sci. Rep.* 17:85.

62. Kelly, R. 1990. "Bombardment-Induced Compositional Change." Chap. 4 in *Handbook of Plasma Processing Technology*, ed. S.M. Rossnagel, J.J. Cuomo, and W.D. Westwood. Park Ridge, N.J.: Noyes Publications.
63. Winters, H.F., and J.W. Coburn. 1992. "Surface Science Aspects of Etching Reactions." *Surface Science Rep.* 14:161.
64. Bradley, R.M., J.M.E. Harper, and D.A. Smith. 1986. "Theory of Thin Film Orientation by Ion Bombardment during Deposition." *J. Appl. Phys.* 60:4160.
65. Petrov, I., L. Hultman, J.-E. Sundgren, and J.E. Greene. 1992. "Polycrystalline TiN Films Deposited by Reactive Bias Magnetron Sputtering: Effects of Ion Bombardment on Resputtering Rates, Film Composition, and Microstructure." *J. Vac. Sci. Technol.* A10:265.
66. Yapsir, A.S., L. You, T.-M. Lu, and M. Madden. 1989. "Partially Ionized Beam Deposition of Oriented Films." *J. Mater. Res.* 4:343.
67. Ohmi, T., T. Saito, M. Otsuki, T. Shibata, and T. Nitta. 1991. "Formation of Copper Thin Films by a Low Kinetic Energy Particle Process." *J. Electrochem. Soc.* 138:1089.
68. Homma, Y., and S. Tsunekawa. 1985. "Planar Deposition of Aluminum by RF/DC Sputtering with RF Bias." *J. Electrochem. Soc.* 132:1466.
69. Oechsner, H. 1975. "Sputtering—A Review of Some Recent Experimental and Theoretical Aspects." *Appl. Phys.* 8:185.
70. Thornton, J.A. 1977. "High Rate Thick Film Growth." *Annual Rev. Mater. Sci.* 7:239.
71. Thompson, M.W. 1968. "The Energy Spectrum of Ejected Atoms during the High Energy Sputtering of Gold." *Phil. Mag.* 18:377.
72. Zalm, P.C. 1986. "Ion Beam Assisted Etching of Semiconductors." *Vacuum* 36:787.
73. Zalm, P.C. 1988. "Quantitative Sputtering." *Surf. and Interface Anal.* 11:1.
74. Hoffman, D.W. 1990. "Intrinsic Resputtering—Theory and Experiment." *J. Vac. Sci. Technol.* A8:3707.

8.9 Recommended readings

Biersack, J.P. 1987. "Computer Simulations of Sputtering." *Nuclear Instr. and Methods in Physics Research* B27:21.

Carter, G., and D.G. Armour. 1981. "The Interaction of Low Energy Ion Beams with Surfaces." *Thin Solid Films* 80:13.

Davies, J.A. 1992. "Fundamental Concepts of Ion-Solid Interactions: Single Ions, 10^{-12} Seconds." *MRS Bull.* (June).

Hubler, G.K., ed. 1992. "Pulsed Laser Deposition." *MRS Bull.* (Feb., special issue).

Lafferty, J.M., ed. 1980. *Vacuum Arcs: Theory and Application*. New York: John Wiley & Sons.

Sanders, D. 1990. "Vacuum Arc-Based Processing." Chap. 18 in *Handbook of Plasma Processing Technology*, ed. S. M. Rossnagel, J.J. Cuomo, and W.D. Westwood. Park Ridge, N.J.: Noyes Publications.

Ziegler, J.F., J.P. Biersack, and U. Littmark. 1985. *The Stopping and Range of Ions in Solids*. New York: Pergamon.

Chapter

9

Glow-Discharge Plasmas

The powerful concept of using a plasma to couple nonthermal energy from an electric field into a film-deposition process was introduced in the last chapter. There, we defined the plasma state and examined the arc type of plasma for metal vaporization. We also examined in detail the use of the ion stream emanating from plasmas to accomplish sputter vaporization and film modification. In this chapter, we will focus on the structure and behavior of the widely used glow-discharge type of plasma. We will further see that plasma energy appears not only as ion bombardment, but also as neutral fragments of gas molecules. These "free radicals" have high chemical reactivity which promotes compound formation at a T much lower than that required for thermal reaction. Additional but smaller amounts of plasma energy appear as light and heat.

There are many ways to couple electrical energy into a plasma and many geometrical arrangements of plasma volume, sputtering target, gas source, and film substrate. These factors greatly influence the generation of ions and free radicals and their delivery to the process, so it is important to understand their behavior to obtain the desired film properties. In the following sections, we will examine energy coupling, plasma configuration, and the chemistry of free-radical reactions.

For all plasma configurations, the stages of energy transfer through the plasma are the same, namely: (1) acceleration of free electrons in the applied electric field; (2) electron-impact reactions with gas molecules to generate ions, electrons, free radicals, and excited-state molecules; (3) diffusion of these energetic particles out of the plasma toward the containment walls and film surface, where they dissipate their energy; and (4) acceleration of ions into the wall and film by the "sheath field" next to it. The steady-state concentration of each type of particle within the plasma represents a balance between its generation rate and its loss rate to the walls. We will first examine the generation of energetic particles by electron impact.

9.1 Electron-Impact Reactions

Consider a free electron of charge q_e immersed in a gas and in an electric field **E** (V/m). The resulting Lorentz force, $q_e\mathbf{E}$, on the electron [Eq. (8.15)] accelerates it in accordance with $F = m_e dv/dt$ until it collides with a gas molecule. There are many possible outcomes of this collision depending on the electron's (translational) kinetic energy, E_e (in electron-volts, or eV). All of these outcomes are important to the behavior of the plasma, and we will examine them in order of increasing E_e. In the lowest E_e range of <2 eV or so depending on the molecule, the collisions are elastic; that is, the energy remains translational. At higher E_e, a variety of inelastic collisions take place, in which E_e is partially converted into internal energy of the target molecule. In the highest E_e range of > 15 eV or so, collisions result in ionization (*potential* energy increase), which sustains the plasma by producing positive ions and new free electrons in accordance with Eq. (8.14).

The key feature of the elastic collision is that the fractional energy transfer, γ_m, is very small, because the electron's mass is only ~10^{-5} of the molecule's mass. For example, in a head-on collision with Ar, $\gamma_m = 5.4\times10^{-5}$ by Eq. (8.20). Thus, if **E** is strong enough, the electrons can continue to gain energy as they drift through the field undergoing many elastic collisions, until some electrons finally gain enough energy to cause the ionization which sustains the plasma. The low γ_m also means that the gas does not get heated very much by the electron flux. In a glow-discharge plasma producing enough free radicals and ions for thin-film processing, the gas T may only reach 200° C or so. This is different from the arc discharge of Sec. 8.3, whose constricted cross section results in such a high electron flux that the gas does become very hot despite the low γ_m. Considerable gas heating can also occur in very intense glow discharges, as we will see in Sec. 9.5.

The ions gain just as much translational kinetic energy as the electrons per unit travel distance along **E** [(see Eq. (6.6)], but they lose much of it in their first collision with a molecule because of a high γ_m. This process also produces some gas heating, but not very much, because, as we will see in the next section, **E** is by far the strongest at the edge of the plasma and is in the direction that accelerates ions into the wall and electrons into the plasma. Thus, the predominant energy transfer paths from the plasma's charged particles are ion bombardment of the walls and inelastic electron collisions with gas molecules.

These inelastic collisions occur at higher electron kinetic energy, E_e, than the elastic ones, and they result first in excitation of an electron in the target atom or molecule. The excitation is stimulated by the electric-field pulse produced in the molecule as the free electron passes through it. The excited molecule can then spontaneously (that is, without further collision) undergo one of the following processes, in order of increasing E_e: (1) electron relaxation back toward the ground state, (2) dissociation, or (3) ionization. We discuss these processes in turn below.

Relaxation of an electronically excited state is practically instantaneous ($\sim 10^{-8}$ s) in most cases and is accompanied by the **emission of a UV or visible photon** whose wavelength corresponds to the excited electron's energy-level drop per Eq. (6.5). These emission lines gives plasmas their glow and also provide for convenient and nonintrusive *qualitative* analysis of the plasma's atomic composition by optical-emission spectroscopy (OES). However, when molecules and free radicals are present along with atoms, the spectra become quite complicated and many lines remain unidentified. Also, quantification of atomic concentration is frustrated by not knowing either the fraction of atoms in the ground state or the electron energy distribution, both of which affect excitation rate. These problem are sometimes but not always eliminated [1–3] by the use of "actinometry" [4], in which an emission line at a nearby energy and from an atom of known concentration is used to calibrate the electron flux at that energy.

We mention in passing that other analytical techniques such as laser-induced fluorescence (LIF) [5] and line-of-sight mass spectrometry [6] provide more quantitative chemical analysis of plasmas and gases, but they are too complex to use for routine film deposition. The two most convenient analytical techniques are optical emission, as discussed above, and mass-spectrometric sampling of the *effluent* gas (Sec. 3.5). A survey of plasma diagnostic techniques is given in Kroesen (1993).

The relaxation of an electronically excited atom by photon emission is sometimes quantum-mechanically impeded, and then the atom can linger in a **"metastable" excited state** for many seconds until it finally radiates a photon or is deactivated in a collision. The three lightest inert gases have the highest-energy metastables [7]: He* at 19.8 eV, Ne* at 16.7, and Ar* at 11.7. Sufficiently energetic metastables can contribute to dissociation and ionization of weaker species by "**Penning**" **reactions:**

$$AB + He^* \to A + B + He \quad (9.1)$$

$$C + He^* \to C^+ + He \quad (9.2)$$

where A and B are atoms or free radicals.

Moving up in electron impact energy, we next come to the **dissociation reactions:**

$$AB + e \to A^+ + B + 2e \text{ (dissociative ionization)} \quad (9.3)$$

$$AB + e \to A + B + e \text{ (dissociation)} \quad (9.4)$$

$$AB + e \to A + B^- \text{ (dissociative electron attachment)} \quad (9.5)$$

The free radicals produced here are very active chemically because of their unsatisfied (dangling) bond, and they are usually the primary reactants in plasma-activated film deposition. Radicals may be molecular fragments or reactive atoms; thus, we have both in

$$SiH_4 \to SiH_3 + H \quad (9.6)$$

In very intense plasmas where fractional ionization is high, positive ions can become the primary reactants instead of neutral radicals. Negative ions are also formed in plasmas [Eq. (9.5)], but they are not reactive when their formation completes an electron shell[8], as with F^- and SiH_3^-.

The reactive-plasma deposition processes are distinguished from each other according to how each source material is supplied. In activated reactive evaporation (ARE), a solid or liquid (say Al) is thermally evaporated into a plasma of reactive gas (say O_2); thus, $2Al + 3O \to Al_2O_3$. Reactive sputtering uses the same plasma for sputter-volatilization of the solid source material and for dissociation of the gaseous one. Plasma-enhanced chemical-vapor deposition (PECVD) uses only gaseous source materials, as does thermal CVD (Chap. 7). Plasma dissociation of gaseous reactants can also be employed in any of the beam techniques of Chap. 8. The various plasma configurations appropriate to these processes will be discussed further in subsequent sections.

The key feature of all plasma-activated reactive deposition is that the film-forming reaction(s) can occur at much lower substrate temperature, T_s, than with thermal processes. This is important in the many cases where high T_s causes undesired diffusion across film interfaces, decomposition of film or substrate, or excessive thermal-mismatch stresses upon cool-down. The T_s for deposition is reduced because electron-impact dissociation of relatively cool gas is providing the deposition reaction's activation energy, E_a. That is, free radicals have highly positive heats of formation because of their dangling bond, so their formation brings the reactants much closer to the top of the activation energy "hill" of Fig. 7.16, or often over it. This was illustrated in Fig. 5.2 for the chemisorption reaction of the gaseous diatomic molecule Y_2 versus its atomic radical Y: for Y_2, there is a positive E_a for chemisorption, but for Y, there is none. The kinetics of free-radical reactions will be discussed further in Sec. 9.6.

The threshold electron kinetic energy to cause dissociation, $E_e°$, lies well *above* the molecular bond strength, E_b, though below the ionization threshold or ion "appearance potential," E_i. The excess of $E_e°$ over bond strength is a consequence of the Franck-Condon principle of quantum mechanics, which observes that electronic transitions occur in a much shorter time than nucleus motion [8]. Thus, electron-impact dissociation becomes a two-step process involving first the excitation of a bonding electron into an antibonding orbital without a change in bond length, as illustrated in Fig. 9.1, followed by separation of the now unbonded radicals and relaxation of their electron energy levels. For example, the SiH_3-H bond strength, E_b, is 376 kJ/mol or 3.9 eV/mc, while the $E_e°$ for SiH_4 dissociation is ≈8 eV (Fig. 9.2). Because of the Fig. 9.1 behavior, $E_e°$ cannot be predicted reliably.

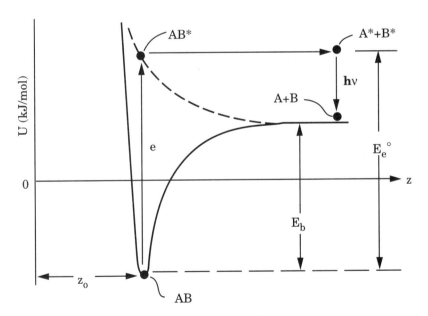

Figure 9.1 Internal energy, U, vs. bond length, z, during electron-impact dissociation of the molecule AB. z_0 is the relaxed bond length, the dashed line corresponds to the antibonding orbital, and **h**ν represents photon emission in relaxation of the excited-state radicals A* and B*.

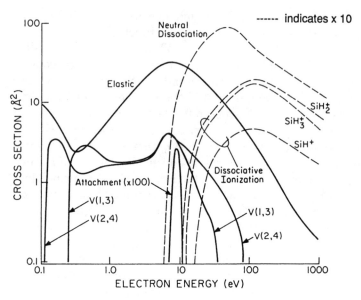

Figure 9.2 Electron-impact reactions of silane gas (SiH$_4$). Vibrationally excited states are denoted by V(i, j). 1 Å^2 = 10^{-20} m^2. (Source: Reprinted from Ref. 12 by permission.)

Despite the high E_e required, even very strongly bonded molecules like N ≡ N (E_b = 9.8 eV) become significantly dissociated in a plasma. As a calibration point, consider [9] plasma CVD from a flowing mixture of 1.6 sccm SiH$_4$ + 45 sccm N$_2$ to form amorphous silicon nitride of composition SiN$_{1.4}$H$_{0.3}$. There, a moderate power of 60 W applied to a plasma confined between 15-cm-diameter electrodes 3 cm apart (0.3 W/cm^2) was found to produce just enough N atoms to consume all of the SiH$_4$ in deposition, implying that 1.1 sccm or 2.5 percent of the N$_2$ was being dissociated. Molecules with more moderate bond strengths will have higher dissociation fractions.

We finally come to the highest-energy electron-impact reaction: **ionization**. The ionization may be simple [Eq. (8.14)] or dissociative [Eq. (9.3)], with the latter having a higher energy threshold for a given molecule. The total ionization cross section by all reactions was plotted versus electron energy for various molecules in Fig. 8.2. Ionization sustains the plasma by supplying new free electrons to replace those being lost to the walls, although electrons are sometimes also supplied by a hot-cathode (thermionic) emitter (Sec. 8.1). The concentration of ions, n_+, and electrons, n_e, in a plasma at steady state—the so-called plasma density—can be quite low and still have a profound effect on film deposition. (Note that n_+ and n_e are about equal except in the thin "sheath" at the plasma boundaries, because a tremendous electrical potential would develop otherwise.) A typical density for the "parallel-plate" glow discharge (Sec. 9.3) is $n_+ \approx n_e \sim 10^{10}$ cm^{-3}, although special electrodeless coupling techniques (Sec. 9.5) can achieve $\sim 10^{12}$ cm^{-3}. Thus, in a parallel-plate discharge operating at 100 Pa, where the molecular concentration at room T is 2.4×10^{16} mc/cm^3 [Eq. (2.10)], the fractional ionization is $<10^{-6}$! The level of n_e is not usually monitored during film deposition. The common techniques for doing so are simple in principle but fraught with practical problems (Kroesen, 1993). One technique involves interpreting the current versus voltage (I-V) curve of a small wire "Langmuir" probe immersed in the plasma [10], but perturbation of the plasma by the probe is always a concern, and interpretation of the curve is not straightforward, especially in the presence of negative ions, rf waves, or magnetic fields. In another technique, the plasma's refractive index, which increases with n_e, is measured with a microwave interferometer coupled through windows [11]. This technique is less intrusive, but beam alignment is difficult and spatial resolution is poor.

Ideally, one would like to know the steady-state concentration of *all* of the active species in the plasma, including metastables, radicals, and ions. In principle, this is obtained by setting the generation rate equal to the loss rate for each species, but in practice these rates are not easily determined. That portion of the loss rate occurring by diffusion to the walls can be estimated reasonably well with the diffusion equation [Eq. (2.27)], and that portion due to fluid flow by the equations of Sec. 3.3. Loss within the bulk of the plasma can take various paths whose rates are more difficult to estimate. Volume recombination of *atomic* ions and radicals requires a third-body collision [Eq. (7.38)] to carry away the energy released, and the probability of such collisions is low at typical glow-discharge pressures. However, *molecular* ions can recombine with electrons dissociatively without a third body:

$$AB^+ + e \rightarrow A + B \tag{9.7}$$

Radicals can also recombine in the gas phase if one of them is molecular and has enough vibrational degrees of freedom to absorb the energy released. In addition, metastables, radicals, and ions can all react with gas molecules to produce new species. A few specific free-radical reactions of PECVD are discussed in Sec. 9.6.4.

Generation rates of the various active species by electron impact are even more difficult to determine. Electron-molecule reactions can be described by a second-order rate equation similar to Eq. (7.34) for thermal reactions:

$$R_{Ai} \text{ (mc/cm}^3 \cdot \text{s)} = k_{Ai} n_e n_A \tag{9.8}$$

Here, subscript i indicates the particular reaction of molecule A to which the rate R and rate constant k refer, and the n terms are the concentrations of electrons and of molecule A per cm^3. The behavior of k does not follow the Eq. (7.35) Arrhenius form of thermal reactions, however, because electron energy rather than thermal energy is supplying the reaction activation energy, E_a. Thus, the E_a extracted from an Arrhenius plot versus gas T or substrate T has little meaning in a plasma reaction, even though such E_a values are often reported. The most one can conclude from such a plot is that if E_a is low (say 0.1 eV), the reaction is plasma activated, while if it is much higher, there is likely to be some degree of thermal activation involved.

It is useful to derive an expression for the k of Eq. (9.8) to see what factors do influence it. An electron moving at speed c_e in the plasma undergoes molecular collisions resulting in the reaction of interest (i) at a frequency given by

$$\nu_{ei} (s^{-1}) = c_e/l_{ei} = c_e(\sigma_{Ai} n_A) \qquad (9.9)$$

where σ_{Ai} is not the *total* collision cross section with molecules but only the cross section for the (i) reaction, and l_{ei} is the corresponding mean free path. The second equality above was derived earlier as Eq. (2.22). Since c_e and σ_{Ai} are both functions of electron energy, E_e, the reaction rate is not simply given by $R_{Ai} = \nu_{ei} n_e$, but must be obtained by integrating over the electron energy distribution: $f(E_e) = (dn_e/dE_e)/n_e$. That is,

$$R_{Ai} = \int_{n_e}^{\infty} \nu_{ei} dn_e = n_e n_A \int_0^\infty \sqrt{\frac{2E_e}{m_e}} \, \sigma_{Ai} f(E_e) \, dE_e \qquad (9.10)$$

where the second equality has been obtained by using Eq. (9.9) for ν_{ei} and Eq. (8.7) for c_e (or v_e) in terms of E_e. Thus, k_{Ai} is given by the second integral.

In only a few cases such as silane (Fig. 9.2) are the $\sigma_{Ai}(E_e)$ functions for the various electron-impact reactions known fairly well, and this is the first hurdle faced in estimating R_{Ai}. In Fig. 9.2, by the way, the one neutral-dissociation curve shown includes all possible neutral fragmentation patterns. The v(1,3) and v(2,4) curves denote vibrational excitation, which is another way for electrons to heat the gas besides elastic collisions. The electron-attachment curve shown is for dissociative reactions of SiH_4 by Eq. (9.5).

The second function needed for Eq. (9.10) is $f(E_e)$. If the electrons gained and lost energy only by colliding elastically with each other—

that is, if they were in kinetic equilibrium—$f(E_e)$ would assume the Maxwell-Boltzmann form given in Eq. (2.1) for molecules. However, e-e collisions dominate only when the plasma's ionization fraction is very high. Generally, the E_e gain and loss mechanisms differ, as discussed above, so that $f(E_e)$ is non-Maxwellian. That is, the electrons continue to gain energy in the electric field until they lose their energy upon reaching the threshold for the high-σ inelastic collisions—dissociation and ionization. On balance, this process tends to produce an $f(E_e)$ that is crudely Maxwellian at low E_e and truncated at high E_e, but $f(E_e)$ cannot presently be estimated accurately enough for calculation of R_{Ai} using Eq. (9.10), except for the simplest plasmas such as the inert gases. Considerable research effort is underway to measure σ values and to predict $f(E_e)$ for practical processing plasmas. For reference, we note that a plasma in which e-molecule collisions dominate is called "Lorentzian," and one in which the ionization fraction is high enough so that e-e collisions dominate is "Coulombic."

Despite the above difficulties with $f(E_e)$, the plasma electrons can be described by a characteristic kinetic energy, which can be extracted from Langmuir probe [10] or other measurements. Recall that in a Maxwell-Boltzmann distribution, the mean translational kinetic energy of a particle is related to temperature by $(3/2)k_BT$ [Eq. (2.11)]. On this basis, it is customary in plasma physics to speak of an electron "temperature" even for a non-Maxwellian $f(E_e)$. However, this temperature is given in energy units of eV for convenience, and the $(3/2)$ is dropped; thus,

$$\tilde{T}_e (eV) = k_B T_e/q_e = 8.63 \times 10^{-5}\, T_e \qquad (9.11)$$

Here, k_B is in J/K, and T_e would be the temperature in K of a Maxwellian distribution. The ($\sim$) over T_e signifies eV units and will be used below as a reminder that the electron temperature is in energy units rather than in K. Because only a small fraction of the electrons need to attain the ionization threshold, E_i, in order to sustain the plasma and because E_i does not vary a great deal among gases, $\tilde{T}_e$ is typically a few eV and is not particularly sensitive to plasma process conditions such as power, pressure, and composition. However, in some plasmas to be discussed later where rf fields and magnetic confinement allow the electrons to gain a great deal of energy before being lost to the walls and where the pressure is too low to provide sufficient collisional cooling, $\tilde{T}_e$ can be much higher. The $\tilde{T}_e$ of a few eV is much higher than the mean energy of the ions in a glow discharge, which are cooled by momentum transfer to approach the gas T of a few hundred °C or $\sim$0.1 eV.

As a final topic in electron-impact phenomena, it is useful to estimate the dc electrical conductivity of a plasma. The conductivity of any substance is equal to its concentration of free charges times the drift velocity, v, of those charges per unit of electric field strength (the charge mobility, μ), as was expressed for electrons in Eq. (6.2). Note that the electron drift velocity, v_e, is the velocity of *net* electron motion in the direction of the field and is not the same as the random thermal speed, c_e, of Eq. (9.9). For low ionization fraction (Lorentzian plasma), the mobility of plasma electrons is limited by the drag produced by their collisions with molecules, which are predominantly elastic collisions. Some electrons will leave these collisions retaining part of their forward momentum, and some will have their momentum reversed, so on average the momentum retained will be zero. Thus, the drift momentum of an electron, $m_e v_e$, is reduced to zero at a frequency of just v_e, the electron elastic-collision frequency with molecules from Eq. (9.9). The drag force so produced must balance the Lorentz force of Eq. (8.15) which is accelerating the electrons; that is,

$$F = d(m_e v_e)/dt = m_e v_e v_e = q_e E \tag{9.12}$$

Solving the last equality for v_e and inserting into Eq. (6.2) using SI units, we have

$$s\left(S/m \text{ or } \Omega^{-1} m^{-1}\right) = \frac{n_e q_e^2}{m_e v_e} = \frac{n_e q_e^2}{\sqrt{2 E_e m_e}\, \sigma_e\, n} \tag{9.13}$$

Here, v_e was evaluated for the last equality using Eqs. (9.9) and (8.7); σ_e (m^2) is a mean value of the elastic cross section; n is the gas concentration (mc/m^3); and E_e is in coulomb-V, not eV. Because the electrons have only 10^{-5} times the mass of the ions and are about the same in concentration, they dominate the conductivity even though their collision frequency is higher than that of the ions due to their higher speed. Equation (9.13) also holds for rf plasmas as long as $\omega_0 \ll v_e$ ("collisional" plasma), where ω_0 is the rf-drive angular frequency (radians/s).

9.2 Plasma Structure

In the above discussion of the electron-impact reactions which couple electrical energy into the plasma gas, we considered only the bulk plasma region. We now move to the boundary region or "sheath" abutting the walls that confine the plasma. Plasma behavior in this region is crucial both to sustaining the plasma and to controlling film deposition. Ion and electron loss occurs mostly at the walls, and electrical energy must pass from the walls through the sheath to sustain the plasma. Plasmas affect thin-film processes principally by providing ion bombardment and free radicals to various surfaces, including substrates, depositing films, and sputtering targets. Plasma-surface interaction is determined largely by the *way* in which electrical energy is coupled to the plasma electrons through surfaces, so we will discuss in subsequent sections the various coupling methods and how they affect thin-film processes. First, however, we need to examine the approach of the plasma to the surface and how this results in the plasma sheath.

Recall that a plasma is a partially ionized gas containing about equal concentrations of positive and negative particles. We assume here, as we have previously, that the negative particles are mostly electrons. On the other hand, plasmas rich in extremely electronegative gases such as the halogens can contain more negative ions than electrons, and this increase in mass of the negative particle alters plasma-sheath behavior considerably [13]. However, such plasmas are generally encountered in plasma etching rather than in film deposition. We also continue to assume here that the gas pressure is low enough that the electrons become much hotter than the ions; that is, we are in the glow-discharge regime. Finally, we assume only singly charged ions, which is valid except in very intense plasmas.

Now envision a volume of plasma suspended within a gas. The free electrons and positive ions are held within this volume by their mutual attraction, and there are two quantities which are very useful in characterizing this attraction. The first describes the screening of an ion charge by the electrons which accumulate around it. The electrical potential surrounding a screened ion falls off exponentially with increasing distance away from it. One would expect the distance needed for effective screening to decrease with increasing plasma density, n_e, and also to increase with electron "temperature" (energy), T_e, since the latter will increase the distance to which the electrons can remove themselves from the attraction of the ion's potential field. The distance within which the ion's potential falls to 1/e of the value it would have without screening (the Coulomb potential of Fig. 8.20) is called the **Debye length** and is derived in Chapman (1980; p. 57) as:

$$\lambda_D \text{ (cm)} = 0.1 \left(\frac{\varepsilon_o \tilde{T}_e}{n_e q_e} \right)^{1/2} = 743 \sqrt{\tilde{T}_e / n_e} \qquad (9.14)$$

where ε_o = electrical permittivity of vacuum = 8.84×10^{-12} F/m

$\tilde{T}_e$ = electron T in eV [Eq. (9.11)]
n_e = plasma density, e/cm^3
q_e = electron charge = 1.60×10^{-19} C

(The 0.1 appears because of the use of cm rather than the SI unit of m in λ_D and n_e.) A typical plasma having $n_e = 1 \times 10^{10}$ e/cm^3 and $T_e = 3$ eV has $\lambda_D = 0.013$ cm.

The second quantity used in characterizing charge attraction relates to the plasma's response to a displacement of electrons from ions. If the electric field causing the displacement is removed, the electrons will be pulled back toward the ions, but they will then overshoot and oscillate about the ions at an angular frequency known as the **plasma frequency,** ω_p (s^{-1}), which is also the rate at which the electrons can traverse λ_D:

$$\omega_p = \frac{\bar{v}_e}{\lambda_D} = 10^3 \left(\frac{n_e q_e^2}{\varepsilon_0 m_e}\right)^{1/2} = 56{,}400 \sqrt{n_e} \qquad (9.15)$$

For the second equality here, the mean electron velocity in the direction of oscillation, $\bar{v}_e$, was evaluated by setting kinetic energy equal to the thermal energy of one dimension, $(1/2) k_B T_e$. (If SI units were used for n_e, the 10^3 would drop out.) The value of ω_p depends only on plasma density, and for our typical n_e of 1×10^{10} e/cm^3, we have $\omega_p = 5.6 \times 10^9$ s^{-1} or, dividing by 2π, 897 MHz or 0.9 GHz (gigahertz, pronounced "jiga," referring to "gigantic"). Observe that the second equality has the form of the equation for a mass-spring oscillator, $\sqrt{k/m}$ [Eq. (4.36)]. Here, the spring constant, k, is the Coulomb restoring force of the plasma charge, $n_+ q_+$ ($= n_e q_e$), acting on an electron charge, q_e. ω_p will be useful later on when we discuss wave propagation of rf electrical energy in plasmas. Because of the restoring force of charge attraction, plasmas are rich in wave phenomena.

We can now proceed with examining our volume of plasma as it approaches the surface. The plasma will diffuse outward toward its containing surfaces, just like any particles diffuse down a concentration gradient. The much higher thermal velocity of the electrons relative to the ions in the bulk of the plasma causes them to lead this diffusion, so that the plasma bulk is left with a net positive charge while the edge collects a negative charge of electrons. These leading electrons are retarded by the positive-ion charge, while at the same time the ions are accelerated out of the plasma by the leading negative charge, so that in steady state, the diffusion velocities of both charges become equal, and the diffusion is said to be "ambipolar." The **ambipolar diffusion velocity,** u_a, can be obtained by setting the electron-plus-ion thermal energy in one dimension equal to the kinetic energy of the diffusing plasma:

$$\frac{1}{2}k_B(T_e + T_+) = \frac{1}{2}(m_e + m_+)u_a^2 \approx \frac{1}{2}m_+ u_a^2 \qquad (9.16)$$

Here, the electron and ion temperatures are in K. Solving for u_a and adding a second term to account for the drag of ion-molecule collisions due to their mean free path, l_+, in collisions with molecules [14], we have

$$u_a(m/s) = \left[\frac{q_e(\tilde{T}_e + \tilde{T}_+)}{m_+(kg)}\right]^{1/2} \left[1 + \frac{\pi\lambda_D}{2l_+}\right]^{-1/2} \rightarrow \left[\frac{q_e\tilde{T}_e}{m_+}\right]^{1/2} \qquad (9.17)$$

where we have converted to $\tilde{T}$ in eV, and where the last expression holds in the low-pressure ($l_+ \gg \lambda_D$) and high-$\tilde{T}_e$ ($\tilde{T}_e \gg \tilde{T}_+$) regime that usually characterizes film-deposition plasmas. A close analogy exists here to the effusion of gas from an orifice. There, the gas reaches a maximum velocity equal to the speed of sound, which has the same m and T dependence as does u_a (see Appendix E). Thus, u_a is known as the "ion sound velocity." Both phenomena involve the conversion of thermal energy into directed energy. For a plasma of Ar ($m_+ = 40$ u $\times$ 1.67×10^{-27} kg/u) with $\tilde{T}_e = 3$ eV, u_a by the last expression of Eq. (9.17) is 2.9×10^5 cm/s.

When the ambipolar diffusion front arrives at the surface, the electron charge that leads it is replaced by a negative surface potential that serves the same purpose; namely, to accelerate the ions and retard the electrons so that, in steady state, the escaping fluxes of ions and electrons are equal. The retardation of electrons produces a sheath of positive space charge against the surface, as shown in Fig. 9.3. In this figure, the surface is shown at ground potential, and the **plasma potential**, V_p, is positive with respect to it by $\Delta V_p = V_p - 0$. The level of ΔV_p needed to repel sufficient electrons increases with $\tilde{T}_e$ and also depends on the electron energy distribution, $f(E_e)$, of Eq. (9.10). This boundary condition leads to a key feature of the glow-discharge plasma, namely that *the plasma bulk is always more positive than the most positive of the containing surfaces* (excluding very small surfaces, such as a fine-wire Langmuir probe). As soon as a substantial surface is introduced which has a more positive potential, say V_s, plasma electrons flow into it until the plasma charges up to ΔV_p above it; that is, until $V_p = V_s + \Delta V_p$. This fact will be important in understanding the behavior of plasmas under ac and rf excitation.

The above considerations also apply to electrically floating surfaces, such as substrates suspended in the plasma for deposition. Since these

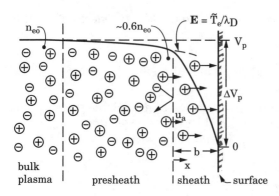

Figure 9.3 Charge distribution and voltage profile near a surface abutting a glow-discharge plasma. ΔV_p is the *minimum* (anode) sheath voltage drop.

surfaces must receive a balanced charge flow in steady state, they develop a **"floating" potential,** V_f (measured from ground), which is negative with respect to the plasma to retard electrons. This means that they also become surrounded by a sheath of positive space charge. For a Maxwellian $f(E_e)$, it can be shown (Chapman, 1980; p. 70) that the voltage drop across this sheath is

$$V_f - V_p = \Delta V_f = -\tilde{T}_e \ln(m_+/2.3m_e) = -\tilde{T}_e (6.7 + \ln M_+) \qquad (9.18)$$

where M_+ is the ion molecular weight. For $Ar(M_+ = 40)$, for example, $\Delta V_f = -10\tilde{T}_e$. Although $f(E_e)$ is generally non-Maxwellian, ΔV_f is still of this order. V_f is easily measured as the potential at which no current flows into or out of a conducting surface or wire (a Langmuir probe [10]) immersed in the plasma.

The presence of ΔV_f and ΔV_p means that *all* surfaces exposed to a plasma have ions accelerated into them through at least this high a potential; that is, at least 10 eV or so. The acceleration potential can be made much higher by negatively biasing the surface, as we will see in the next section, but it cannot be made lower. Fortunately, 10 eV lies in the desirable ion-bombardment-energy window in which surface processes are activated but subsurface damage is avoided, as was discussed in Sec. 8.5. To estimate and control ion-bombardment energy and flux, we need to examine more closely the structure of the sheath and its interface with the bulk plasma.

The interface between the sheath and the plasma occurs roughly where the charge neutrality of the bulk plasma breaks down. According to the conventional "Bohm criterion," this is where the ion velocity begins to exceed the u_a of Eq. (9.17). A more recent definition [15], which provides a smoother mathematical transition across the interface, places it where the electric field reaches a level of

$$E = \tilde{T}_e/\lambda_D \tag{9.19}$$

The region of ion acceleration between the bulk plasma and the sheath edge is known as the **presheath** and is shown in Fig. 9.3. Acceleration in the presheath causes a drop in plasma density to a fraction of the bulk value, n_{eo}, which is 0.6 if one makes the approximations of a Maxwellian $f(E_e)$ and thermal equilibrium (Chapman, 1980; p. 69). Using this 0.6 factor and the Bohm criterion, the ion current density or **ion flux** injected into the sheath is given by the conventional expression for flux as a product of particle concentration and velocity [as in Eqs. (2.5), (4.35), and (6.2)]:

$$j_+ \text{ (A/cm}^2) \approx 0.6 n_{eo} q_e u_a \tag{9.20}$$

with n_{eo} and u_a in cm units. For our earlier example [after Eqs. (9.14) and (9.17)] of a 10^{10} e/cm^3 Ar plasma at $\tilde{T}_e = 3$ eV, we have $j_+ = 0.28$ mA/cm^2. Measurement of n_{eo} requires special techniques (Sec. 9.1 under **ionization**), but we will see later that j_+ can be roughly estimated just from plasma power density for both dc (Sec. 9.3.1) and rf (Sec. 9.4.3) plasmas.

We next need to know the **sheath width,** shown as b in Fig. 9.3. This width is determined by the necessity to maintain a positive field, $E = -dV/dx$, to draw ions to the surface despite the intervening positive space charge, which reduces E at the sheath-presheath boundary. In the low-pressure limit where the ions undergo free fall across the sheath without encountering collisions with molecules, this situation is governed by the Child-Langmuir law of space-charge-limited current flow, which was derived as Eq. (8.11) for thermionic electron emission from hot surfaces. Substituting m_+ for m_e in that equation, we have

$$\boxed{j_+ = \frac{4\varepsilon_o}{9}\sqrt{\frac{2q_e}{m_+}}\frac{V_b^{3/2}}{b^2} = 5.5\times 10^{-8}\left(\frac{V_b^{3/2}}{M_+^{1/2}b^2}\right)} \tag{9.21}$$

where M_+ is in atomic mass units, and where the units of b determine the units of j_+. The sheath voltage drop, V_b, will be equal to ΔV_p of Fig. 9.3, or it will be larger if a negative bias is applied to the surface. Solving for the sheath width, b, and setting the current injected from the plasma [Eq. (9.20)] equal to the current extracted by the sheath field [Eq. (9.21)], we have

$$b^2 = 1.05 \times 10^{-2} \frac{\varepsilon_o V^{3/2}}{n_{eo} q_e \tilde{T}_e^{1/2}} \qquad (9.22)$$

for units of cm and e/cm^3. Observe that b has the same n_{eo} dependence, as does λ_D [Eq. (9.14)], but a different $\tilde{T}_e$ dependence. For this collisionless or **Child-Langmuir sheath,** the ions bombard the surface at their maximum energy, which is the sum of their kinetic energy upon entering the sheath [(1/2) $\tilde{T}_e$ from Eq. (9.17)] and the potential-energy drop across the sheath (q_+V_b). In rf sheaths, however, bombardment energy is less, as we will see in Sec. 9.4.2.

At higher pressures, or for higher V_b where b is larger, various types of collisions occur within the sheath, which both reduce the ion-bombardment energy and increase the space charge. In the **charge-exchange collision,** an electron from a gas atom or molecule jumps over to an ion passing nearby, leaving the atom ionized and the ion neutralized. The latter continues on its course toward the surface without further acceleration, and the former accelerates in the sheath field. When the ion and molecule have the same nuclei, the collision cross section, σ_m, is very large; for example, it is about 4×10^{-15} cm^2 for Ar$^+ \to$ Ar in the 10-eV range. Ions can also be deflected by molecules without charge exchange occurring, as they are in solids (Fig. 8.21), but the effective σ_m values are much lower for these collisions. Ionization can also occur in the high-pressure sheath by the impact of secondary electrons which are emitted from the surface because of ion bombardment [Eq. (8.13)] and are accelerated toward the plasma by the sheath field.

The effects of charge exchange and electron-impact ionization on b have been calculated [16] for Ar plasma and are shown in Fig. 9.4 for various values of V_b and of injected j_+. Also shown is the mean free path of Ar$^+$ in room-T Ar based on its charge-exchange σ_m quoted above and using Eq. (2.22) [which applies here instead of Eq. (2.24) because the Ar$^+$ is traveling much faster than the Ar]. We see that b begins to decrease when pressure, p, increases to the point where the mean free path is less than b; finally, b∝1/p in the "ionization limit" where electron-impact ionization dominates the space charge. The increase in space-charge density with p causes b to decrease because it increases the electric-field gradient in accordance with Eq. (8.4). A more recent sheath model [15] uses the presheath-interface criterion of Eq. (9.19) and accounts only for charge exchange, which was found

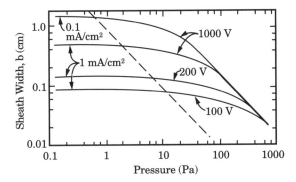

Figure 9.4 Calculated effect of pressure, sheath voltage, and injected-current density, j_+, on the sheath width in an Ar dc plasma. The dashed line is the mean free path of Ar^+ between charge-exchange collisions. (Source: Adapted from a figure in Ref. 16.)

to have a relatively small effect on b. The graphical results reported can be approximated by

$$b^2 \approx 10^{-2} \frac{\varepsilon_o V_b^{4/3}}{n_e q_e \tilde{T}_e^{1/3}} \qquad (9.23)$$

where n_e (e/cm^3) is the value at the presheath interface, and b is in cm. Observe that this equation is similar to Eq. (9.22) with some differences in the exponents. It is not easy to determine b experimentally, because it is perturbed by probing and is often smaller than 1 mm. One possible method is observing the position at which gas-phase macroparticles accumulate (see Sec. 9.6.2).

It is important to know the extent to which the sheath is collisional in order to estimate ion-bombardment conditions at the surface. Collisions reduce the energy of the bombardment and also increase its flux by adding to it fast charge-exchanged neutrals and new ions made by electron impact. Both modeling and measurement of these energy and flux distributions are difficult and are not usually available for film-deposition processes, but it is useful to be aware of the qualitative effects of plasma parameters on ion bombardment when designing and operating such processes. Figure 9.4 and Eq. (9.21) together can be used as a guide to determine whether the maximum ion bombardment energy will be reduced by sheath collisions. When subsurface film damage is a concern, it is best to keep the sheath voltage drop at its minimum at the film surface by avoiding negative bias, rather than depending on sheath collisions to reduce ion energy. On the other hand, when sputtering is the objective, one wants to minimize sheath collisions and maximize the voltage drop across the sheath in front of the sputtering target. We will see in the following sections how to control the electric fields driving the plasma so as to accomplish these objectives.

9.3 DC Excitation

In the following three sections, we will examine the ways in which electrical energy can be coupled to the plasma and the consequences for thin-film processing. These ways include: (1) applying a dc voltage across electrodes within the plasma vacuum chamber; (2) applying ac or rf voltage across these electrodes; and (3) coupling electromagnetic energy through an insulating chamber wall (electrodeless excitation). DC excitation is the oldest and also the simplest of these in concept, so we will address it first. However, its use is limited to electrically conducting materials on the electrodes. RF excitation avoids this restriction because of capacitive coupling, and it also reduces the drive voltage required. Coupling of electromagnetic energy through the vacuum wall is receiving increased attention because of its avoidance of electrodes within the chamber and because of its ability to achieve extremely high plasma density without developing excessive voltages within the chamber.

Recall from previous discussion that plasmas can enhance a film-deposition process in two ways: ion bombardment and free-radical reactions. Ion bombardment can be used for sputter-volatilization of source material or for structural modification of the depositing film, as was discussed in Sec. 8.5. Free radicals made by electron-impact dissociation of gas molecules are useful for their enhanced reaction rates with sputtered source material (reactive sputtering), evaporated source material (activated reactive evaporation—ARE), or each other (plasma-enhanced chemical vapor deposition—PECVD), as was discussed in Sec. 9.1. In this chapter, these various ion-bombardment and free-radical processes will be analyzed in the context of each particular electrical coupling technique rather than process by process.

9.3.1 Parallel-plate configuration

This most widely used plasma configuration and the potential distribution between its two electrode plates are shown in Fig. 9.5. It is often called a "planar diode" by analogy to the vacuum-tube diode, although its electrical behavior is actually rather different. In the figure, the anode is at ground and the cathode is driven negative by the power supply. The bulk of the plasma floats above ground by the plasma

9.3.1 Parallel-plate configuration 471

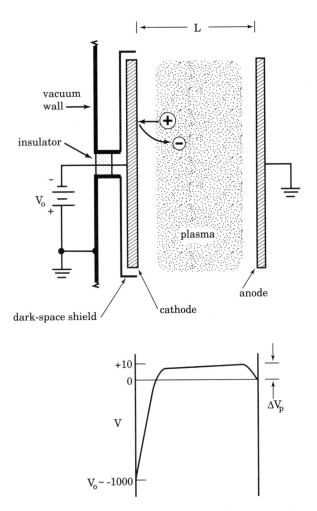

Figure 9.5 Geometry and typical voltage profile of the parallel-plate dc glow discharge. For good coating uniformity in sputtering or PECVD onto stationary substrates, the ratio of electrode diameter to L would be much larger than shown.

potential, ΔV_p, and has little voltage drop across it because of its high conductivity relative to that of the sheaths (see Exercise 9.2). This means that essentially all of the applied voltage, V_o, appears across the cathode sheath. This voltage drop results in high-energy ion bombardment and sputtering of the cathode. The polarity could be reversed, with the cathode at ground and the anode (and the plasma!) driven positive. However, then the grounded chamber walls would also act as part of the cathode, and their sputtering would contaminate the process.

The cathode voltage drop also *sustains* the plasma by accelerating secondary electrons emitted from the cathode [Eq. (8.13)] into the plasma where they initiate a cascade of ionizing collisions, as was shown in Fig. 8.2. The *dc* parallel-plate glow discharge operates at a pressure in the 3–300 Pa range and at a V_0 of about 1000–2000 V. The pressure limits are approximate and will shift with electrode spacing and gas composition. At pressures below the limit, not enough collisions occur before the electrons reach the anode. At higher pressures, the discharge tends to switch to the concentrated and low-voltage arc mode shown in Fig. 8.7, especially at high power. The high V_0 of the dc glow discharge is required so that each cathode secondary electron can produce enough ionizing collisions before losing its energy. A small increase in V_0 results in a large increase in current because of the cascade effect, so for good power control a current-regulated power supply is used. To "strike" (initiate) the discharge, it is often necessary to supply a spike of higher voltage or to adjust pressure to the minimum of the Paschen curve (Fig. 8.3) so that the gas will break down at the voltage available.

As we follow the secondary electrons out from the cathode, they first cross the "dark space," which is dark because not enough inelastic collisions with molecules have occurred yet for the glow from their excited states to be visible. The dark-space width may be smaller than the sheath width at high pressure and low plasma density, or it may be larger in opposite case, and it may *not* in general be assumed equal to the sheath width. Since the electrons follow the sheath field, which is perpendicular to the cathode surface, they travel in a broad parallel beam and are known as "beam" electrons. After acceleration, they pass into the "negative-glow" region, where they ionize gas molecules and also lose their directionality by scattering. If the electrode gap, L, is smaller than the width of the negative glow, the beam electrons are likely to reach the anode before undergoing an ionizing collision. Such a discharge is said to be "obstructed," and any further decrease in L causes a sharp rise in V_0 and ultimately extinction of the plasma. The region of plasma beyond the negative glow is known as the "positive column."

The width of the negative glow is roughly the mean free path for ionizing collisions as given by Eq. (2.22). Using the ideal-gas law [Eq. (2.10)] and room T to obtain the molecular concentration, n, in Eq. (2.22), and taking the ionization cross section for small molecules at 1 keV from Fig. 8.2 ($\sigma_m \approx 1\times 10^{-16}$ cm^2), we have

$$\boxed{Lp \sim 40 \text{ Pa} \cdot \text{cm}} \qquad (9.24)$$

for the minimum product of electrode gap and gas pressure in a dc discharge. This minimum is useful in designing plasma reactors. It is also used to advantage in preventing undesired discharges along the back of the electrode and its voltage lead, by installing a grounded "dark-space shield" along these surfaces at a spacing well under the L of Eq. (9.24), as shown in Fig. 9.5. A similar minimum L applies with rf excitation (Sec. 9.4).

One would like to estimate the ion-bombardment flux or current density, j_+, at the electrodes to estimate its effect on film processing. Note that j_+ will be about the same at both electrodes, because it is determined by the rate of ion injection into the sheath from the bulk plasma and not by the sheath voltage drop [see Eq. (9.20)]. In dc and low-frequency-ac plasma (< 1 MHz), the current passing across the sheath is mostly ion current, since the secondary-electron yield at the cathode is only about 0.2 [Eq. (8.13)], so j_+ is roughly estimated by dividing current in the external circuit by the electrode area. For the grounded anode, this area needs to include all grounded surface reached by the plasma, not just the anode itself.

9.3.2 Chemical-activation applications

Both dc and rf parallel-plate plasmas can be used to activate CVD reactions at low T, such as the silane decomposition reaction of Eq. (9.6), which is one of the reactions leading to amorphous Si deposition. These processes are known as **plasma-enhanced CVD** (PECVD). We will address the chemistry of PECVD reactions in Sec. 9.6 and concentrate for now on the plasma configuration. As with thermal CVD, plasma-CVD films deposit wherever the reaction energy is being supplied, including wherever the plasma touches a surface and sometimes downstream of the plasma as well. Usually, substrate-T control is important to film quality, and therefore the substrate is generally placed on one of the electrodes rather than being suspended in the plasma. This maximizes heat transfer by gas conduction (Sec. 5.8.2) between the substrate and the electrode, whose T can be controlled. The choice of anode or cathode for the substrate position depends on the desired energy of film ion bombardment. In amorphous Si films, for example, termination of dangling Si bonds by H is needed to remove electron trap states from the band gap, and excessive bombardment dislodges the H. There, "anode material" is a better semiconductor than "cathode material." On the other hand, when the tensile stress of a film is excessive, high bombardment energy can counteract it by the ion-peening effect (Sec. 8.5.3). For good plasma uniformity across the substrate, both electrodes should be larger in radius than the substrate by several times the electrode gap to avoid the plasma-density dropoff region at the edges, which results from ambipolar diffusion (Sec. 9.2).

When a large substrate needs to be coated, the reactant gases are distributed evenly over it through a "showerhead" array of holes in a hollow counterelectrode into which the gases are metered. The holes should have a diameter much less than the sheath width so that they do not create "hot spots" in the plasma. This problem results from the hollow-cathode effect illustrated in Fig. 9.6. Penetration of the sheath edge into the hole creates a region of plasma that is more intense than elsewhere, because beam electrons are being injected into it from all around rather than just from a plane. This localized region of enhanced conductivity preferentially attracts the applied power, which further enhances the local current of ion bombardment and beam electrons. This is an unstable, "runaway" situation which results in arc plasmas (Sec. 8.3) developing at some or all of the holes and much less power being dissipated where one usually wants it—in the uniform glow discharge. Moreover, the arcs quickly erode away the edges of the holes. The same situation develops in rf plasmas. Because sheath width decreases with increasing plasma density [Eq. (9.22)], arc breakdown sets an upper limit to plasma density and power in glow discharges and is one of the reasons why electrodeless discharges (Sec.

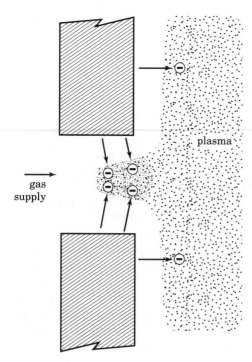

Figure 9.6 Hollow-cathode intensification effect at a hole in the plasma cathode.

9.3.2 Chemical-activation applications

9.5) are attractive. Of course, one could specifically design a showerhead to operate in the hollow-cathode mode as a way to achieve high localized plasma density. Even in the absence of holes in the electrodes, high enough plasma density leads to arc breakdown at cathode surface asperities due to field emission [Eq. (8.12)].

A glow discharge can also be used to activate a gaseous reactant in the presence of a thermally evaporated reactant in the process known as **activated reactive evaporation** (ARE) [17]. The evaporant is usually a metal, and the gas can be, for example, O_2, NH_3, or CH_4, for the formation of metal oxide, nitride, or carbide films, respectively. In ARE, a thermionic (hot) cathode (Fig. 8.1) is preferred to a simple plate cathode, for the following reason. The parallel-plate glow discharge is a "cold-cathode" discharge in which the sustaining electrons are generated mostly by gas ionization. This places a lower limit on gas pressure, p [Eq. (9.24)], and this limit can cause various problems. In subsequent discussion, we will see how excessive p scatters ions and vaporized source material, depletes the kinetic energy of sputtered species, and causes macroparticle formation. There are several ways in which a glow discharge can be made to operate at lower p, including electrodeless excitation (Sec. 9.5), the magnetron (Sec. 9.3.4), and the thermionic cathode which we presently consider.

One geometry for ARE is shown in Fig. 9.7. Electrons thermionically emitted from the hot-filament cathode are accelerated toward the pos-

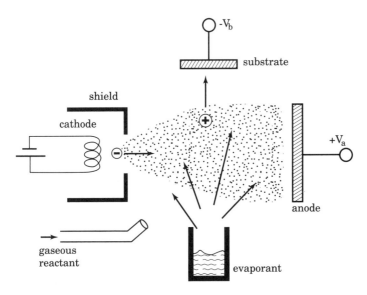

Figure 9.7 Typical geometry for compound-film deposition by plasma-activated reactive thermal evaporation.

itively biased anode plate. Even without a plasma, this current dissociates and ionizes intervening gas by electron impact, as it does in the ion gauge (Fig. 3.6c). With increasing p, additional current flows because of the resulting plasma. However, because of the thermionic emission, there is no lower p limit. Thus, p can be low enough that evaporant can be transported across the plasma to the substrate without being excessively scattered by intervening gas. Some of this evaporant will also become ionized as it crosses the electron beam (see Exercise 9.6). The ions of gas and vapor can be accelerated into the film if so desired by applying a negative potential to the substrate. Finally, the film-forming reaction takes place on the surface between adsorbing vapor and gas radicals. The filament may need to be shielded from the vapor to prevent alloying, and in oxidizing gases it needs to be made of Ir rather than W. Filament design is discussed in Appendix D. ARE can also be done with an electron-beam evaporation source (Fig. 8.4). There, the plasma which is always generated over the source in the presence of a gas background can be enhanced by adding an anode collector plate or ring above the source. Electrodeless sources of gas radicals and ions can also be used for ARE, and these will be discussed in Sec. 9.5.

The third mode of plasma chemical activation uses a sputtered source material along with a gaseous one. The gas becomes dissociated in the sputtering plasma and reacts to form a compound film, such as $Ti + N_2 \rightarrow TiN$. This **reactive sputtering** technique will be discussed in the next section.

9.3.3 Sputtering

The parallel-plate plasma of Fig. 9.5 is widely used to supply vapor for film deposition by sputter-erosion of the cathode, or "target." Often, the plasma is magnetized using the "magnetron" cathode to be discussed in the next section. In either case, the cathode is bombarded by plasma ions having energies approaching the externally applied voltage, although ion energy is distributed downward by scattering in the sheath (Fig. 9.4). The mechanism of sputtering was discussed in Sec. 8.5.4. Target purity and cooling are discussed later with reference to Fig. 9.10 for the magnetron cathode. Here, we examine three effects of the *plasma* on sputtering process behavior; namely, scattering of particles by the plasma gas, negative-ion ejection from the target, and reactive sputtering. A fourth effect involves the acceleration of plasma ions into the substrate using the negative bias shown in Fig. 8.25. The resulting "resputtering" of the depositing film can produce effective planarization of rough topography (Fig. 8.31), and the bombardment can modify film structure in various ways (Sec. 8.5.3).

9.3.3 Sputtering 477

Even at the lowest operable pressure of the dc-diode plasma [Eq. (9.24)], there is considerable **gas scattering** of sputtered particles as they cross the plasma, with consequent loss of their desirable kinetic energy and loss of deposition rate by backscattering. Magnetic confinement is widely used to reduce minimum pressure and thus avoid these problems, but for now we examine the simple planar diode. By Eq. (9.24), minimum diode pressure is about 10 Pa for a typical electrode gap of L = 4 cm. At this pressure, the mean free path between atomic collisions, l, is ~0.1 cm per Eq. (2.24), so a sputtered particle will undergo many collisions in crossing the gap. The number of collisions needed to "thermalize" *any* energetic particle to the mean kinetic energy of the surrounding gas atoms depends on the efficiency of energy transfer as expressed in Eq. (8.20) and is therefore a minimum for equal masses of particle and gas atom. Ultimately, one wants to know the distance, d, traveled toward the substrate before **thermalization,** and this is shown in Fig. 9.8 in terms of number of mean free paths (d/l) versus the ratio of gas to energetic-particle mass (M_g/M_s) and for initial translational energies corresponding to that of a typical

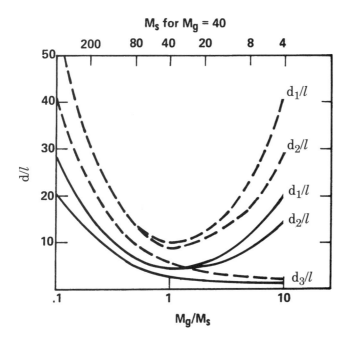

Figure 9.8 Distance, d, traveled before thermalization of a gas of mean free path l, for particles of initial translational energy of 5 eV (solid line) or 1000 eV (dashed line) using three calculational models (d_1, d_2, d_3) vs. the mass ratio of the gas atom (M_g) to the energetic particle (M_s). (Source: Reprinted from Ref. 18 by permission.)

sputtered particle (5 eV) and that of a reflected plasma ion or ejected negative ion (1000 eV) coming from the target. These calculations also take into account the increase in mean free path with decreasing mass and with increasing energy [18]. The thermalization distance is minimum for $M_g = M_s$, as expected. For our example above, the substrate was 40 mean free paths from the target, so even for large mass differences, sputtered particles will be thermalized before reaching the substrate. On the other hand, the more energetic reflected (positive) or ejected (negative) ions can arrive still retaining some energy. Figure 9.8 can be used to estimate the conditions needed either to preserve the desirable kinetic energy of the sputtered particles or to thermalize the much higher and usually undesirable kinetic energy of the reflected or ejected ions. At high sputtering power, however, local rarefaction of the gas in front of the target due to "sweep-out" by the sputtered particle flux increases the l in the electrode gap above that calculated from gauge pressure [19].

The fractional **deposition-rate loss** due to a thermalization distance, d, which is less than the electrode gap, L, can be estimated as follows. The sputtered particles are essentially being "implanted" into the plasma gas to a depth d from their source at the target surface, just as ions are implanted into a solid to a depth determined by the stopping power of their binary collisions with atoms in the solid (Sec. 8.5.2.2). Once thermalized, the particles have lost their directed energy and then proceed to diffuse by Fick's law [Eq. (2.27)] both forward toward the substrate and backward toward the target, from a planar source situated at distance d from the target. Since both surfaces are particle sinks, the concentration gradients driving the diffusion depend only on distances to the two surfaces. Thus, the fraction of thermalized particles reaching a substrate at distance L from the target is just d/L, and the rest is redeposited on the target. Deposition efficiency in planar-diode sputtering has recently been analyzed in more detail along these lines [20].

Scattering of the sputtered particles also broadens their spread of incident angles at the substrate. Thermalization and spreading together cause a generally undesirable shift in **film microstructure** from the bombardment-compacted Zone T to the more porous and weakly bonded Zone 1, as was discussed in Sec. 5.4.1 and later in connection with Fig. 5.38 on film stress. Operation at lower plasma pressure using the magnetron avoids this problem.

We now turn to the second plasma effect on sputtering behavior: **negative ions** ejected from the target. In a compound target, when one element has a low ionization potential, E_i, (say 6 eV) and the other has a high electron affinity (say 2 eV), so that the difference between the two becomes small, it is likely that the latter element will be sputtered as a negative ion rather than as a neutral atom. Elements of high electron affinity include the halogens (F, Cl,...) and the chalcogens (O, S,...) to the right of the periodic table, and some transition metals such as Au. Elements of low E_i include the alkali metals (Li, Na,...) and alkaline earths (Be, Mg,...) to the left, and the rare earths (La,... Sm,...). Indeed, the negative-ion yield from a Au target was found [21] to increase by 10^4 when the Au was alloyed with Sm. Negative ions get accelerated into the plasma along with the beam electrons by the cathode sheath field. For pressures above 1 Pa or so, they will be stripped of the extra electron in the plasma [22]; but unless the Lp product [Eq. (9.24)] is very high, they can still cross to the depositing film and bombard it with enough energy to damage it or erode it. This will occur more in the direct beam path than off to the side, especially in magnetron sputtering at low pressure. Negative-ion-beam effects have even been seen [22] in ZnO, despite the relatively large difference between the appearance potential of Zn^+ and the electron affinity of O (9.4–1.46 eV). When negative-ion flux is substantial, one is faced with the dilemma of operating either at low pressure to retain the desirable sputtered-particle kinetic energy while suffering negative-ion damage, or at high pressure to dissipate negative-ion energy while losing the Zone T film structure due to thermalization and scattering. This is a fundamental problem in *glow-discharge* sputtering of compounds and has been particularly troublesome in the deposition of high-T_c superconducting films [23] such as YBCO (Y-Ba-Cu oxide). On the other hand, in *ion-beam* sputtering (Fig. 8.25a), there is no cathode sheath over the target to accelerate the negative ions into a high-energy beam. However, ion-beam sputtering is restricted to small substrate areas due to its awkward geometry.

The third plasma effect to be discussed here is **reactive sputtering**. In this technique, a reactive gas is added to the usual Ar sputtering plasma to shift compound-film stoichiometry in sputtering from a compound target or to deposit a compound film from a metallic target. The former application arises from a tendency for the electronegative element to become depleted in compound sputtering. Although both elements must leave the surface at the target composition ratio in steady state [see discussion following Eq. (8.38)], the sticking coefficient of the electronegative element at the substrate is likely to be lower, resulting in a substoichiometric deposit. This shortfall can be compensated by addition of an appropriate amount of that element to the supply flow of sputtering gas. Thus, the transparent conducting film indium-tin oxide (ITO) is sputtered from a compound target in Ar + O_2. The O_2 flow is raised enough to reduce film optical absorption by metal precipitates but not enough to produce excessively resistive films.

Compound deposition by reactive sputtering from a metal target generally lowers target fabrication cost and increases target purity as compared to using a compound target, but it complicates process control when film composition is critical. The process-control problem is illustrated in Fig. 9.9. This **hysteresis loop** can be plotted using a quartz-crystal deposition-rate monitor (Sec. 4.8.1) or by monitoring an optical-emission line from the metal vapor in the plasma. When the metal target is being sputtered in pure Ar (point A), the sputtering rate is relatively high. As the reactive-gas supply flow is increased, rate remains high until some point, B, at which the gas adsorption rate on the target exceeds the sputtering rate. Then, the target becomes "poisoned" with adsorbed gas, and the sputtering rate drops to a much lower level (C) because much of the ion bombardment now goes into sputtering away the continuously adsorbing gas layer rather than into sputtering the underlying metal. If the reactive-gas flow is now decreased, the poisoned condition persists until point D, where the sputtering rate begins to exceed the adsorption rate so that the target can clean itself and return to a high sputtering rate.

To describe the hysteresis loop of Fig. 9.9 quantitatively and to determine the relationship between mass flow and film composition, it is necessary to think in terms of the partial pressure, p_i, of the reactive gas. Equation (3.8) relating p_i to pumping speed needs to be modified for this situation as follows:

$$p_i = Q_i/(C_o + C_s) \tag{9.25}$$

Here, Q_i (Pa·l/s) is the mass flow rate of the reactive gas, C_o (l/s) is the pumping speed of the chamber vacuum pump reduced by its connecting piping and throttle valve (Fig. 3.3), and C_s is the pumping speed of the *depositing metal* for the reactive gas, which adsorbs on it

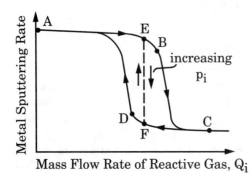

Figure 9.9 Typical process behavior of metal sputtering in reactive gas i, with the mass flow rate of Ar fixed.

and becomes buried ("getter" pumping). If the metal is depositing much faster than the gas, C_s is the same as the conductance of a molecular-flow orifice: 11.6×A l/s for the case of room-T air [Eq. (3.5)], where A is the adsorption area in cm^2. That is, all of the gas arriving at A is pumped by burial. For slower metal-deposition rate or higher p_i, C_s will be reduced by the fractional surface coverage of reactive gas, Θ, in accordance with the Langmuir adsorption model discussed prior to Eq. (6.20). Returning now to Fig. 9.9, consider an attempt to operate at point E on the hysteresis loop, where the ratio of metal deposition flux to p_i happens to produce the desired stoichiometry in the compound being deposited. Now if Q_i should drift up by the slightest amount, p_i will increase and cause both Θ and target poisoning to increase. These effects both reduce C_s and cause p_i to increase further per Eq. (9.25). This *positive feedback* produces an inherently unstable situation which results in the process drifting to point F, where for the same Q_i, metal deposition rate is much lower and p_i is much higher, thus reducing process throughput and shifting film composition. From point F, a slight downward drift in Q_i could cause a return to point E, so film composition is not controllable within the hysteresis-loop area.

The width of the loop can be reduced by increasing C_o in Eq. (9.25), because that lessens the effect of a fluctuating C_s. Also, the loop can be shifted to the right by introducing the reactive gas near the substrate to produce a higher p_i there than at the target [24], thus reducing the degree of poisoning for a given Q_i. Operating at high sputtering rate further reduces poisoning by causing gas rarefaction in front of the target [19]. However, it may still be desirable to operate at point E to obtain the desired stoichiometry at high deposition rate. This can be done if sufficient *negative* feedback is imposed to counteract the positive feedback inherent to the process. To do this, one needs to monitor p_i or some related signal and inject it into a feedback control loop to drive Q_i. (The same technique was applied to T control in Sec. 4.5.3.) That is, if p_i starts to drift up from its programmed set point, Q_i needs to be driven down to restabilize it. Convenient signals include the p_i reading on a downstream mass spectrometer, an optical-emission line from the reactive gas in the plasma, or the voltage across the plasma under constant-power control, since plasma impedance changes with gas composition. Extensive mathematical modeling of reactive sputtering has been done [24], including the case of *two* reactive gases for the deposition of compounds such as oxynitrides, where the control problem is much more difficult [25]. In reactive deposition by the narrow-beam evaporation processes of Secs. 8.2–8.4, source poisoning does not occur, because the evaporation flux is always much higher than the adsorption flux of reactive gas.

9.3.4 Magnetrons

The lower limit of operating pressure in the planar-diode sputtering plasma (Fig. 9.5) was imposed by the need for the beam electrons ejected from the cathode to undergo enough ionizing collisions with the gas to sustain the plasma before they reach the anode and are removed there. The magnetron has been a major advance in sputtering technology, and greatly improves upon this situation. Basically, it incorporates a crosswise magnetic field over the cathode, which traps the beam electrons in orbits in that location and thus greatly increases their path length before they finally escape to the anode by collisional scattering. Because the electron's travel path is now much longer than the electrode gap, L in Eq. (9.24), the minimum pressure to sustain the plasma is much lower for the magnetron than for the planar diode—typically 0.1 Pa instead of 3 Pa. At 0.1 Pa, the sputtered particles retain most of their kinetic energy upon reaching the substrate, so one obtains the beneficial effects of this energy on film structure, as discussed in Sec. 8.5.1.3. Also, deposition rate is increased because of reduced scattering and redeposition of sputtered particles on the cathode, although redeposition is still measurable [19]. Finally, the increased efficiency of electron usage means that lower applied voltage (typically 500 V) is needed to sustain a plasma of a given density, n_e, and that the voltage increases even less steeply with power than it does in the planar diode [26]. Unfortunately, the magnetic field cannot be made strong enough to deflect the problematic cathode negative ions that were discussed in the previous section, although their influence is reduced by the lower cathode-sheath voltage of the magnetron. Another problem is that the erosion pattern of a magnetron target is highly nonuniform across the surface, as we will see below. This pattern becomes imprinted on films deposited on stationary substrates when negative ions are affecting the film, because of the beam nature of these ions. Deposition-rate nonuniformity is less sharply imprinted, because the sputtered particles are neutral and are emitted in more or less the broad cosine distribution (Sec. 8.5.4.2). Both magnetrons and planar diodes can also be operated using rf excitation when one needs to couple power through insulating targets.

Many configurations of magnetic field and cathode shape have been developed for various applications [26, 27], but we will discuss magnetron operating principles with reference to the planar, circular configuration of Fig. 9.10. The overall setup for magnetron sputtering was discussed in connection with Fig. 8.25b, while here we examine the magnetron cathode in more detail. The target material to be sputtered is a disc 3 to 10 mm thick which is bonded for good thermal contact to a water-cooled Cu backing plate. Bonding is best done by soldering, al-

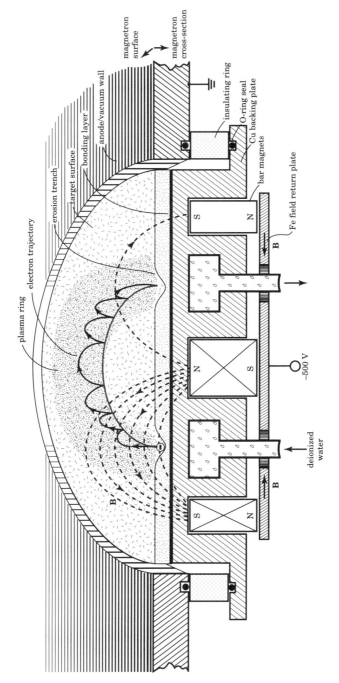

Figure 9.10 Planar-magnetron structure and behavior. The electron-orbit radius is shown much larger than actual size for clarity.

though epoxy bonding or clamping onto a coating of thermally conductive vacuum grease can be employed instead at moderate plasma power densities. Target-material selection was discussed in Sec. 8.5.4.1. The cooling water is best deionized to prevent electrolytic corrosion between the electrically biased backing plate and the grounded water supply. The entire cathode assembly is floated off ground by a ceramic insulator ring which can also form part of the vacuum wall by employing O-ring seals. The adjoining grounded metal vacuum wall then acts as the anode, although grounded shields are often added to confine the sputtered material. The anode is spaced closely enough to the edge of the cathode so that a plasma cannot ignite between them, in accordance with the discussion following Eq. (9.24). The crosswise magnetic field is established by a ring of bar magnets plus one central one, and these are connected on the back by an Fe "field-return" plate to complete the magnetic circuit and to confine the field. Using the strongest (Nd-Fe-B) magnets, the field over the target can approach one kilogauss, or 0.1 T in SI units.

Upon igniting the plasma, beam electrons emitted from the cathode become accelerated into the plasma by the cathode-sheath electric field, **E**, just as in the case of the planar diode of Fig. 9.5. However, the presence of the magnetic field, **B**, causes them to also curve into orbits as shown in Fig. 9.10 as a result of the Lorentz force given by Eq. (8.15). The gyratron-orbit radius depends on **B** and on the electron velocity component perpendicular to **B** in accordance with Eq. (8.17). For example, an electron accelerated vertically to a kinetic energy of 500 eV by **E** and immersed in a **B** of 0.03 T would have a gyratron radius of 2.5 mm. For **B** to affect the electrons, pressure must be low enough that the electron mean free path is not significantly less than the gyratron radius; that is, it must be less than a few Pa. The plasma electrons are then said to be "magnetized," although the ions are not. The magnetron will still operate as a sputtering source at much higher pressure, but gas scattering will dominate electron behavior rather than **B**.

At low pressure, the magnetron behaves as follows, with reference to Fig. 9.10. Electrons emitted from the target surface or created by ionization in the sheath field are accelerated vertically by **E** but, at the same time, forced sideways by **B**, so they eventually reverse direction and return toward the target, decelerating in **E** as they proceed until their direction is again reversed and the cycle repeats. The net motion is a clockwise drift around the circle of the target, in the so-called **E**×**B** direction using the vector notation of Eq. (8.15). If there were *no* collisions and if **E** were uniform, an electron starting at the target with zero kinetic energy would follow a "cycloidal" path, which is the same as that of a point on the periphery of a wheel rolling along the ground, and the time-average $\mathbf{E} \times \mathbf{B}$ drift velocity would be (Cecchi, 1990)

$$u_d\,(\text{m/s}) = \frac{|\mathbf{E}|\,(\text{V/m})}{|\mathbf{B}|\,(\text{tesla})} \quad (9.26)$$

Actually, the path is more complicated because of collisions and because $\mathbf{E}$ decreases with distance from the target (see Fig. 9.3). Also, u_d measured in magnetrons [27] has been found to be much larger than that predicted by Eq. (9.26), suggesting that plasma *waves* may be driving the motion. In any case, good magnetron design requires that the $\mathbf{E} \times \mathbf{B}$ drift path close on itself so that the electrons do not pile up somewhere. The post magnetron of Fig. 9.11 is another common closed-path configuration. It offers very efficient target-material utilization in coating numerous small parts placed all around the cylindrical substrate holder as shown.

The planar magnetron of Fig. 9.10 suffers from poor target-material utilization because of the trenched erosion pattern shown. The radial narrowness of this trench results from radial compression of the plasma by the magnetic-mirror effect illustrated in Fig. 9.12. This effect is also important in other magnetized plasmas to be discussed in Sec. 9.5. Consider first the electron a, which is orbiting at some velocity $\mathbf{v}_e$ in a plane perpendicular to the z axis due to $\mathbf{B}$, which is becoming stronger with increasing z as shown by the converging field lines. By Eq. (8.15), the resulting force $\mathbf{F}$ exerted on this electron will be in the $\mathbf{v}_e \times \mathbf{B}$ direction, perpendicular to $\mathbf{B}$. Because $\mathbf{B}$ at the orbit periph-

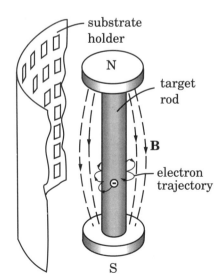

Figure 9.11 Geometry of the cylindrical post magnetron.

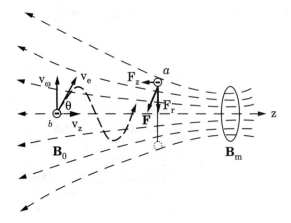

Figure 9.12 Behavior of electrons in a magnetic mirror.

ery is not parallel to z, **F** resolves into an axial component, F_z, as well as a radial one, F_r, and F_z accelerates the electron *away from the converging field*. Note that this would also be true for reversed polarity of field or charge. Since **F** is always perpendicular to $\mathbf{v}_e$, the electron's kinetic energy remains unchanged in this process, and therefore the axial acceleration has to be accompanied by a corresponding reduction in the circumferential (orbit) velocity. Conversely, consider electron b, whose velocity $\mathbf{v}_e$ has a z component, v_z, as well as a circumferential one, v_ω. As this electron spirals toward the converging field, the directed velocity, v_z, decreases, and the "thermal" one, v_ω, increases. If v_z passes through zero before the electron reaches the position of maximum field strength, $\mathbf{B}_m$, the electron will be reflected by the magnetic mirror and accelerated back in the diverging direction, with v_ω decreasing (cooling) as it proceeds. This cooling and acceleration is analogous to molecular behavior in the supersonic nozzle of Fig. 4.11, although the energy transfer mechanism is different. Whether electron b gets reflected depends on its approach angle, θ, at the z position where **B** has some lower value, $\mathbf{B}_0$. When the fractional change in **B** per orbit cycle is small, conservation of energy and momentum lead to a simple formula (Cecchi, 1990) for the angle of the "escape cone" within which the mirror is not strong enough to reflect the electron:

$$\sin^2\theta = |\mathbf{B}_0/\mathbf{B}_m| \tag{9.27}$$

Returning now to Fig. 9.10 to relate the magnetic-mirror effect to magnetron operation, we see that plasma electrons will be forced away from both the small and the large magnetron radii where **B** converges toward the magnet pole pieces. The plasma electrons will be

9.3.4 Magnetrons

compressed by these mirrors toward an intermediate radius where **B** is uniform, and that is where the plasma and the ion bombardment of the target will be most intense. This effect can be reduced somewhat by designing a flatter **B** or by mechanically scanning the magnets back and forth during sputtering, although the latter solution is awkward.

The film-thickness nonuniformity that results from plasma compression in magnetrons can be avoided by moving the substrates around during deposition, as was done in dealing with point sources of evaporant in Sec. 4.6. One common deposition geometry utilizes the rectangular magnetron, a variation of the Fig. 9.10 design in which the disc is stretched to an arbitrary length in one direction so that the electrons follow an oblong "racetrack" $\mathbf{E} \times \mathbf{B}$ path above the target. When a rectangular magnetron is used in the Fig. 4.13 substrate-transport configuration, ±2 percent uniformity can be obtained over very large areas such as rolls of plastic film or sheets of architectural glass.

Localization of the plasma over the target by the magnetron's transverse magnetic field results in a much lower plasma density over the substrate than in the case of the planar diode (Fig. 9.5), and ion bombardment flux to the substrate is reduced proportionately per Eq. (9.20). This is desirable when the neutral sputtered particles alone carry sufficient kinetic energy to optimize film structure or when one wants to minimize the substrate heating that results from ion bombardment. In other cases, however, one may want to further increase film bombardment while retaining the low operating pressure of the magnetron. One way to do this is by "unbalancing" the magnets [28, 29] as shown in Fig. 9.13. There, the central bar magnet has been replaced by a smaller one that cannot pull in all the field lines emanating from the magnets in the ring. Some of these lines then curve away toward the substrate. Since electrons traveling parallel to **B** are not

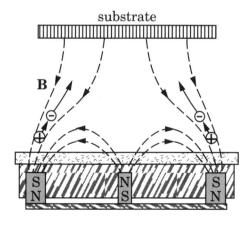

Figure 9.13 Plasma flow toward the substrate along magnetic field lines in an unbalanced magnetron.

acted on by **B**, they can escape along these field lines and toward the substrate, pulling positive ions along with them by ambipolar diffusion [Eq. (9.17)] and thus increasing ion flux to the substrate. The bombardment energy may then be increased if necessary by negatively biasing the substrate.

Another way to increase ion-bombardment flux to the film is to ionize the mostly neutral sputtered-particle flux during its transport to the substrate. This can be done as shown in Fig. 9.14, where an rf-powered coil is used to couple energy inductively into a secondary plasma downstream of the magnetron plasma [30]. The inductive coupling mechanism is discussed in Sec. 9.5.3. By increasing pressure to 4 Pa to increase sputtered-particle transport time and by using an 8- to 10-cm deep secondary plasma, 85 percent ionization of sputtered metal can be achieved [30]. This is much higher than the ionization fraction of the background Ar, because the ionization potential of metals is much lower. The ionized metal can then be accelerated into the substrate using negative bias. Metal-ion acceleration has an advantage over just Ar^+ acceleration in that it causes the depositing metal to arrive more perpendicularly to the substrate, thus reducing the problem of film buildup on the top corners of topography, which was illus-

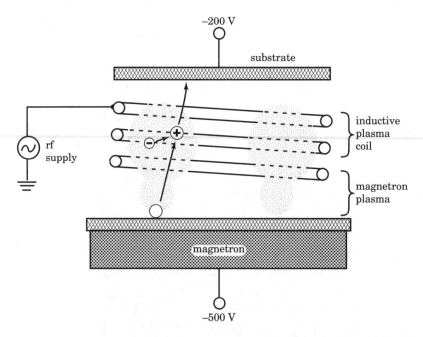

Figure 9.14 Ionization of sputtered particles using a secondary, inductively coupled plasma.

trated in Fig. 8.30. Cathodic-arc evaporation (Sec. 8.3) shares the advantage of high ionization fraction, but it suffers from the macroparticle problem.

9.4 Frequency Effects

We have so far focused on plasmas driven by dc electrical power, such as the vacuum arc (Sec. 8.3), the dc planar diode (Fig. 9.5), the thermionically supported diode (Fig. 9.7), and the dc magnetron (Fig. 9.10). DC operation requires that the surfaces of both electrodes be electrically conductive, including any sputter-target materials, substrates, or thin films that may be placed on them. The presence of insulating layers on the electrodes deflects dc plasma current into any surrounding conductive areas and thus leads to gross plasma nonuniformity or to plasma extinction. Insulating substrates can be placed elsewhere, parallel to the current flow path, but this tends to be an awkward and limited mode of operation. Therefore, when insulating materials are involved, ac power is usually employed so that power may pass through the insulator by capacitive coupling.

The effect of electrical drive frequency on capacitive coupling may be understood by considering the ignition of a planar-diode plasma when an insulating substrate is covering the driven electrode. For analytical simplicity, we will use a square-wave voltage symmetrical about ground, although the voltage would normally be sinusoidal. Let the voltage start on the negative swing of the cycle at $-V_0$, as shown at the left of Fig. 9.15. Initially, the substrate surface also floats to $-V_0$ so that most of the drive voltage is dropped across the gas space to ground. Then, a dc plasma ignites, and ion current starts to be drawn to the surface of the substrate across the cathode sheath just as in Fig. 9.5. How quickly the surface charges up positively due to this ion current is determined by the capacitor operative equation,

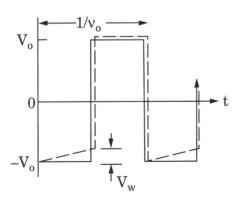

Figure 9.15 Voltage on the surface of an insulating substrate (dashed line) which is sitting on a planar-diode electrode being driven by a square-wave voltage (solid line) of frequency v_0 Hz.

$$I = j_+A = C(dV_w/dt) \tag{9.28}$$

where, using SI units,
- I = current, A
- j_+ = ion current density, A/m^2
- A = area of the insulating substrate and of the electrode, m^2
- C = capacitance, F
- V_w = voltage drop across the capacitor, V
- t = time, s

For a parallel-plate capacitor such as the substrate in question,

$$C = \varepsilon_0 \varepsilon_d A/w \tag{9.29}$$

where
- ε_0 = electrical permittivity of vacuum = 8.84×10^{-12} F/m
- ε_d = dielectric constant of the capacitor medium
- w = plate spacing (substrate thickness here), m

During the initial chargeup, when $V_w \ll V_o$, dV_w/dt is constant and may be replaced by $\Delta V_w/\Delta t$. Then we may combine Eqs. (9.28) and (9.29) and rearrange to give the time to charge up to some ΔV_w:

$$\Delta t = 8.84 \times 10^{-12} \varepsilon \Delta V_w / j_+ w \tag{9.30}$$

Taking the case of a 1-mm-thick quartz substrate ($\varepsilon = 3.8$) and the typical j_+ of 0.28 mA/cm^2 [which was calculated following Eq. (9.21)] and converting these to SI units, we find that ΔV_w rises to 100 V in only 1.2 µs. This means that if we wish to restrict substrate charging to a maximum of 100 V, we must reverse the voltage within 1.2 µs by operating at a drive frequency of 400 kHz. When V_o becomes positive, electrons immediately flow out of the plasma until the bulk plasma floats up to $+V_o$ plus the plasma potential, ΔV_p, of Fig. 9.3. These electrons neutralize the positive charge residing on the surface of the substrate so that it again starts from $\Delta V_w = 0$ when the next dose of ions arrives, as shown in Fig. 9.15. That is, the electrons *displace* the positive charge, and the capacitive substrate is said to be passing "displacement" current.

The above situation may also be analyzed using the concept of circuit impedance. The capacitive impedance or reactance, X_C, to current flow decreases with increasing *sinusoidal* drive frequency, v_o (Hz), according to

$$X_C (\Omega) = 1/2\pi v_o C = 1/\omega_o C \tag{9.31}$$

where ω_o is the angular drive frequency, $v_o/2\pi$ (radians/s or s^{-1}), and C is in farads (F). The impedance of our substrate for unit area at 400 kHz is thus 1.2×10^5 $\Omega \cdot$ cm^2. This may be compared to the impedance of the plasma, which is largely in the cathode sheath and is resistive at this v_0. For a typical sheath voltage of 1000 V, the sheath impedance is (1000 V)/(2.8×10^{-4} A/cm^2) = 3.6×10^6 $\Omega \cdot$ cm^2. Considering the substrate and sheath as a voltage divider, this says that the *time-average* voltage drop across the substrate will be 3 percent of V_0, which is consistent with our Eq. (9.30) calculation in which the *maximum* substrate voltage drop was 10 percent of V_0. Operation at lower v_0 would result in too much of V_0 being dropped across the substrate, causing plasma extinction, or causing the plasma to seek lower-impedance current paths, or causing plasma nonuniformity across the substrate due to variation in the substrate-electrode gap and therefore in capacitive coupling. When the depositing film itself is the largest capacitive impedance, as in the case of reactive sputtering or PECVD of insulating compound films onto conducting substrates, ΔV_w may have to be held to an even lower value to avoid electrical-breakdown damage across the film (see Exercise 9.8).

The above example illustrates the motivation to operate glow discharges at high frequency. Plasma drive voltage dropped across insulating layers can cause film damage, loss of ion-bombardment energy, plasma nonuniformity, or plasma extinction, and this voltage drop decreases with increasing frequency. However, other factors also influence the selection of drive frequency. We will see below that there is a distinct change in plasma behavior in crossing 1 MHz or so. The <1 MHz regime includes 60 Hz ac up through the audio range and through the low-frequency rf range and will be termed the low-frequency regime here. Above 1 MHz is the high-frequency regime. Later, in Sec. 9.5, we will examine operation in the microwave or UHF regime (> 300 MHz) as a means to drive an electrodeless discharge.

9.4.1 Low-frequency regime

At drive frequencies below 1 MHz or so, the planar-diode plasma may be thought of as a dc plasma that is periodically reversing polarity, as shown in Fig. 9.16. To the left is the sinusoidal waveform of the voltage being applied to the driven electrode, and to the right is the electrical potential versus position across the gap, L, from the driven electrode to the grounded electrode at the far right. At t = 0, the plasma looks identical to the dc plasma of Fig. 9.5: the driven electrode is at its negative peak voltage, $-V_0$, and is acting as the cathode, and the potential of the bulk plasma is floating ΔV_p above ground. A half cycle later, the driven electrode is at its positive peak voltage and has become the *anode*. Meanwhile, the plasma potential has floated up to a value of ΔV_p above $+V_0$ in accordance with the principle of Sec. 9.2

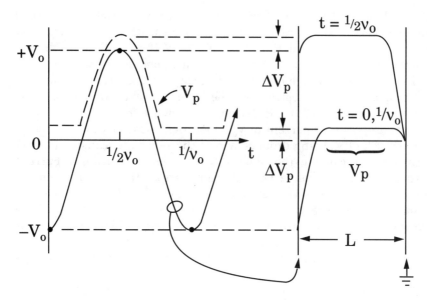

Figure 9.16 Voltages vs. time (left) and vs. position across the electrode gap, L (right), in a low-frequency planar-diode glow discharge. On the left, the solid waveform is the voltage applied to the electrode, and the dashed waveform is the plasma potential relative to ground, V_p.

that the bulk plasma is always more positive than the most positive of its containing surfaces. The plasma potential measured from ground, V_p, tracks the positive-going drive voltage essentially instantaneously as shown on its Fig. 9.16 waveform, because the mobile plasma electrons are quickly drawn into the positive-going electrode. Thus, at $+V_o$ the cathode sheath has shifted from the driven electrode to the grounded electrode. Then, as drive voltage decreases, V_p drops to maintain the ΔV_p criterion, by the flow of ions to ground. V_p eventually remains pinned at ΔV_p above ground when the drive voltage returns to the negative swing of its cycle. The resulting V_p waveform is essentially that of a half-wave rectified sinusoid because of the rectifying action of the sheaths, which pass electrons much more easily than ions.

What are the effects of the above behavior on film deposition? The low-frequency plasma is most often applied to plasma-enhanced CVD, with the substrate residing on one or the other electrode. First of all, for a drive waveform symmetrical about ground (no dc component), the experience of a film on either electrode will be the same. With regard to ion bombardment, either film will receive alternating bursts of low-energy ions across the anode sheath and high-energy ions across the cathode sheath, as the sheaths periodically swap places. This leads to the commonly observed "bimodal" time-average ion-energy distribution at the electrodes of low-frequency plasmas [31]; that is, a low-energy hump at ΔV_p and a smaller high-energy hump near V_0. However, at high pressure, sheath collisions will distribute the high-energy hump to lower energies as discussed at the end of Sec. 9.2. Meanwhile during this oscillation, the plasma is being sustained by bursts of high-energy "beam electrons" (Sec. 9.3.1) which are emitted from the cathode and accelerated into the plasma across the cathode sheath. Thus, the beam-generated region of high electron-impact reaction rates (excitation, dissociation, ionization; see Sec. 9.1) also alternates between electrodes, and the degree of reaction localization at each electrode increases with pressure. These effects can be seen using time- and position-resolved monitoring of optical emission from electronically excited molecules [32], which relax rapidly ($<10^{-7}$ s) compared to the drive-cycle time, $1/\nu_0$.

Because of the rapid optical-emission decay time, the plasma *appears* to extinguish in the drive-voltage crossover region between the V_0 peaks, where little excitation occurs because of the low energy of the beam electrons. However, electron temperature, $\tilde{T}_e$, and the free-radical population decay less rapidly, so whether a drop is seen in them depends on drive frequency, ν_0 (and on pressure). The $\tilde{T}_e$ decay time may be estimated as follows. The electrons are cooled by elastic collisions with molecules, as discussed in Sec. 9.1, and this collision frequency is given by

$$\nu_e = c_e/l_e \tag{9.32}$$

where c_e and l_e are the electron mean speed and mean free path. At the typical $\tilde{T}_e$ of 3 eV, the mean electron kinetic energy is 4.5 eV [see discussion preceding Eq. (9.11)], and c_e is 1.3×10^8 cm/s by Eq. (8.7). By Eq. (2.22), $l_e \approx 4/p$ cm, where p is the pressure in Pa. However, the large electron-molecule mass disparity results in a low fractional energy transfer per head-on collision, which is $\gamma_m = 5.4 \times 10^{-5}$ for Ar, for example [(by Eq. (8.20)]. Thus, the characteristic time for electron thermalization to the gas T is

$$\tau_T \text{ (s)} \approx 1/\nu_e\gamma_m = l_e/c_e\gamma_m = 6\times10^{-4}/p \tag{9.33}$$

or 6 μs at a typical PECVD operating pressure of 100 Pa. At drive frequencies low enough that the voltage crossover interval is longer than this time, the loss of electron energy causes the sheaths to collapse and thus causes ion bombardment to cease. Sheath collapse also means that any negative ions or negatively charged macroparticles that were trapped by the sheath's retarding field can now diffuse to the film surface. The negative-particle problem will be discussed further in Sec. 9.6.3.

The population of the free radicals, some of which are film precursors, decays much less rapidly because these usually must reach a surface before reacting away. The characteristic time, τ_D, for diffusion across a length Λ with a diffusivity D may be found by rearranging Eq. (5.25). Taking $\Lambda = 1$ cm as typical, and using the D of Ar in Ar from Table 2.1 with a 1/p pressure dependence, we have

$$\tau_D = \Lambda^2/4D \approx 1.3 \times 10^{-5} p \qquad (9.34)$$

or 1.3 ms at 100 Pa. Thus, the free-radical supply to the surface will be unmodulated except at very low drive frequencies or pressures. Note that the pressure dependence of this decay time is opposite to that of the T_e decay time in Eq. (9.33).

9.4.2 Transition to high frequency

Two significant changes in plasma behavior occur as drive frequency, v_o, crosses into the high-frequency regime above 1 MHz or so, although there appears to be no coupling between these changes. One is that plasma ions injected into the sheath no longer have enough time to cross to the surface within a half-cycle time, $1/(2v_o)$. The other is that a new electron-energy gain mechanism becomes operative, which involves reflection of bulk electrons off the oscillating sheath. These phenomena and their implications for film deposition will be discussed in turn below.

9.4.2.1 Ion transit time. Because ions are so much more massive than electrons, the electron-related assumption of instantaneous response to changes in sheath voltage at rf frequencies is not valid for ions. The time for an ion to cross a collisionless (low-pressure) sheath may be estimated by setting the force exerted on it by the sheath electric field, **E** [see Eq. (6.6)], equal to the force of accelerating its mass:

$$q_+\mathbf{E} = m_+(dv_+/dt) = m_+ d(dx/dt)/dt \qquad (9.35)$$

Actually, **E** increases (more or less linearly) with distance from the plasma edge to the electrode surface, as seen in Fig. 9.3, but it will suffice here to give it a constant, average value of V_b/b across the sheath width, b. If one also neglects the injection velocity at the sheath edge [Eq. (9.17)] compared to the subsequent acceleration, integration of Eq. (9.35) gives for the characteristic ion transit time across the sheath (Exercise 9.9), in SI units:

9.4.2 Transition to high frequency

$$\tau_+ = \sqrt{\frac{2m_+ b^2}{q_+ V_b}} \qquad (9.36)$$

For Ar$^+$ crossing a typical cathode sheath having V_b = 1000 V and b = 1 cm (see Fig. 9.4), we find that τ_+ = 0.29 µs. For some ions to bombard the surface with energy corresponding to nearly the full sheath voltage, they must cross the sheath while the drive voltage is near its peak, or within a radian or so of the voltage cycle; that is, $v_o \leq$ 0.6 MHz. At significantly higher v_o, the ions will take many cycles to cross, experiencing a burst of acceleration during each cathodic phase and then coasting during each anodic phase, so that they arrive with an energy corresponding to the *time-average* sheath voltage drop. For the V_p waveform of Fig. 9.16, this average is V_o/π. However, we will see in the next section that in the high-frequency regime, the waveform becomes more sinusoidal so that the average becomes $V_o/2$. We will also see that a lower V_o is required in the high-frequency regime. The result of the time averaging and the lower V_o is that *the maximum ion-bombardment energy is much lower in the high-frequency regime.*

The effect of the ion-energy transition is dramatically reflected in the intrinsic stress of silicon nitride films deposited by PECVD over a range of drive frequencies, as shown in Fig. 9.17. These films are tensile at >1 MHz due to a chemical mechanism which will be discussed in Sec. 9.6.4.2. However, at <1 MHz, the "ion-peening" effect of high-energy bombarding ions, as expressed in Eq. (8.33), overcompensates for the chemical effect and results in compressive stress. In principle, one could operate at an intermediate frequency to obtain zero stress. However, this is an awkward frequency regime due to the changing plasma behavior (Sec. 9.4.2.2) and also due to power-coupling problems (Sec. 9.4.4). Instead, one can use a mixture of a high and a low

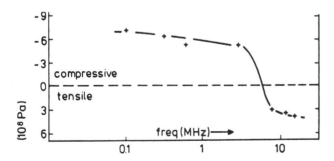

Figure 9.17 Effect of drive frequency on stress of PECVD silicon nitride films deposited from SiH$_4$ + NH$_3$ + N$_2$. (Source: Reprinted from Ref. 33 by permission.)

frequency in the power proportion appropriate to give zero stress (see Sec. 9.4.4).

9.4.2.2 Electron energy gain. We now turn to the second change in plasma behavior that occurs upon crossing into the high-frequency regime. Recall that in the dc and low-frequency regimes of plasma drive power, the plasma is sustained by secondary electrons that are generated by ion impact on the cathode and accelerated across the sheath (the "beam" electrons). While this mechanism is still operative at high frequency, another electron-energy gain mechanism becomes more important: sheath oscillation. At any frequency, the sheath thickness expands as the driven electrode swings negative, because of the need to expose more volume of positive space charge to accommodate the increased voltage drop [(see derivation of Eq. (8.11)]. Then it contracts again on the positive swing and proceeds to oscillate in thickness in a roughly sinusoidal manner, with some harmonic content due to various nonlinearities [14]. Figure 9.18 shows the ion and electron concentrations, n_+ and n_e, versus position across the sheath at an intermediate point of time in the cycle where the sheath has some intermediate thickness, $b(t)$, between the anodic and cathodic limiting thicknesses, b_a and b_c. On the plasma side of $b(t)$, we have n_e

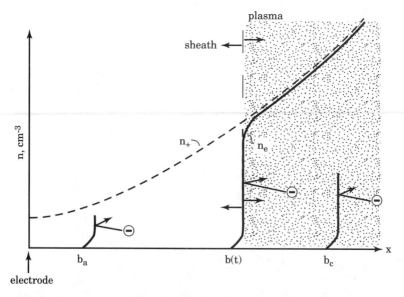

Figure 9.18 Ion and electron concentrations across the rf sheath edge, $b(t)$, which is oscillating between b_a and b_c limits and from which plasma electrons are reflected. (Source: Adapted from a figure in Ref. 14.)

9.4.2 Transition to high frequency

$\approx n_+$ in accordance with the principle of plasma charge neutrality, while at b(t), n_e drops precipitously as we enter the positive-space-charge region of the sheath. Meanwhile, n_+ decreases slowly toward the electrode as the ions become accelerated toward it [Eq. (4.35) at constant flux].

Now, recall from the discussion of Fig. 9.3 that the sheath develops in the first place because of the high velocity of the electrons relative to the ions. The sheath provides a potential barrier for electrons, which reduces their escape flux so that it becomes equal to that of the ions. Thus, all but the fastest electrons in the tail of the electron-energy distribution are reflected from the sheath edge and returned to the plasma. If the sheath happens to be expanding when a particular plasma electron reflects off it, that electron will receive a "swat" analogous to that imparted to a ball by a tennis racket, and it will return to the plasma *hotter*. At the usual rf frequency of v_0 = 13.56 MHz, the energy gain per swat is only 0.1 eV or so (see Exercise 9.10), but the electron soon bounces off a gas molecule and returns for another swat. Statistically, some of the electrons will gain enough energy through successive favorable collisions to ionize molecules, and the plasma is thus sustained by this *volume* energy-gain mechanism without the need for beam electrons from the surface. This is considered to be "ohmic" heating because it is dominated by collisions. The advancing-sheath mechanism is analogous to the heating of a gas by an advancing piston during gas compression. The electrons being swatted are sometimes described as "wave-riding" on the advancing sheath, but this is misleading because it does not account for the multiple bounces which are necessary for reaching ionizing energy.

The energy gained from sheath oscillation is proportional [34] to the square of the sheath velocity and thus to v_0^2. For v_0 < 1 MHz or so, the electron energy gained in this way is insufficient to sustain the plasma, while for increasing v_0 above 1 MHz, the drive voltage, V_0, required becomes monotonically lower because the sheath-oscillation mechanism is becoming more efficient. At 13 MHz, the V_0 required is around one-third of that required to sustain a low-frequency plasma of the same density. The lower V_0 and the ion transit time together are responsible for the much lower ion-bombardment energy represented by the high-frequency results of Fig. 9.17. Because ion-bombardment energy decreases monotonically with increasing v_0, a smaller fraction of the applied electrical power is dissipated in ion bombardment of the electrodes, leaving a larger fraction available for dissociating molecules. This may be one reason why plasma chemical reactions are often observed to increase in efficiency with v_0, as in the case of amorphous Si PECVD rate [35]. Another difference in V_0 behavior in the high-frequency regime is that V_0 rises much more steeply with power than at low frequency, being roughly proportional to plasma current.

Other characteristics of the high-frequency sheath are of interest as well. Calculations of the collisionless (low-pressure), high-frequency, rf **sheath thickness,** b(t) in Fig. 9.18, show it to be roughly the same [14] as in the dc case [Eq. (9.23)] for a given voltage drop. When the collision rate in the sheath is high and the voltage drop is also high, b(t) decreases by at most 50 percent. In contrast to the dc plasma, here the time-average sheath thickness is the *same* as that of the "dark space" observable at the edge of the glow, since the energetic electrons which produce the glow are being generated at the sheath edge. Electrically, this sheath looks much like a capacitor, because at high rf frequencies the capacitive impedance is much less than the resistance represented by the flux of ions across it, as may be seen by examining the calculation following Eq. (9.31). In other words, the displacement of electrons as the capacitor's plate spacing, b(t), oscillates represents a **displacement current** that is much higher than the dc ion current flowing in parallel to it. Across a capacitor, the current is at its peak when the sinusoidal applied voltage is crossing its midpoint and is therefore at its maximum rate of rise, as seen in Eq. (9.28). That is, current *leads* voltage by a quarter cycle or 90°. This is different from the dc and low-frequency plasmas, where the sheaths were resistive because they mostly passed ion current in phase with the sheath voltage. Note that there is no power dissipation in a capacitor. Power is dissipated only by the in-phase component of the $I \times V$ vector product. Therefore, there is a large excess of current flowing through the high-frequency rf plasma—the displacement current—which is not carrying power but which still must be handled by the power circuit. This fact is important in driving these plasmas, as we will see in Sec. 9.4.4. Another consequence of the relative immobility of the ions at high frequencies is that the **plasma potential,** V_p, does not track the drive voltage in the half-wave rectified manner of Fig. 9.16, as it did at low frequencies. Instead, it oscillates in a roughly sinusoidal manner with the sheath capacitance [31], between limiting values of ΔV_p above ground and ΔV_p above the most positive excursion of the driven electrode. The maximum (collisionless) ion-bombardment energy is then determined by the time-average value of this oscillating V_p.

The sheath-edge and the beam mechanisms of electron energy gain are actually both operative at high frequency [36], and the plasma is said to be in the "α" or "γ" mode when the first or second mechanism is dominant, respectively, since α and γ are the historical symbols for the volume and surface secondary-electron emission coefficients. The beam mechanism increases in relative importance at higher power where sheath voltage is higher.

9.4.3 RF bias

The low impedance of capacitors in the high-frequency regime [see Eq. (9.31)] allows rf power to be coupled to the driven electrode through a series capacitor as shown in Fig. 9.19a. This is an important and widely used technique which greatly improves plasma control. Consider the typical "asymmetric" plasma geometry of Fig. 9.19a, in which the area of the grounded "electrode" is much larger than that of the driven electrode because it includes not only the substrate and its

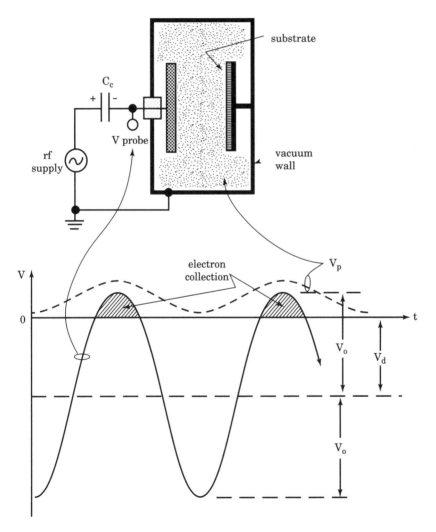

Figure 9.19 DC self-biasing of a capacitively coupled rf waveform due to plasma-electrode asymmetry.

platform but also all of the vacuum-wall surface that is reached by the plasma. Now, recall that the ion flux or current density, j_+, injected into the sheath and arriving at the surface is determined only by plasma density and $\tilde{T}_e$ through Eq. (9.20). If these two quantities are reasonably uniform throughout the plasma, then the electrode of smaller area, A, will receive less ion current, $I_+ = j_+ A$, than will the larger electrode. Conversely, electron current to a surface is determined by the plasma potential, V_p, relative to that surface. When the driven electrode tries to become more positive than V_p on its positive swing, the plasma instantly floats above it by discharging electrons into that electrode. Similarly, when the driven electrode tries to pull V_p below ground on its negative swing, the plasma remains above ground by discharging electrons into the substrate and the wall. Thus, electron current is not governed by electrode area, whereas ion current is. This means that the driven electrode in the asymmetric reactor of Fig. 9.19a receives less ion current than electron current, averaged over time. Because the coupling capacitor blocks this net dc current, the electrode and capacitor proceed to charge up negatively. Without the capacitor (dc coupling), this dc current would flow around the circuit, and the plasma would behave as a dc plasma in parallel with an rf one. A dc plasma leads to many problems when insulating surfaces are in the circuit, as was discussed at the beginning of Sec. 9.4.

As the coupling capacitor of the asymmetric rf plasma continues to charge negatively, the rf waveform at the electrode shifts downward as shown in Fig. 9.19b; that is, it develops a negative dc self-bias voltage, V_d. This bias reduces both the potential across which electrons are being collected from the plasma and the length of the time intervals during which this collection takes place, as represented by the shaded portion of the biased waveform. Within a few cycles, a steady state is reached, at which the time-average electron current to the driven electrode just balances the ion current. At steady state, the bias voltage often reaches its maximum possible value, $V_d \approx V_0$, where the waveform is essentially all below ground potential. Meanwhile, V_p remains above ground, following a sinusoid [37] as shown in Fig. 9.19b rather than the low-frequency waveform of Fig. 9.16, because of the capacitive behavior of the sheaths. That is, the sheaths may be thought of as a capacitive voltage divider for the applied voltage, with V_p being between the series capacitors. Now, the maximum ion-bombardment energy in the high-frequency regime is given by ion charge (unity) times the *time-average* of the sheath voltage drop, $\bar{V}_b$. At the driven electrode, this energy is the time-average difference between the two sinusoids of Fig. 9.19b, which is $V_0/2$ for an applied waveform symmetrical about ground and V_0 for the fully biased case. For intermediate biases, then, we have

9.4.3 RF bias

$$V_b = (V_o - V_d)/2 \qquad (9.37)$$

where V_o is always positive and V_d negative. At maximum bias, ion-bombardment energy elsewhere has been reduced to a minimum equal to the potential drop of a dc anode sheath, ΔV_p from Fig. 9.5. This is a very desirable situation for sputtering, because it concentrates the ion energy on the target where it is wanted and eliminates unwanted sputtering elsewhere. It is the standard configuration for rf sputtering of insulating materials using either the planar diode or the magnetron. The driven electrode is always referred to as the cathode because of its negative bias.

RF bias may also be applied to insulating *substrates* using the same circuitry. This is useful for ion-bombardment cleaning of the substrate ("backsputtering") prior to film deposition, or for increasing the energy of film ion bombardment during deposition to produce structural modification. When the substrate is being transported past the target or other vapor source to obtain good deposition uniformity, the rf must be coupled to it through a metal backing plate which must carry the very high rf displacement current mentioned in the previous subsection. Sliding or rolling electrical contact is usually insufficient, but the current can be *capacitively* coupled from an rf-driven plate spaced closely behind the moving backing plate [38]. In PECVD applications of the rf-biased planar-diode reactor, the substrate can be placed on either the cathode or the anode depending on how much ion-bombardment energy is desirable for the process at hand.

Considerable modeling and measurement of the rf-bias effect has been done [37]. However, it is difficult to predict the degree of bias which will be developed in a given reactor under particular operating conditions, because many factors contribute to determining the *effective* area of the grounded electrode (that is, the area actually reached by the plasma). Nevertheless, it is useful to keep in mind some trends. This area increases with plasma power, because the plasma spreads out more. It decreases with increasing pressure, decreasing electrode gap, and increasing gas electronegativity, because these factors all interfere with the diffusional spreading of plasma electrons. Electronegative gases immobilize free electrons by dissociative attachment [Eq. (9.5)] or by simple attachment as follows:

$$A + e \rightarrow A^- \qquad (9.38)$$

A magnetic field applied perpendicular to the electrodes also reduces effective area by inhibiting the radial diffusion of electrons across the field lines. In the low-frequency regime, plasma spreading is further reduced because the perpendicularly directed beam electrons dominate plasma excitation [39].

RF bias may be hard to predict, but it is easy to measure using a high-impedance, high-voltage rf oscilloscope probe on the cathode vacuum feedthrough as shown in Fig. 9.19a. It is useful to measure both V_o and V_d in this way so that the fractional bias may be determined. Alternatively, V_d alone can be measured with the circuit of Fig. 9.20. Here, the inductor or choke, L, blocks the rf and passes the dc, since the impedance of an inductor is given by

$$X_L (\Omega) = 2\pi v_o L = \omega_o L \qquad (9.39)$$

where L is the inductance in henrys (H). Meanwhile, the capacitor shunts any remaining rf around the resistive voltage divider because of its low rf impedance given by Eq. (9.31). The resistive divider reduces V_d to a conveniently measured level. At the conventional rf frequency of 13.56 MHz, the impedances of the L and C shown are 3.4 kΩ and 12 Ω, respectively, so that only 0.4 percent of the rf voltage appears across the capacitor and resistive divider.

For maximum bias where $V_d \approx V_o$, an upper limit on ion current, I_+, to the biased electrode can be estimated by assuming that all of the

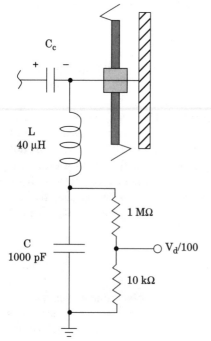

Figure 9.20 DC-bias probe with L and C sized for 13-MHz operation.

applied power, P, goes into ion acceleration across the cathode-sheath voltage drop, V_b; that is,

$$P = I_+ V_b \approx I_+ V_d \approx I_+ V_o \qquad (9.40)$$

This assumption is most reasonable for monatomic gases at low pressure and high power [40], where the fraction of power dissipated in other ways is smallest.

9.4.4 Power coupling

The efficient and accurate delivery of power to a plasma load requires special techniques that vary considerably with the drive-frequency regime. It is important to understand these techniques when designing or tuning plasma-deposition equipment and when attempting to measure or reproduce the power delivered to a process. The various techniques will be outlined below, beginning with the simplest (dc) case and then proceeding to the low-frequency, high-frequency, and microwave regimes.

The **dc power** delivered to a load is simply the product of the current through the load and the voltage drop across it:

$$P(W) = IV_o \qquad (9.41)$$

To maximize the power delivered at *any* frequency, the load impedance, X_l, must be matched to the source impedance of the power supply, X_s. In the dc case, X_l is just a resistance, $X_l = R = V_o/I$, because the plasma is a resistive load. At any frequency, X_s is the V/I slope of the power supply's output. For example, a 1-kW dc supply rated at 1000 V and 1 A has $X_s = 1000 \, \Omega$. If X_l is also 1000 Ω, the full rated power of the supply can be delivered. For $X_l < X_s$, the power supply will reach its current limit before full voltage is developed, whereas for $X_l > X_s$, it will deliver less than its rated current at maximum voltage. Good dc power supplies designed for driving plasmas are impedance-matched to the expected load and also have constant-power and constant-current control, arc protection, high-voltage pulses for plasma ignition, and safety interlocks.

RF power supplies for both the low- and high-frequency regimes have a standard source impedance of 50 Ω, which is much lower than the typical plasma X_l. For example, the low-frequency, resistive sheath discussed after Eq. (9.31) has R = 36 kΩ for a 100 cm^2 area. A high-frequency, *capacitive*, rf sheath 3 mm thick and 100 cm^2 in area has C = 2.9×10^{-11} f (29 pf or "puffs") by Eq. (9.29) and $X_C = 400 \, \Omega$ at 13.56 MHz by Eq. (9.31). This impedance mismatch is remedied in the

low-frequency case by inserting between the power supply and the load an impedance-matching transformer with a 50 Ω primary winding and a secondary winding matched to the X_l of the plasma.

In the **high-frequency** regime, two complications arise with this simple approach. One is that transformers do not work well, because their inductive coils develop too high an impedance [Eq. (9.39)], and because there is too much capacitive shunting across the coils [Eq. (9.31)]. Also, the plasma is now largely capacitive, and the resulting large displacement current which was discussed in Sec. 9.4.2.2 must be blocked from the power supply. This out-of-phase current represents "reflected" power, which would overload the supply and prevent it from delivering its rated power. Both of these problems can be solved by inserting an L-C matching network [41] such as the "π" network shown in Fig. 9.21. In a crude analysis of the operation of this network, variable capacitor C_1 is tuned to the 50-Ω source impedance, or about 200 pf at 13 MHz [Eq. (9.31)], and C_2 is tuned to the 400-Ω load impedance, or about 25 pf. In the inductor, L, V leads I by 90°, just the opposite of a capacitor, so when $X_L = X_C$, the I-V phase angle is restored to zero. Then, the network-plus-plasma looks like a resistive load from both ends, so the reflected power drops to zero. The L needed to match $(C_1 + C_2)$ is (0.7 + 5.6) μH at 13 MHz [Eq. (9.39)], and an additional amount is required to account for the C of the plasma and C_c in series. A coil suspended in air makes a suitable rf inductor, and

$$L (\mu H) = 0.024 j^2 \phi \qquad (9.42)$$

where j is the number of turns and ϕ is the diameter in cm. In actual network operation, there is considerable coupling between C_1 and C_2, so they must be tuned together to the point of zero reflected power. Forward and reflected power are usually measured separately on 50-Ω,

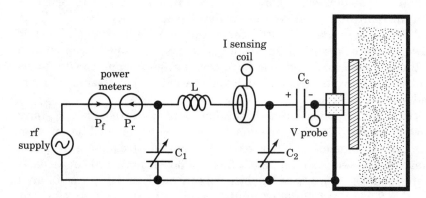

Figure 9.21 Typical matching network for high-frequency rf plasma coupling.

in-line power meters such as those manufactured by Bird and shown in Fig. 9.21. Reflected power may also be fed back to motor-driven capacitors for automatic tuning. This is useful when plasma conditions are being changed often, since that affects tuning.

L-C networks are operable from 1 to 100 MHz or so, but 13.56 MHz is generally used for high-frequency rf plasmas, because it is one of the frequencies assigned by the United States Federal Communications Commission (the FCC) for industrial use. Stray radiation from equipment operating at other frequencies can interfere with radio communications. Radiation in the low-frequency regime is less of a problem, because the wavelength is so much longer than typical equipment dimensions that the equipment does not make a very good radiating antenna. **Radio-frequency interference** (RFI) at high frequency can be a problem for electronics being used with the plasma process, such as vacuum gauges and mass flow controllers. This can be minimized [42] by first enclosing all rf-hot equipment in grounded metal boxes or screens (Faraday cages), which is good safety practice, anyway. Also, all electronics units and the rf supply should have only one ground lead each, and these leads should be brought to a common point of known good ground to eliminate rf ground loops. At high powers, the ground leads are best made of Ag-plated Cu sheet to give the largest surface area of high conductance, since at rf frequencies, current is carried near the surface of a conductor due to the "skin effect."

The **skin effect** is a result of the exponential attenuation of electromagnetic radiation upon entering a conductive medium. The transverse **E** field of the radiation transfers energy to the conducting charges so that the oscillation of **E** becomes damped. The effect limits the depth to which rf or microwave fields or light can penetrate any conductor, including *plasmas*, as we will see in Sec. 9.5. Light attenuation is expressed in Beer's law [Eq. (4.49)]. For rf, the characteristic depth at which **E** has been attenuated to 1/e is known as the "skin depth" [43],

$$\delta_s (m) = \sqrt{\frac{\varepsilon_0 c_0^2}{\pi s v_0}} = \sqrt{\frac{2.54 \times 10^5}{s v_0}} \qquad (9.43)$$

where c_0 is the speed of light, s is the conductivity of the medium (S/m), v_0 is the radiation frequency (Hz), and all units are SI. For the most conductive metal, Ag, s = 6.3×10^7 S/m at room T, so δ_s = 17 µm at 13 MHz.

In the matching network of Fig. 9.21, there can be substantial **power losses** in the inductor and in the lead wire connecting it to the plasma electrode. These losses are due to the skin effect coupled with the fact that the circulating displacement current may be ~100 A. The inductive coil and the capacitive plasma sheath essentially form a resonant or "tank" circuit in which a large amount of electrical energy sloshes back and forth, at a resonant angular frequency of $1/\sqrt{LC}$, between storage in the magnetic field of the coil when it is at peak current and storage in the electric field of the sheath when it is at peak voltage. When the absolute power delivered to the plasma needs to be known for purposes of process transfer, scaling, or modeling, these coupling losses must be subtracted from the net forward power measured on the upstream power meters of Fig. 9.21. A qualitative evaluation of whether power loss is significant and where it is occurring can be made by running the plasma for awhile and then turning off the power, *disconnecting* the power supply for safety, and quickly feeling around the matching network and vacuum feedthrough for hot spots.

For quantitative evaluation of power losses, the wiring's series resistance, R_s, at v_0 can be determined [44] by measuring its current with the sensing coil shown in Fig. 9.21 (a "Rogowsky" coil or "current donut"). Calibrated current-sensing coils are manufactured for various v_0 ranges. When the circuit is driven by the power supply *without* a plasma (gas evacuated), power (P) losses can occur only in the wiring, so that R_s is readily determined from

$$P = P_f - P_r = I_r^2 R_s \qquad (9.44)$$

where P_f and P_r are the forward and reflected power, and I_r is the root-mean-square current, which is $\sqrt{2}$ times the peak current in the case of a sinusoid. Then, with the plasma operating, the power loss is $I_r^2 R_s$, and the difference between that and the power measured on the meters, $(P_f - P_r)$, is the power dissipated in the plasma.

Sometimes, one wants to use **mixed drive frequencies,** v_0; that is, to apply low-v_0 and high-v_0 power to the plasma simultaneously, such as for controlling the proportion of high-energy ion bombardment in PECVD, as discussed in Sec. 9.4.2.1. To do this, one can connect the matching networks of both power supplies to the same electrode, provided that the two power sources are sufficiently decoupled from each other [45]. The coupling capacitor, C_c, of Figs. 9.19 through 9.21 will suffice to keep the low-v_0 power from feeding back into the high-v_0 source because of its high X_C at the lower v_0. Similarly, a series inductor placed between the low-v_0 matching transformer and the electrode will block high-v_0 power because of its high X_L at the higher v_0.

In the **microwave frequency regime** (above 300 MHz), still other power-coupling techniques must be employed. At such high frequencies, the L-C network of Fig. 9.21 does not behave well, because the X_C

between inductor coils becomes as low as the coil's X_L, and the X_L of the capacitor lead wires becomes as low as the capacitor's X_C. That is, inductors start to look like capacitors and vice versa. Also, radiation is increased because of the short wavelength,

$$\lambda = c_0/\nu_0 \tag{9.45}$$

which is 12 cm at the FCC-assigned industrial microwave ν_0 of 2.45 GHz (2450 MHz). Therefore, it is important to transmit microwave power in tightly shielded and firmly connected coaxial cables or hollow metal waveguide tubes. The latter have lower transmission loss but are more awkward because of their rigidity. Both phase and impedance matching to the plasma load are necessary just as in the high-ν_0 rf regime. Phase is tuned with a slider-adjustable length of rigid transmission line, and impedance is tuned by adjusting the proximity of a ground plane (a "tuning stub"). Because of the short λ, it becomes easy to launch a microwave beam into the plasma through a dielectric vacuum wall such as glass or ceramic. This can be done from a waveguide or from an antenna, and some specific examples will be given in the next section.

9.5 Electrodeless Excitation

We have seen in the previous two sections that glow-discharge-plasma excitation using electrodes always requires a high-voltage sheath to accelerate electrons to ionizing energy. While this sheath is useful when high ion-bombardment energy is desired at the surface, it also harbors some problems; namely, it can cause undesired sputtering, it consumes much power in accelerating ions, and it strongly limits the power and plasma density achievable. As pointed out in connection with the hollow-cathode effect of Fig. 9.6, high power density ultimately leads to arc breakdown at the cathode. Conversely, electrodeless excitation can avoid the high-voltage sheath and can therefore achieve ~100 times higher plasma density in the glow-discharge regime, thus providing much higher concentrations of active species for film deposition, with gas ionization fractions approaching unity when pressure is low enough. Because there is no high-voltage sheath, these ions bombard the film with low energies of ~10 eV, which is within the desirable energy window for rearranging surface atoms without doing subsurface damage (Sec. 8.5). Since many electrodeless techniques can also operate at very low pressures approaching the molecular-flow regime (Kn > 1), they can provide a highly directed flux of ionic depositing species, and this substantially reduces self-shadowing when coating surfaces having rough topography.

We will examine below several ways of accomplishing electrodeless excitation. All of these involve coupling rf or microwave energy through a dielectric vacuum wall from an external coil or antenna. Since the plasma is most intense nearest to this external coupler, the uniform distribution of active species over large film areas is often a problem. The equipment also tends to be more expensive than that of the parallel-plate reactor. However, since electrodeless plasmas are relatively new in thin-film deposition, their various pros and cons compared to more traditional plasmas are still in the process of assessment.

9.5.1 Microwaves

The wavelength of electromagnetic radiation at the industrial microwave frequency of 2.45 GHz is only 12 cm [Eq. (9.45)], so microwave energy can easily be beamed into a plasma from an external antenna or waveguide. The intervening vacuum-wall material needs to be an electrical insulator (a dielectric) for microwave transparency, and it is typically borosilicate glass, quartz, sapphire, or ceramic. One of the simplest of plasma couplers is the metal **quarter-wave cavity** [46] shown in Fig. 9.22. When the cavity is $(1/4)\lambda$ long, grounded on one end, and driven from the coaxial-cable voltage at the other end, a standing wave develops as shown in the plot of the wave's transverse electric field, **E**. This is similar to the standing wave that develops in an organ pipe. Because this is a resonant situation, **E** can become quite large, and when it is large enough within the glass tube, the gas therein breaks down. Active species formed in the resulting plasma are transported downstream in the flowing gas and can be used in film-forming reactions with other species supplied downstream either by evaporation or as gases. Another way to activate a microwave plasma in a tube involves the launching of **"surface waves"** in the plasma using external electrodes [47]. The waves propagate along the tube, dissipating power in the plasma. Plasmas are rich in wave phenomena, as we will see below. Surface-wave devices may be operated over a wider range of drive frequency than resonant cavities, and they are less sensitive to changes in gas pressure or composition. They can also be operated at very high plasma density of $n_e > 10^{12}/cm^3$, but at most, one-half of the applied power can be coupled into the plasma, which is a distinct disadvantage.

The pressure effect on microwave plasma excitation can be understood by first considering the microwave **E** field acting on a free electron in *vacuum*. Figure 9.23a shows the **E** field of a microwave oscillating as $\mathbf{E} = \mathbf{E}_0\cos(\omega_0 t)$, as did the light wave of Eq. (4.44). The force exerted on the electron by **E** accelerates it parallel to **E** in accor-

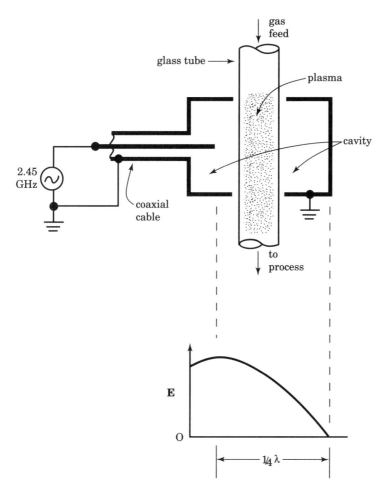

Figure 9.22 Typical quarter-wave microwave cavity for plasma coupling. (Tuning adjustments not shown.)

dance with Eq. (9.35) (written for ions), and the resulting electron velocity versus time, $v_e(t)$, is shown in Fig. 9.23b. When **E** crosses zero and reverses, v_e reaches its maximum negative value and begins decelerating, finally reversing direction and reaching another maximum when **E** again crosses zero. Thus, **E** and v_e are phase-shifted by 90° ($\pi/2$), and there is *no net energy gained* by the electron. This is analogous to the case of inductors and capacitors, where voltage leads or lags current by 90° so that there is no power dissipation (Sec. 9.4.4). When gas is present in the microwave field, however, electrons change their directions in collisions with molecules, and some of them will shift their phase so as to become further accelerated in **E**. The elec-

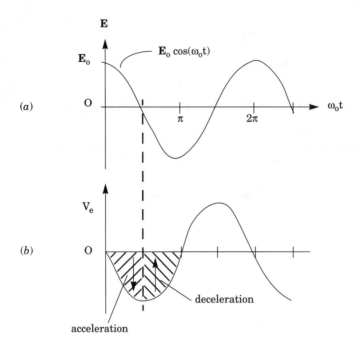

Figure 9.23 Response of a free electron's velocity in vacuum (b) to the oscillating electric field of a microwave (a), showing the 90° ($\pi/2$) phase shift.

tron-energy distribution thus continues to spread upward until some electrons reach ionizing energy. Power coupling by this mechanism is most efficient when ω_0 is equal to v_e, which is the electron-molecule collision frequency for momentum transfer [Eq. (9.9) and Exercise 9.11] (Cecchi, 1990; p. 44). However, with proper tuning, microwaves can be coupled to plasmas with little reflected power over a wide range of v_e/ω_0 (that is, over a wide range of pressure for a given ω_0). In the low-pressure limit of this range, there are not enough collisions, and in the high-pressure limit there are so many that too much energy is lost in momentum transfer. This pressure behavior is similar to the Paschen curve (Fig. 8.4) for gas breakdown in a dc field, although in the present case the low-pressure limit involves the electron's phase rather than its mean free path.

Collisions are not needed to change the electron phase if a strong enough magnetic field, **B**, is established parallel to the direction of the microwave beam and thus perpendicular to the **E** field. Figure 9.24 shows microwaves being beamed through a dielectric window into a plasma chamber containing a **B** field supported by an external solenoid coil. Electrons accelerated perpendicular to **B** by **E** will orbit in a

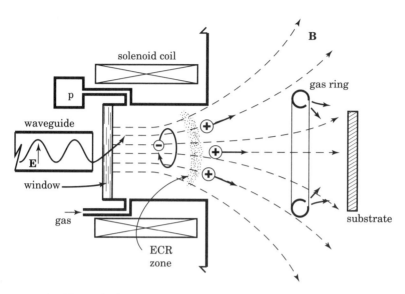

Figure 9.24 Mirror-field ECR plasma source driven by a waveguide and configured for downstream film deposition.

plane perpendicular to **B** as shown and with a cyclotron radius, r_c, given by Eq. (8.17). If the microwave angular frequency, ω_0, is the same as the angular frequency of the orbit, $v_\perp/r_c$, then the electron will be in phase with the field at both $+\mathbf{E}_0$ and $-\mathbf{E}_0$. This phenomenon is known as **electron cyclotron resonance** (ECR), and from Eq. (8.17) it can be seen to occur when

$$|\mathbf{B}| = \omega_0 m_e/q_e = 2\pi v_0 m_e/q_e = 3.58 \times 10^{-11} v_0 \qquad (9.46)$$

or when $|\mathbf{B}| = 0.0875$ T (875 Gauss) for $v_0 = 2.45$ GHz. (Due to other plasma interactions, the resonance point can be shifted somewhat from this prediction.) At pressures low enough so that many orbits occur between collisions, the electrons continue to gain energy and spiral outward with increasing r_c until they finally collide with a gas molecule or with the side wall. To sustain any plasma, it is only necessary to have enough ionizing collisions to balance the electron loss that occurs by diffusion out of the plasma. Here, the loss is mostly at the end faces ($\perp \mathbf{B}$), since lateral diffusion is inhibited by **B**. Consequently, the low-pressure limit is only 10^{-2} to 10^{-3} Pa. At the other extreme, when pressure is high enough that there are many electron collisions per orbit, the resonance is lost (Exercise 9.12) and **B** has little effect. In other words, at high pressure, the electrons are no longer "magnetized," and the plasma behaves more like a conventional microwave discharge. In principle, ECR can be operated at any v_0 as long as

Eq. (9.46) is satisfied; but as v_0 is reduced, r_c increases, so the high-pressure limit is reduced.

Because the microwave beam becomes much more strongly absorbed by the plasma once it arrives at the **B** corresponding to resonance, the zone of intense plasma in ECR tends to form a sheet as shown in Fig. 9.24, and its position can be adjusted by the solenoid current. The efficient coupling of microwaves to this sheet requires careful source design, and to understand this problem it is necessary to first review the interaction of electromagnetic radiation with matter. We begin with the case of an *un*magnetized plasma, where we are dealing with a conducting medium: that is, one containing free charges (electrons), which collide with atoms and thus dissipate energy. This describes a metal, too, and in fact the interaction of radiation is similar for unmagnetized plasmas and for metals. Metals absorb light and longer-wavelength radiation by the motion of free electrons, and the characteristic absorption depth or "skin depth," δ_s, becomes shallower with increasing conductivity in accordance with Eq. (9.43). The shallower that δ_s becomes, the larger is the fraction of radiation that gets reflected before it can be absorbed, which is why Ag makes the best mirrors. However, metals also have a plasma frequency, ω_p, which is given by Eq. (9.15). If the radiation angular frequency, ω_0, exceeds ω_p, then the electrons respond much less to the **E** oscillations of the radiation, and the medium becomes much more transparent. Thus, alkali metals become transparent in the UV, and plasmas become transparent when their density, n_e, is low enough that $\omega_p < \omega_0$, although there will still be some absorption due to electron collisions with molecules. For a $\omega_0/2\pi$ ($=v_0$) of 2.45 GHz, we can see from Eq. (9.15) that transparency occurs below $n_e = 7.45 \times 10^{10}$ cm^{-3}. Plasmas are designated "underdense" and "overdense" when n_e is below or above this point.

ECR plasmas are usually run overdense so as to achieve the desired level of gas activation for film deposition. However, the presence of a strong enough **B** field between the window and the ECR zone allows microwaves to propagate across this distance without being absorbed within the skin depth. This is because a highly magnetized plasma behaves like a *dielectric* medium rather than like a metal. Recall that the electrons in a dielectric are localized about individual atoms and experience a restoring force when displaced (polarized). An electromagnetic wave induces local oscillation in this polarization, but no conduction and therefore no energy dissipation except at high amplitude (nonlinear effects). The atomic oscillators essentially re-radiate the wave, and it is not absorbed. The wave is just slowed down or "loaded" by the oscillator interaction, and the wavelength is shortened, as expressed in the index of refraction [Eq. (4.40)]. The effect of a strong **B** field on a plasma is to localize free electrons into cyclotron orbits about the field lines. This inhibits the conductivity and thus allows certain waves to propagate without absorption. In particular, when **B** is above the ECR value of Eq. (9.46), a right-hand circularly polarized (RHP) wave (see Fig. 4.19) is known to propagate [48]. A linearly polarized wave may be viewed as the sum of right-hand and left-hand (LHP) components, so part of that wave can propagate as well. Wave propagation along **B** also occurs in helicon plasmas (Sec. 9.5.2). The physics of wave behavior in magnetized plasmas is quite complex and not fully understood, but the point here is that the microwave entering the window of Fig. 9.24 can be made to propagate into the ECR zone and thus couple efficiently into the plasma there if the **B** field is made higher upstream as shown (Lieberman, 1993). In this case, the ECR zone of absorption is said to be a "magnetic beach" by analogy to ocean waves hitting the shore. The converging **B** upstream also acts as a magnetic mirror (Fig. 9.12) to reduce plasma losses at the window, so this design is known as a mirror-field ECR. In an alternate ECR design where **B** is lower outside of the ECR zone, the wave is made to penetrate sufficiently into the ECR zone despite the skin-depth absorption by using a resonant cavity to achieve very high **E** [49].

The behavior of an ECR plasma changes dramatically in crossing from the underdense to the overdense region, which is done by increasing the microwave power or the gas pressure [50]. Microwave absorption and plasma density increase abruptly, and the plasma moves outward radially within the ECR zone. This and other mode shifts and nonuniformities can be a problem for process control, so it is important to characterize the behavior of an ECR source and to establish stable and reproducible operating conditions before running a process.

Typical plasma density in an ECR source is 10^{12} cm^{-3}, much higher than the 10^{10} cm^{-3} of the parallel-plate glow discharge. Toward the low end of the operating pressure range, this density becomes comparable to the gas concentration itself, which is 2.4×10^{12} cm^{-3} at 10^{-2} Pa and room T and still lower at the elevated gas T present in such an intense discharge. Thus, the **ionization fraction** can approach *unity* in ECR and other high-density, low-pressure plasma sources, whereas in the parallel-plate plasma it is more like 10^{-6}. This difference fundamentally changes the chemistry of plasma-enhanced film-deposition processes. Instead of a process dominated by neutral radicals as the depositing species and supplemented by ion bombardment, we have one in which ions carry both a depositing species and bombardment energy. One effect of this change is that the deposition flux is now directed, which is useful for filling the deep via holes encountered in integrated-circuit fabrication while avoiding pileup of deposit on the sidewalls and resultant cavity formation, as shown in Fig. 9.25a and b. Another effect is that nearly every depositing atom carries at least several eV of kinetic energy, thus making the most of the beneficial effects of energetic deposition on film properties. However, the effects of this added energetic flux on film properties as compared to films deposited with energetic flux from conventional plasmas have not yet been adequately evaluated.

Two other effects result from high-power, low-pressure operation. One effect is **depleted gas pressure** in the source relative to that measured downstream, even if the gas is fed through the source as shown in Fig. 9.24. This happens when ionization fraction is high, because ambipolar diffusion of ions out of the source [Eq. (9.17)] is much faster than thermal diffusion of neutral particles. Thus, for accurate characterization of source behavior, pressure should be sampled within the source as shown. Note that a thermal-transpiration correction must then be applied [Eq. (3.18)] because of the high gas T. The second effect is **thermal runaway** of electron temperature, $\tilde{T}_e$. Without enough gas collisions, the $\tilde{T}_e$ of ECR electrons would continue to increase until it was so large that the orbit radius would approach the source radius. Thus, when source-pressure depletion sets in at high power, $\tilde{T}_e$ and ionization fraction rise, causing still more depletion. This is a positive-feedback, unstable situation. Further instability arises when $\tilde{T}_e$ exceeds 100 eV or so, because then ionization cross section begins to decrease with increasing $\tilde{T}_e$ (Fig. 8.2), and thus less electron energy is lost to collisions for a given pressure (Lieberman, 1993). High $\tilde{T}_e$ defeats one purpose of electrodeless excitation, namely elimination of the high-voltage sheath, because sheath voltage increases with $\tilde{T}_e$, as discussed prior to Eq. (9.18). Thus, in ECR sources run at too low a pressure, undesirable sputtering occurs where the ECR zone intersects the sidewall. Too high a $\tilde{T}_e$ can also result in ex-

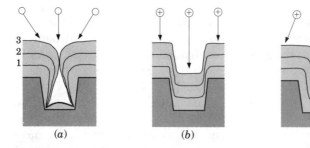

Figure 9.25 Characteristic film-thickness profiles at three successive times (1, 2, 3) during deposition into a via or trench on the substrate, when the depositing species are (a) randomly directed neutrals, (b) perpendicularly directed ions, and (c) obliquely directed ions.

cessive ion-bombardment energy at the film because of electron acceleration in the diverging **B**, as we will see below.

The diverging **B** field downstream of the solenoid of Fig. 9.24 is a reverse magnetic mirror: it is a "**magnetic nozzle**," which accelerates orbiting electrons in the diverging direction as discussed prior to Eq. (9.27) (Cecchi, 1990). In the limit as **B** → 0 downstream, all of the orbiting (thermal) kinetic energy of the electron can be converted to kinetic energy directed along **B**. Moreover, ions are accelerated along **B** together with the electrons by the ambipolar-diffusion effect (Sec. 9.2), even though **B** is not strong enough to magnetize the ions. Thus, the higher the $\tilde{T}_e$, the higher the ion-bombardment energy on a downstream substrate. Ion acceleration along the **B** lines shown in Fig. 9.24 also causes ions to impinge at an oblique angle, which destroys the advantage of directed deposition for via filling, as shown in Fig. 9.25c. The ion flux can be collimated and the electron acceleration reduced at the same time by adding a downstream solenoid coil to straighten out the field lines [51] as shown in Fig. 9.26a.

If *higher* ion bombardment energy is desired than is supplied by the magnetic nozzle, negative bias can be applied to the substrate. When the substrate or film is a dielectric, this needs to be **rf bias** (Fig. 9.19). The application of bias also provides a convenient method of measuring ion current to the substrate, by assuming that the bias power applied is the product of the bias voltage times the ion current [Eq. (9.40)]. This assumption is valid at low pressure and at high power density in the source plasma, in which case most of the applied bias power goes into ion acceleration across the cathode sheath. The power input to the substrate from the high ion-bombardment flux can heat it significantly, even when bias is not applied. Recall from Sec. 5.8 that thermal contact between a substrate and its platform at low pressure is poor even if clamping is employed. Since platform T is almost always measured instead of actual substrate T, much of the work in which "room-T" deposition of high-quality films is reported actually involved a much higher substrate T. It may be that high-density plasma allows deposition at lower substrate T, but one has to be very careful in determining this T. These same considerations regarding substrate biasing and heating also apply to other high-density sources to be discussed below.

Plasma confinement downstream of a high-density plasma, or wall-loss reduction around any plasma, can also be achieved using a "**magnetic bucket**." One common configuration is shown in Fig. 9.26b. The array of permanent magnets surrounding the chamber presents to approaching electrons **B** fields that are crosswise between the magnets and thus deflect the electrons, and that are converging at each magnet and thus act as magnetic mirrors. Electrons can escape through the

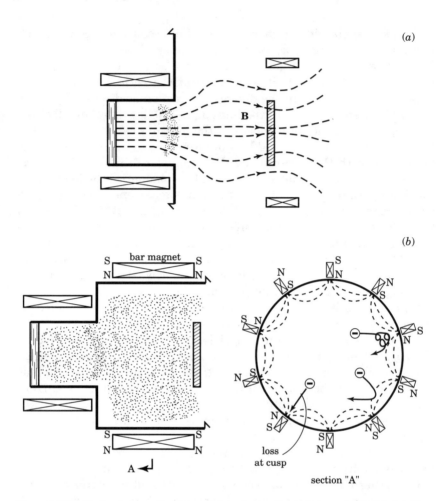

Figure 9.26 Methods of magnetically confining plasma downstream of a high-intensity plasma source such as that of Fig. 9.24: (a) solenoid field shaping and (b) magnetic bucket.

mirror only if they have velocities nearly parallel to and aligned with the "cusps" in **B** parallel to the chamber radius at the center of each magnet [Eq. (9.27)]. Figure 9.27 is a photograph of a plasma being confined by a magnetic bucket, looking from downstream. Bright plasma extends outward toward the wall at the ten cusp points. The magnet spacing in a bucket is always a compromise between too much loss at the cusps if the spacing is too small and too much **B** penetration toward the center of the chamber if the spacing is too large.

Gas distribution is the final aspect of ECR design for film deposition to be discussed. We have mentioned injection of gas upstream in the ECR source, but if one were to introduce a *depositing* species

Figure 9.27 Photograph of plasma confined by a magnetic bucket, viewed as in Section A of Fig. 9.26b. (Source: Reprinted from Ref. 52 by permission.)

there, most of the film deposition would occur within the source rather than on the downstream substrate. Therefore, the depositing species is instead injected just upstream of the substrate, through the "gas ring" shown in Fig. 9.24. The ECR plasma supplies the gaseous species which either activates or reacts with the depositing species to form the film. For example, SiH_4 from the gas ring and activated O_2 from the plasma will deposit SiO_2, and SiH_4 plus activated He will deposit amorphous Si. The chemistry of film deposition downstream of a plasma is discussed further in Sec. 9.6.3. The gas ring contains an array of effusion orifices directed at the substrate. These need to be positioned so that the flux of gas arriving at the substrate is reasonably uniform. They also need to be small enough, compared to the inside diameter of the ring tubing, that the fractional pressure drop through the ring tubing is small, to ensure uniform effusion rate from orifice to orifice. The pressure drop due to fluid flow in tubing can be found from Eqs. (3.2) and (3.7), and flow rate versus pressure drop for orifices in the sonic-flow limit is given in Appendix E.

9.5.2 Helicons

Electromagnetic waves of *rf* frequency can propagate through an overdense plasma when it is magnetized, just as could the microwaves of the ECR plasma of Fig. 9.24. Recall from that discussion that a plasma is overdense when its plasma frequency, ω_p [Eq. (9.15)], exceeds the angular drive frequency, ω_0, of the applied power. In an overdense plasma without magnetization, free electrons can respond to the electromagnetic field to absorb and reflect it. However, magnetization of the electrons inhibits their conductivity so that the plasma behaves more like a dielectric medium. In particular, when the magnetic field, **B**, is strong enough so that the electron cyclotron angular frequency,

$$\omega_c = (q_e/m_e)|\mathbf{B}| \tag{9.47}$$

[compare Eq. (9.46)], is larger that the drive frequency, ω_0, right-hand circularly polarized waves (see Fig. 4.19) can propagate along **B** [48]. These are the "whistler" waves discovered in the early days of radio, which are generated in lightning storms and then propagate through the ionospheric plasma along the Earth's magnetic field. They are more generally known as "helicon" waves because of the helical path traced by the tip of the rotating **E** vector as it propagates.

Interaction with the orbiting electrons of a magnetized plasma slows down a helicon wave just as interaction with the oscillating dipoles in a dielectric medium slows down a light wave. Thus, the wave speed in either medium is

$$c = \frac{c_o}{\tilde{n}} = \nu_o \lambda = \nu_o \left(\frac{\lambda_o}{\tilde{n}}\right) = \frac{\omega_o}{2\pi/\lambda} = \frac{\omega_o}{\mathbf{k}} \tag{9.48}$$

where Eq. (9.45) provides the second equality, and where
 c_o = speed of light in vacuum or air = 3×10^8 m/s
 $\tilde{n}$ = index of refraction of the medium at ω_o
 λ = wavelength in the medium
 λ_o = wavelength in vacuum or air
 ν_o = frequency of the wave, Hz

9.5.2 Helicons

ω_0 = angular frequency of the wave = $2\pi\nu_0$, s^{-1}
k = wave vector or propagation vector = $2\pi/\lambda$ [see text following Eq. (4.44)]

For plane waves, c is also the "phase velocity," which is the speed at which a point of given phase advances in the direction of propagation. Now, ñ increases with ω_0, especially as one approaches the resonant frequency of the medium and the interaction becomes very strong. This accounts for the *dispersion* of white light into its constituent colors by a prism. Dispersion relationships can be derived from Maxwell's equations for any electromagnetic situation, and for a right-hand circularly polarized wave propagating parallel to **B** in a plasma, it is found [48] that

$$\tilde{n}^2 = 1 - \frac{\omega_p^2}{\omega_0^2 - \omega_0\omega_c} \approx \frac{\omega_p^2}{\omega_0\omega_c} \qquad (9.49)$$

The last equality holds for the present case of interest, which is when $\omega_0 \ll \omega_c \ll \omega_p$ (drive, cyclotron, and plasma frequencies, respectively). That is, our plasma is strongly magnetized for the drive frequency being applied, and it is dense enough that the plasma frequency is very high. Note that in the more general form of Eq. (9.49), when $\omega_0 \to \omega_c$, ñ $\to \infty$, so the wave is stopped dead in its tracks. This is the *resonant* situation of the ECR plasma [Eq. (9.46)]. However, we return now to the situation at lower ω_0.

By inserting into the approximation at the end of Eq. (9.49) the expressions for ω_p [Eq. (9.15)] and ω_c [Eq. (9.47)], we obtain (in SI units)

$$\tilde{n}^2 = n_e q_e / \varepsilon_0 B \omega_0 \qquad (9.50)$$

Then, using Eq. (9.48) to eliminate ñ, we have for the wavelength of the helicon wave in the plasma,

$$\lambda = \sqrt{\frac{2\pi\varepsilon_0 c_0^2}{q_e}} \sqrt{\frac{B}{n_e \nu_0}} = 5.59 \times 10^{12} \sqrt{\frac{B}{n_e \nu_0}} \qquad (9.51)$$

Taking typical conditions of ν_0 = 13.56 MHz, **B** = 0.01 T (100 Gauss), and $n_e = 10^{18}$/m^3 (10^{12}/cm^3), we find that ñ = 147 and λ = 15 cm. The very high ñ produces a very *slow wave* whose resulting short λ means that 13.56-MHz power can now be coupled conveniently into the plasma wave from an *antenna* rather than through the electrodes discussed in Sec. 9.4. This arrangement is shown in Fig. 9.28. The entire

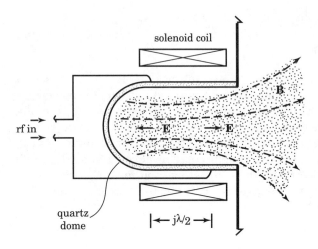

Figure 9.28 Helicon plasma source. The external antenna loop (not shown) is structured so as to drive the helicon wave to the right along **B**.

plasma container needs to be a dielectric here, not just the end window as in ECR, so a glass or quartz dome is usually used. The antenna loop (not shown) is wrapped around the dome and is driven through a matching network such as the one shown in Fig. 9.21. As in the case of ECR, the plasma electrons are accelerated out of the source by the magnetic nozzle effect of the diverging solenoid **B** field, and the ions follow by ambipolar diffusion. The magnetic confinement techniques of Fig. 9.26 can be used downstream. Since **B** is about 10 times smaller than in ECR, however, the solenoid requirements are much less severe.

Helicon behavior is strongly affected by the power level applied. At low power, the voltage on the antenna can be over 1 kV and can therefore act as an electrode to couple power capacitively through the quartz wall into the plasma sheath, as with the parallel-plate plasma of Sec. 9.4.2. As power is increased, there is a discontinuity [53] at which the helicon-wave coupling mode activates and becomes dominant, with an accompanying decrease in antenna voltage and a ×30 increase in plasma density, n_e. However, the antenna voltage may still be high enough to cause local capacitive coupling with accompanying high-energy ion bombardment and sputtering of the quartz dome. As power continues to increase in the helicon mode, there are further smaller discontinuities in n_e. These probably correspond to jumps in the number, j, of half wavelengths that best match the antenna length [53] [see Fig. 9.28 and Eq. (9.51)]. For stable process operation, one should adjust power or **B** to a point midway between these jumps.

The mechanism of plasma electron energy gain in the helicon is different from that in either the rf parallel-plate or the ECR plasma, and it is believed to be "Landau damping," which operates as follows. In the regions of time and space where the **E** field of the helicon wave points partly upstream (to the left in Fig. 9.28), plasma electrons are accelerated downstream. When an electron's velocity downstream is close to the wave's phase velocity, c [Eq. (9.46)], it can "surf" on the wave and receive considerable energy from it, damping (attenuating) the wave as it does so. In the example after Eq. (9.51) where ñ = 147, we have c = 2×10^6 m/s. From Eq. (8.6), we find that this corresponds to 12 eV of electron energy, which is in the range of typical ionization potentials (Fig. 8.3) and thus provides efficient coupling from the wave to plasma ionization. This mechanism of electron energy gain does not require collisions with molecules to operate, in contrast to the sheath-edge mechanism of the high-frequency parallel-plate plasma (Fig. 9.18) and the phase-shift mechanism of the unmagnetized microwave plasma (Fig. 9.22). Collisionless electron energy gain also occurs in the ECR plasma of the last section. In addition, confinement of the electrons by **B** in both of these plasmas greatly increases their path length before they are lost to the walls. Collisionless coupling and magnetic confinement together allow these plasmas to be operated down to very low pressures of 10^{-2} Pa or so where the molecular mean free path exceeds the plasma diameter.

9.5.3 Inductive coils

The third and last method of electrodeless plasma excitation to be discussed is inductive coupling [54], which has the simplest concept and equipment. If a highly conductive helical coil of j turns is wrapped around a dielectric containment tube or dome as shown in Fig. 9.29 and operated in resonance with applied rf power, the very large rf current circulating in the coil, I_{rf}, generates an axial rf magnetic field, B_{rf}, within the tube, which in turn induces a circulating rf electron current in the plasma, jI_{rf}, once the plasma is lit. This is basically a transformer, with the plasma itself acting as the single secondary winding. Since the electrons are confined to orbits by B_{rf}, this plasma can operate down to 10^{-1} Pa or so, although some collisions are required for electron phase shifting between rf cycles, as in the microwave plasma of Fig. 9.22. Since high-voltage sheaths are not involved, power coupling to the electrons is efficient and power density can be high, so plasma densities of ~10^{12} cm^{-3} are achievable. The inductive plasma can also operate at 1 atm as a "plasma torch," which has been explored for deposition of diamond films from methane but will not be discussed further here.

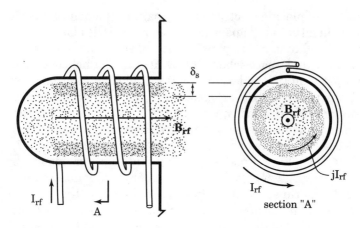

Figure 9.29 Inductive coupling to a plasma from a helical coil. [δ_s is the skin depth of Eq. (9.43).]

Coil resonance can be established using the coil's inductance in conjunction with a capacitor in an L-C circuit similar to that of the Fig. 9.21 matching network. However, the helical-resonator coil of Fig. 9.30 achieves a sharper and stronger resonance (referred to as a higher "Q") and therefore a higher current. This type of coil [55] is grounded on one end and floating on the other, and is $(1/4)\lambda$ from end to end along the helix, so that it resonates like an antenna. The length required is found from Eq. (9.45) and is 5.5 m at 13.56 MHz. The grounded enclosure provides some parallel capacitance, and imped-

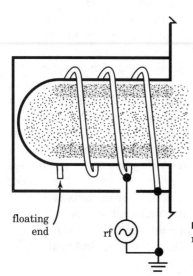

Figure 9.30 Geometry of a helical resonator.

ance matching to the power supply is accomplished with a supplemental parallel capacitor and by the position of the rf input on the coil. In either of the above resonators, the voltage from end to end of the coil is many kV, so *capacitive* coupling to the plasma through the dielectric wall is also possible, as with the helicon antenna. This is generally undesirable because of the accompanying high-voltage sheath and consequent sputtering and elevation of the plasma potential above ground. However, the capacitive mode can be blocked effectively by inserting a cylindrical metal electrostatic shield between the coil and the dielectric wall. The shield needs to have periodic slots oriented parallel to the cylinder axis to prevent circulating currents from developing and robbing power from the plasma. This shield cannot be used with the helicon, because its antenna couples through the *electric* field.

The plasma of Fig. 9.29 is most intense in the annular region where the current is circulating, since this is where the electrons are being accelerated to ionization potential. This region extends in from the wall by about the skin depth, δ_s, of Eq. (9.43). For a given applied frequency, v_0, δ_s depends on the plasma conductivity, s. We can use Eq. (9.13) to find s so long as the pressure is high enough that the electron collision frequency, v_e, is much larger than $2\pi v_0$ ($= \omega_0$). Into Eq. (9.9) for v_e, we insert the following:

- the speed of a 4.5 eV ($\tilde{T}_e$ = 3 eV) electron, 1.3×10^6 m/s from Eq. (8.7)
- a typical collision cross section of 10^{-19} m^2 (0.1 nm^2)
- the concentration of a room-T gas at 10 Pa

For this case, we find $v_e = 3 \times 10^8$ s^{-1}, which is sufficiently larger than the ω_0 of 9×10^7 s^{-1} at 13.56 MHz. Under these conditions, and for a plasma density of 10^{18} m^{-3} (10^{12} cm^{-3}), we find $\delta_s = 1.4$ cm. Note that δ_s decreases with increasing s and therefore with increasing power (or n_e) and decreasing pressure. For the most efficient and uniform power coupling into the plasma volume, one wants δ_s to be about the same as the radius of the containment cylinder. If δ_s is larger, the impedance of the "secondary winding" represented by the plasma becomes too high, and inductive coupling dies out. If δ_s is smaller, the intense region of the plasma becomes more concentrated toward the cylinder periphery. However, the center region still "fills in" with plasma to some extent by diffusion.

As in the cases of ECR and the helicon, the inductive plasma diffuses out the end of the source and can be used downstream in conjunction with a gas ring for film deposition, as shown in Fig. 9.24 for ECR. All three sources share the ability to achieve high plasma density of $>10^{12}$ cm^{-3} at low pressures without developing high sheath voltages. This combination gives ionization fractions near unity and high fluxes of low-energy, directed ions that are particularly well suited to reaction activation and structural modification at the surface of a depositing film. The main difference with inductive coupling is that there is no solenoid **B** field with its accompanying problems of ion steering and acceleration along the diverging field lines. Although inductive coupling is an old concept, large, high-intensity inductive sources have become commercially available only recently, so they have been explored very little for film deposition.

A few other inductive configurations deserve mention. One is the *immersed* coil of Fig. 9.14, which was used there to enhance the ionization of sputtered particles. Another is the spiral coil of Fig. 9.31, which has the potential advantages of generating a reasonably uniform plasma over a large diameter and of coupling it closely to the depositing film. An electrostatic shield with radial slots can be used here to block capacitive coupling as in the case of the helical coil. Finally, Fig. 9.32 shows a ferrite-core transformer driven at frequencies as low as 60 Hz, with a plasma loop acting as the secondary winding [54]. This is probably not scalable to large areas, but it could provide an efficient and inexpensive source of free radicals and ions for use in downstream reactions on small substrates, as did the microwave cavity of Fig. 9.22.

9.6 Plasma Chemistry

This chapter has thus far focused on the *physics* of the glow-discharge plasma; that is, the generation of charged particles and the electric fields and currents which these charges produce. The principal features observed are the presence of energetic electrons (T_e of several eV) throughout the bulk of the plasma, and the bombardment of all

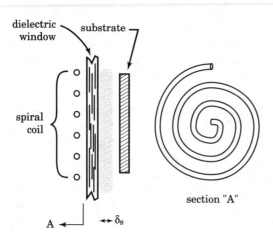

Figure 9.31 Inductive coupling from a spiral coil.

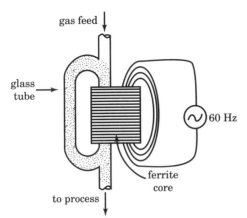

Figure 9.32 Transformer coupling to a plasma loop.

surrounding surfaces by energetic ions (10 to 1000 eV). Chapter 8 dealt with the physical effects of this ion bombardment on sputtering targets and depositing films. We now turn to the *chemistry* which is induced by interaction of charged particles with gaseous and adsorbed reactants.

Recall from Sec. 6.5.4 that most elements can be gasified by reacting them with terminating radicals: for example, Si becomes SiH_4 (silane) gas, and Al becomes $Al(CH_3)_3$ (trimethyl Al) vapor. This allows film-forming elements to be transported by fluid flow to the reaction zone on or above the substrate, where the terminating radicals are eliminated and the remaining elements combine with each other to deposit the film. In *thermal* CVD, the energy needed to surmount the reaction-activation-energy barrier (Fig. 7.16) comes from the elevated substrate T. In *plasma* CVD, it comes from the charged particles. We have established the concept of plasma activation of chemical reactions at low T in previous discussion at the beginning of this chapter and Chap. 8. Below, we will examine in more detail how this occurs and how it can be controlled.

As in the case of thermal CVD, the plasma-deposition process consists of a series of steps, any one of which may be the rate-controlling one, and there are likely to be parallel pathways as well. Usually, the chemical kinetics are not known well enough that we may be certain of the dominant reaction pathway through this web. The plasma case can be even more complex than the thermal one, because charged particles tend to be less selective than thermal energy in activating reactions. This makes reaction modeling a major challenge, and it also can increase film contamination. For example, unwanted decomposition of the methyl radical in $Al(CH_3)_3$ leads to several percent of C incorporation in Al film, although this can be controlled by using a minimum of plasma power [56]. Background gases such as CO and H_2O also become activated, and this increases their reaction rates over what they would be in thermal CVD. These problems of plasma CVD are accepted in return for the ability to deposit just about any material at low T and to simultaneously supply ion bombardment for film structural improvement.

9.6.1 Kinetics

What happens to a gas molecule when it dares to enter a plasma? There are various possible fates, as shown schematically in the flow diagram of Fig. 9.33, and it is important to know their nature and rates to design and optimize reactive plasma processes. The primary chemical-activation event is electron-impact dissociation to produce free radicals, as shown at (*a*). Ionization also occurs, of course, but at a much lower rate because of the higher energy required, except in very intense low-pressure plasmas. The free radicals may diffuse directly to the substrate surface (*b*), or they may first react with other radicals or molecules (*c*) to produce gaseous precursor molecules which then diffuse to the surface (*d*). Some of the source gas reaches the surface without activation (*e*), and some is pumped away directly (*f*). The species arriving at the surface have a certain adsorption probability that depends on conditions there, and the adsorbate then reacts (*g*) to form the film and volatile by-products. The surface reactions may be activated by electron and ion bombardment and by the substrate T. Reactive plasma processes other than plasma CVD involve a solid source material as well as a gaseous one, those processes being reactive sputtering, activated reactive evaporation, laser ablation, and arc evaporation. In those cases, the vaporized solid will be adsorbing and reacting too. Here, however, we concentrate on plasma CVD, since it involves the richest gas-phase chemistry, so that the other processes are subsets of plasma CVD in that regard. We proceed below to make rough estimates of the rates of the various gas-phase steps shown in Fig. 9.33 and their dependence on the primary plasma parameters of pres-

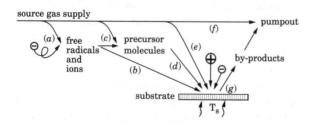

Figure 9.33 Plasma-CVD reaction steps as defined in the text.

sure and power. This exercise will show which steps are going to dominate under what ranges of conditions. In Sec. 9.6.4, we will give specific examples of plasma-CVD processes.

The formal expression for the rate of an electron-impact reaction is given by the integral of Eq. (9.10); but as pointed out in the accompanying text, the functional forms of $\sigma(E_e)$ and $f(E_e)$ needed to solve this integral are not usually available. Nevertheless, the simplifications to Eq. (9.10) offered below will give us a good *qualitative* sense of plasma chemical behavior. In this pursuit, we first need to know the **plasma density, n_e**. Measurements on parallel-plate plasmas indicate that at typical pressure (p) and power density, $n_e \sim 10^{10}$ cm^{-3}. The p and power dependence of n_e may be estimated by writing the following equation for the dissipation of electrical input power, P (in W), in the plasma (Lieberman, 1993):

$$P = j_+ A\varepsilon_L = 0.6 n_{eo} q_e u_a A\varepsilon_L \approx n_e q_e u_a A\varepsilon_L \tag{9.52}$$

where A is the effective area of plasma contact with its container, and the second equality is given by Eq. (9.20) for the ion flux, j_+, injected into the sheath. In the last equality, we are neglecting gradients in n_e across the bulk of the plasma. The term ε_L (in eV, $\equiv$ V) is the total energy dissipated in the plasma per ion injected into the sheath, and it consists of three terms:

$$\varepsilon_L = \varepsilon_c + 2\tilde{T}_e + \varepsilon_i \tag{9.53}$$

which are, respectively, the total electron energy lost in collisions in the plasma per ion-electron pair created by ionization, the mean energy of an electron escaping the sheath for a Maxwellian electron-energy distribution, and the ion energy gained in crossing the sheath. $\tilde{T}_e$ decreases slowly with increasing p, but it is largely "pinned" to a few eV by the thresholds for the inelastic collisions with gas molecules (dissociation and ionization), except at <1 Pa in ECR plasmas, where "thermal runaway" can occur due to insufficient collisions (Sec. 9.5.1). Meanwhile, ε_c increases slowly with increasing p, so ε_L is only a weak function of p. u_a also varies little with p, since it is determined by $\tilde{T}_e$ through Eq. (9.17). Finally, A decreases somewhat with increasing p due to electrons diffusing less far from their generation region. The net effect of these compensating behaviors of the Eq. (9.52) factors is that n_e is a weak (usually increasing) function of p, and we will therefore assume for present purposes that n_e is independent of p. We can also see from Eq. (9.52) that n_e will increase with power. By the way, Eqs. (9.52) and (9.53) also show why the electrodeless plasmas of Sec. 9.5 are more efficient at generating high n_e, because the absence of a high-voltage sheath makes ε_i lower and thus n_e higher for a given power input.

Given n_e, we now estimate the **dissociation rate**: the generation rate of a typical free radical, B, from molecules of AB. To do this, we first replace the cross section function $\sigma(E_e)$ in Eq. (9.10) with an abrupt dissociation threshold of 9 eV. This threshold is just above the low-energy edge of the total-neutral-dissociation curve for SiH_4 in Fig. 9.2, where the dissociation-collision cross section is 2×10^{-16} cm^2. We also need the speed of a 9-eV electron, which is $c_e = \sqrt{2 E_e q_e / m_e} = 1.8 \times 10^8$ cm/s. For the electron-energy-distribution term, $f(E_e)$, we assume the Maxwellian form of Fig. 2.4, where we can see by eyeball integration that about 10% of the distribution lies above the c_e corresponding to

$$2\overline{c_e^2}$$

—that is, above twice the mean energy. Thus, for a typical plasma with a mean electron energy of 4.5 eV ($\tilde{T}_e = 3$ eV), about 10 percent of n_e is energetic enough to dissociate SiH_4 or some other typical reactant molecule. Putting this all together in Eq. (9.10), we have for a dissociation rate of AB per unit volume,

$$R \text{ (mc/cm}^3\cdot\text{s)} = n_e n_{AB} c_e \sigma \cdot 10\% = 3.6 \times 10^{-9} \, n_e n_{AB} \quad (9.54)$$

for n_e and n_{AB} in cm^{-3}. From this, the *frequency* at which an individual molecule of AB becomes dissociated is

$$\nu_d = R/n_{AB} = 3.6 \times 10^{-9} \, n_e = 36 \text{ s}^{-1} \quad (9.55)$$

where the last equality applies when $n_e = 10^{10}$ cm^{-3}. We are going to compare this frequency with the frequencies of the other events that can befall a molecule entering the plasma, and they are all plotted versus pressure in Fig. 9.34.

For the frequency of the second event—**reaction with a free radical**, we first need an estimate of radical concentration. Since dissociation energy thresholds are lower than ionization thresholds, dissociation rate is much higher than ionization rate. Also, radical diffusion to the wall is much slower than electron diffusion because of the much lower particle speed. Consequently, the steady-state dissociation fraction of molecular plasmas is much higher than the ionization fraction or plasma density, n_e. In the Sec. 9.1 example of N_2, whose triple bond makes it one of the most difficult molecules to dissociate, this fraction was found experimentally to be 0.025 at a moderate power level of 0.3 W per cm^2 of parallel-plate cross section. We

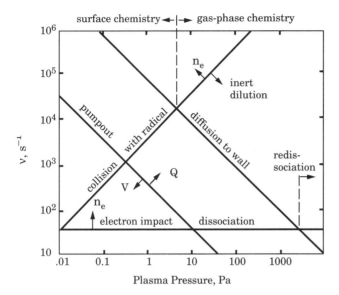

Figure 9.34 Approximate frequencies of various events that can befall a molecule in a glow-discharge plasma, for conditions and assumptions given in the text. (Source: Adapted from a figure in Ref. 72.)

will assume for present purposes a typical value of 0.1n for free-radical concentration, where n is the total gas (molecule) concentration. To the molecule-radical reaction, we can now apply the rate equation for a conventional (thermal) second-order reaction [Eq. (7.34)]. Indeed, free radicals are involved as intermediate species in many thermal chain reactions such as combustion and polymerization. As in the thermal case, the rate constant here is exponential in T per Eq. (7.35) unless the activation energy, E_a, is zero. E_a is in fact always zero in the case of *radical*-radical reactions, because then no bonds need to be broken to activate the reaction—but E_a may be nonzero when one reactant is a complete molecule. In the limit of $E_a = 0$, reaction proceeds at "collision rate" as expressed by Eq. (7.39), where the room-T rate constant is $k_c = 1.6 \times 10^{-10}$ cm^3/mc·s. Thus, for the collision *frequency* and upper-limit reaction frequency of an individual molecule, say CD, with B radicals, we have

$$v_r = R/n_{CD} = k_c n_B = 1.6 \times 10^{-10} \cdot 0.1n = 3.8 \times 10^3 \, p \qquad (9.56)$$

where the last equality uses n given by the ideal-gas law, Eq. (2.10), with p in Pa. This frequency is also plotted in Fig. 9.34. It will increase with n_e because of increased dissociation fraction, and it will decrease if the process-gas mixture is diluted with a gas that does not provide reactive radicals.

A comment is needed here regarding **positive-ion chemistry**. Most positive ions have incomplete outer electron shells just like free radicals do, and they are therefore very reactive [8]. For example, H abstraction by noble gas ions,

$$Ar^+ + H_2 \rightarrow ArH^+ + H \tag{9.57}$$

or

$$Ar^+ + SiH_4 \rightarrow ArH^+ + SiH_3 \tag{9.58}$$

is very favorable thermodynamically, because the ArH^+ product has a filled electron shell and is therefore very stable (large negative $\Delta_f G$). Ion-molecule reactions such as these are likely to be a parallel route to free-radical generation in most plasmas. However, their rates are limited by the electron-impact generation rate of reactive ions, which is much smaller than the electron-impact generation rate of radicals because of the higher electron-energy threshold. Ion-molecule reactions may be more important downstream of the plasma, where the electrons will have dissipated their kinetic energy (more in Sec. 9.6.3).

The remaining two events befalling a molecule entering the plasma are the physical processes of **diffusion to the wall and pumpout**. The frequency at which an individual molecule reaches the wall is the inverse of its characteristic diffusion time, which was estimated in Eq. (9.34) for a typical diffusion length of $\Lambda = 1$ cm. Thus,

$$\nu_D = 1/\tau_D = 4D/\Lambda^2 = 7.7 \times 10^4/p \tag{9.59}$$

Finally, the pumpout frequency is the inverse of the molecule's residence time in the plasma:

$$\nu_p = W/V = 1.7 \times 10^3 \, Q/Vp \tag{9.60}$$

where W = total volume flow rate of gas, cm^3/s
 V = plasma volume, cm^3
 Q = mass flow rate, sccm
 p = total pressure, Pa

For a typical flow rate of 100 sccm through a 500 cm^3 plasma, we have $\nu_p = 330/p$.

Comparison of these four order-of-magnitude frequency estimates in Fig. 9.34 gives a good qualitative idea of which events will dominate under various plasma operating conditions. In the ~100 Pa regime of

9.6.1 Kinetics

conventional plasma CVD, the fastest event is collision with a radical. Therefore, if a given molecule's or radical's reaction rate is fast with the available radicals, it will react to product before diffusing to the surface. The figure also shows that an incoming molecule is likely to encounter a surface before being dissociated by electron impact. However, if it cannot chemisorb in its unactivated state, it will just reflect or quickly desorb (see Fig. 5.1) until it finally does become dissociated in the plasma. Then, because the radical collision frequency is orders of magnitude higher than the electron-impact dissociation frequency, *thermochemically driven reactions are likely to dominate the plasma chemistry despite the presence of fast electrons.* This makes it possible to form quite complicated precursor molecules in plasmas, and it also suggests that those formed will be selectively the most stable ones (most negative $\Delta_f G$), although there is little information available to confirm the latter point. At very high pressure, however, diffusion becomes so slow that electron-impact redissociation of these precursor molecules becomes likely before they reach the surface and deposit, especially at higher power (higher n_e). The high reactivity of the resulting precursor *radicals* may contribute to the gas-phase macroparticle formation (homogeneous nucleation) which is often encountered at high pressure or power in plasma CVD. Macroparticle formation will be discussed further in the next subsection.

Moving down in pressure to <4 Pa in Fig. 9.34, we reach a crossover beyond which a radical, once formed, will diffuse to the surface before colliding with another radical. In this regime, the film-forming reaction sequence involves less gas-phase precursor formation and is instead dominated by radical reactions on the surface. At about the same pressure, depending on n_e and mass flow rate, reactant molecules begin to be pumped away faster than they can become dissociated, so that reactant utilization fraction (η in Fig. 5.1) decreases. This is an example of a shift in the rate-limiting step in a series of steps: reactant dissociation has become limiting here rather than gas supply. However, η can be made high at low pressure by using one of the high-power plasma sources discussed in the previous section. These have an n_e about 100× higher than that assumed for Fig. 9.34, and this raises the dissociation frequency so that it intersects the pumpout frequency at 100× lower a pressure. Figure 9.34 and the procedures used to construct it are useful in determining which of the gas-phase processes is going to dominate a given plasma-CVD process.

We now turn to the **surface chemical processes** by which the film is formed from the adsorbed reactants and precursors (step *g* in Fig. 9.33). Little is known about the surface chemistry of plasma CVD, largely because of the difficulty of analyzing a surface that is immersed in a plasma. Because of the transient nature of many important surface species, analysis really has to be carried out *during* deposition and not after the plasma has been extinguished. New techniques in infrared-absorption spectroscopy and infrared ellipsometry are promising in this pursuit (see Sec. 7.3.3 and Kroesen, 1993). In addressing the surface chemistry, we first note that the sequence of steps which was discussed extensively in the context of thermal CVD in Sec. 7.3.3 also applies to plasma CVD. That is, reactants adsorb, they react to form the film, and by-products desorb. The added factor in the present context is plasma activation of the surface chemistry, which appears both as predissociation of the reactants before they adsorb, and as electron and ion bombardment of the surface. The relative influence of free radicals versus charged particles on film deposition rate and quality can be distinguished by experimentally comparing direct plasma deposition with downstream deposition, since in the latter case the charged particles are selectively attenuated (more in Sec. 9.6.3). *Thermally* activated surface processes may also be occurring, and they can be identified by a change in film properties with increasing substrate T.

One can learn much about the surface processes of both thermal and plasma CVD by examining the profile of films deposited onto structured substrates. A particularly useful structure [57] is shown in Fig. 9.35. This can easily be fabricated by (1) depositing SiO_2 and polycrystalline Si layers onto a Si(100) wafer, (2) defining a trench etch mask by photolithography, (3) etching through the polycrystalline Si, and (4) undercut-etching the SiO_2 in HF solution (which etches Si much more slowly). The trenches should be oriented perpendicular to a (110)

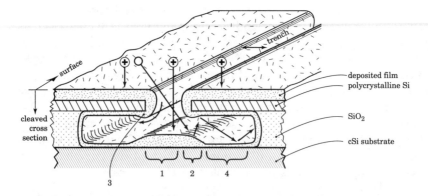

Figure 9.35 Hypothetical thickness profile of a film deposited into the test-structure cavity described in Ref. 57. Deposition in numbered regions is due mainly to (1) ion activation, (2) neutral species of high sticking coefficient, S_c, (3) surface diffusion, and (4) low-S_c species.

cleavage plane, so that after film deposition the substrate can be cleaved to expose the film-thickness profile within the etched cavity. The profile can be viewed and measured by scanning electron microscopy (SEM). Even the profile at incremental depths through the film, as drawn in Fig. 9.25, may be seen by interrupting or modifying the deposition briefly at intervals to produce SEM contrast. The hypothetical film profile shown in Fig. 9.35 illustrates four surface processes that can be distinguished by this technique. Deposition activated by or carried by ion bombardment appears only directly under the trench (region 1), because ions impinge perpendicularly from the plasma, provided that the trench width is much smaller than the plasma-sheath thickness. On the other hand, neutral depositing species impinge over a broad solid angle and therefore also deposit in region 2. Surface diffusion would produce a thickening in region 3 due to surface species diffusing around the trench lip from the top face. If the sticking coefficient, S_c, of a particular depositing species is very low, its deposition will be uniform out to the edge of the cavity in region 4. An intermediate S_c will result in a dropoff in thickness moving outward both on the bottom surface (region 2) and on the underside of the top surface (region 3). The latter effect may be distinguished from surface diffusion by increasing the height of the cavity, which will cause thinning and spreading of an intermediate-S_c profile but will not affect a surface-diffusion profile.

9.6.2 Macroparticles

Plasma-deposition processes are notorious generators of macroparticles, otherwise known as particulates, dust, or powder. For now, we will refer to these simply as particles, although in general, the term "macroparticle" is more precise, because atoms and molecules are particles, too. When plasma-deposition processes are used in dust-free clean rooms for integrated-circuit fabrication or other critical applications, great care must be taken to minimize particle generation and to keep particles from being transported to the substrate or out into the room. Particle transport is minimized by using the vacuum techniques of Sec. 3.4.3. Particle generation occurs in several ways. One involves the flaking off of deposits built up on surfaces surrounding the substrate. This buildup is controlled by periodic system cleaning using plasma etching, wet etching, or abrasion and washing. Particles can also be ejected from various process-material vapor sources such as electron-beam hearths, cathodic arcs, or sputtering targets, and their minimization was discussed in connection with those sources. Remaining generation is due to gas-phase (homogeneous) nucleation of film material, which was discussed as a thermal-CVD problem in

Sec. 7.3.2. There, the remedies were to minimize reactive-gas partial pressure and residence time within the reaction zone, and the same remedies apply to plasma CVD. In plasma CVD, however, homogeneous nucleation and particle growth are both greatly increased by the presence of the plasma, as explained below.

A particle immersed in a plasma behaves like any other electrically floating solid in a plasma in the sense that it develops a negative charge, q, determined by the floating potential, ΔV_f, given by Eq. (9.18). Upon doing so, the particle becomes surrounded by a sheath of positive space charge and thus behaves like a spherical capacitor [58], as shown in Fig. 9.36. The charge required to raise a spherical capacitor to a given potential is proportional to its radius [43], r, so q grows with the particle. Now, the peripheral sheath field surrounding the plasma, which retards electron escape, will far more efficiently retard the escape of these much more massive negative particles, because they have far less translational energy with which to surmount the field barrier. Consequently, massive particles become *trapped* in the plasma and can linger there for minutes or even hours! The longer they linger, the bigger they grow, because vapor atoms and free radicals are continually colliding with them and depositing just as they do on the film surface.

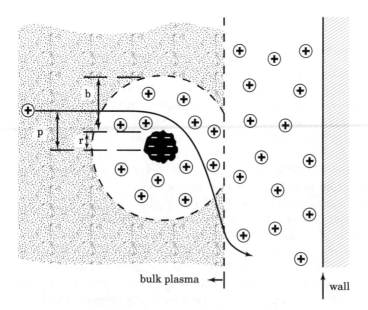

Figure 9.36 Coulombic collision of a plasma ion with a negatively charged macroparticle of radius r which is electrostatically trapped at the edge of the plasma sheath.

9.6.2 Macroparticles

In the meantime, numerous other forces are acting on the suspended particles, including the gas flow, gravity, thermophoresis (Sec. 7.4.3), and the "ion wind." One or the other of these can eventually push the particles out of the plasma, sometimes onto the depositing film where they are most unwelcome. Ion wind refers to the flux of ions from the bulk plasma into the peripheral sheath as described by Eq. (9.20), and it interacts with the particles as follows. When one of these ions passes within a particle's sheath, it begins to feel the particle's negative charge, q, and is deflected as shown in Fig. 9.36, thereby transferring momentum to the particle. The smaller the impact parameter, p, the less that q is screened by the particle's sheath, and the larger is the deflection. This is a screened coulombic collision like the ones between ions and solid atoms which were described in Sec. 8.5.2.2, except that here the coulombic interaction is attractive rather than repulsive. The collision cross section may be expressed as $\sigma = \pi p_0^2$, where the characteristic interaction distance, p_0, is proportional to q and therefore to particle radius, r, by the spherical-capacitor relation discussed above. This cross section is much larger than the physical cross section of the particle, since r is much less than the particle's sheath width, b. The drag force produced on the particle by these collisions [59] increases with ion flux and therefore with plasma density, and it increases as r^2, since $\sigma \propto p_0^2 \propto q^2 \propto r^2$ in accordance with the above discussion. This drag force pushes the particle toward the containing wall or depositing-film surface against the electrostatic retarding force of the plasma's peripheral sheath.

The other forces acting on the suspended particles also increase rapidly with particle radius, r: gas-flow drag is proportional to surface area [Eq. (2.28)] and therefore to r^2, thermophoresis is proportional to cross section (r^2), and gravity is proportional to mass (r^3). On the other hand, the electrostatic trapping force is proportional only to the *first* power of r. Thus, the growing particle eventually reaches a critical size at which the trapping is overcome and the particle is pushed out of the plasma. *Which* force overcomes the trapping and whether the particle lands on the depositing film depends on conditions of ion flux [$\propto n_e$ by Eq. (9.20)], gas flow velocity, T gradients, and substrate orientation relative to gravity. In thermal CVD, the charge-related forces are absent, but homogeneously nucleated particles can still be suspended in a layer above a heated substrate by the thermophoretic force acting against gravity and gas flow (Sec. 7.4.3). In plasmas, the behavior of suspended particles and how it depends on process conditions is more complex and is just beginning to be understood.

Suspended particles can be observed easily by the scattering of a laser beam directed into the plasma [60]. They are often, though not always, observed to accumulate at the edge of the sheath as shown in

Fig. 9.36. Indeed, their boundary might be a good marker of the edge of the sheath, not an easy point to measure otherwise. The particles also segregate laterally when there are steps in the surface that are bigger than the sheath width or when a change in surface material changes plasma coupling, presumably because of the lateral field components induced thereby. The particles segregate toward cooler areas due to thermophoresis. They flow laterally along the direction of gas flow and can be swept downstream once they overcome the trapping sheath at the edge of the plasma.

Among the plasma-deposition processes, plasma CVD has the largest potential for particle generation, because of homogeneous nucleation. The nucleation rate is determined by the second-order gas-phase reaction between film precursors. When a third-body collision is required to remove reaction energy and thereby prevent redissociation of the reaction product, the pressure dependence of the reaction rate can even approach third-order [see discussion preceding Eq. (7.38)]. Particle growth therefore increases rapidly with process pressure and plasma density. Indeed, there appears to be a threshold in both of these parameters below which no particles are observed, and in one case it was determined by laser scattering at $\lambda = 532$ nm to be ~10 Pa in pressure for typical parallel-plate plasma densities [61]. Of course, threshold sensitivity depends on the probe, being proportional to laser intensity and to the $1/\lambda^4$ Rayleigh-scattering factor.

Even in a plasma full of particles, these particles will not become incorporated into the film if final film thickness is reached before they have time to grow to the critical size at which they can escape the sheath. When this condition cannot be achieved for a particular process, it is still possible to reduce film contamination by deliberately allowing the particles to escape periodically while they are still too small to degrade film homogeneity. Recall that when plasma power is turned off, electron T and the sheath field decay with a time constant given by Eq. (9.33) and amounting to about 6 µs at 100 Pa. Without the sheath field, the particles are free to escape, and a clean plasma can then be reignited. Thus, 100 percent modulation (on-off) of input power at a switching frequency of a few kHz is very effective in reducing the concentration of observable particles in the plasma [62]. This can be done easily using an in-line electronic rf switch triggered by a square-wave generator.

9.6.3 Downstream deposition

Films can also be deposited "downstream" in the gas flow stream emanating from a plasma, instead of at the very edge of the plasma. This can reduce undesired film deposition within the plasma source if one of

9.6.3 Downstream deposition

the reactants is injected through a downstream gas ring as shown in Figs. 9.24 and 9.37. Downstream deposition also reduces the variety of energetic species present at the deposition surface and may consequently improve film quality. The price paid for these potential advantages is in deposition rate, since the species that activate the deposition reaction become deactivated as they proceed downstream.

To determine the dominant deposition mechanism of a particular downstream thin-film process, one must first determine what active species are available there. That is, one must assess the degree to which the deposition region is *decoupled* from the plasma. Coupling occurs in both the forward and backward directions: energetic species are transported downstream, and reactants that are injected downstream can "backstream" up into the plasma source against the flow to become activated there. Forward coupling of UV light, which can cause activation of adsorbed species and radiation damage within dielectric films, is easily blocked by the optical baffle shown in Fig. 9.37. For the other active species, the degree of forward coupling depends on their lifetime relative to the time it takes them to be transported downstream.

The transport of ions and electrons is initially at the ambipolar diffusion velocity of Eq. (9.17): $u_a \approx 3 \times 10^5$ cm/s for $\tilde{T}_e = 3$ eV. But once the electrons are no longer receiving energy from the plasma power source, the drag of ions colliding with molecules thermalizes both species to the gas T. The mean free path for molecular collisions was estimated after Eq. (2.24) to be l (cm) $\approx 1/p$ for p in Pa, and since ion-molecule momentum transfer is efficient for similar masses [Eq. (8.20)], the thermalization time constant is

$$\tau_+ \approx l/u_a = 3 \times 10^{-6}/p \qquad (9.61)$$

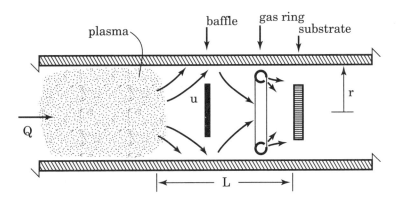

Figure 9.37 Geometry for downstream plasma CVD.

Note that this is 200× faster than the time constant for thermalization of electrons by direct momentum-transfer collisions within the bulk of the plasma when the power is shut off [Eq. (9.33)]. The travel distance during τ_+ is only $u_a\tau_+ = l$, so in downstream transport, the ions and electrons may be considered fully thermalized. Consequently, these species are carried along with the neutral active species at the mean velocity of the gas flow stream, $\bar{u}$ (cm/s). This velocity depends on mass flow rate, Q (sccs), and flow cross section, A (cm^2), in accordance with Eq. (3.14), so that the time for the stream to travel the distance L (cm) downstream from the plasma to the substrate in Fig. 9.37 is

$$t = L/\bar{u} = ALp/10^5Q \qquad (9.62)$$

[Note that the numerator represents just the quantity of gas within volume AL, so that t amounts to a residence time within a volume as in Eq. (7.41).] If we want to operate in the fluid-flow regime to minimize backstreaming, we need to have the Knudsen number, Kn, be ≤0.01, where Kn = $l/r \approx 1/pr$ [Eq. (2.25)] and r is the flow-tube radius. Taking r = 5 cm, we need p = 20 Pa (or more). For a typical Q of 100 sccm or 1.7 sccs and an L of 20 cm, we then find from Eq. (9.62) that t = 0.2 s. This time might be made ×100 or so faster by operating at very high Q or low p. Against this time, we must now compare the survival times of the various active species carried by the flow stream.

During transport, the active species are going to be diffusing to the wall with a time constant τ_D given by Eq. (9.59). Taking an average diffusion length Λ = 2 cm for our 5-cm-radius tube, we have τ_D = 5×10^{-5}p or 10^{-3} s at 20 Pa, which is much faster than the transport time of 0.2 s. Raising p slows the diffusion but also slows the transport, so we conclude that all active species are going to experience many wall collisions before reaching the substrate, unless the tube is much wider or the mass flow rate much higher. The active species will also experience many collisions with other species in the gas phase, and we now must assess the likelihood of their survival at the substrate given this situation.

Table 9.1 lists the possible outcomes of three types of collisions involving each of the active species: ions, metastable atoms [as described before Eq. (9.1)], and free radicals. [The electrons can no longer activate film deposition, because they have been thermalized per Eq. (9.61).] For making a trial assessment of survival probability, consider a single-component flow stream such as He or O_2, with the other (depositing) reactant, say SiH_4, being injected downstream so as to minimize upstream reaction and deposition. We now examine the active species of Table 9.1 one by one.

TABLE 9.1 Fates of Active Plasma Species in Common Collisions

Collision with	Active species		
	B^+	B^*	B (radical)
Molecule	Deactivated by Eq. (9.58) if molecule contains H	Penning reactions [Eqs. (9.1) and (9.2)]	Can react or can recombine in third-body collision
Electron	Recombination needs third body [Eq. (7.38)]		Can become unreactive B^- by Eq. (9.38)
Wall	Neutralized if electron available	Energy lost to wall [7]	Reaction rate increases with adsorption time and wall reactivity (Sec. 5.1)

1. The B^+ ions can react into an inert ion by H abstraction [Eq. (9.58)] only to the extent that the downstream reactant is backstreaming or another H-bearing molecule is available. Recombination with an electron is unstable if pressure is low enough (such as 20 Pa) that third-body collisions are not available to carry away the energy of recombination [see discussion preceding Eq. (7.38)]. Ions are always neutralized at conducting walls, but they are neutralized at insulating walls only if electrons are supplied. Since there are always electrons in the flow stream, wall neutralization is likely.

2. The metastables, B^*, can transfer energy to other molecules by Penning reactions [Eqs. (9.1) and (9.2)]. Even if no such reactions are available, they will generally be deactivated in their first collision with the wall [7].

3. The free radicals can be kept from reacting with other molecules if backstreaming is avoided, and from recombining in the gas phase if pressure is held low enough to avoid third-body collisions. They can also survive wall collisions if the wall is unreactive toward them (no chemisorption) and if their adsorption lifetime is short enough that they do not find each other and recombine before they desorb. (The kinetics of surface reactions were discussed in Sec. 5.1.) Fluorocarbon coatings such as Teflon provide the best passivation against free-radical deactivation, with dielectrics next and metals worst in general, due to the relative adsorption energies and lifetimes on these surfaces. The radicals can also form inactive negative ions by electron attachment [Eq. (9.38)], but there are so many more radicals than electrons in the stream that depletion will be negligible.

In summary, metastables and ions are unlikely to survive unless wall collisions can be avoided, but free radicals can survive if care is taken. In cases where downstream deposition has been reported using a noble-gas plasma such as He along with downstream injection of a reactant such as SiH_4, it is therefore most likely that backstreaming and activation of the reactant is dominating the process.

In the molecular-flow regime of Kn > 1 (p < 0.2 Pa for r = 5 cm), backstreaming is very fast, because the counterflowing streams do not collide with each other. In the fluid-flow regime, the backstreaming problem was solved in connection with pump-oil backstreaming [Eq. (3.13)], where the concentration of the downstream species was found to decrease exponentially with distance upstream, with a 1/e attenuation distance of $L_0 = D/\bar{u}$. We take the diffusivity, D, to be $1.9 \times 10^4/p$ cm^2/s for Ar-Ar (Table 2.1) and find the mean flow-stream velocity $\bar{u}$ from Eq. (3.14). At Q = 100 sccm and p = 20 Pa, this pressure being chosen at the bottom end of the fluid-flow regime for maximum $\bar{u}$, we find that L_0 = 9 cm. A 99 percent attenuation would take $4.6 L_0$ or 40 cm, so we conclude that the elimination of backstreaming reactant is going to require a very high mass flow rate or long transport distance. Reactant that does backstream can be dissociated by active species that are still surviving just downstream of the plasma, and those of the resulting free radicals that are nondepositing can then be swept back to the gas-ring region where they can dissociate depositing reactant. For SiH_4 backstreaming into a He plasma, for example, we expect Penning dissociation [Eq. (9.1)] into SiH_3 + H. Then, when the H is swept downstream, the following H-abstraction reaction occurs:

$$H + SiH_4 \rightarrow SiH_3 + H_2 \qquad (9.63)$$

for which the rate constant is known [63, 64]: $k = 4 \times 10^{-13}$ cm^3/mc·s. The SiH_3 deposits in both regions. This is the most likely mechanism for deposition downstream of noble-gas plasmas, but careful experimental analysis of each situation is needed.

When a high-intensity, low-pressure plasma source is used in the Fig. 9.24 configuration with the depositing reactant being injected from a downstream gas ring, one might expect activation by backstreaming to easily dominate the deposition, since Kn is high. However, a surface activation mechanism also becomes important, whereby the high flux of energetic ions and electrons impinging on the substrate can dissociate reactant which becomes adsorbed there following downstream injection, as discussed in Sec. 8.5.1.2. The dominance of this mechanism was indicated for Si deposition from an ECR H_2 plasma plus downstream SiH_4 when a deposition-thickness pattern was seen on the substrate which replicated that of a grid placed in the flow stream [65]. Free radicals emanating from the plasma are not as collimated as are electrons and ions confined to magnetic field lines, so the pattern from radical deposition would have been much less distinct. The surface activation mechanism accounts for the directional deposition illustrated in Fig. 9.25b which can be obtained downstream of these plasmas. Backstreaming will still occur to the extent that reactant can reflect or desorb from downstream surfaces before reacting, but for ionization fraction approaching unity in a high-intensity, low-pressure source, much of this backstreaming reactant will become ionized and thrown back at the substrate rather than depositing upstream. This direct deposition by ions is also directional.

One might expect that a high enough ion flux emanating from a plasma would sweep backstreaming reactant downstream even for Kn > 1, but it may be argued as follows that the ion flux is not high enough to do so. Figure 2.6b showed an ion passing into the plane of the paper through an array of gas molecules, and the ion's mean free path was calculated as the distance the ion had to travel before the collision cross sections with the molecules it passed, σ_m, "filled" the total cross section. Using similar reasoning, we now consider the reverse case of a stream of ions passing *out* of the plane of the paper past a relatively stationary thermal molecule. The mean time that it takes for this stream to fill the cross section and thus collide with the molecule is

$$t = 1/J_+ \sigma_m \qquad (9.64)$$

where J_+ is the ion flux in ions/cm$^2 \cdot$s (versus j_+ for A/cm^2). Using Eq. (9.20) for the ion flux, we find that for $n_e = 10^{12}$ cm^{-3} and $\tilde{T}_e = 3$ eV, $t = 2 \times 10^{-3}$ s. During this time, a room-T molecule travels 80 cm, so backstreaming is not significantly impeded by the ion flux.

At higher pressures and flow rates where backstreaming can be avoided, and when the plasma generates chemically active free radicals such as O or N atoms, downstream deposition occurs by reaction of these radicals with a film-forming gas injected downstream, such as SiH_4 to form SiO_2 or SiN_xH_y films, respectively. At still higher pressures including atmospheric, O reacts with O_2 to form O_3 (ozone), which is also an active downstream oxidant. For example, its reaction with $Si(OCH_2CH_3)_4$ (tetra-ethoxysilane or TEOS) forms SiO_2 which is known for excellent deposition conformality over substrate topography.

9.6.4 Case studies

Only a few plasma-CVD chemical systems have been studied enough to describe their deposition mechanisms. Chemical-kinetics studies require elaborate apparatus and tedious measurement. For process development, it is easier to just keep depositing films and looking at them until the desired results are achieved. Indeed, most (all?) technologically important plasma-CVD processes have been developed in this manner. However, a deeper understanding of the chemistry can explain various puzzling behaviors, point the way toward even better film properties, and help develop new processes. Hydrogenated amorphous Si (aSi:H or just aSi) and Si nitride (SiN_xH_y) have been studied the most, and their mechanisms will be discussed here as examples of the plasma principles presented above. In this discussion, keep in mind the sequence of steps shown in Fig. 9.33. When more than one of the sequences shown are operating in parallel, the fastest one dominates the process, and within that sequence, the slowest step controls the deposition rate.

9.6.4.1 Amorphous silicon. Plasma-CVD aSi is a semiconductor of surprisingly respectable electronic properties. It is widely used for solar cells and photosensors and for thin-film transistors in active-matrix, liquid-crystal, flat-panel displays (AMLCDs) for portable televisions and computers. It is impossible for the four bonds of each Si atom to fully coordinate with each other in an amorphous network. That is, the network is "overconstrained." The resulting dangling bonds would trap all charge carriers and render the material insulating were it not for the termination of these bonds by H amounting to about 10 at.% of the film composition. The keys to effective H incorporation are to (1) start with a H-saturated Si source such as silane gas (SiH_4), (2) use low plasma power to prevent excessive dissociation, and (3) use enough substrate T to develop the Si network without driving out too much H (about 280° C). Electronic-quality aSi cannot be deposited by thermal CVD, because too high a T is required to activate the deposition reaction. The mechanism by which SiH_4 transforms into aSi is quite complex and still somewhat controversial, involving parallel gas-phase reactions and many surface and subsurface processes. The following is a distillation of the current understanding.

The process begins with electron-impact dissociation of SiH_4 into various products. The most likely reactions are

$$SiH_4 \xrightarrow{e} SiH_4^* \rightarrow SiH_3 + H \quad (\Delta_r H = +4.3 \text{ eV/mc}) \quad (9.65)$$

$$\rightarrow SiH_2 + H_2 \quad (\Delta_r H = +2.6) \quad (9.66)$$

$$\rightarrow SiH_2 + 2H \quad (\Delta_r H = +7.2) \quad (9.67)$$

One might expect the reaction requiring the least energy input ($\Delta_r H$) to dominate, but actually there is more than enough energy for any of these reactions available in the electronically excited species, SiH_4^*, whose internal-energy content is well above that of the products in accordance with the Franck-Condon principle illustrated in Fig. 9.1. Its energy content has to be ≥ 8 eV, in fact, because that is the observed electron-energy threshold for SiH_4 dissociation (Fig. 9.2). The portion of this energy not used in bond-breaking appears as either internal or translational energy of the products. It has been argued that Eq. (9.67) should dominate [66] because it has been determined to do so when SiH_4 excited by 8.5-eV photons dissociates, but it is likely that all three reactions occur to some extent. This is different from the thermal-CVD situation, where the vertical electronic excitation of Fig. 9.1 does not apply. Thermal excitation instead involves *vibrational* excitation through molecular collisions, as discussed in Sec. 7.3.2. Since the occupancy of vibrational energy levels is exponentially distributed in accordance with Boltzmann statistics as expressed in Eq. (5.13), the low-$\Delta_r H$ reaction of Eq. (9.66) is strongly favored in thermal CVD. This is a crucial difference between plasma chemistry and thermochemistry.

Thus, three principal radical species are produced in some proportion by the above reactions, and these "primary" radicals then proceed to react with neighboring species in ways which *are* governed by thermodynamic favorability. The insertion reactions

$$SiH_2 + SiH_4 \to Si_2H_6^* \xrightarrow{M} Si_2H_6 \tag{9.68}$$

and

$$SiH_2 + Si_nH_{2n+2} \to Si_{n+1}H_{2n+4}^* \xrightarrow{(M)} Si_{n+1}H_{2n+4} \tag{9.69}$$

are known by direct measurement [67] to proceed at collision rate; that is, with the fastest possible rate constant of k $\approx 2 \times 10^{-10}$ cm^3/mc·s [see Eq. (7.39)]. De-activation of the $Si_2H_6^*$ requires a third-body (M) collision, without which it will just fall apart again. However, k reaches its high-pressure limit with only 130 Pa of He, and this would be lower for other gases. Polysilane species with n > 2 in Eq. (9.69) have enough vibrational degrees of freedom to absorb the reaction energy *without* the aid of a third body, so they are fast at arbitrarily low pressure. These latter reactions are responsible for gas-phase powder formation in both thermal and plasma CVD of Si when pressure is too high, as shown in Fig. 7.15. The other two primary radicals react less quickly. H produces additional SiH_3 by H abstraction [Eq. (9.63)] with k = 4×10^{-13} cm^3/mc·s as measured by several labs [63, 64]. SiH_3 has no thermodynamically favorable reactions with SiH_4, but at high enough concentration and pressure, it will recombine with itself to form Si_2H_6 at collision rate with the aid of a third body.

All of these gas-phase products impinge on the deposition surface along with the SiH_4, with their relative fluxes depending on how far from the surface the primary radicals were generated and how many reactive collisions they encountered before reaching the surface. That is, moving away from the generation zone one would expect to see mostly SiH_3 and Si_2H_6, and farther away mostly Si_2H_6. The aSi surface will be almost completely covered with chemisorbed H because of the continual arrival of H-laden precursors and because H is known to effectively passivate the surface of Si, as discussed in Sec. 7.3.3. Because of the passivation, the sticking coefficients, S_c, of SiH_3 and Si_nH_{2n+2} are low, and this accounts for the uniform conformality obtained for aSi deposited over surface topography at low power [68]. On the other hand, SiH_2 is expected to have a very high S_c because of its ability to insert into the Si-H bond as in Eq. (9.68). Fortunately, it never reaches the surface unless pressure is very low or the SiH_4 is highly diluted. The fractional surface coverage by H, Θ, may be roughly modeled by the Langmuir isotherm, Eq. (7.46), which shows that Θ will decrease as the H-desorption rate constant, k_d, increases. It is on the bare Si sites, of fraction $(1 - \Theta)$, that SiH_3 and Si_nH_{2n+2} can adsorb and react, and therefore H desorption is believed to be the deposition-rate-limiting step [69].

There are several possible H-desorption mechanisms, including thermal desorption, reaction with impinging H, and ion bombardment. Thermal desorption cannot be important, however, because the aSi deposition rate varies little with substrate T. The relative importance of the other two mechanisms is not clear. As the SiH_4 is diluted with any of the noble gases, film conformality into trenches degrades, and TEM images begin to show a microcolumnar structure [68] characteristic of "Zone 1" quenched growth (Sec. 5.4). Both of these trends suggest that S_c is becoming high. One would indeed expect a shift to a lower steady-state Θ upon increasing the ratio of nondepositing-ion flux to H-bearing precursor flux if ion-induced desorption were important. However, dilution also increases the probability of the high-S_c SiH_2 reaching the surface before reacting with SiH_4, as has been observed in mass spectrometry [68]. The message here is that it is difficult to do an unambiguous experiment when there are so many things going on, and this is one of the challenges of plasma-CVD analysis.

In addition to its possible role in desorbing the passivating H, impinging H is known to etch Si. Selective etching of strained Si-Si bonds by H can account for the transition of film structure from amorphous to "microcrystalline" that occurs with increasing H_2 dilution [68] of the SiH_4 or with increasing exposure of the film to H_2 plasma [70] between intervals of deposition in pure SiH_4. The stronger, unstrained bonds occur where a cluster of Si atoms happens to be approaching its crystalline arrangement, so these crystalline nuclei are selectively left unetched. Now, the microcrystalline Si has far fewer dangling bonds upon which to attach H than does aSi, since they reside only at the grain boundaries instead of throughout the solid. Thus, we are presented here with a curious situation in which a 10× or so dilution of the SiH_4 plasma in H_2 results in a 10× *decrease* in the H content of the deposited film! Similar etching selectivity is believed to be responsible for the ability to deposit diamond films from CH_4-H_2 plasma. Even though graphite is the thermodynamically more stable form of C (except at very high pressure), a higher etch rate for graphite would cause kinetics to win out over equilibrium considerations.

Selective etching can also be used to obtain substrate-selective deposition, because the bonding of depositing Si precursors is weaker to dissimilar substrates such as SiO_2 than it is to Si itself. Since F atoms etch Si, an SiH_4-SiF_4 mixture can be used to obtain substrate-selective deposition [71]. Within a certain window of SiH_4/SiF_4 ratio, deposition is obtained on Si but not on SiO_2, while outside of the window, deposition is obtained on both or neither of the two surfaces. Selective deposition was discussed more fully in the context of thermal CVD toward the end of Sec. 7.3.3.

9.6.4.2 Silicon nitride. Plasma-CVD silicon nitride, SiN_xH_y, deposited from SiH_4 + NH_3, is widely used in the microelectronics industry as a diffusion-barrier coating and as the gate dielectric in aSi field-effect transistors. Like aSi, an amorphous network of Si and N having insufficient H is overconstrained by the presence of four bonding electrons on Si and three on N. However, during SiN_xH_y plasma deposition, considerable H (~30 at.%) becomes incorporated and serves to reduce the overconstraint by terminating Si and N bonds. By contrast, SiO_2 is not overconstrained, because the O, with only two bonding electrons, can more easily adjust itself in the network to find Si neighbors. This is why bulk SiO_2 can form a glass (fused quartz) as well as a crystal and why plasma-CVD SiO_2 incorporates relatively little H (~2 at.%). The H in plasma SiO_2 appears as OH, which decreases with increasing substrate T as it combines and evaporates in the form of H_2O, leaving the remaining O bonded to Si. Bond termination in dielectric films is important for many electronics applications, because dangling bonds can trap injected charge and thereby produce internal electric fields which alter device properties.

SiN_xH_y is a material of widely variable composition, unlike crystalline Si_3N_4, and this adds a new complication to deposition-process control. If a plasma-deposited (amorphous) SiN_xH_y film *happened* to have the 4/3 "stoichiometric" N/Si ratio corresponding to crystalline silicon nitride, then the ~30 at.% H would be equally distributed on Si and N atoms, because each broken bond of the overconstrained Si_3N_4 would produce one Si and one N dangling bond. However, the N/Si ratio can in practice vary continuously from zero (aSi:H) to almost two [$Si(NH)_2$, silicon di-imide], depending on deposition conditions, and the H shifts from the Si to the N with increasing N/Si, as seen by infrared absorption of the H bonds [6]. It turns out [72] that the trapping rate of injected electrons drops by ~10^3 as composition shifts from N/Si $\approx$ 4/3 to the N-saturated limit where all of the H is bonded to N, so the traps apparently are Si dangling bonds or Si-H bonds broken upon encountering an energetic electron. In the N-saturated limit, most of the H appears as NH groups and some as NH_2 groups. NH has two bonding electrons, like O, so it may be thought of as acting like O does in SiO_2 to tie together the Si atoms into a glassy network. By a similar analogy, NH_2 makes only one bond and is therefore equivalent to OH. With increasing substrate T, NH_2 and NH progressively combine and evaporate [6] as NH_3. The electron trapping rate of N-saturated SiN_xH_y is about the same as that of plasma-CVD SiO_2, which further supports the analogy between the two amorphous networks [72].

SiN_xH_y composition depends both on the NH_3/SiH_4 reactant ratio and on plasma power density. Even with a large NH_3/SiH_4 ratio of ~10, Si-rich nitride is obtained if power is too low [6], which indicates that NH_3 requires more electron energy to dissociate than does SiH_4. Note that this result does not *necessarily* follow from the fact that the N-H bond is stronger than the Si-H bond, because of the electronic-excitation factor illustrated in Fig. 9.1. The Si-rich deposit in excess NH_3 also indicates that SiH_n radicals do not readily react with NH_3 gas. To obtain N-saturated deposition, power must be sufficient to *dissociate* enough NH_3 to completely react with all of the SiH_n so as to saturate the depositing Si with bonds to N. Under conventional parallel-plate conditions of 10–100 Pa where gas-phase chemistry is important (see Fig. 9.34), this process has been found to involve the formation of the precursor molecule tetra-aminosilane [6], $Si(NH_2)_4$. This is an unstable molecule that cannot be detected downstream and mostly loses the first amino (NH_2) group in the plasma to form $Si(NH_2)_3$. This radical adsorbs on the surface and then undergoes a "condensation" reaction involving NH_3 gas evolution to form the SiN_xH_y network, as shown in Fig. 9.38. The continuation of this condensation beneath the surface where the network is becoming more rigid would result in stretched Si-N bonds and can thus account for the high tensile stress which is

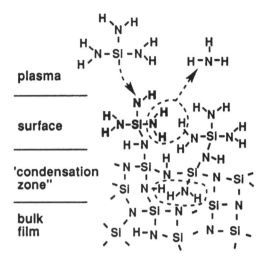

Figure 9.38 Surface and subsurface reactions in the deposition of SiN_xH_y from $SiH_4 + NH_3$. (Source: Reprinted from Ref. 6 by permission.)

characteristic of this material unless deposited at low frequency (see Fig. 9.17). SiN_xH_y deposited this way also exhibits good film conformality over topography, indicating that $Si(NH_2)_3$ has a low sticking coefficient, S_c. The condensation reaction is thermally activated, because with increasing substrate T, stress increases, and N and H content decrease. The endpoint of this reaction would be the formation of Si_3N_4, but at 530° C the film still contains 20 at.% H. For precursor decomposition, we can write the following overall balanced reaction:

$$3Si(NH_2)_4 \rightarrow Si_3N_4 + 8NH_3 \tag{9.70}$$

The reaction of SiH_n with NH_m represents a chemical "oxidation" of the Si. The plasma generates the NH_m oxidant from NH_3 in proportion to the power applied, and this oxidant is then consumed in reacting with the SiH_n. The same behavior has been observed [72] in the reaction of SiH_4 with N generated from N_2 to form SiN_xH_y and with O from N_2O to form SiO_2. In all cases, complete oxidation of the Si requires the presence of excess oxidant. As SiH_4 mass flow rate is increased, the oxidant becomes depleted as shown in Fig. 9.39a until an endpoint (F_c) is reached where the oxidant is completely consumed. At higher flow rates, the excess SiH_n begins to react with *itself* to form Si_2H_6 gas, which can be detected by mass spectrometry downstream, and a smaller amount of the SiH_n deposits as aSi which becomes an undesired part of the nitride or oxide film.

The Fig. 9.39a experiment is essentially a "titration" of the oxidant using SiH_4, with the appearance of Si_2H_6 being the endpoint signal. SiH_4 plasma-oxidation processes may be "tuned" in this way to control

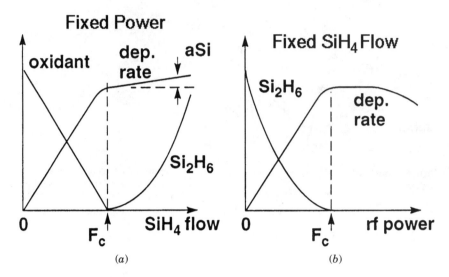

Figure 9.39 Characteristic process behavior of SiN_xH_y and SiO_2 plasma CVD. (Source: Reprinted from Ref. 72 by permission.)

composition and produce good electronic material. The amount of available oxidant increases with power, and the amount needed increases with SiH_4 flow, so the end-point condition is represented by a critical ratio,

$$F_c = (\text{plasma power})/(SiH_4 \text{ mass flow rate}) \tag{9.71}$$

which will be a characteristic of a given reactor and may vary somewhat with reactor geometry and plasma frequency. F_c also varies with the ease of oxidant dissociation, being largest for N_2, smallest for N_2O, and intermediate for NH_3. The F_c point may instead be approached by varying power at a given SiH_4 flow as shown in Fig. 9.39b. Downstream Si_2H_6 disappears, and deposition rate levels out, when there is enough power to oxidize all of the SiH_4. In this level region, most of the SiH_4 is incorporated into the depositing film, so the approximate deposition rate expected may be found from the SiH_4 flow rate and is controlled by that rate. In making this calculation, remember that the film deposits on all accessible surfaces, not just on the substrate.

Although the behavior of these three SiH_4-oxidation processes is similar in the aspects discussed above, it is different in other ways. $SiH_4 + N_2$ does not form a gas-phase precursor molecule, deposits poorly over topography, and produces the undesirable microcolumnar "Zone 1" structure [9, 72] (Sec. 5.4). These characteristics indicate that the deposition precursors are high-S_c SiH_n ($n \leq 2$) radicals and N atoms. $SiH_4 + N_2O$ is widely used to deposit SiO_2. N_2O is preferable to O_2 as an O-atom source because it does not spontaneously react with the SiH_4 in the gas-supply manifold to form powder like O_2 does. Upon electron impact, N_2O dissociates into O and N_2, and negligible N is incorporated into the film except at excessive power. Considerable $Si(OH)_4$, which is the analog of $Si(NH_2)_4$, is seen in the gas phase along with other product molecules, but they are unlikely to be the dominant film precursors, because their subsurface condensation reaction would produce tensile stress as in the case of nitride, whereas plasma SiO_2 actually has compressive stress [72]. This stress reversal is used to advantage in depositing oxynitride, $SiO_xN_yH_z$, from $SiH_4 + NH_3 + N_2O$. By adjustment of the NH_3/N_2O ratio, film stress can be made exactly zero, so that films many microns thick can be deposited without cracking or buckling.

9.7 Conclusion

The glow-discharge plasma is a convenient and versatile source of energetic particles for activating film-deposition processes. It is characterized by a low gas T and by a much higher electron T which is driven by electrical power input. Free radicals generated by electron impact can activate CVD reactions at much lower T than is required for thermal CVD, thus allowing processing of T-sensitive materials. The positive-space-charge sheath surrounding the plasma accelerates positive ions into surrounding surfaces at kinetic energies of 10 to 1000 eV. These ions are useful for source-material sputtering and for film structural modification during deposition. Indeed, the sputter-magnetron is the most widely used technique for volatilizing solid material in thin-film deposition.

Free-radical chemistry and ion energy vary considerably with substrate location and with plasma conditions, especially gas pressure, electrical frequency, and power level. Therefore, the choices of reactor geometry and plasma operating conditions are critical to achievement of the desired film properties. Higher pressure allows more gas-phase chemistry to take place and often produces macroparticle contamination if it is too high. It also scatters ions and sputtered particles and thus reduces their kinetic energy at the surface. Electrical frequency at or above the standard 13.56 MHz is preferred for coupling power through dielectric materials. Increasing frequency reduces ion energy, through a transition in sheath behavior at about 1 MHz and through a steady decrease in sheath voltage at higher frequencies.

One may identify two main categories of plasma-CVD reactors: the conventional parallel-plate discharge, and electrodeless discharges. Electrodeless discharges can be operated at much higher power density without causing arc breakdown. When the electrons are magnetically confined, electrodeless discharges can also be operated at much lower pressure without causing extinction. This shift in power and pressure causes a change in plasma chemistry from control by free radicals to control by the ion stream. The potential advantages of this new chemistry for film deposition are presently being explored.

9.8 Exercises

9.1 What are the ways in which free electrons can be lost from a plasma?

9.2 (a) Estimate the conductivity, s (S/cm or $1/\Omega \cdot$cm), of a 10-Pa SiH_4 plasma of density 1×10^{10} e/cm^3 and $\tilde{T}_e$ = 3 eV. How does this compare with the conductivity of (b) Al metal? and (c) the sheath, of width b, at an electrically floating substrate, taking s $=j_+/\mathbf{E} \approx j_+/(\Delta V_f/b)$ per Eq. (6.2)?

9.3 Show that the three expressions for the plasma frequency in Eq. (9.15) are equivalent.

9.4 In the low-pressure limit, (a) what is the maximum ion-bombardment flux at the electrodes of a dc Ar plasma operating at 1200 V and 100 W between parallel electrodes 15 cm in diameter and 3 cm apart? (b) What is the plasma density, assuming $\tilde{T}_e$ = 3 eV?

9.5 Name four reasons for preferring an electrodeless discharge in PECVD.

9.6 Activated reactive evaporation of ZnO is being carried out from Zn and O_2 in the geometry of Fig. 9.7. The O_2 pressure is low enough that Zn-vapor scattering during transport can be neglected. The electron current across the evaporant stream is 1 A and is distributed over an area of 10 × 10 cm. Estimate the probability of Zn ionization.

9.7 Using the mean free path calculated following Eq. (2.24) and using Fig. 9.8, estimate the fraction of sputtered Pt that reaches the substrate in a planar-diode plasma having an 8-cm gap and operating at the minimum pressure given by Eq. (9.24).

9.8 Silicon nitride (ε = 6.5) is being deposited by PECVD in a symmetrical, parallel-plate discharge. Assuming square-wave drive voltage for simplicity, at peak voltage V_o = 800 V and 1 mA/cm^2 of ion flux across the sheaths, what is the minimum drive frequency that will keep the film from charging up to beyond half of its breakdown strength of 6 MV/cm?

9.9 (a) Derive Eq. (9.36). (b) What is the transit time of H^+ across a 1000-V, 1-cm collisionless sheath? (c) What is the functional dependence of τ_+ on V_b?

9.10 Consider a plasma sheath to be oscillating sinusoidally with a peak-to-peak amplitude equal to the sheath thickness of Eq. (9.23). For 13.56 MHz drive power at a peak voltage of 300 V and for $n_e = 10^{10}$ cm^{-3} and $\tilde{T}_e = 3$ eV in the plasma, what is the maximum energy gain in eV of an electron reflected off the oscillating sheath? (Hint: use a moving coordinate system.)

9.11 Estimate the pressure for most efficient power coupling in a 2.45 GHz plasma having $\tilde{T}_e = 3$ eV. (Sec. 9.5.1)

9.12 Assuming a typical electron-molecule collision cross section for momentum transfer, calculate the pressure at which a 10-eV electron in a 2.45-GHz ECR plasma encounters one collision per orbit.

9.13 An rf-bias power of 400 W is needed to achieve a dc bias of -300 V on a 150-cm^2 substrate platform downstream of an ECR source. (a) What is the ion flux to the substrate, and (b) what is the plasma density over the substrate?

9.14 Derive Eq. (9.51) from Eq. (9.49).

9.15 (a) What two operating conditions in high-power, electrodeless glow discharges result in ionization fractions near unity? (b) What allows each of these operating conditions to be achieved?

9.16 Show that a molecule can indeed travel about 80 cm upstream against the plasma-source ion flux in the example following Eq. (9.64).

9.17 Describe at least three deposition mechanisms operable in the apparatus of Fig. 9.24.

9.18 Describe three ways to reduce macroparticles in plasma CVD and how they work.

9.19 (a) How many sccm of SiH$_4$ are required to deposit SiN$_{1.7}$H$_{1.1}$ of density 2.3 g/cm^3 at 500 nm/m in a 30-cm-diameter parallel-plate reactor, assuming 60 percent SiH$_4$ utilization? (b) What is the minimum mass flow rate of NH$_3$ needed to achieve N saturation of the SiH$_4$, assuming 20 percent NH$_3$ dissociation?

9.9 References

1. Collart, E.J.H., J.A.G. Baggerman, and R.J. Visser. 1991. "Excitation Mechanisms of Oxygen Atoms in a Low Pressure O$_2$ Radio-Frequency Plasma." *J. Appl. Phys.* 70:5278.
2. Aydil, E.S., and D.J. Economou. 1992. "Theoretical and Experimental Investigations of Chlorine RF Glow Discharges." *J. Electrochem. Soc.* 139:1406.
3. Krogh, O., T. Wicker, and B. Chapman. 1986. "The Role of Gas Phase Reactions, Electron Impact, and Collisional Energy Transfer Processes Relevant to Plasma Etching of Polysilicon with H$_2$ and Cl$_2$." *J. Vac. Sci. Technol.* B4:1292.
4. Coburn, J.W., and M. Chen. 1980. "Optical Emission Spectroscopy of Reactive Plasmas: A Method for Correlating Emission Intensities to Reactive Particle Density." *J. Appl. Phys.* 51:3134.

5. Donnelly, V.M. 1989. "Optical Diagnostic Techniques for Low Pressure Plasmas and Plasma Processing." Chap. 1 in *Plasma Diagnostics*, vol. 1, ed. O. Auciello and D.L. Flamm. Orlando, Fla.: Academic Press.
6. Smith, D.L., A.S. Alimonda, C.-C. Chen, S.E. Ready, and B. Wacker. 1990. "Mechanism of SiN_xH_y Deposition from NH_3-SiH_4 Plasma." *J. Electrochem. Soc.* 137:614.
7. Kolts, J.H., and D.W. Setser. 1979. "Electronically Excited Long-Lived States of Atoms and Diatomic Molecules in Flow Systems." Chap. 3 in *Reactive Intermediates in the Gas Phase*, ed. D.W. Setser. Orlando, Fla.: Academic Press.
8. Libby, W.F. 1979. "Plasma Chemistry." *J. Vac. Sci. Technol.* 16:414.
9. Smith, D.L., A.S. Alimonda, and F.J. von Preissig. 1990. "Mechanism of SiN_xH_y Deposition from N_2-SiH_4 Plasma." *J. Vac. Sci. Technol.* B8:551.
10. Hershkowitz, N. 1989. "How Langmuir Probes Work." Chap. 3 in *Plasma Diagnostics*, vol. 1, ed. O. Auciello and D.L. Flamm. Orlando, Fla.: Academic Press.
11. Meuth, H., and E. Sevillano. 1989. "Microwave Diagnostics." Chap. 5 in *Plasma Diagnostics*, vol. 1, ed. O. Auciello and D.L. Flamm. Orlando, Fla.: Academic Press.
12. Kushner, M.J. 1988. "A Model for the Discharge Kinetics and Plasma Chemistry during Plasma Enhanced Chemical Vapor Deposition of Amorphous Silicon." *J. Appl. Phys.* 63:2532.
13. Gottscho, R.A. 1986. "Negative Ion Kinetics in RF Glow Discharges." *IEEE Trans. on Plasma Science* 14:92.
14. Godyak, V.A., and N. Sternberg. 1990. "Dynamic Model of the Electrode Sheaths in Symmetrically Driven RF Discharges." *Phys. Rev. A* 42:2299.
15. Godyak, V.A., and N. Sternberg. 1990. "Smooth Plasma-Sheath Transition in a Hydrodynamic Model." *IEEE Trans. on Plasma Science* 18:159.
16. Pennebaker, W.B. 1979. "Influence of Scattering and Ionization on RF Impedance in Glow Discharge Sheaths." *IBM J. Res. Develop.* 23:16.
17. Bunshah, R.F. 1983. "Processes of the Activated Reactive Evaporation Type and Their Tribological Applications." *Thin Solid Films* 107:21.
18. Westwood, W.D. 1978. "Calculation of Deposition Rates in Diode Sputtering Systems." *J. Vac. Sci. Technol.* 15:1.
19. Rossnagel, S.M. 1988. "Deposition and Redeposition in Magnetrons." *J. Vac. Sci. Technol.* A6:3049.
20. Stutzin, G.C., K. Rózsa, and A. Gallagher. 1993. "Deposition Rates in Direct Current Diode Sputtering." *J. Vac. Sci. Technol.* A11:647.
21. Cuomo, J.J., R.J. Gambino, J.M.E. Harper, and J.D. Kuptsis. 1978. "Significance of Negative Ion Formation in Sputtering and SIMS Analysis." *J. Vac. Sci. Technol.* 15:281.
22. Tominaga, K., S. Iwamura, Y. Shintani, and O. Tada. 1982. "Energy Analysis of High-Energy Neutral Atoms in the Sputtering of ZnO and $BaTiO_3$." *Jap. J. Appl. Phys.* 21:688.
23. Venkatesan, T., X.X. Xi, Q. Li, X.D. Wu, R. Muenchausen, A. Pique, R. Edwards, and S. Mathews. 1993. "Pulsed Laser and Cylindrical Magnetron Sputter Deposition of Epitaxial Metal Oxide Thin Films." In *Selected Topics in Superconductivity*, ed. L.C. Gupta and M.S. Multani. Singapore: World Scientific, 625.
24. Berg, S., M. Moradi, C. Nender, and H.-O. Blom. 1989. "The Use of Process Modelling for Optimum Design of Reactive Sputtering Processes." *Surface and Coatings Technol.* 39/40:465.
25. Carlsson, P., C. Nender, H. Barankova, and S. Berg. 1993. "Reactive Sputtering using Two Reactive Gases: Experiments and Computer Modelling." *J. Vac. Sci. Technol.* A11:1534.
26. Thornton, J.A. 1982. "Coating Deposition by Sputtering." Chap. 5 in *Deposition Technologies for Films and Coatings*, ed. R.F. Bunshah. Park Ridge, N.J.: Noyes Publications.

27. Rossnagel, S.M. 1991. "Glow Discharge Plasmas and Sources for Etching and Deposition." Chap. II-1 in *Thin Film Processes II*, ed. J.L. Vossen and W. Kern. Orlando, Fla.: Academic Press.
28. Window, B., and G.L. Harding. 1990. "Ion-assisting Magnetron Sources: Principles and Uses." *J. Vac. Sci. Technol.* A8:1277.
29. Window, B., and G.L. Harding. 1992. "Characterization of Radio Frequency Unbalanced Magnetrons." *J. Vac. Sci. Technol.* A10:3300.
30. Rossnagel, S.M., and J. Hopwood. 1994. "Metal Ion Deposition from Ionized Magnetron Sputtering Discharges." *J. Vac. Sci. Technol.* B12:449.
31. Köhler, K., D.E. Horne, and J.W. Coburn. 1985. "Frequency Dependence of Ion Bombardment of Grounded Surfaces in RF Argon Glow Discharges in a Planar System." *J. Appl. Phys.* 58:3350.
32. Gottscho, R.A. 1987. "Glow-Discharge Sheath Electric Fields: Negative-Ion, Power, and Frequency Effects." *Phys. Rev. A* 36:2233.
33. Claassen, W.A.P., W.G.J.N. Valkenburg, M.F.C. Willemsen, and W.M.v.d. Wijgert. 1985. "Influence of Deposition Temperature, Gas Pressure, Gas Phase Composition, and RF Frequency on Composition and Mechanical Stress of Plasma Silicon Nitride Layers." *J. Electrochem. Soc.* 132:893.
34. Godyak, V.A., R.B. Piejak, and B.M. Alexandrovich. 1991. "Ion Flux and Ion Power Losses at the Electrode Sheaths in a Symmetrical RF Discharge." *J. Appl. Phys.* 69:3455.
35. Howling, A.A., J.-L. Dorier, and Ch. Hollenstein. 1992. "Frequency Effects in Silane Plasmas for Plasma Enhanced Chemical Vapor Deposition." *J. Vac. Sci. Technol.* A10:1080.
36. Godyak, V.A., and A.S. Khanneh. 1986. "Ion Bombardment Secondary Electron Maintenance of Steady RF Discharge." *IEEE Trans. on Plasma Science* PS-14:112.
37. Köhler, K., J.W. Coburn, D.E. Horne, and E. Kay. 1985. "Plasma Potentials of 13.56 MHz RF Argon Glow Discharges in a Planar System." *J. Appl. Phys.* 57:59.
38. Smith, D.L., and A.S. Alimonda. 1994. "Coupling of Radio-Frequency Bias Power to Substrates without Direct Contact, for Application to Film Deposition with Substrate Transport." *J. Vac. Sci. Technol.* A12:3239.
39. Gottscho, R.A., G.R. Scheller, D. Stoneback, and T. Intrator. 1989. "The Effect of Electrode Area Ratio on Low-Frequency Glow Discharges." *J. Appl. Phys.* 66:492.
40. Godyak, V.A., R.B. Piejak, and B.M. Alexandrovich. 1991. "Ion Flux and Ion Power Losses at the Electrode Sheaths in a Symmetrical RF Discharge." *J. Appl. Phys.* 69:3455.
41. Hall, G.L. et al. (eds.). 1984. *The Radio Amateur's Handbook*, Chaps. 2 and 19. Newington, Conn.: American Radio Relay League.
42. Alcaide, H.D. 1982. "RFI Prevention in RF Plasma Systems." *Solid State Technol.* (April).
43. Feynman, R.P., R.B. Leighton, and M. Sands. 1964. *The Feynman Lectures on Physics*, vol. 2. Reading, Mass.: Addison-Wesley.
44. Suggestion by R.B. Piejak, May, 1991.
45. Hey, H.P.W., B.G. Sluijk, and D.G. Hemmes. 1990. "Ion Bombardment: a Determining Factor in Plasma CVD." *Solid State Technol.* (April):139.
46. Fehsenfeld, F.C., K.M. Evenson, and H.P. Broida. 1965. "Microwave Discharge Cavities Operating at 2450 MHz." *Rev. Sci. Instr.* 36:294.
47. Moisan, M., J. Hubert, J. Margot, G. Sauvé, and Z. Zakrzewski. 1993. "The Contribution of Surface-Wave Sustained Plasmas to HF Plasma Generation, Modeling and Applications: Status and Perspectives." In *Microwave Discharges: Fundamentals and Applications*, ed. C.M. Ferreira and M. Moisan. New York: Plenum.
48. Chen, F.F. 1974. *Introduction to Plasma Physics*, Chap. 4. New York: Plenum.

49. Asmussen, J. 1989. "Electron Cyclotron Resonance Microwave Discharges for Etching and Thin Film Deposition." *J. Vac. Sci. Technol.* A7:883.
50. Gorbatkin, S.M., L.A. Berry, and J.B. Roberto. 1990. "Behavior of Ar Plasmas Formed in a Mirror Field Electron Cyclotron Resonance Microwave Ion Source." *J. Vac. Sci. Technol.* A8:2893.
51. Matsuoka, M., and K. Ono. 1988. "Magnetic Field Gradient Effects on Ion Energy for Electron Cyclotron Resonance Microwave Plasma Stream." *J. Vac. Sci. Technol.* A6:25.
52. Berry, L.A., and S.M. Gorbatkin. 1995. "Permanent Magnet Electron Cyclotron Resonance Plasma Source with Remote Window." *J. Vac. Sci. Technol.* A13(2).
53. Perry, A.J., D. Vender, and R.W. Boswell. 1991. "The Application of the Helicon Source to Plasma Processing." *J. Vac. Sci. Technol.* B9:310.
54. Hopwood, J. 1992. "Review of Inductively Coupled Plasmas for Plasma Processing." *Plasma Sources Sci. and Technol.* 1:109.
55. Cook, J.M., D.E. Ibbotson, P.D. Foo, and D.L. Flamm. 1990. "Etching Results and Comparison of Low Pressure Electron Cyclotron Resonance and Radio Frequency Discharge Sources." *J. Vac. Sci. Technol.* A8:1820.
56. Masu, K., K. Tsubouchi, N. Shigeeda, T. Matano, and N. Mikoshiba. 1990. "Selective Deposition of Aluminum from Selectively Excited Metalorganic Source by the RF Plasma." *Appl. Phys. Lett.* 56:1543.
57. Cheng, L.-Y., J.P. McVittie, and K.C. Saraswat. 1991. "New Test Structure to Identify Step Coverage Mechanisms in Chemical Vapor Deposition of Silicon Dioxide." *Appl. Phys. Lett.* 58:2147.
58. Kilgore, M.D., J.E. Daugherty, R.K. Porteous, and D.B. Graves. 1994. "Transport and Heating of Small Particles in High Density Plasma Sources." *J. Vac. Sci. Technol.* B12:486.
59. Kilgore, M.D., J.E. Daugherty, R.K. Porteous, and D.B. Graves. 1993. "Ion Drag on an Isolated Particulate in a Low-Pressure Discharge." *J. Appl. Phys.* 73:195.
60. Selwyn, G.S., J.H. Heidenreich, and K.L. Haller. 1991. "Rastered Laser Light Scattering Studies during Plasma Processing: Particle Contamination Trapping Phenomena." *J. Vac. Sci. Technol.* A9:2817.
61. Yoo, W.J., and Ch. Steinbrüchel. 1992. "Kinetics of Particle Formation in the Sputtering and Reactive Ion Etching of Silicon." *J. Vac. Sci. Technol.* A10:1041.
62. Verdeyen, J.T., J. Beberman, and L. Overzet. 1990. "Modulated Discharges: Effect on Plasma Parameters and Deposition." *J. Vac. Sci. Technol.* A8:1851.
63. Becerra, R., and R. Walsh. 1987. "Mechanism of Formation of Tri- and Tetrasilane in the Reaction of Atomic Hydrogen with Monosilane and the Thermochemistry of the Si_2H_4 Isomers." *J. Phys. Chem.* 91:5765.
64. Johnson, N.M., J. Walker, and K.S. Stevens. 1991. "Characterization of a Remote Hydrogen Plasma Reactor with Electron Spin Resonance." *J. Appl. Phys.* 69:2631.
65. Nakayama, Y., M. Kondoh, K. Hitsuishi, M. Zhang, and T. Kawamura. 1990. "Behavior of Charged Particles in an Electron Cyclotron Resonance Plasma Chemical Vapor Deposition Reactor." *Appl. Phys. Lett.* 57:2297.
66. Gallagher, A. 1988. "Neutral Radical Deposition from Silane Discharges." *J. Appl. Phys.* 63:2406.
67. Jasinski, J.M. 1994. "Gas Phase and Gas Surface Kinetics of Transient Silicon Hydride Species." In *Gas-Phase and Surface Chemistry in Electronic Materials Processing*, Proceedings vol. 334. Pittsburgh, Pa.: Materials Research Society.
68. Street, R.A. 1991. *Hydrogenated Amorphous Silicon*, Chap. 2. Cambridge, U.K.: Cambridge U. Press.
69. Veprek, S., and M. Heintze. 1990. "The Mechanism of Plasma-Induced Deposition of Amorphous Silicon from Silane." *Plasma Chem. and Plasma Processing* 10:3.

70. Boland, J.J., and G.N. Parsons. 1992. "Bond Selectivity in Silicon Film Growth." *Science* 256:1304.
71. Baert, K., P. Deschepper, J. Poortmans, J. Nijs, and R. Mertens. 1992. "Selective Si Epitaxial Growth by Plasma-Enhanced Chemical Vapor Deposition at Very Low Temperature." *Appl. Phys. Lett.* 60:442.
72. Smith, D.L. 1993. "Controlling the Plasma Chemistry of Silicon Nitride and Oxide Deposition from Silane." *J. Vac. Sci. Technol.* A11:1843.

9.10 Suggested Readings

Cecchi, J.L. 1990. "Introduction to Plasma Concepts and Discharge Configurations." Chap. 2 in *Handbook of Plasma Processing Technology*, ed. S.M. Rossnagel, J.J. Cuomo, and W.D. Westwood, eds. Park Ridge, N.J.: Noyes Publications.

Chapman, B. 1980. *Glow Discharge Processes: Sputtering and Plasma Etching*. New York: John Wiley & Sons.

Kroesen, G.M.W., and F.J. de Hoog. 1993. "In-Situ Diagnostics for Plasma Surface Processing." *Appl. Phys. A* 56:479.

Lieberman, M.A., and R.A. Gottscho. 1993. "Design of High Density Plasma Sources for Materials Processing." In *Physics of Thin Films: Advances in Research and Development*, vol. 3, ed. J. Vossen. Orlando, Florida: Academic Press.

Thornton, J.A. 1982. "Plasmas in Deposition Processes." Chap. 2 in *Deposition Technologies for Films and Coatings*," ed. R.F. Bunshah. Park Ridge, N.J.: Noyes Publications.

Chapter

10

Film Analysis

At last we come to final step in the film-deposition process sequence of Fig. 1.1, which is to actually have a look at the film. In many cases, some of this will have been done already during deposition using various *in-situ* monitoring techniques discussed earlier; but after film removal, a host of other analytical techniques become available. There are two levels of film analysis, as pointed out in Sec. 1.2. The applications level involves measuring behavioral properties (Table 1.1), meaning how the film interacts with its environment: light, electric and magnetic fields, chemicals, mechanical force, and heat. The deeper level involves probing structure and composition, which together *determine* film behavior and thus provide a bridge of understanding between the deposition step and the film behavior.

Below are described the common techniques for analyzing film structure, composition, and properties. Since film analysis is a vast topic, these descriptions are brief and intended mainly to make the reader aware of the techniques available and aware of their principal features and shortcomings. More information appears in Brundle (1992) and still more in books dealing with individual techniques. Film analysis is performed by various service laboratories, and equipment for purchase may be examined at the equipment shows accompanying major film-related conferences. Analytical laboratories and equipment manufacturers are listed in the annual Buyers' Guides published by the journals *Physics Today* and *Semiconductor International*.

Three qualities are used to evaluate an analytical technique: sensitivity, accuracy, and repeatability or precision. Sensitivity describes how small a quantity can be detected. In terms of fractional content of an element in a material, 1 ppm (part per million) would be considered high sensitivity and 1 percent would be low. Sensitivity is important when looking for trace impurities. Accuracy is the degree to which the measured quantity corresponds to the actual quantity. It is defined in terms of the estimated error, such as 250 mg ±25 mg or ±10 percent. Often, the repeatability from one measurement to the next is better than the absolute accuracy, especially among samples analyzed in rapid succession. Repeatability or precision determines the ability of the analysis to detect small differences among samples. In other words, it describes to how many *meaningful* decimal places a measurement can be made. Weighing is one of the most precise analytical tools: a weight of 183.583 mg ±0.002 mg (±10 ppm) would be considered very precise. A measurement can be very precise without being accurate if the instrument is not calibrated carefully. Calibration samples having accurately known characteristics are useful here, such as the standard crystals with known lattice constants which are used in x-ray diffraction. Accurate analysis for elemental composition may require calibration samples having composition close to that of the "unknown" sample to be measured, because of "matrix" effects; that is, changes in the calibration for one element due to the presence of other elements. Compositional-analysis techniques differ greatly in degree of matrix effect.

10.1 Structure

Structure may be examined at different levels of spatial resolution. We will proceed from the coarsest—film thickness and surface topography—down through polycrystalline microstructure and porosity within the film and, finally, to the atomic symmetry of crystals and to chemical bonds.

10.1.1 Thickness

If a shadow mask is employed during deposition to define an edge to the film either on the actual substrate or on a "dummy" substrate, thickness may be determined directly with a **stylus profiler**. This instrument drags a stylus across the film while measuring stylus vertical deflection to about 1 nm. The shadow of the mask edge must be sharp enough that substrate height variation across the shadow region is much less than the film thickness (see Fig. 10.1). Therefore, very thin films are difficult to measure this way, as are CVD films because of their tendency to penetrate crevices. A Si wafer provides a very flat substrate, and a sharp shadow-mask edge can be made by cleaving a second Si wafer along a (111) plane, which forms an acute angle to one face. Alternatively, a sharp edge for thickness measure-

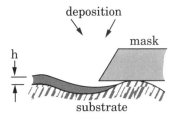

Figure 10.1 Problem of measuring the thickness, h, of a very thin film on a non-flat substrate by stylus profiling across a tapered edge generated by a shadow mask.

ment can be generated *after* deposition by chemically etching [1] through the film using a drop of photoresist or wax as a mask. However, it first must be confirmed using similar masking that the etchant does not also etch the substrate; that is, high etch "selectivity" is needed so as to accurately delineate the substrate interface. If the film is a polymer or other soft material, the stylus can deform it so as to give an erroneously low step height at the film edge. If the deformation is plastic and therefore permanent (Fig. 5.30), it will be observable in an optical microscope. Elastic deformation, which springs back and is therefore *not* observable afterwards, can be detected as an increase in measured step height upon decreasing the stylus force setting of the profiling instrument.

Transparent films can be measured optically without having to generate a step. If the reflectivity of the substrate and the refractive index of the film are known, and if the film is not too much thinner than one optical thickness period [Eq. (4.43)], thickness can be calculated from the **interference oscillations** in reflected monochromatic light intensity as wavelength is scanned. The Nanospec instrument performs this calculation automatically and also directs the spectrally scanned light through a microscope so that very small areas can be probed, such as mesas or trench bottoms on patterned substrates. Transparent films viewed in white light often show color, which is especially vivid if the substrate reflectivity and the film's index of refraction are high. This is due to constructive interference at the particular wavelength for which the film thickness corresponds to an integral number of optical periods [Eq. (4.43)]. Color also changes with viewing angle by the same equation. If the film thickness is nonuniform, the wavelength for constructive interference varies across the surface, so that spectral bands or fringes are seen, usually in a blue-gold-magenta sequence which can be quite beautiful and is some consolation when one does not achieve the desired film uniformity. The degree of nonuniformity may be estimated by counting fringes, with each fringe representing a thickness change of about 250 nm ÷ (index of refraction).

Ellipsometry can measure both the thickness and refractive index of the film once the substrate polarization ellipse is measured (Fig. 4.20). It can measure films as thin as ~1 nm, but a large flat area of a few mm^2 is needed for probing. Even multilayer films can be measured if additional data is gathered by scanning the probe-light wavelength over a range using a spectroscopic ellipsometer. The resulting plots of optical constants versus wavelength are computer-fitted to an optical model of the multilayer film. The ability to calculate a fit and the reliability of that fit depend on how many of the thicknesses and refractive indices in the multilayer stack are known at the outset and how accurately they are known. A good fit then reveals the remaining thicknesses and indices. In this technique, the flexibility and convenience of the software are very important. Bear in mind that the refractive index of a film may differ from that of the same material in bulk form because of structural or compositional differences. In particular, porosity reduces refractive index.

Metal films on electrically insulating substrates can be measured for thickness by the **eddy current** induced in the film by an rf-driven coil placed close to the surface. This is similar to the inductive coupling of Fig. 9.31. The resulting impedance change of the coil can be converted to film thickness if the resistivity of the film material is known. Note that this may be higher than the bulk resistivity of the same material due to porosity or impurities or, in the case of films <10 nm thick or so, due to surface and interface scattering of the conducting electrons [2].

The more tedious procedure of **cross-sectioning** is required for examining thickness variation over topography or thicknesses in complex multilayers such as superlattices. Cross-sectioning is greatly simplified by using a cleavable, single-crystal substrate such as a Si wafer. Otherwise, sawing and polishing must be done. For examining cross-sectional thicknesses >100 nm, scanning electron microscopy (SEM) is sufficiently accurate, but for thinner films, transmission electron microscopy (TEM) is required, and this involves thinning the cross section to <100 nm so that the beam can pass through it (more in Sec. 10.1.3). When examining layers of similar materials in SEM, "staining" of the cross-section surface with a selective chemical etchant [1] may be required to improve contrast and thus reveal the interfaces.

10.1.2 Surface topography

The degree of overall surface roughness may be qualitatively evaluated by observing the intensity of diffuse scattering under green-light illumination. Short-λ light is best, because scattering intensity $\propto 1/\lambda^4$ for features smaller than λ. The sensitivity and the film appearance also depend on the *spatial* wavelength of the roughness, λ_s. For $\lambda_s \ll \lambda$, a rough film appears shiny; for $\lambda_s \sim \lambda$, it appears matte; and

10.1.2 Surface topography

for $\lambda_s \gg \lambda$, it looks grainy. Imaged light scattering can be used to count particles on a relatively smooth surface. Surface monitoring by laser-beam scattering *during* deposition was discussed in Sec. 5.4.1.

To examine *individual* roughness features, microscopy is needed. Standard optical microscopy can reveal prominent features such as cracks and bubbles (Fig. 5.35) and hillocks (Fig. 5.36), but **Nomarski microscopy** can reveal much shallower topography, down to ~1 nm if the spatial wavelength of the topography is of the order of the microscope's resolution. In the Nomarski microscope (Fig. 10.2), polarized white light is reflected from a beamsplitter plate down through a Wollaston prism. This prism is made from a "birefringent" crystal, the refractive index of which varies with the rotational orientation of the light's polarization relative to a crystal axis. For diagonal orientation, the beam becomes split in the prism into two beams polarized along the directions of maximum and minimum index. The degree of splitting is designed so that these beams reflect from the sample surface at a spacing slightly less than the microscope's resolution, so that no image doubling occurs. After reflection, the beams recombine in the prism and then pass through the beamsplitter and a second polarizer (the "analyzer") to the eyepiece. If the sample surface is sloped as shown, one returning beam will have traveled a longer path than the other, so that the two are now partly out of phase and their sum is therefore elliptically polarized (Fig. 4.19) upon encountering the analyzer. If the analyzer's polarizing direction is rotated 90° to that of the first polarizer, total extinction occurs over flat areas of the sample, but in sloped areas some light reaches the eyepiece due to the elliptical polarization. This technique greatly enhances the contrast produced by shallow surface topography, giving the appearance of illuminating the film with a grazing light. The Wollaston prism should be rotated to give black-and-white rather than colored contrast, and then the degree of contrast is adjusted by rotating the analyzer. By comparison, the scanning electron microscope (SEM), despite its 100× higher reso-

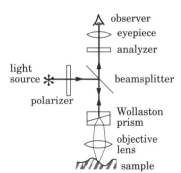

Figure 10.2 Elements of the reflection Nomarski microscope.

lution, is generally less revealing of surface topography because of its poorer contrast.

Topography can be *quantified* either by **optical interferometry** or by stylus profiling. The Michelson interferometer (Fig. 10.3) combines monochromatic beams reflected from the sample and from a flat reference mirror which is slightly tilted so that parallel, equally spaced interference fringes appear across the sample. A fringe-line undulation amounting one fringe spacing corresponds to a bump or depression $\lambda/2$ high in the surface, and features $>0.1\lambda$ or so in height can be detected. Distinguishing bumps from depressions requires a calibration sample of known topography. The Fizeau interferometer (Fig. 10.4) produces much sharper fringes because of the multiple reflections which occur between the Fizeau plate and the sample. Consequently, it is sensitive to much smaller heights of a few nm, but the Fizeau plate must be pressed against the sample and may damage it. Interferometry is good for scanning large areas quickly, but stylus profiling is more sensitive and has better lateral resolution.

The stylus profiler used in the previous section to measure the step height of a film edge can also measure more general topography. The best instruments can take successive scans in the x direction while incrementing in y after each scan to trace a "raster" pattern, so that a 3D image of area xy and height z is obtained. The lateral resolution is determined by the stylus radius. For example, Fig. 10.5 shows a blunt stylus failing to reach the bottom of a narrow trench and therefore giving an erroneously low depth reading. The **atomic-force microscope** or scanning force microscope is a miniaturized, rastering stylus profiler which uses piezoelectrically driven scanning and a microlitho-

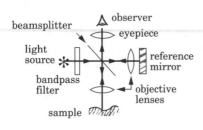

Figure 10.3 Elements of the Michelson interferometric microscope.

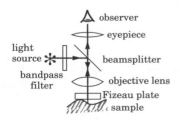

Figure 10.4 Elements of the Fizeau interferometric microscope.

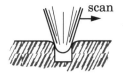

Figure 10.5 Trench-profiling error due to finite stylus radius.

graphically etched stylus and supporting cantilever [3]. It has a height sensitivity of ~0.1 nm and a surprisingly good lateral resolution such that individual atoms can be resolved under good conditions. The area scanned is much smaller than for the stylus profiler: ~100 μm square versus >1 mm square. The cantilever's vertical deflection is sometimes measured directly as it is dragged across the sample surface; however, less force is applied by using the "tapping mode" in which *incipient* contact of a vertically vibrating stylus to the surface is detected by the accompanying drop in cantilever resonant frequency. The atomic-force microscope is similar in design to its predecessor, the scanning tunneling microscope (STM). The difference is that the STM detects incipient contact by the exponential rise in tunneling current between an electrically biased tip and the surface. The STM is therefore restricted to measuring conducting and semiconducting films, but it can capture remarkable images of surface atomic structure, such as that of Fig. 6.29.

10.1.3 Bulk inhomogeneity

Common features in this category include grain boundaries in polycrystalline films, dislocations in epitaxy, voids in the microcolumnar "Zone 1" structure (Sec. 5.4.1), and precipitate nodules. Surface topography that interferes with the observation of grain boundaries can be removed by mechanical polishing with a very fine abrasive ("lapping"). If this is done at a slight angle to the surface, the grain structure down through the film is revealed under the optical microscope as in Fig. 5.22. Brief **crystallographically selective etching** in a reagent that preferentially attacks certain facets [1] is used to "stain" the surface so as to develop contrast from grain to grain. The same technique can be applied to cross-sectional surfaces, although grain structure can often be seen in cleaved cross sections without polishing and etching. Other etchants [1] selectively attack the points where dislocations intersect the surface of single crystals, as discussed in Sec. 6.3, producing pits which can be counted under the optical microscope, or under the SEM if their areal density is very high. The resulting **etch-pit density** is a convenient and quantitative measure of epitaxial quality.

Observation of smaller features such as microcolumnar voids and patterns of dislocations in the bulk requires **transmission electron**

microscopy (TEM) [4]. For a typical electron-beam energy of 400 keV, the wavelength is only 0.002 nm [Eq. (6.7)], so atomic resolution, in principle, can be obtained. The beam's diameter is much larger than atomic spacing, but *phase coherence* over this diameter allows high-resolution imaging. The sample must be thinned to a ~10 nm wedge by ion milling a crater into it. This allows the electrons to pass through the sample without being scattered more than once. Under these conditions, diffracted beams are produced as shown in Fig. 10.6. When these beams are refocused by a magnetic lens, the interference between them produces a phase-contrast image which reveals the lattice periodicity and crystallographic defects within the sample volume through which the beam passed. Surface contamination and native oxide degrade the image somewhat, so it is best to remove these if possible by *in-situ* cleaning under the TEM's vacuum.

The average film porosity over a large area can be determined by measuring film **density**. The film and substrate are weighed on a microbalance; then the film is dissolved off in a selective etchant, and the substrate is reweighed. Alternatively, the substrate can be weighed before deposition. Accuracy is improved by using a thick film and a thin, low-density substrate such as a Si wafer or a sheet of mica. From film thickness and weight, density can be calculated. Determining porosity from density requires knowing the film's crystallographic phase and the bulk density of that phase.

10.1.4 Crystallography

Diffraction induced by the lattice periodicity is the basic technique for examining crystallography, and its principles were discussed in detail in Sec. 6.4.2 in the context of electron diffraction for *in-situ* monitoring

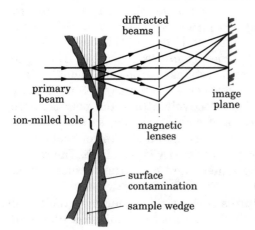

Figure 10.6 Greatly simplified schematic of TEM optics.

of epitaxial film growth. The low-energy beams used there for LEED and RHEED only penetrate the top few monolayers, so these technique are not practical for examining films after air exposure, since the resulting oxide growth and contaminant adsorption are usually difficult to remove. The TEM discussed above generates diffraction patterns from the bulk film if it is very thin, but the sampled region is only a tiny volume ~10 nm across. X-ray diffraction, on the other hand, samples a much larger volume of a few μm deep by ~1 mm across. When the film is thinner than this and the substrate is not amorphous, care must be taken to distinguish film from substrate diffraction features; or alternatively, **grazing-incidence diffraction** can be used. When the grazing angle is small enough that total external reflection of the x-ray beam occurs, only the few tens of nm within the x-ray wave's "evanescent tail" beneath the surface are probed. (Compare the waveguiding effect of Fig. 6.5.) This technique is particularly useful for accurately measuring the atomic spacing of planes oriented perpendicular to the surface, from which the mechanical strain of a film can be calculated [5, 6]. Note that film strain is biaxial, so that tensile strain produces lattice expansion parallel to the surface but compression perpendicular to it, per Fig. 5.31b. Grazing x-rays can also probe elemental composition, as we will see in Sec. 10.2.1.

Many modes of x-ray diffraction are used to examine the bulk of both epitaxial and polycrystalline films, as discussed in Sec. 6.4.2.1 and summarized briefly here. For polycrystalline films, the **2θ scan** identifies crystal phases by the θ-angle positions of the diffraction peaks and also gives a qualitative evaluation of the degree of preferred crystal-plane orientation ("texturing") parallel to the film surface by comparison of the relative peak heights from various planes with those from a randomly oriented reference sample consisting of the powdered bulk material. Many of these "powder patterns" are available in x-ray databases. The widths of the peaks in θ give the average crystallite size by Eq. (6.9). In the **pole-figure** mode, the intensity of a selected peak is measured versus angular orientation of the sample to determine degree of texturing. A powder sample would give a spherical pole figure.

For epitaxial films, the **x-ray topograph** presents a two-dimensional map of dislocations and other defects that locally distort the lattice periodicity and thus attenuate the diffraction. Nonlocalized disorder, such as that due to low deposition T or impurity alloying, can be measured by the width of the **rocking curve** obtained by tilting the film while the diffractometer is fixed on a particular Bragg peak in 2θ. Disorder can also be measured by the backscattering of small projectiles such as He^{++} (alpha particles) from lattice nuclei. This is a form of **Rutherford backscattering** spectrometry (RBS), which is also used for elemental analysis (Sec. 10.2.3). Here, the epitaxial film is aligned so that the projectiles travel along a channeling direction through the crystal lattice and therefore penetrate much more deeply before being backscattered than if they were traveling in a random direction. The more perfect the crystal lattice, the deeper the backscattering and the lower the projectile's escape probability back through the surface. Thus, the backscattering yield relative to that from a randomly oriented sample of the same crystal, a ratio known as χ, is a quantitative measure of the degree of lattice disorder. The film orientation is adjusted until the minimum ratio, χ_{min} ("khi-min"), is obtained. For a good crystal, χ_{min} can be as low as 2 percent or so.

10.1.5 Bonding

Molecular bonds are well modeled as mass-spring oscillators, and as such they have resonant frequencies described by Eq. (4.36):

$$\nu_k = \frac{1}{2\pi}\sqrt{\frac{k}{m_r}} \tag{10.1}$$

Subscript k denotes a particular stretching or bending vibrational mode of the bond. The spring constant, k in the numerator, represents the bond stiffness. The mass is the "reduced" mass of the bonded pair, A-B:

$$m_r = \frac{m_A m_B}{m_A + m_B} \tag{10.2}$$

Bonding is analyzed by measuring ν_k, usually by infrared absorption (IR) or Raman spectroscopy. IR is more useful for determining how the various elements in a material are bonded to each other, and Raman is more useful for determining the degree of bonding *order* within a material.

We first discuss **infrared-absorption** spectroscopy. Bonds are either polar or polarizable, so they can couple to and absorb electromagnetic radiation at the ν_k of Eq. (10.1). (The absorbed energy then gets dissipated into other modes and becomes heat.) The term ν_k is related to the absorption wavelength by the velocity of light: $\nu_k = c_o/\lambda_o$ [Eq. (4.39)]. The subscripts on c_o and λ_o denote measurement in vacuum (or dry air); these quantities both decrease upon moving into a solid (or liquid), but ν_k remains the same. Typically, $\nu_k \sim 10^{14}$ s^{-1}, so $\lambda_o \sim 3$ μm, which is in the near infrared. The spectroscopic frequency scale is conventionally calibrated in "wavenumbers" = $1/\lambda_o$

(cm^{-1}), so a ν_k of 10^{14} s^{-1} ≡ 3300 cm^{-1}. The ν_k values of most chemical bonds are tabulated in IR-spectroscopy databases. When there are many absorption bands in a spectrum and bond identification consequently becomes difficult, film deposition can be carried out with a source material enriched in an isotope of the element in doubt, thus shifting the ν_k of its bonds in accordance with Eqs. (10.1) and (10.2). If a film's substrate is also absorbing at ν_k, it must be dissolved off before analysis.

IR absorption can also be used for *quantitative* analysis once a bond has been identified. Radiation passing through an absorbing material is attenuated exponentially in intensity, I, with the quantity of absorbing material within its path in accordance with Beer's law, [Eq. (4.49)]; so for quantitative IR it is convenient to rearrange this equation and define an "absorbance," A, which is unitless and linearly related to material quantity:

$$A = -\log_{10}(I/I_o) = \varepsilon_k n_k h \quad (10.3)$$

For present purposes, the path length for attenuation is the thickness, h, of the film being analyzed. I_o and I are the radiation intensities at ν_k upon entering and leaving the film, n_k is the concentration (cm^{-3}) of bonds absorbing at ν_k, and ε_k is the "oscillator strength" of the bond's kth vibrational mode. The value of ε_k varies widely from bond to bond and mode to mode, so calibration is necessary using a material of known n_k.

Infrared spectroscopy was originally done by "dispersing" a source of blackbody radiation into its spectrum with a prism or diffraction grating and then selecting wavelength with a slit and sweeping it mechanically. However, much better performance is obtained using **fourier-transform infrared spectrometry** (FTIR) [7], especially for the weak absorbances available from thin films. In FTIR, the blackbody source radiation is passed through a Michelson interferometer (similar to Fig. 10.3), whose path length is varied sinusoidally by wiggling the mirror at ~1 Hz, so that the wavelength of constructive interference is varied. The waveform of the radiation intensity in the interferometer will be nonsinusoidal, because the absorption by the intervening sample varies with wavelength. The fourier transform of this waveform is the IR spectrum: A versus $1/\lambda_o$. The FTIR spectrometer is essentially collecting the entire spectrum with every mirror cycle, and signal collection may be continued for hours if necessary until an acceptable signal-to-noise ratio is achieved for small A values.

Bond frequencies and lattice-vibration (phonon) frequencies can both be measured by **Raman spectroscopy**. Recall that light passing through transparent materials is weakly scattered from individual atoms by dipole interaction. The light's transverse electric field polarizes the atoms, and the oscillating atomic dipoles re-radiate the light in all directions. Most of the light undergoes Rayleigh scattering, which radiates at the incident frequency, but some undergoes Raman scattering, which is shifted down (Stokes shift) or up (anti-Stokes) in frequency by the transfer of a quantum of energy into or out of a vibrational mode of the molecule or a phonon of the lattice to which the scattering atom is bonded. Usually, the sample is illuminated with a high-intensity Ar^+ laser beam (green), and the spectrum of the scattered light is collected in the perpendicular direction. Single crystals with perfect bonding order exhibit sharp Raman-scattering peaks, whereas amorphous materials have broad peaks because of the variation in bonding among the scattering atoms. For example, Raman spectroscopy has been useful in determining the fraction of diamond versus graphite or amorphous C in the development of diamond film-deposition processes. Film strain can also be measured, because the asymmetry of bond potential wells (Fig. 9.1) causes a shift in phonon frequency with bond length.

10.1.6 Point defects

This category of structure includes for single crystals the native point defects of Table 4.1, and for amorphous materials other native defects such as weak or dangling bonds. The category also includes impurity atoms, whose identity and concentration can be determined directly by sufficiently sensitive elemental analysis as discussed in Sec. 10.2. Here, we discuss the detection of both native and impurity point defects in semiconducting and dielectric films by their effects on the film's optical and electrical properties. These properties are affected because the point defects usually produce electronic states within the band gap, which can trap or detrap electrons or holes (more in Sec. 6.2). Unfortunately, this indirect detection makes it difficult to unambiguously identify the particular type of defect involved.

Optical properties are affected because the energy for detrapping can be supplied by photon absorption, and the energy released upon trapping can cause photon emission (luminescence). In **photoluminescence spectroscopy**, the sample is cooled with liquid N_2 to reduce *thermal* detrapping and is illuminated with intense laser light of sufficient photon energy to activate the states under study. The spectrum of emitted light contains peaks whose wavelengths can be related to the positions of defects within the band gap [8] and whose intensities are proportional to defect concentration.

The very small optical absorption caused by point defects can be detected by **photothermal-deflection spectroscopy** [9], shown schematically in Fig. 10.7. The intense, wavelength-tunable, pump-laser beam is partly absorbed by the film, causing it and the air just over it to heat up. Meanwhile, the reflected beam from the He-Ne probe laser is being monitored with a position-sensitive photodetector. The expansion of the heated film toward the beam and the refractive-index reduction of the heated air deflect the probe beam in proportion to the amount of pump-beam absorption. Note that excess absorption by surface electronic states will also occur, but this can be separated from bulk absorption because it produces a smaller decrease in the modulation of the detected signal with increasing pump-beam chopping frequency on account of the lower associated thermal mass [9]. Photothermal deflection is better than the FTIR of the previous section for detecting very small absorbance, because it is measuring the response of the material directly rather than measuring a very small fractional change in the probe-light intensity. Measuring a small difference between large signals is always difficult, because it requires high precision in the measurement.

Point defects that produce "shallow donor" states (close to the conduction-band edge) cause electron conduction in semiconductors, and shallow acceptor states (close to the valence-band edge) cause hole conduction (see Sec. 6.2). The concentrations of these shallow states can be calculated from **conductivity and mobility** measurements, as discussed in Sec. 10.3.2, provided that the measurement T is high enough that all of the states are ionized. Two additional pieces of information can be obtained by making measurements at reduced sample T. First, from the T at which the carriers start to "freeze out," the depth of the level from the band edge can be calculated using Fermi statistics. Second, the maximum mobility reached at cryogenic T is a measure of crystalline perfection, since carriers are scattered by any kind of disorder, such as point defects or dislocations. (At room T, mobility is limited by phonon scattering.)

The **depth profile** of carrier concentration through a film can be obtained by a "C-V" measurement. Here, a metal-semiconductor junction

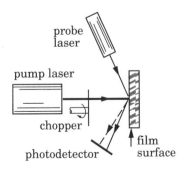

Figure 10.7 Geometry of photothermal-deflection spectroscopy.

is made, where the metal is most conveniently a mercury-droplet probe. Reverse-biasing the junction produces a depletion region whose width is proportional to V and is inversely related to average carrier concentration, n_e. Depletion width is measured by junction capacitance as determined from the parallel-plate formula, Eq. (9.29). Differentiation of the C-V profile yields n_e versus depth.

Electrical states located deeper within the band gap of a semiconductor can be probed by **deep-level transient spectroscopy** (DLTS) of a reverse-biased metal/semiconductor junction [8]. As the edge of the depletion region sweeps across a region containing deep states, trapping and detrapping occur at a rate whose T dependence gives the depth of the states from the band edge.

10.2 Composition

Here, we first need to perform *qualitative* analysis; that is, to identify the elements present in the film, including both the major constituents and trace elements such as dopants and contaminants. Then, we usually want to analyze *quantitatively* for the atomic fraction of each element and, often, also its spatial distribution, especially in the direction of depth through the film(s). The techniques discussed below all have the advantages of being sensitive to most elements and of being able to analyze the very thin layers encountered in thin-film work. However, they vary greatly in their sensitivity to trace amounts, their quantitative accuracy, and their depth resolution. X-ray fluorescence and mass spectrometry are the most sensitive. Auger, XPS, SIMS, and RBS have the best depth resolution (a few nm). None of the techniques has better than about 1 percent repeatability for reasonable signal-collection times, and this unfortunately precludes the detection of slight variations in stoichiometry within the single-phase field of most compounds (see Fig. 4.6b), variations which can have large effects on material properties. Traditional **wet chemical analysis** techniques have much higher repeatability when a sufficient quantity of material is available, so they can be applied to this problem when necessary. However, the procedures are different for each element, as well as being generally tedious, and one must beware of interferences from other elements. These techniques will not be discussed further here.

The **instrumental analysis** techniques discussed below are divided into three categories according to the physics involved: those using electron or x-ray probe beams and signals, those performing mass analysis on volatilized atoms, and those using a backscattered atomic probe beam. Finally, analysis for H is treated separately, since it has special problems and solutions.

10.2.1 Electrons and x-rays

There are four techniques in this category, and they are characterized by the probe-beam species and the detected species as outlined in Table 6.1 and discussed in Sec. 6.4.1. **X-ray photoelectron spectroscopy** (XPS) and **Auger spectroscopy** detect emitted electrons and therefore analyze only the first few nm beneath the surface because of the shallow escape depth of the electrons. For bulk film analysis, this feature provides good depth resolution but also complicates quantification, as we will see below. **Depth profiling** of composition is achieved by sputtering away the film during analysis, using an ion beam raster-scanned over an area larger than the electron collection area to ensure a spatially uniform sputtering rate. Even when depth profiling is not needed, surface adsorbates and native oxides must still be sputtered away before analysis in order to determine bulk composition.

The sputtering process degrades both depth resolution and compositional accuracy. Depth resolution decreases because surface roughness increases with the depth sputtered, even on homogeneous films, due to statistical roughening, which was discussed in Sec. 5.4.1 in the context of deposition. Roughening is still faster when the sputtering yield, Y_s [Eq. (8.37)], is spatially nonuniform. This occurs in polycrystalline films due the dependence of Y_s on crystal plane, or in any film where contaminants having lower Y_s are present in islands on the surface or in precipitate nodules in the bulk. In addition to roughening effects, abrupt interfaces in multilayer structures become smeared out to ~5 nm by knock-on implantation, ion mixing, and radiation-enhanced interdiffusion (see Sec. 8.5.3). Sputtering also reduces compositional accuracy, because the surface becomes enriched in elements of lower Y_s and in elements that thermodynamically prefer to segregate to the surface and can do so with the help of radiation-enhanced diffusion.

Other factors also limit XPS and Auger accuracy. The dependence of electron escape depth on film composition results in a matrix effect, which can be corrected for computationally to some degree. Adsorption of vacuum background gas during analysis often increases C and O readings, especially in Auger spectroscopy, where the intense electron beam activates adsorbate chemisorption. Absolute accuracy for the other elements is ~10 percent, and typical sensitivity is ~1 at.%. XPS can also determine the degree of electronegativity in an element's bonding environment by the **chemical shift** of the photoelectron peaks in energy, although one must beware of bonding perturbation by sputtering and of peak shifts due to charge-up in the case of insulating films. As for *lateral* resolution, Auger spectroscopy's 15 nm is far better than that of any other elemental-analysis technique.

The other two techniques of Table 6.1 detect emitted *x-rays*, whose escape depth of >1 µm is much larger than that of electrons. The **elec-**

tron microprobe is convenient because it can be attached to a scanning electron microscope. Its lateral-resolution dimension is much larger than the probe-beam diameter, however, because of beam scattering and spreading beneath the surface. The depth of the analyzed volume is determined by the hundreds-of-nanometer penetration depth of the ~10-keV probe beam. When the film is thinner than this, elements in the substrate will also be seen, while the sensitivity to film elements will be reduced. For accurate quantitative analysis of such films, one needs a calibration sample having the same thickness as well as a similar composition to that of the unknown. On the other hand, **x-ray fluorescence** can be made to analyze only the top 10 nm or so by directing the x-ray probe beam at grazing incidence so that total external reflection occurs. This technique has very high sensitivity exceeded only by mass spectrometry, but it requires a large sample diameter of 10 mm and is not sensitive to elements lighter than S. Moreover, if the surface is not absolutely flat, reflection is not total and the ratio of surface to subsurface sensitivity drops.

10.2.2 Mass spectrometry

Here, vapor is ionized and then separated according to particle mass-to-charge ratio (q/m) using a **quadrupole** electric field or a **magnetic** field. For analysis of sputter-volatilized solids such as thin films, sensitivity is better than 1 ppb (part per billion, or 10^{-9}) for some elements, so it is very good for detecting trace contaminants, although it is less accurate than other techniques having less sensitivity.

In **secondary-ion mass spectrometry** (SIMS), an ion beam sputters away the film, and the small fraction that is sputtered as ions rather than as neutrals is analyzed. This fraction varies by orders of magnitude depending on ion-beam element, sputtered element, and film composition (the matrix effect). Even with calibration samples, it is difficult to obtain accuracy of better than a factor of two, although sensitivity is very high. Depth resolution is the same as in XPS and Auger and for the same reasons. Lateral resolution of 1 µm can be obtained by imaging the pattern of sputtered ions onto a position-sensitive ion detector.

In **glow-discharge mass spectrometry**, a plasma over the sample is used to sputter it (see Sec. 9.3.3). Sensitivity is even higher than in SIMS because of larger sample area and ionization of the vapor in the plasma, but depth resolution is poor.

10.2.3 Rutherford backscattering

Rutherford backscattering spectrometry (RBS) [10] is basically billiard-ball physics and is therefore the most quantitative of the elemental-analysis techniques discussed here. Relative sensitivity from element to element and matrix effects can be calculated accurately from first principles without the use of calibration samples. In RBS, a beam of He^{++} (alpha particles) is directed at the sample at high enough energy ($\approx$2 MeV) so that the particles scatter from the sample's atomic nuclei in binary Coulomb collisions unscreened by the surrounding electron clouds (Rutherford scattering). Screened collisions of ions at lower energy are more complicated and were discussed in Sec. 8.5.2.2. From the energy spectrum of the backscattered particles, elemental concentrations and their depth profiles can be calculated. A target atom scattering the beam withdraws from the He^{++} an amount of energy determined only by the target-atom mass, m_t, and given by Eq. (8.20) corrected for the scattering angle, so that m_t is easily calculated from the kinetic energy remaining in the backscattered beam. Heavy elements withdraw little energy, so they are more difficult to resolve from each other, but the sensitivity for them is much higher because the scattering cross-section increases as atomic number squared.

When the scattering atom is at some depth beneath the surface, additional energy is lost by the ingoing and outgoing beam because of the stopping power of the material for the beam as expressed in Eq. (8.22). Thus, the thicker the film, the broader in energy becomes the backscattered peak from an element that is evenly distributed through it, and variation in peak height with backscattering energy can be used to calculate **depth profile** with a resolution of about 10 nm. Thick films having multiple elements of similar masses will encounter peak-overlap problems, however. Meanwhile, the substrate elements generate a background signal in the energy spectrum which extends all the way from the Eq. (8.20) energy down to zero energy because of the stopping effect. Thus, when a major element in the substrate is heavier than one in the film, the peak from that film element will sit on top of the substrate background, greatly reducing its signal-to-noise ratio. Substrate m_t can be minimized by depositing the film on pyrolytic graphite sheet.

10.2.4 Hydrogen

Hydrogen gives no signal in any of the electron or x-ray spectroscopies, because it has no inner-shell electrons. Also, it is too light to backscatter the beam in RBS. Its bonds do produce absorption peaks in infrared spectroscopy (Sec. 10.1.5), but quantification requires a calibration sample of known H content in the *same bonds* to determine oscillator strength [Eq. (10.3)]. H can be measured by mass spectrometry, but accuracy is poor. The following techniques are useful for quantifying H.

In a modification of RBS sometimes called **forward-recoil scattering**, a 2-MeV He^{++} beam is directed at a grazing angle to the sample surface, so that the H which is scattered forward by the beam has a high probability of escaping from the surface. The sensitivity is about 0.01 at.%.

Hydrogen can be **volatilized** from the sample by sufficient heating. When this is done in a vacuum chamber having a known pumping speed and a calibrated pressure gauge, the amount evolved can be found by integrating Eq. (3.2) over the evolution period. It is important to also monitor the evolution with a mass spectrometer to determine the molecular form of the evolving H and to determine the fraction of the pressure burst that does not contain H. For example, H evolves as H_2 from metals and Si, but as H_2O from silicon dioxide and as H_2 and NH_3 from plasma-deposited silicon nitride. Some of the evolving H may be coming from water desorbed either from the surface or from internal porosity, but this typically evolves at a lower T (<500 K) than the H from within the solid. The sample may have to be heated to as much as 1200 K or so to release all the H from the solid.

Nuclear-reaction analysis can be both sensitive and accurate for H and other light elements (Brundle, 1992), but it requires access to a particle accelerator as well as careful calibration. For example, a beam of the ^{15}N isotope at the specific resonance energy shown here undergoes the reaction

$$H + {}^{15}N(6.385 \text{ MeV}) \rightarrow {}^{12}C + {}^{4}He + \gamma(4.43 \text{ MeV}) \qquad (10.4)$$

where γ represents gamma rays whose flux can be calibrated for H concentration. This technique has been used to determine H in silicon nitride [11], for example.

10.3 Properties

Film properties refer to the film's interactions with its environment. These interactions determine the film's performance in its application: it is where "the rubber hits the road." Properties are governed by structure and composition, so analysis of all three kinds gives insight into how to modify the deposition process to improve properties. The effects of various deposition parameters on film structure and composition have been a major theme throughout the book.

Film properties were categorized in Table 1.1. In the subsections below, we address optical, electrical, and mechanical properties. **Magnetic properties** are determined by film interaction with a coil such as that in the read/write head of a memory disc. Also, magnetic domains can be mapped with the atomic force microscope by using a magnetic tip to sense magnetic fringe fields. **Chemical properties** mainly involve the film's dissolution rate in various etchants [1].

10.3.1 Optical behavior

A film may reflect, absorb, transmit, or scatter light, and these properties are functions of wavelength. All of the impinging light undergoes one of these four fates. **Reflection and scattering** are measured by the intensity of a light beam returned at the specular angle and at nonspecular angles, respectively. Scattering increases with surface roughness and particulate contamination (Sec. 10.1.2) and, for transparent materials, with bulk inhomogeneity. **Absorption and transmission** from the infrared (IR) to the ultraviolet (UV) regime can be determined by spectroscopy, as discussed for the infrared in Sec. 10.1.5.

The general optical behavior of solids is as follows. Metals reflect and absorb light strongly, because their free electrons prevent electromagnetic-wave propagation. (Compare microwaves in plasmas, Sec. 9.5.1.) The higher the electrical conductivity, the higher is the reflected fraction versus the absorbed fraction because of the shallower "skin depth" [Eq. (9.43)], which is why Ag makes the best mirrors. Nonmetals absorb light strongly by electron/hole-pair generation at photon energies above the band-gap energy or "fundamental absorption edge," which is in the UV for insulators and in the visible or near-IR for semiconductors. Additional absorption occurs at various wavelengths below the band gap due to defects, impurities, and bond resonances.

The amounts of reflection, absorption, and transmission occurring in nonscattering materials are determined uniquely by the fundamental optical constants of Eq. (4.47): the index of refraction, $\tilde{n}$, and the extinction coefficient, κ. A transparent film's $\tilde{n}$ can be measured by interference oscillations if its thickness is known, and its $\tilde{n}$, κ and thickness can all be determined by ellipsometry. These techniques were discussed in Sec. 10.1.1.

10.3.2 Electrical behavior

Here, the analytical techniques and the properties measured differ for metals, semiconductors, and insulators (dielectrics). For metals, the basic electrical property is the **resistivity**, ρ (Ω-cm), of the bulk material, which is defined as the resistance, R (Ω), of a 1-cm cube of material between opposite faces. With thin films, it is more convenient to think in terms of the resistance of a square of film between opposite edges, as shown in Fig. 10.8. This resistance is known as the "sheet resistivity" and is given in ohms per square,

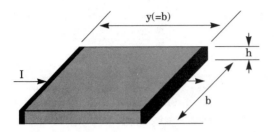

Figure 10.8 Geometry defining the sheet resistivity of a film of thickness h.

$$\rho_s \, (\Omega/\square) = \rho y/A = \rho/h \tag{10.5}$$

where h is the film thickness and A is the cross-sectional area of the conduction path, bh. Note that ρ_s is independent of the size of the square, so that the R of a conducting thin-film line is just proportional to the number of squares represented by the y/b ratio of the line. The value of ρ_s is conveniently measured with the linear four-point probe shown in Fig. 10.9, provided that the substrate has high enough ρ so that most of the current, I, passes through the film. By measuring the voltage drop, V, with a different pair of probes than the pair used for current flow, the voltage drop associated with current flow through the contacts is removed from the measurement, and only the voltage drop across a distance d of film is measured. The input impedance of the voltmeter must be much higher than the film resistance across the distance d so that it does not provide a shunt path for current flow. The four-point arrangement makes the quality of the probe contacts noncritical, so long as they can pass enough current to generate a measurable V. Thus, they can usually be just point contacts placed under pressure by a spring-loaded probe jig. For equal probe spacing, ρ_s is given simply by [12]

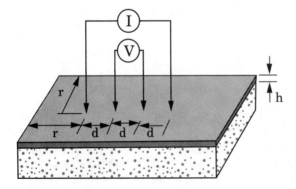

Figure 10.9 Linear four-point probe for sheet-resistivity measurement of a film on an insulating substrate.

$$\rho_s = \frac{\pi}{\ln 2}\frac{V}{I} = 4.53\frac{V}{I} = 4.53\,R \tag{10.6}$$

For good accuracy, the contact diameter should be much smaller than d, and the distance, r, from the edge of the probe array to the edge of the film should be much larger than d. For example, when r/d = 15, measured ρ_s is 1 percent higher than actual ρ_s, and when r/d = 6, it is 10% higher [12]. When film thickness is also known, ρ can be calculated from ρ_s by Eq. (10.5) and compared with the known ρ of the bulk material as a measure of film quality. However, very thin films of h < 10 nm or so have ρ_s increased by surface and interface scattering of the conducting electrons [2].

The ρ_s of semiconductor films can also be measured with the linear probe array, but it is more common to use the square "van der Pauw" array [13] of Fig. 10.10a or c. This way, one can separately obtain the **electron mobility and concentration**, μ_e and n_e (or μ_h and n_h for hole conduction), whose product determines the conductivity, s = $1/\rho$ [see Eq. (6.2)]. For ρ_s measurement, the current and voltage contacts are configured in parallel as in Fig. 10.10a, so that a resistance $R_{AB,CD} = V_{DC}/I_{AB}$ is obtained. Then, the array is rotated 90° to obtain $R_{BC,AD}$. If pattern asymmetry is small enough that these two R values are within 30 percent of each other, then the average R is related to ρ_s by Eq. (10.6) (to within 1 percent), just as for the linear probe. The contacts need to be at the periphery of the film and much smaller in diameter than the distance between them to avoid measurement errors. The geometry of Fig. 10.10c is less susceptible to these errors, but it requires delineation by etching. The film periphery may actually be of any shape that has reasonable four-fold symmetry.

The "Hall mobility," μ_H, is obtained from the same contact array using the perpendicular current and voltage configuration of Fig. 10.10b. Upon application of a magnetic field, **B**, perpendicular to the film, charge carriers making up the current I_{AC} are deflected sideways by

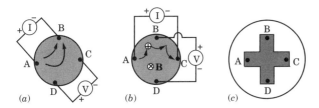

Figure 10.10 Van der Pauw geometries and circuits: (a) resistivity configuration, (b) mobility configuration with the **B** field pointing into the plane of the figure, and (c) patterned film for error reduction.

the Lorentz force, Eq. (8.15), so that they pile up at one or the other voltage contact, depending on the carrier sign, and thus cause a sideways voltage drop V_{BD}. This is the "Hall effect." The deflection in Fig. 10.10b is shown for positive charge carriers (holes). Upon reversing **B**, the Hall voltage inverts, and the voltage difference between reversals, ΔV_{BD}, is measured so as to reduce asymmetry errors. To further reduce error, a second measurement is made with the array rotated 90°, so that two "resistances" are obtained, $\Delta R_{AC,BD}$ (= $\Delta V_{BD}/I_{AC}$) and $\Delta R_{BD,CA}$. Their average, ΔR, is related to μ_H by

$$\mu_H \text{ (cm}^2/\text{V·s)} = 10^8 \Delta R/2\mathbf{B}\rho_s \qquad (10.7)$$

for **B** in Gauss, where a field of ~2000 Gauss is typically used. Note that the 10^8 would drop out for μ_H and **B** in SI units. The μ_H can be as much as 30 percent larger than the actual carrier mobility—the μ_e or μ_h given by Eq. (6.2)—depending on material and carrier concentration [14]; but to first order, the two μ values can be assumed to be the same if there is only one carrier type. On the other hand, in "compensated" semiconductors, where both electrons and holes are conducting, the Hall voltage will be partly cancelled out, and the calculation of carrier mobility from μ_H becomes more difficult. Finally, at least for uncompensated material, the carrier concentration, n_e or n_h, can be obtained from μ_e or μ_h using Eq. (6.2). Additional information about the material can be obtained by making ρ_s and μ_H measurements at reduced T, as discussed in Sec. 10.1.6.

We now turn to insulating (dielectric) materials. A dielectric film sandwiched between conducting substrate and overlayer materials forms a parallel-plate capacitor, and the application of an electric field across it can produce a wide range of phenomena in the film, including: charge injection from the conducting layers, charge trapping within the dielectric, ion migration, passage of leakage current, ac power dissipation, expansion and contraction of piezoelectric crystalline films such as ZnO, and permanent polarization of ferroelectrics such as $BaTiO_3$. Here, we focus on capacitor behavior and charge motion. If the overlayer material in our sandwich is patterned into a known area and if the dielectric-film thickness is known, the **dielectric constant**, ε, can be found from Eq. (9.29) upon measuring the sandwich capacitance, C, with a capacitance bridge or an "LCR" meter. The latter measures ac impedance and the phase angle between applied voltage and current, and from these it calculates C and R (as well as inductance, L). In an *ideal* capacitor, the phase angle is 90° and there is no power dissipation, but power dissipation can occur in dielectrics due to leakage current and also, at high frequency, due to a phase-angle shift, δ, resulting from dissipative charge oscillation within the dielectric. The latter is expressed in terms of the dissipation factor or **loss tangent**, tan δ. By the way, when high C is desired in a thin film, there are available some easily deposited materials of very high ε, such as Ta_2O_5 (ε = 22).

Application of fields of more than a few MV/cm across a dielectric film sandwiched between conducting electrode layers causes charge injection (tunneling) through the interfacial potential barrier from the Fermi level, E_f, of the electrode into the conduction band, E_c, of the dielectric, as shown in the energy diagram of Fig. 10.11 (compare Fig. 6.4). This charge consists of electrons from the negative electrode and sometimes also holes from the positive electrode tunneling into the valence band. The charges may either pass through to the other electrode, causing leakage current, or they may become trapped in defect states deep within the band gap of the dielectric. (*Porous* dielectric films may have conductive pathways through their internal surface area at much lower fields.) At still higher fields, the injected electrons gain enough kinetic energy to cause impact ionization of atoms within the dielectric, and this results in an electron avalanche and dielectric breakdown. This is analogous to the breakdown of gas to form a plasma, as discussed after Eq. (8.14). Leakage behavior and the onset of breakdown is conveniently measured by the **ramp I-V characteristic** of the sandwich. The voltage, V, across the electrodes is increased in steps of ~1 V at intervals of ~1 s, and the leakage current, I, is measured at each step once I recovers from the V step. The stepwise V ramp is preferable to a continuous ramp, because it removes from the measurement the capacitive displacement current that accompanies

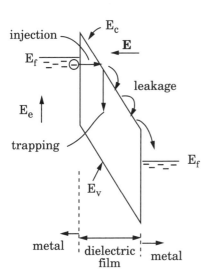

Figure 10.11 Electron potential-energy (E_e) diagram and charge behavior for a dielectric film under high electric field, **E**, between two conductors.

the increasing V [Eq. (9.28)]. In a typical I-V plot, I remains at noise level for awhile and then increases exponentially but smoothly until a sudden large jump in I occurs at a particular V step, which is defined as the breakdown voltage. Good SiO_2 has a breakdown voltage of ≈11 MV/cm, for example. Although the I-V characteristic is a convenient measurement, its shape and the breakdown V can vary with the ramp rate and with the film's electrical-stressing history when charge trapping is occurring, and then it is important to also measure the charge trapping directly.

The amount of charge residing in a dielectric film can be measured if the film is deposited on a lightly n-doped Si wafer and then overcoated with a metal film patterned into mesas to make MIS (metal/insulator/semiconductor) capacitors as shown in Fig. 10.12a. When the metal electrode is biased positively with respect to the Si, conduction electrons in the Si accumulate at the dielectric interface so that the capacitance, C, is determined by the dielectric thickness, h. However, under negative bias, electrons are repelled from the interface to a "depletion depth," b, which increases with bias and with decreasing Si doping level. Then, C is determined by the thickness (b + h) and is therefore lower, in accordance with Eq. (9.29). Thus we obtain the characteristic C-V curve shown in Fig. 10.12b. [This is the shape when high-frequency modulation of the bias (>10 kHz or so) is used to measure C; at low frequency, C instead rises again under high negative bias in the "inversion" regime where minority-carrier holes are accumulated, because they can respond to slow modulation and thus act as an electrode.] When there is no *net* charge in the dielectric, the transition from electron depletion to accumulation occurs near zero bias. The transition V is known as the "flat-band voltage," because it is where the conduction and valence bands of the Si are bent neither up (deple-

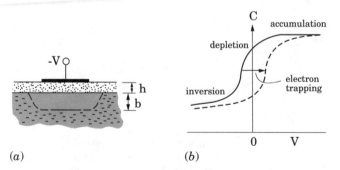

Figure 10.12 MIS-capacitor behavior: (a) n-doped semiconductor depletion to depth b under negative metal-electrode bias, and (b) high-frequency C-V curve with no charge (solid line) and with negative charge (dashed line) trapped in the dielectric film.

tion) nor down (accumulation) at the interface. When negative charge is trapped in the dielectric, it acts like negative bias on the electrode, causing depletion even with no bias, so that a more positive electrode bias is required to accumulate the Si, and the C-V curve shifts to the right as shown in Fig. 10.12b. In quantitative terms, the trapped charge causes an electric-field change across it which is given by Eq. (8.4). Conversely, positive trapped charge causes a C-V shift to the left.

From the amount of the **C-V shift**, one can calculate the concentration of trapped charge as well as its distribution across the dielectric film in the field direction [15]. This technique is very sensitive for measuring the amount of charge trapped. The charge may have been (1) embedded during plasma deposition, (2) diffused in as ionic contaminants during or after deposition, or (3) injected during high-field stressing. Fixed charge in the deposited film (case 1) is revealed by a nonzero flat-band voltage of the initial C-V curve. Ionic contamination (case 2) is often due to Na^+. The presence of minute quantities of this ubiquitous ion in the gate oxide of field-effect transistors frustrated the development of these key devices and thus delayed the establishment of the integrated-circuit industry for many years. Mobile-ion contamination can be distinguished from fixed charge in the dielectric by **"bias-temperature" stressing**, whereby the ions are made to drift at elevated T in a moderate electric field, causing a C-V shift. The tendency of the film to trap electrons injected into it (case 3) is determined by **avalanche injection** of a metered pulse of charge, followed by measurement of the resulting C-V shift [15]. The result can be characterized in terms of the product of the concentration of traps per cm^3 times their cross section in cm^2, $n_t \sigma_t = \alpha_e$ (cm^{-1}), where α_e is essentially an electron absorption coefficient analogous to the one for light defined by Eq. (4.49). When the film is thinner than $1/\alpha_e$, part of the injected charge passes through to the opposite electrode, whereas when it is thicker, all of the charge becomes trapped.

10.3.3 Mechanical behavior

Here, we address four properties: stress, adhesion, hardness, and stiffness.

The level of a film's tensile or compressive **stress** can be determined by the resulting concave or convex curvature in the substrate as given by Eq. (5.57) and discussed in detail there.

Adhesion of a film to its substrate is the most difficult of the four to measure. Simple qualitative tests include determining whether or not the film peels off along with a strip of sticky tape pressed onto it and then pulled off, or whether it chips off upon being scratched with a stylus. A more quantitative test is illustrated in Fig. 10.13. A metal stud of known cross-section, A, is glued to the film, and then a force ramp (F) is applied until something breaks—either the substrate or the glue or the film/substrate interface. Thin substrates can be supported by gluing on a backing plate as shown. A strong glue such as epoxy should be used, since its strength sets the upper limit of measurable adherence strength. Soldering or brazing provides a stronger bond, but the heating may cause interdiffusion or structural change at the interface to be tested. When the film/substrate interface breaks first, the interfacial tensile strength is given directly by F/A. The tensile strength of well cured epoxy is about 80 MPa or 12 kpsi. With multilayer films, the weakest interface will be the one that breaks, of course, and which interface broke can be identified by compositional analysis or sometimes by microscopic inspection. The pull force must be exactly perpendicular to the film surface so as to distribute the tension evenly over A and thereby maximize the breaking force.

Hardness of thin films can be determined by "nanoindentation" [16], a miniaturized version of the Vickers hardness test for bulk materials, in which the point of a pyramidal diamond stylus is pressed into the surface with a specified force. The size of the resulting dimple is inversely related to hardness. The film should be thicker than about 10 times the depth of the dimple so as to avoid substrate influence.

The stiffness of a material is characterized by its **elastic modulus**, Y (Pa), as defined in Eq. (5.50). Y varies with microstructure, so its value in a thin film is not necessarily the same as in the bulk material. The Y of a thin-film material can be found from the resonant frequency of a suspended section of film. The cantilever structure of Fig. 10.14 can be fabricated using microlithography, first to pattern the film into a finger shape and then to undercut-etch a well in the substrate using a second photomask, so as to leave the end of the finger freely suspended. The fundamental resonant frequency of this rectangular cantilever beam of cross section A = bh and uniform mass density ρ_m is related to Y and geometry by the following equation [17]:

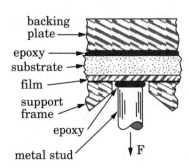

Figure 10.13 Epoxied-stud pull test for film adherence.

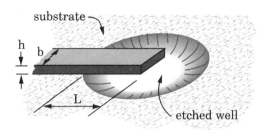

Figure 10.14 Etched-cantilever structure for determining film stiffness.

$$v_o = \frac{1.875^2}{2\pi L^2}\sqrt{\frac{YI_z}{\rho_m A}} = 0.162\frac{h}{L^2}\sqrt{\frac{Y}{\rho_m}} \tag{10.8}$$

in SI units (Pa, kg, m, Hz), where $I_z = bh^3/12$ is the moment of inertia of the beam [compare Eq. (5.56)]. Note that this is a $\sqrt{k/m}$ expression characteristic of mass-spring resonance [Eq. (4.36)]; here, the spring constant $k \propto YI_z/L^3$, and the mass $m \propto \rho_m AL$. The value of v_o can be found by mounting the Fig. 10.13 structure on a piezoelectric transducer driven by a variable-frequency voltage and watching under a microscope for a blurring of the cantilever upon reaching resonance. The lowest of the various frequencies at which blurring will be seen is the fundamental resonance.

10.4 Conclusion

We have now completed the film-deposition process sequence outlined in Fig. 1.1: source-material supply, transport to the substrate, deposition reactions, and film analysis. The results of analysis contain clues as to how to modify the operation of the first three steps to improve film properties for the desired application. These clues will be more apparent and more numerous if structural and compositional analyses are carried out along with property analysis. Many of the linkages between deposition-process parameters and the nature of the resulting film have been discussed in Chaps. 4 through 9. Still, there are many more linkages to be discovered and new processes to be invented. Hopefully, the information and principles presented in this book will prove to be a useful platform and guide for this endeavor.

10.5 References

1. Walker, P., and W.H. Tarn (eds.). 1991. *CRC Handbook of Metal Etchants*. Boca Raton, Fla.: CRC Press. (Also covers nonmetals.)
2. Berry, R.W., P.M. Hall, and M.T. Harris. 1979. *Thin Film Technology*, Chap. 6, "Electrical Conduction in Metals." Huntington, N.Y.: Krieger Publishing.

3. Sarid, D., and V. Elings. 1991. "Review of Scanning Force Microscopy." *J. Vac. Sci. Technol.* B9:431.
4. Gibson, J.M. 1991. "High Resolution Transmission Electron Microscopy." *MRS Bull.* (March):27.
5. Segmüller, A. 1991. "Characterization of Epitaxial Thin Films by X-ray Diffraction." *J. Vac. Sci. Technol.* A9:2477.
6. Clemens, B.M., and J.A. Bain. 1992. "Stress Determination in Textured Thin Films using X-ray Diffraction." *MRS Bull.* (July):46.
7. Back, D.M. 1991. "Fourier Transform Infrared Analysis of Thin Films." In *Physics of Thin Films* 15, "Thin Films for Advanced Electronic Devices," ed. M.H. Francombe and J.L. Vossen. Orlando, Fla.: Academic Press.
8. Herman, M.A., and H. Sitter. 1989. *Molecular Beam Epitaxy: Fundamentals and Current Status*, Chap. 5. Berlin: Springer-Verlag.
9. Olmstead, M.A., and N.M. Amer. 1983. "A New Probe of the Optical Properties of Surfaces." *J. Vac. Sci. Technol.* B1:751.
10. Chu, W.K., and G. Langouche. 1993. "Quantitative Rutherford Backscattering from Thin Films." *MRS Bull.*, (January):32.
11. Lanford, W.A., and M.J. Rand. 1978. "The Hydrogen Content of Plasma-Deposited Silicon Nitride." *J. Appl. Phys.* 49:2473.
12. Smits, F.M. 1958. "Measurement of Sheet Resistivities with the Four-Point Probe." *Bell System Tech. J.* (May):711.
13. van der Pauw, L.J. 1958. "A Method of Measuring Specific Resistivity and Hall Effect of Discs of Arbitrary Shape." *Philips Res. Rep.* 13:1.
14. Look, D.C. 1990. "Review of Hall Effect and Magnetoresistance Measurements in GaAs Materials and Devices." *J. Electrochem. Soc.* 137:260.
15. Nicollian, E.H., and J.R. Brews. 1982. *MOS (Metal-Oxide-Semiconductor) Physics and Technology*, Chap. 11. New York: Wiley-Interscience.
16. Pharr, G.M., and W.C. Oliver. 1992. "Measurement of Thin Film Mechanical Properties Using Nanoindentation." *MRS Bull.* (July):28.
17. Rothbart, H.A. (ed.) 1964. *Mechanical Design and Systems Handbook*, Table 6.6b. New York: McGraw-Hill.

10.6 Recommended Reading

Brundle, C.R., C.A. Evans, Jr., and S. Wilson (eds.) 1992. *Encyclopedia of Materials Characterization: Surfaces, Interfaces, Thin Films*. Boston: Butterworth-Heinemann.

Appendix A
Units

A.1 SI (Système International) Units

These units are self-consistent; that is, all conversion factors are unity. They should be used whenever possible to avoid errors in calculation and to facilitate comparison with other peoples' results. The SI units used in this book include the following:

Quantity	Units	Symbol
Capacitance	farads	F
Charge	coulombs	C
Conductance	siemens	S
Current	amperes	A
Distance	meters	m
Electrical potential	volts	V
Energy	joules	J
Force	newtons	N
Frequency	hertz	Hz
Inductance	henrys	H
Magnetic flux density	tesla	T
Mass	kilograms	kg
Power	watts	W
Pressure	pascals	Pa
Resistance	ohms	W
Temperature	kelvin	K
Time	seconds	s

A.2 Conversion Factors

To convert: From To	To From	Multiply by Divide by
Å	nm	0.1
atm	Pa	1.01×10^5
bar	Pa	10^5
dynes/cm	N/m (J/m^2)	10^{-3}
dynes/cm^2	Pa	0.1
eV	J	1.60×10^{-19}
eV/mc	kJ/mol	96.3
gauss	tesla (webers/m^2)	10^{-4}
g-cal	J	4.186
psi	Pa	6895
sccm	mc/s	4.48×10^{17}
sccm	Pa·l/s	1.84
stones/acre	Pa	0.0154
torr	Pa	133 (=400/3)
u	kg	1.66×10^{-27}

A.3 Equivalent Units

Charge	$C = A \cdot s$
Energy	$J = kg \cdot m^2/s^2 = N \cdot m = W \cdot s = V \cdot A \cdot s = C \cdot V$
Force	$N = kg \cdot m/s^2$
Molecular dose	1 L (langmuir) = 10^{-6} torr·s
Pressure	$Pa = N/m^2$; μm (microns of Hg) = millitorr; mm Hg = torr

Appendix B

Vapor Pressures of the Elements

The following figures have been reproduced by permission of the present copyright owner, General Electric Corp., from R. E. Honig and D. A. Kramer (1969), "Vapor Pressure Data for the Solid and Liquid Elements," *RCA Review* 30:285. To cover a wide range of T, log p_V has been plotted versus log T rather than versus 1/T, even though the latter would be more linear in accordance with Eq. (4.13). For each element, available data considered reliable have been fitted to a formula consisting of Eq. (4.13) plus some corrective terms. The authors estimated the reliability of the raw data to be ±20 percent, but it always covered a range narrower than that of the figures (and given in the above reference), so reliability may be less in the extrapolated regions, particularly at very low pressure. If accuracy is critical, more recent literature should be compared.

The first two figures include the more common elements, while the third includes element numbers 21 (Sc), 43 (Tc), 56–72, and >84. Elements X having more than one significant vapor species are labeled "ΣX." Note that the red and white solid phases of P have different p_V values. For the p_V values of alloys and compounds, see the discussion of Ch. 4.

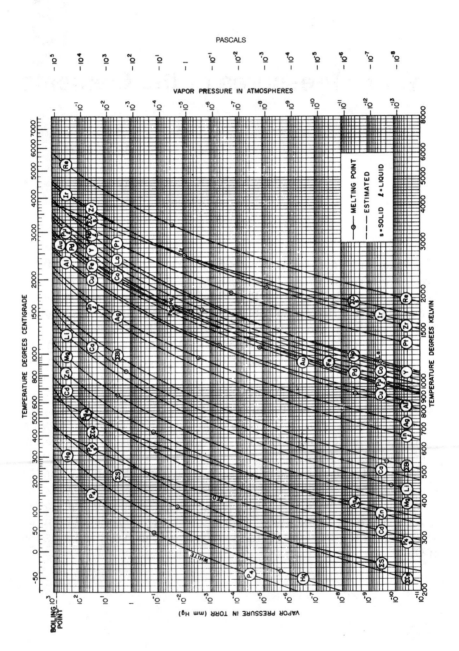

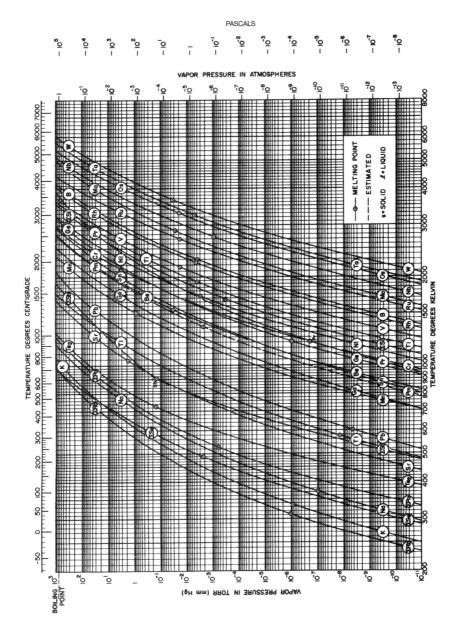

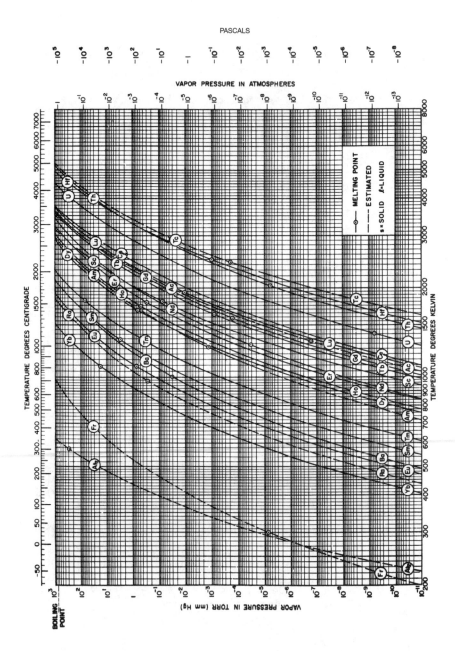

Appendix C

Sputtering Yields of the Elements

Data for the following figure were mostly taken from curves of sputtering yield, Y_S, versus Ar^+ impingement energy at perpendicular incidence, these curves having been calculated as the best fit to available data and reported by N. Matsunami et al. (1984), "Energy Dependence of the Ion-Induced Sputtering Yields of Monatomic Solids," *Atomic and Nuclear Data Tables* 31:1. Only values of Y_S for Ar^+, the usual sputtering gas, are plotted here, although Y_S for many other ions is reported in the above reference. For Mg, Mn, Zn, and In, the Y_S values plotted are those calculated for 1-keV Ar^+ by P. C. Zalm (1988), "Quantitative Sputtering," *Surface and Interface Analysis* 11:1. Elements not plotted are too volatile, low-melting, reactive, or nasty to be suitable for sputtering. The Y_S values given are for pure elements under background pressures of reactive gases that are low enough to prevent significant adsorption, which would reduce Y_S. For other Y_S values, see the discussion of Sec. 8.5.4.2.

For all but four elements, Y_S is plotted at three ion energies increasing in factors of two: 0.5, 1.0, and 2.0 keV. This spans the typical range of ion energies encountered in sputtering for film deposition. Note that Y_S increases less rapidly than does ion energy over this range, except for Be and C between 0.5 and 1 keV. This means that for a given plasma power, the sputtering rate, Q_S of Eq. (8.38), will be higher for lower ion energy and higher ion current.

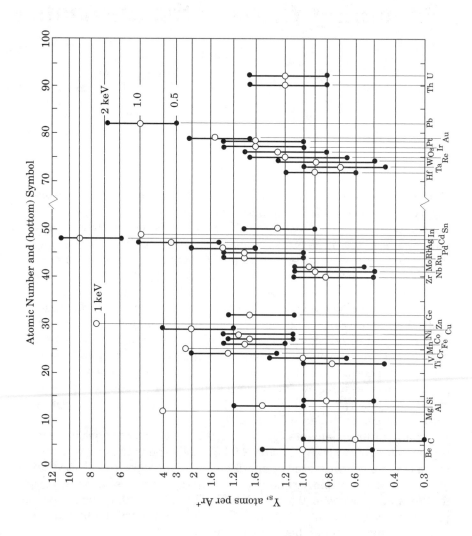

Appendix D

Characteristics of Tungsten Filaments

Hot filaments are frequently needed in film-deposition work as heaters or electron emitters. Since W has the lowest vapor pressure of any element at a given T, it is usually the element of choice, although Re is sometimes alloyed in to reduce brittleness after heating, and Ir is used instead in oxidizing gases to avoid etching.

The following procedures are useful in fabricating W-wire filaments. The wire should first be wiped with fine sandpaper and then with a methanol-soaked, lint-free cloth to remove native oxide. A coil can be made by winding the wire around a rod having a diameter somewhat smaller than the coil diameter desired. The coil will be uniform if the windings are laid against each other and if the wire is not allowed to twist during the winding. This is most easily done if the rod is held in a lathe chuck and rotated slowly. After winding, the coil can be stretched to the desired pitch. Beware that if W wire is bent too sharply or more than once in the same spot, it will fracture lengthwise into strands, forming a sometimes-unnoticeable weak spot which will break or burn out during use. Tungsten wire can be spot-welded with a fast, high-current pulse to other metals used as lead-in and support rods, Ni being the best for weldability. The W around the weld becomes very brittle, however, so the wire should be wrapped around the rod a few times and then spot-welded at the outside end to provide support to the weld.

Tungsten filaments can be operated up to a T of 2200 K or so before their life becomes inconveniently shortened by evaporation. The accompanying figure plots versus T the four characteristics listed in the table below for a typical W filament of diameter $\phi = 0.01$ cm and length $L = 10$ cm. The scaling factors to filaments of other dimensions are also listed. Filament T is conveniently measured with an optical pyrometer.

Characteristic	Symbol	Units	Scaling factor
Electron emission	I_e	mA	$L\phi$
Evaporation rate	dh/dt	$nm/10^6$ s	1
Heating current	I_h	A	$\phi^{3/2}$
Voltage drop	V	V	$\dfrac{L}{\sqrt{\phi}}$

Data for the figure were mostly taken from H. A. Jones and I. Langmuir (1927), "The Characteristics of Tungsten Filaments as Functions of Temperature," *General Electric Review* 30:310. For the band shown for dh/dt indicating its uncertainty, the upper bound was taken from this source, while the lower bound was calculated from vapor pressure (Appendix B) using Eqs. (2.19) and (4.17) with an evaporation coefficient assumed to be maximum at $\alpha_v = 1$. The solid line for I_e is the level thermodynamically determined by Eq. (8.1), and the data from the above source fit this equation with B = 60 and ϕ_W = 4.50. However, if the electric field sweeping emitted electrons away from the filament is insufficient, the space-charge limit on emitted current density sets in at a level determined by Eq. (8.11) and suggested by the dashed line in the figure.

Characteristics of Tungsten Filaments

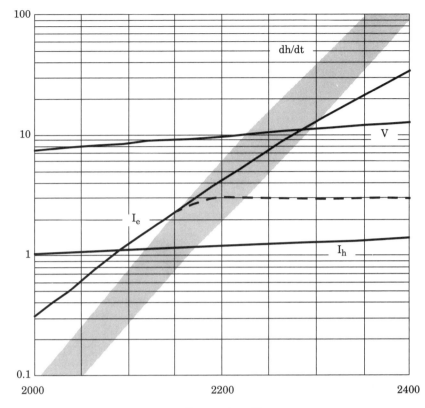

Temperature, K

Appendix E

Sonic Orifice Flow

Sonic or "choked" flow of a gas through an orifice occurs when the gas velocity in the orifice constriction has reached the speed of sound. It represents the maximum mass flow rate, Q* moles/s, through the orifice at a given upstream pressure no matter how low the downstream pressure. This situation occurs in various places in thin-film deposition, including gas flow into vacuum chambers (Sec. 3.3), supersonic-nozzle evaporation sources (Sec. 4.5.4), and safety flow restrictors on hazardous gas cylinders (Sec. 7.1.1). The maximum value of Q* (that is, neglecting orifice end losses) is given by the following equation, which has been adapted from that given in V. L. Streeter (1962), *Fluid Mechanics*, McGraw-Hill Book Co., New York, p. 256:

$$Q^*(\text{mol/s}) = \frac{A^* C^*}{V_m^*} = \frac{A^* p_o}{\sqrt{MRT_o}} \left[\gamma \left(\frac{2}{\gamma+1} \right)^{(\gamma+1)/(\gamma-1)} \right]^{1/2}$$

$$= \frac{A^* p_o}{\sqrt{MRT_o}} f(\gamma)$$

where A^* = cross-sectional area of the orifice restriction, m^2
 c^* = speed of sound under conditions in the orifice, m/s
 V_m^* = molar volume under conditions in the orifice, m^3/mol
 M = molecular weight, kg/mol
 R = gas constant = 8.31 J/mol·K
 p_o = upstream pressure, Pa
 T_0 = upstream temperature, K
 γ = heat capacity ratio = c_p/c_v

The gas heat capacities c_p and c_V measured at constant pressure and volume, respectively, are given by Eqs. (2.13) and (2.12). Their ratio reaches its maximum value of $\gamma = 5/3$ for monatomic gases, and it approaches unity for complex molecules because of their many internal degrees of freedom. The function $f(\gamma)$ is plotted versus γ in the figure below and may be taken to have a value of 0.65 ± 0.03 for complex molecules.

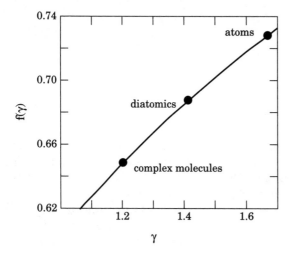

Index

A
"A" face vs. "B" face 279
Abnormal glow discharge 388
Abnormal grain growth 177
Absorbance 567, 569
Absorption vs. adsorption 23
Absorptivity 73, 111
Acceptor dopants 230, 274
Accuracy, defined 558
Acoustic impedance 103
Actinometry 455
Activated complex 339
Activated reactive evaporation 387, 456
Activation energy
  adsorption 123
  determination 339
  reaction 125
Activity coefficients 75, 276
Adhesion 145, 197
  enhancement by ion mixing 428
  failure 193
  measurement 581
Adiabatic expansion 92
Adsorbate-inhibition mechanism in ALE 272
Adsorption 120
  competitive 348
  dissociative 122
  into polymers 52
  isotherm 346
  saturation 272
  self-limiting 269
  vs. absorption 23

Adsorption-limited deposition 349
Al vacuum chambers 53
Al_2O_3 184
Al-Cu 77
(AlGa)As 74, 113, 229
Alkali metals 479
Alkaline earths 479
Alkaline-earth fluorides 240
Alloys
  definition 74
  deposition 262
  evaporation 74
  ordering 229, 263, 270
  semiconductor 228
    composition control 262
Alloys vs. compounds in deposition 260
Alpha mode of plasma drive 498
Al-Si 182
Ambipolar diffusion velocity 464
Amorphitization by ion bombardment 431
Amorphous films 177
Amorphous Si 178, 542
Amphoteric dopants 274
  in GaAs 279
Angle of incidence, light 107
Angular frequency 109
Anions
  and surface charge 142
  defined 268
Anisotropy
  defined 140
  in deposition rate 153

in epitaxy   308
in surface tension   142
Annealing   3, 119, 288
  during deposition   191
Anodic arc   390
Antiphase-domain boundaries   295
Antireflection coating   108
Antisite point defects   79
Appearance potential   379
AR coating   108
Arc breakdown in glow discharges   474
Arc plasmas   387
Arrhenius equation   125
Arrhenius plot
  in a plasma reaction   460
  in CVD   357
Arsenic passivation of Si   241
Aspect ratio   349
Atomic orbital notation   243
Atomic terraces   154, 293
Atomic-force microscope   562
Atomic-layer epitaxy   269
  gas control in   314
Atomic-layer growth   155
Au film growth   172
Auger electron   243
Auger emission   379
Auger spectroscopy   571
Avalanche injection   581
Avalanche multiplication   379
Axisymmetric flow   322
Azimuthal angle   265
Azimuthal axis   157

B

Backsputtering   432, 501
Backstreaming
  oil   46
  against an ion flux   541
  in downstream plasma deposition   540
Baffle, oil   46
Baking and outgassing   51
Ballistic aggregation   164
Ballistic deposition   161
Ballistic transport of electrons in
    semiconductors   234
Band gap   230
Band-gap engineering   230
Basal plane   141
Bayard-Alpert gauge   57
Beam electrons   472, 493
Beam-equivalent pressure   97

Beer's law   111
Bending moment   190
BET isotherm   347
Biaxial elastic modulus   188
Biaxial stress   187
Bicrystal   281
Bimodal ion-energy distribution   493
Bimolecular reaction rate   339
Birefringence   179
Bistable flow pattern   329
Black films   166
Blackbody   73
  radiation   202
  well   74, 215
BN, pyrolytic   53, 84, 188
Bohm criterion   466
Bolometers   179
Boltzmann factor   132
Boltzmann transport equation   422, 438
Bond types   142, 144, 152, 199, 222
Bonding layer   145, 199
Bonding determination in films   566
Boundary layer
  chemical   344
  concentration   353, 356
  laminar   319
  velocity   322
Bragg condition   247
Bragg-Brentano scan   248
Bravais lattice   140
Breakdown voltage   580
Brewster's angle   107
Bruggeman approximation   114
Buckling of films   193
Buckminster-Fuller ene   93
Buckyballs   93
Buffer layers   188
  in epitaxy   298
Burgers vector   284

C

CaF_2 199, 227, 259, 297
Calorimeter   211
Calorimetry of energetic particles   405
Capacitance
  diaphragm gauge   57
  manometer   57
  measurement   578
  parallel-plate   490
Capacitive coupling
  of plasma power   489
Capacitive impedance   490

Index

Capacitive plasma sheath 498
Capillarity model 144
Carbon contamination
   in Auger spectroscopy 246
   in MOCVD 335
Carbonyl 267
Carrier concentration measurement 577
Catalyst 351
Cathode sheath 471
Cathode spot behavior 391
Cathodic arc 390
Cations
   and surface charge 142
   defined 268
Centrifugal force on electrons 383
Cermets 179
$(CH_3)_3Ga$ 335
   decomposition 271
Chalcogens 261, 479
Charge injection 579
Charge neutrality at interfaces 266
Charge trapping in dielectrics 580
Chemical activation in plasmas 473
Chemical boundary layer 344
Chemical equilibrium 330
Chemical potential 68, 137
   in mixtures 331
Chemical shifts in XPS 245, 571
Chemical sputtering 441
Chemical vapor deposition 4, 307
Chemisorption 121
Chemisorption-reaction probability 126
   in CVD 345
Child-Langmuir law 377, 467
Choked flow in nozzles 91, 597
Clausius-Clapeyron equation 69
Cluster ions 92
Coble creep 194
Co-evaporation 75, 97
Coherence length 110
Coherency strain 281
   and growth morphology 298
Coherent epitaxy 281
Cold-cathode emission 378
Collimated sputtering 444
Collimation of vapor 169
Collision cascade of atoms 411
Collision cascade of electrons 379
Collision rate constant 340
Collisional plasma 462
Collisionless electron energy gain 521
Collisions, third-body 93, 340
Columnar structure 160

Commensurate growth 279
Compensation of charge carriers 578
Competitive adsorption 348
Composites 179
Composition control in compounds 259
Composition control in alloys 74
Compound, definition 74
Compound semiconductors, doping
   of 274
Compounds, evaporation 77
   II–VI 228, 260
   III–V 228, 262
   IV–VI 78, 228
Compressive stress and ion
   bombardment 429
Concentration boundary layer 353
Condensation
   coefficient 73
   in vacuum pumps 40
   reaction 546
   theory 72
   water 50
Conductance, gas-flow 42
Conduction band 230
Conductive compounds 184
Conductivity, electrical; see Electrical conductivity
Confinement energy of electrons 235
Conformal coating 307, 348
Congruent evaporation 80, 395
Congruent vaporization 373
Contamination
   adsorbed 198
   desorption by energetic particles 373
   during evaporation 84
   from source gases 267
   in CVD 315
   in evaporants 85
   in sputtering 434
   particle 54
   sources 45
Continuity equation 41
Continuous growth mode 156
Control
   PID 85
   proportional 85
Convection
   forced vs. free 317
   free 327
Corrosion resistance 177, 193
Corrugation ratio 134
Cosine flux distribution 96
Coulomb potential 417

602 Index

Coulomb's law 417
Coulombic plasma 461
Coupling capacitor in rf plasma 500
Cracking furnace 80, 267
Cracking of films 193
Creep under stress 187
Critical damping 89
Critical flow in nozzles 91
Critical nucleus concentration 149
Critical nucleus radius 147
Critical point of vapor 11
Critical thickness for epitaxy 300
Cr-SiO 179
Crucible materials, 84
Crystallographic texturing by compression 430
Crystallization and entropy 222
Crystallographic
  defects 279
  directions 141
  disorder measurement 249, 565
  nomenclature 140
  planes 140
  symmetry 223
  texturing 156
    by compression 430
    by sputtering 441
Crystals, bulk growth 226
Cubic lattice 140
$CuInSe_2$ 75, 229
Curvature
  and surface energy 175
  measurement 190
Cusp magnetic field 516
C-V measurement 569, 580
CVD 4, 307
CVD phase diagram 335
CVD reactor models 358
Cyclotron radius 384

D

Dangling bonds 142, 178
Dark space in plasma 472
Dark-space shield 473
DC plasma 470
DC power supplies 503
Dead spots in gas flow 53, 316
De-adherence 198
deBroglie wavelength 233
Defect types (point) 79
Defects in crystals 279
Defects in semiconductors 282

Deformation-mechanism maps 196
Degeneracy of quantized energy levels 14, 66, 131
Degreasing of polymers 52
Delta doping 235
Density measurement of films 564
Depletion layer 232, 580
Deposition
  directional; see Directional deposition
  into trenches 165, 169, 349, 445, 514, 532
  flux monitoring 100
Deposition rate
  anisotropy 153
  in planar-diode sputtering 478
  monitoring with RHEED 258
  vs. flux 22
  vs. frequency in PECVD 497
Depth profiling of carrier concentration 569
Depth profiling of composition 571
Desorption 50
  by ion bombardment 407
  rate 127
Devitrification 178, 180
Dew point 316
Diamond film nucleation 128
Diamond structure 223
Dichroic materials 112
Dielectric breakdown 579
Dielectric constant 112, 490
Differential measurement 57
Differentially pumped seals 98
Diffraction 246
  electron 251, 254
  grating 251
Diffusion
  barrier films 177, 183
  coefficient, gas 27
  effect of vacancies on 181
  in CVD 353
  length 134, 137
  of oil on surfaces 47
  of point defects 410
  thermal 363
  velocity 277
Diffusion-limited deposition 355
Diffusivity, gas 27
Digital control of thickness 270
Dimensionless numbers 25, 320
Dimer rows on surfaces 265
Diode lasers 236
  ZnSe 275

Directional deposition
   in sputtering   488
   in plasma CVD   541
Discommensurate growth   282
Dislocations   279
   climb   283
   edge-type   282
   epithreading   291
   etch pits   238, 563
   glide   283
   glissile   283
   nucleation   289
   pinning   289, 291
   screw-type   284
   sessile   283
   threading   290
   velocities   289
Disorder, thermodynamics of   67, 156
Dispersion   519
Displacement current   490
   in rf plasma   498
Displacement reaction   309
Dissipation factor
   in dielectrics   579
Dissociation rate by electron impact   528
Dissociation reactions   456
Dissociation threshold   457
DLTS   570
Donor dopants   230, 274
Dopants   273
   amphoteric   274, 279
   compensation   278
   fractional activation   278
   in semiconductors   230
   lattice-site control   270
   out-diffusion   277
   shallow implantation   277
Doping superlattices   237
Dose control of vapors   272
Dose in ALE   270
Downstream plasma deposition   517, 536
Drift velocity   462
Dry lubricants   142
Dust; see also Macroparticles
   contamination   54
   toxic   55
Dynamic interactions in RHEED   259

E
E × B drift   485
Edge energy   154, 295
Edge growth   155

Effective electron mass   233
Effective heat of formation   185
Effective-medium models   114
Elastic modulus   186
   biaxial   188
   measurement   582
Elastomers   52
Electrical breakdown of dielectrics   579
Electrical conductivity
   of a plasma   462
   of a solid   232
Electrical properties of films   575
Electromagnetic radiation   105
   blackbody   73, 202
   heat transfer   202
   interaction with conducting media   512
   microwave   508
   vaporization by   394
   wave behavior   105
Electromigration   178, 181
Electron affinity and negative ions   479
Electron attachment   456
Electron cooling
   in magnetized plasmas   486
   in plasmas   493
   in thermionic emission   392
Electron
   collisions   454
   conduction in semiconductors   232
   conduction in plasma   462
   cyclotron radius   384
   cyclotron resonance   511
   diffraction   251
   emission from surfaces   374
   energy gain in helicons   521
   energy gain in rf plasma   496
   escape depth   245, 297
   heating at sheath edge   497
   in magnetic field   382
   mass   233
   mean free path   24
   microprobe   244, 571
   tunneling   378
   spectroscopy   242, 571
   terminal velocity   233
Electron temperature   461
   in ECR plasma   514
Electron trapping
   in dielectrics   581
   in Si nitride   546
Electron wavelength, free   247
Electron-beam evaporation   382
Electron-beam excitation of vapor   99

Electronegative gas in plasmas   463
Electronic stopping power   416
Electron-impact dissociation   526
Electron-impact reactions   454
Electron-molecule reaction rates   459
Electron-volt, defined   90
Electrostatic shield   523
Eley-Rideal reaction mechanism   124
Ellipsometry   109, 559
Emission spectrometer, electron-impact   99, 243, 572
Emissivity   73
　in substrate heating   205
Endothermic reaction   68, 338
Energetic bombardment   197, 371
Energetic condensation   90, 401, 408
Energetic irradiation   145
Energetic neutrals
　and film compression   429
　in sputtering   405
Energies of particles, typical   90
Energy
　dislocation   285
　distribution of sputtered particles   444
　enhancement of evaporants   90
　free   67
　internal   64
　minimization in epitaxy   287
　transfer in collisions   401
Energy-enhanced deposition   166
Energy-enhanced epitaxy   300
Enthalpy   20, 67
Entropy   65, 139
　configurational   180
　configurational vs. thermal   67
Epitaxy
　at low T   300
　atomic-layer   269
　defined   221
Epithreading dislocations   291
Equation of state   17, 317
Equilibrium
　chemical   330
　constant   138, 333
　defined   69
　deviation in epitaxy   227
　vapor-liquid   64
Equipartition theorem   19, 131
Equivalent beam pressure   264
ESCA   244
Etch selectivity   559, 563
Etch-pit density   563
Euler's formula   110

Eutectic   179
　Al-Si   182
Evanescent tail   234
Evaporation
　coefficient   73
　congruent   80
　cooling in   68
　electron-beam   382
　GaAs   80
　of alloys   74
　of compounds   77
　of SiO_2 77
　rate   71
　sources   81
　thermodynamics   63
Evaporative cooling   68
Ewald sphere   250
Excess-flow valve   311
Exothermic reaction   338
Exponential relaxation   77
Extension alloys for thermocouples   213
Extinction coefficient, optical   111

F

Face-centered cubic lattice   141
Facets   141
Facetting   152
Faraday cage   505
Fe film growth   173
Fermi level   232, 374
　and electron emission   375
　in metals   212
　pinning   278
Fermi-Dirac distribution   374
Ferromagnetic films   157
Field-effect transistor   232
Filaments, W   593
Filtered arcs   393
Finite-size effect   164
Flat-band voltage   580
Floating potential   466
Flow
　control   312
　of gases in tubes   43
　pattern, bistable convective   329
　sonic orifice   597
　visualization   330
Fluence of laser energy   396
Fluid-flow regime   25
Fluorites   226, 297
Fluoroptic thermometer   214
Flux monitoring of vapors   98

Flux, molecular impingement 15
Forward-recoil scattering 574
Forward sputtering 444
Fowler-Nordheim equation 378
Fractional-order LEED spots 254
Franck-Condon principle 457
Frank-van der Merwe growth 143
Free convection 327
Free energy 67
Free radicals
  deactivation downstream of
    plasma 539
  defined 453
  reaction rate 528
  reactivity 456
  recombination 459
Freezing-out of chemical processes 150
Frenkel pair 411
  recombination 416
Frequency factor 125
Frequency, angular 109
Fundamental absorption edge 575

G
GaAs
  "chloride" CVD 351
  composition control 261
  deposition kinetics 262
  dopants 276
  evaporation 80
  polar faces 279
  surface structure 264
Galling 1
Gamma mode of plasma drive 498
Gamma rays 574
Gas
  2D 131, 137
  2D, nucleation of 155
  adsorption 120
  evolution 50
  condensable 12
  condensation in pumps 40
  contamination 315
  evolution from films 198
  evolution from surfaces 50
  flow control 312
  hazardous 41, 267, 311
  heat transfer 29, 208
  heating in plasmas 454
  kinetics 9
  law, ideal 17

  pressure reduction by sputtering
    flux 478
  residence time 330
  ring 517
  supply modulation 269
  throughput 41
  toxic 41
  vs. vapor 9
Gaseous source material for PVD 266
Gasification radicals 267
Gas-phase nucleation 342, 365
Gauss's law 376
Gaussian distribution 134, 162
Ge epitaxy 92
Gettering of impurities
  from inert gas 316, 481
  in CVD reactors 61
  in semiconductors 237
Gibbs free energy 67
Gibbs free energy of reaction 138
Gibbs' phase rule 79
Gigahertz 464
Glow discharge, dc 387
Glue layer 145, 199
Grain
  boundaries 157
  growth 150, 176, 195
  growth, abnormal 177
  size, x-ray determination 249
  texturing 152
Grain-boundary
  migration 176
  problems 177
  stuffing 177
Graphite 142, 280
  pyrolytic 84
Graphoepitaxy 158
Grashof number 329
Gray-body assumption 87, 202
Grazing-incidence x-ray diffraction 565
Growth shape of crystals 153
Gyration radius, in magnetrons 484

H
H abstraction from silane 540
H analysis 573
H in amorphous Si 542
H passivation of Si 240, 345
Hall mobility 577
Halogens 261
Hamel flow 321
Hardness measurement 582

He leak detector 59
Heat
  capacity ratio of gases 596
  capacity, gas 19
  conduction, gas 29, 208
  conduction, radiative 202
  convection 327
  input determination 211
  of evaporation 69
  of formation 184
  of reaction 338
  transfer coefficient gas 30
  transfer coefficient, radiative 88
Heat-control coatings 98
Heat-energy balance 205
Heat transfer; see Heat conduction
Heat-transfer gases 210
Helical resonator 522
Helicon plasma 518
Helmholtz free energy 331
He-Ne laser 108
Heteroepitaxy 222
Heterogeneous nucleation 93, 139
Heterogeneous reaction 309, 344
Heterojunction 230
Heterostructure design 229
Hexagonal crystal symmetry 141
High-pressure limit of reaction rate 341
High-vacuum regime 25
Hillocks 194
Hole, conducting 232
Hollow-cathode effect 474
Homoepitaxy 222
Homogeneous nucleation 93, 180
  in CVD 342, 365
  in plasmas 534
Homogeneous reaction 309, 336
Hooke's law 187
  in three dimensions 281
Hydrolysis reaction 309
Hysteresis
  in reactive sputtering 480
  in thermal cycling 195

I
Ideal-gas law 17
Impact mobility 167
Impact parameter 419, 535
Impedance matching 504
Impedance, acoustic 103
Implantation of inert gas 423
Impurity atoms in crystals 274
Implantation theory 416

Impurity "gettering," see Gettering
Impurity problems; see Contamination
Impurity rejection during
  crystallization 276
Incommensurate growth 280
Incubation time for nucleation 351
Index of refraction 106, 111, 512
  in plasma 512
Indium-tin oxide 263, 479
Induced-dipole interaction 120
Inductance of coils 504
Inductive plasma 521
Inelastic collisions of electrons 454
Inert-gas purification 405
Inficon 99
Infrared absorption in CVD 348
Infrared electro-optic devices 230
Infrared absorption spectroscopy 566
Inhomogeneity in films 563
Insulators for vacuum use 53
Interdiffusion at film interfaces 181
Interdiffusion minimization in CVD 352
Interface grading 200
Interface width 163
Interfaces 180
Interfacial compounds 183, 240
Interference
  constructive and destructive 108
  fringes 191, 559, 562
  holography 330
  oscillations in RHEED 258
  oscillations in laser monitoring 106
Interferometry 562
Intermixing of interfaces 114
Intermixing, ion-induced; see Ion mixing
Interstitial defects 79
Interstitial impurities 274
Interstitial production by ion
  bombardment 428
Intrinsic resputtering 446
Invar, thermal expansion 191
Inversion layer 232
Ion acceleration in ECR plasma 515
Ion activation of surface reactions 540
Ion bombardment 400
  and atomic displacement 411
  and deposition activation 533
  and desorption 407
  and film microstructure 408
  and film stress 196, 429
  and planarization 409, 444
  and radiation damage 412
  and surface chemistry 406

and surface diffusion 409
and surface point defects 410
energy vs. plasma frequency 495
flux in dc plasmas 473
cleaning 239
damage threshold 414
for suppressing 3D nucleation 299
Ion
  activation of reactions 406, 540
  atom cluster 92
  channeling 402
  collisions with negative particles 535
  deactivation downstream of
    plasma 539
  dissociation upon impact 406
  energy distribution 493
  flux out of a plasma 467
  gauge 57
  gun 432
  implantation depth 416
  implantation threshold 413
  mean free path 24
  mixing 200, 428
    defined 411
  negative; see Negative ions
  peening 197, 429
  plating 386
  reactions 530
  reflection 402
  sound velocity in plasma 465
  source, electron-impact 92
  transit time across plasma sheath 494
  wind 535
Ion-atom collision trajectory 419
Ion-induced
  amorphitization 431
  atomic displacement, 407
  compression 196, 429
  desorption 407
  electron emission 378
  radiation damage 424
Ionization
  atmospheric-pressure 59
  by electron impact 379
  cross section vs. energy 379
  fraction in plasmas 513
  gauge 57
  in arcs 390
  in plasmas 458
  limit, in sheaths 468
  of sputtered metal 488
  potential of molecules 379
Isentropic process 92

Isotherm, adsorption 389

J
Jellium model 144
Joule heating 82, 389

K
Kalrez 52
KI 77
Kinchen-Pease formula 425
Kinematic interactions in RHEED 258
Kinetic
  barriers to doping 276
  emission of electrons 379
  inhibition 299
  limitations 150
Kinetics
  chemical reaction 336
  gas 9
  in selective deposition 350
  of adsorption 124
  plasma reaction 526
  vs. thermodynamics 150
Kink sites 155
Kinks and dislocation nucleation 289
Kirchoff's law 74, 202
Kirkendall voids 185
Knock-on effect 199, 277
Knock-on implantation 411, 424
Knudsen cell design 83
Knudsen number 25

L
Laminar flow, in ducts 318
Landau damping 521
Langmuir
  adsorption isotherm 346
  adsorption model 272
  probe 459
  unit of gas dose 23
Langmuir-Hinshelwood mechanism 124
Lapping 563
Larmor radius 384
Laser ablation 394
Laser-induced fluorescence 342, 455
Latent heat of vaporization 69
Lattice, crystal 140
Lattice mismatch 279
  critical 288
  defined 223
Laue pattern 248
Layered structures 142, 188
L-C matching network 504
LCR meter 578

Lead-salt alloy semiconductors 230, 263
LEED 251
Light
  attenuation in absorbing media 111
  emission from plasmas 455
  intensity 111
  physics of 105
  scattering 560
  trapping 179
Light-emitting diodes 236
Limited-reaction CVD 352
Line tension of dislocation 286
Liquid-phase epitaxy 227
Load-lock 35
Lobe-shaped flux distribution 96
London dispersion force 120
Lorentz force 382
Lorentzian plasma 461
Loss tangent 579
Low-frequency plasma 491
LSS theory 416

M

Macroparticles
  in laser ablation 397, 399
  in plasmas 533
  in Si PECVD 543
  in arcs 392
  in CVD 342
  in electron-beam evaporation 385
  in evaporation 84, 180
  in sputtering 435
  in T gradients 364
Magnetic beach 513
Magnetic properties of films 574
Magnetic mirror 485
Magnetization of electrons 484
  in ECR plasma 511
  in magnetrons 484
Magnetron, unbalanced 487
Magnetron sputtering 482
Mandrel 84
Mass balance
  equation 41
  in chemical reactions 334
Mass flow controller 312
Mass flow rate 21
Mass spectrometry 59, 572
Material safety data sheet (MSDS) 312
Matrix effects in analysis 558
Matthews criterion 285
Maxwell-Boltzmann distribution 12

Mean free path 23
Mechanical polishing 238
Mechanical properties of films 581
Meissner trap 51
Membrane films 198, 199
Metal compounds, conductive 184
Metallic glasses 178
Metalorganic compounds 267
Metamict phase 431
Metastable excited state 455
Metastables, deactivation downstream of plasma 539
Mg doping of GaAs 278
MgF_2 77
Microcrystalline Si 545
Microwave interferometer 459
Microwave plasma excitation 508
Microwave power coupling 506
Migration-enhanced epitaxy 272
Miller indices 140
Mirror-field ECR 513
Miscibility gap 182
Misfit dislocations 282
Mn contamination 52
Mobile-ion contamination 581
Mobility 233
Mobility measurement 577
Mode-hopping of quartz crystals 104
Moiré pattern 280
Moisture-barrier layers 98
Mole 9
Molecular
  beam chopping 100
  dynamics modeling 167
  flow regime 25
  impingement flux 15
  mean free path 24
  scattering from surfaces 29, 96
  weight 15
Molecular-beam epitaxy 241
  sources for 83
Molecular-dynamics calculations 167, 406, 419
Moment of inertia 190
Monolayer 22
Monomers, defined 146
Monte-Carlo calculation 403
MoS_2 142, 280
Mosaic crystal 157
MSDS (material safety data sheet) 312
Multiphoton absorption 398
Multiple steady states of reaction 348

N

N_2O 549
NaCl   259
Nanoindentation   582
Nanoparticles   93
Native point defects   79
   as dopants   274
   in epitaxy   263
   types   79
Negative glow   472
Negative ions   463
   in magnetrons   482
   in sputtering   478
Nomarski microscopy   561
Normal incidence   108
Nuclear stopping power   416
Nuclear-reaction analysis   574
Nucleation   127
   2D   154, 258
      in epitaxy   293
   3D   145
   and ion damage   431
   barrier   148
   classical   145
   coarseness   149
   defined   119
   heterogeneous   93, 139
   homogeneous; see Homogeneous nucleation
   of clusters   93
   of dislocations   288
   sites   127
Nucleus mobility   158
Nutrient depletion   165, 350

O

Oblique incidence of vapor   165, 515
Obstructed discharge   472
Off-Bragg condition   253, 258
Ohm's law   233
Ohmic heating of electrons   497
Oil
   backstreaming   46
   outgassing   50
Optical
   absorption depth   111, 395
   behavior of solids   575
   confinement in diode lasers   237
   interference filters   1
   interference in films   107, 559
   lever   191
   monitoring of deposition   105

   path length   106
   properties of films   575
   pyrometers   215
Optical-emission spectroscopy   455
Optically absorbent films   179
Organo-metallic compounds   267
Orifice conductance   43, 597
Orifice, ideal   72
Oscillation, temperature   89
Oscillator
   mass-spring   102
   molecular-bond   566
   quartz crystal   102
Oscillator strength of molecule
      bonds   567
Ostwald ripening   150, 176
Outgassing   50
   in sputtering   434
   ion-gauge   58
Overconstrained amorphous
      network   542, 545
Oxidation reaction   309

P

Parabolic velocity profile   320
Parallel-plate plasma reactor   470
Parasitic reaction   342
Particle
   energy, typical   90
   motion in T gradients   364
   spitting   267
Particles, contamination by   54; see also
      Macroparticles
Parting layer   199
Partition function   131
Paschen curve   381
Passivation of Si
   by As   241, 299
   by H   240, 345
Passivation of surfaces   143
Peach-Kohler force   291
Péclet number   277
Peierls lattice resistance   289
Penning reactions   455
Period, oscillation   102
Period, thickness   107
Phase angle   109
Phase diagrams   78
   binary   182
   in CVD   335
   ternary   184
Phase shifts upon reflection   108

Index

Phase velocity 519
Phasor 112
Phonon scattering 233, 569
Photochemical activation 353
Photoelectron 243
Photoemission 378
Photoluminescence 568
Photolysis 342
Photothermal-deflection
    spectroscopy 568
Physical vapor deposition 3
Physisorption 121
Pi (π) network 504
Piezoelectricity 101
Pirani gauge 55
Planar diode 470
Planarization 409, 444
Planck's law 74
Plane of incidence 106
Planetary fixture 97
Plasma
    analysis by titration 547
    conductivity 462
    chemical activation 473
    chemistry 524
    chemistry vs. thermochemistry 543
    CVD 526
    dc 470
    defined 5, 373
    density 458
    diffusion 464
    downstream 536
    ECR 511
    etching 407
    frequency transition 494
    helicon 518
    inductive 521
    ionization fraction in 513
    low-frequency 491
    magnetized 512
    microwave 508
    overdense 512
    parallel-plate 470
    potential 465
    potential in rf plasma 498
    power coupling 503
    power modulation 536
    sheath 462
    sustaining mechanisms 381
    torch 521
    underdense 512
    voltage vs. frequency 497
Plasma-enhanced CVD

chemistry 525
defined 456
reactors 473
Plug flow in CVD 359
Point defects 279, 568; see also Native
    point defects
Poiseuille flow 320
Poisoning of sputtering target 480
Poisson's equation 376
Poisson's ratio 187
Polar faces in zinc-blende
    compounds 279
Polarization of light 106
Polarizing angle 107
Polaroid sheet 112
Polishing of crystals 238
Polycrystalline films 157
Polycrystalline Si 178
Polymorphism 225
Porosity 166
    measurement 104
Positive column in plasmas 472
Post magnetron 485
Potential emission of electrons 379
Potential energy, molecular 64
Potential flow 323
Powder formation; see Macroparticles
Powder patterns 565
Power dissipation in plasmas 527
Prandtl number 329
Precipitates 260
Precision, defined 558
Precursor adsorption state 121
Precursor molecules in plasma CVD 531
Pre-exponential factor 125, 133
Presheath 467
Pressure control 36
    in CVD 314
Pressure vs. deposition rate in CVD 358
Proportional bandwidth 86
Proximity doping 233
Pseudomorph 225
Pseudomorphic growth 281
Puffs (pF) 503
Pump
    ballasting 40
    molecular drag 38
    purging 40
    roughing 35
    selection 36
    throttling 44
Pumpdown, slow 55
Pumping speed 42

Pumps, condensation in   40
Purging of pump forelines   48
PVD; see Physical vapor deposition
Pyrolysis   3, 84
Pyrolytic BN   53, 84 188

Q
Quadrupole mass spectrometer   59
Quantum state   66, 131
Quantum well   235
Quantum wires   236
Quarter-wave coating   108
Quartz, thermal expansion   191
Quartz crystal microbalance   101
Quaternary allows   230
Quenched growth   143, 161

R
Radiation damage   412
Radiation-enhanced diffusion   412, 424, 428
Radiative heat transfer   202
Radicals, free; see Free radicals
Radicals, terminating   267
Raman spectroscopy   567
Random-walk problem   134
Raoult's law   75
Rapid thermal processing   181, 352
Rare earths   479
Rate constant measurement   151, 342
Rate of strain   195
Rate-limiting step   338
Rayleigh number   329
Rayleigh scattering factor   536
Reactant
  depletion in cavities   349
  utilization fraction   356
      in plasma CVD   531
Reaction
  endothermic   338
  exothermic   338
  first-order   124
  in gas phase   336
  heterogeneous   309, 344
  homogeneous   309, 336
  kinetic modeling   341
  kinetics   339
    in plasma   526
  order   338
  rate
    bimolecular   339
    on surfaces   346
    saturation (surfaces)   347

  theory   129
  rate-constant determination   151, 342
  reversible   138, 150
  steps   337, 526
Reactive deposition using arcs   394
Reactive evaporation   387
Reactive sputtering   479
  defined   456
Reactive sticking coefficient   307
Reactivity of free radicals   456
Reactor modeling in CVD   360
Reciprocal space   249
Recoil energy   401
Recrystallization   195
Rectification in plasma sheaths   492
Reduced mass   340, 566
Reducing agent   77
Reduction reaction   309
Reference junction, thermocouple   212
Reflected power   504
Reflectivity oscillation   259
Refraction   106
Refractive index; see Index of refraction
Refractory metals   52
Replacement collision   425
Residence time of gas   343, 530
Residual gas analyzer   59
Resistivity measurement   575
Resistor films   179
Resonant cavity   508, 513
Resonant frequency   102
Resonant frequency of cantilever beam   582
Resputtering
  and planarization   444
  defined   432
  intrinsic   446
Reynolds number   320
RF
  bias   499
  choke   502
  heating   318
  power losses   505
  sputtering   501
RFI   505
RGA; see Residual gas analyzer
RHEED   254
RHEED oscillations   258
  in migration-enhanced epitaxy   272
  T window for   295
Richardson-Dushman equation   375
Right-hand-screw rule   383
Rocking curve   249

612 Index

Rock-salt structure 228
Rod-feed evaporation 385
Rogowsky coil 506
Roll cells 327, 344
Roughening in sputter-erosion 571
Roughening, statistical 299
Roughness exponent 163
Roughness monitoring 163
Rutherford backscattering
  spectrometry 565, 572
Rutherford scattering 422

S

Sacrificial barrier 185
Saddle point 129
Sapphire 226
Sb doping of Si 278
Scanning tunneling microscopy 129, 293, 563
Scattering of charge carriers in
  solids 233
Schlieren photography 330
Screened Coulomb potential 417
Secondary electron 243, 374
Seeded flow stream 92
Segregation coefficient 276
Selective etching in Si and diamond
  deposition 545
Selective-adsorption mechanism 272
Selective-area deposition 127, 350
  of W 334
  using ALE 270
Self-absorption of light 236
Self-diffusion 28
Self-limiting adsorption 269
Self-limiting interfacial reaction 184
Self-shadowing 163, 443
Semiconductor devices 230
Semiconductors, lattice constants and
  band gaps 228
Sensitivity, defined 557
Shadow mask 558
Shallow donors and acceptors 230
Shear modulus 286
Shear stress
  in crystals 282
  in gases 28
Sheath collapse 493
Sheath collisions 468
Sheath edge 466
Sheath edge determination 536
Sheath width 467
  vs. dark-space width (dc) 472

vs. dark-space width (rf) 498
Sheet doping 235
Sheet resistivity 575
Showerhead electrode 474
SI Units 20, 585
Si
  alloy with Ge 289, 292
  amorphous 178, 542
  crystal structure 223
  microcrystalline 545
  polycrystalline 178
  Sb doping 278
  surface passivation; see Passivation
  wafer properties 191
Si_2H_6 345, 543
Si-Ge 289, 292
SiH_4 reactions 342
Silicon nitride
  plasma CVD 545
  stress 196, 495
SIMS 572
Single-phase field 79, 261, 263
SiO 77
SiO_2 178, 180, 545
Skimmers 92
Skin depth (rf) 512
  in inductive plasma 523
Skin effect (rf) 505
Slag 385
Slip in crystals 178, 282
Sn 225
Snell's law 107
Soak period for evaporants 75, 85
Soft pumpdown and venting 55
Solid-on-solid assumption 163
Solution, ideal 331
Sonic flow, in nozzles 91
Sonic orifice flow 597
Soret diffusion 363
Space-charge-limited current flow 376
Spalling 54
Spatial frequency 163
Spatial wavelength 163
Specific heat 20
Spectroscopic ellipsometry 114
Speed, molecular 15
Spike doping 235
Spiking at interfaces 183
Spinel 226
Spitting of macroparticles; see Macroparticles
Spontaneous ordering of alloys 263
Sputter-deposition rate estimation 439

Sputtered particle energy 444
Sputtering 431, 476
 chemical 441
 collimated 444
 defined 411
 directed deposition 488
 effect on composition 441
 magnetron 482
 planar-diode 476
 rate measurement 105
 reactive 479
 superconducting films 479
 target bonding 482
 target poisoning 480
 threshold 435
 yield 436
 yield vs. angle 442
 yields of the elements 591
Staining, metallographic 560, 563
Standard conditions of T and p 21
Standard deviation 134, 162
Standard state of elements 331
Standing wave in microwave cavity 508
Statistical mechanics 13, 66, 131
Statistical roughening 162, 299
Stefan flow 354
Stefan-Boltzmann radiation law 73
Stefan-Maxwell equations 354
Step coverage by resputtering 442
Step-flow growth 295
Steric factor 340
Sticking coefficient
 defined 6, 121, 125
 determination 99, 533
 for compounds 259
 in CVD 345
 of silane + radicals 544
 reactive 307
 vs. coverage 272
Stoichiometric coefficients 333
Stoichiometry 74
 deviation 79
  and precipitates 261
  measurement 570
 deviation and point defects 264
 in compound sputtering 479
Stokes shift 568
Stopping power for ions 416
Straggling of ion range 422
Strain 186
 effects in epilayers 291
 energy 225
 in epitaxial films 281

rate 195
 tetragonal 281
Strained-layer superlattices 237, 289
Stranksi-Krastanov growth 143, 297
Stress 185
 concentration at edges 193
 in films vs. plasma frequency 495
 intrinsic 196
 intrinsic vs. extrinsic 188
 measurement 190
Stress-strain curve 186
Structural zones 159
Structure development 159
Sublimation 11
Sublimation temperature 70
Substitutional impurities 274
Substrate
 biasing 501
 cleaning 198, 239
 curvature 190
 heating in high-density plasma 515
 orientation 54
 polishing 238
 rotation in CVD 325
 temperature calibration 209
 temperature control 200
 temperature transients 210
 transport 98
  in sputtering 487
Superconducting films
 grain boundaries 158
 laser ablation of 395
 sputtering of 479
Superlattices 1, 237
Supersaturation 10
Supersaturation ratio 147
 2D 155
Supersonic expansion in laser
 ablation 398
Supersonic nozzle 90
Surface activation 145, 199
Surface analysis
 by electron spectroscopy 242
 by IR absorption 348
Surface
 carbonization by electron
  bombardment 246
 charging in plasmas 489
 contamination 198, 239
 diffusion 129, 533
  and film morphology 170
  coefficient 137
  energy 139

passivation; see Passivation
phase diagram 266
reactions in plasmas 531
reconstruction 142
   observed by LEED 254
   vs. composition 264
roughening and RHEED 256
segregation 276
smoothness enhancement in ALE 270
stress 140, 192
structure monitoring by RHEED 265
structure notation 254
tension 139
topography measurement 560
waves in plasmas 508
Surface-energy minimization 171
Surface-reaction rates 346
Surface-reaction-limited deposition 350
Surfactant effect 145, 299
Susceptor 318
Symmetry, crystallographic 223

T

Ta_2O_5 579
Tangent rule 165
Tank circuit 506
TE polarization 106
TEM 563
Temperature control
  of evaporation sources 85
  of substrates 200
Temperature
  controller tuning 89
  electron 461
  gradients in CVD 363
  measurement 211
  oscillation 89
  window for ALE 270
  window for deposition 261
Templates for graphoepitaxy 158
Tensile strength 187
Tensile stress in PECVD films 546
TEOS 541
Terminal velocity of electrons 233
Terminology for vacuum deposition using gases 268
Ternary alloys vs. compounds 229
Tetragonal strain 281
Tetrahedral bonding 223
Texturing, crystallographic 156
  by compression 430
  in sputtering 441

x-ray determination 248
Thermal
  accommodation coefficient 31
  accommodation factor 30
  barrier coatings 166
  conductivity, gas 29
  cycling 195
  diffusion 363
  diffusion length 395
  diffusivity 392
  expansion 188
  grooving 174
  load 87
  mass 87
  runaway of electron temperature 514
  shock 198, 400
  spike 167, 412
  time constant 86
  transpiration 59
Thermal vs. electronic excitation of molecules 543
Thermal-contact materials 201
Thermalization of energetic particles in gases 477
Thermally activated growth 170
Thermionic arc 389
Thermionic cathode 475
Thermionic emission 374
Thermionically enhanced field emission 389
Thermocouples 212
Thermodynamics
  first and second laws 65
  of crystallization 222
  of disorder 67, 156
  of evaporation 63
  of interfacial compound formation 184
  vs. kinetics 150
Thermophoresis 364
Thermophoresis in tracer smoke 330
Thick-film technology 2
Thickness measurement of films 558
Thickness period 107
Third body and reaction rate 340
Thompson distribution 444
Threading dislocations 290
Threshold current in diode lasers 236
Throwing power 394
$TiAl_3$ 185
TiC 348
$TiCl_4$ 348
TiC-Mo 179
TiF_4 115

Time constant 77
TiN 1, 184, 394
Titration in plasma analysis 547
Ti-W 183
TM polarization 106
Tool-bit coatings 193
Torque 190
Total internal reflection 237
Toxic dust 55
Tracer smoke 327
Transformer-coupled plasma 521
Transition regime in gas flow 25
Transition state in reactions 129
Translational energy of molecules 13
Transmission electron microscopy 563
Transport
    high-vacuum 94
    molecular-flow 94
    properties of gases 25
Trap
    liquid nitrogen 46
    Meissner 51
    molecular-sieve 48
    states in semiconductors 230, 263, 274
Trapping probability, gas 121
    in CVD 345
    oscillation 259
Trenches, deposition into; see Deposition
Tribological coatings 2
TRIM code 404
Triple point of vapor 11
Tungsten filaments 593
Tuning stub 507
Tunneling of electrons 378
Turbulent flow 319
Twinning in crystals 153

U

Ultra-high vacuum 22
Unbalanced magnetron 487
Undersaturated deposition
    conditions 260
Undersaturated vapor 147
Uniformity of deposition in high
    vacuum 97
Unimolecular reaction rate 341
Unit conversions 585
Utilization fraction of vapor 122, 345

V

Vacancy defects 79
Vacancy production by ion
    implantation 427

Vacuum gauges 55
Valence 222
Valence band 230
Van der Pauw array 577
Van der Waals epitaxy 280
Van der Waals force 120
Vapor
    bubblers 314
    condensable 12
    condensation in pumps 40
    flux feedback control 100
    flux modulation 269
    flux monitoring 98
    pressure 10
    pressure estimation 70
    pressure measurement of
        evaporants 103
    pressures of the elements 587
    utilization fraction 122, 345
    vs. gas 9
Vapor-liquid/solid equilibrium 64
Vapor-phase epitaxy 227
Vegard's law 229
Velocity boundary layer in axisymmetric
    flow 323
Velocity
    ambipolar diffusion 464
    electron 377
    E×B drift 485
    fluid-flow 49
    ion sound 465
    molecular 15
Venting, slow 55
Vicinal plane 295
Virtual leak 53
Viscosity
    gas 28
    kinematic 321
Viton 52
Void formation 185, 194
Volmer-Weber growth 143, 296
Vortices 321

W

Water
    condensation 50
    outgassing 50
    vapor pressure 71
Wave-riding electrons 497
Wave vector 109
Wavenumbers 566
Wear coatings 179

Web coating process   98
Wetting   143
Wetting angle   153
WF_6 334
Whiskers   159
Whistler waves   518
Wien's displacement law   202
Window glass, coating of   98
Work function   212, 374
Wulff shape of crystals   153
Wulff theorem   152
Wurtzite structure   228

X
XPS   243
X-ray
   diffraction   246, 565
   fluorescence   243, 572
   limit in ion gauge   58

mirrors   237
photoelectron spectroscopy   243, 571

Y
YBCO   241, 395
Yield point   186
Young's equation   153
Young's modulus; see Elastic modulus

Z
Z-correction factor   103
Zero-point energy   132
Zinc-blende structure   223
ZnO   157
ZnS   269
ZnSe   275
Zone refining   276
Zones of film structure   159
ZrO_2 166